MUSIC DATA ANALYSIS
Foundations and Applications

Chapman & Hall/CRC
Computer Science and Data Analysis Series

The interface between the computer and statistical sciences is increasing, as each discipline seeks to harness the power and resources of the other. This series aims to foster the integration between the computer sciences and statistical, numerical, and probabilistic methods by publishing a broad range of reference works, textbooks, and handbooks.

SERIES EDITORS
David Blei, Princeton University
David Madigan, Rutgers University
Marina Meila, University of Washington
Fionn Murtagh, Royal Holloway, University of London

Proposals for the series should be sent directly to one of the series editors above, or submitted to:

Chapman & Hall/CRC
Taylor and Francis Group
3 Park Square, Milton Park
Abingdon, OX14 4RN, UK

Published Titles

Semisupervised Learning for Computational Linguistics
Steven Abney

Visualization and Verbalization of Data
Jörg Blasius and Michael Greenacre

Design and Modeling for Computer Experiments
Kai-Tai Fang, Runze Li, and Agus Sudjianto

Microarray Image Analysis: An Algorithmic Approach
Karl Fraser, Zidong Wang, and Xiaohui Liu

R Programming for Bioinformatics
Robert Gentleman

Exploratory Multivariate Analysis by Example Using R
François Husson, Sébastien Lê, and Jérôme Pagès

Bayesian Artificial Intelligence, Second Edition
Kevin B. Korb and Ann E. Nicholson

Published Titles cont.

Computational Statistics Handbook with MATLAB®, Third Edition
Wendy L. Martinez and Angel R. Martinez

Exploratory Data Analysis with MATLAB®, Second Edition
Wendy L. Martinez, Angel R. Martinez, and Jeffrey L. Solka

Statistics in MATLAB®: A Primer
Wendy L. Martinez and MoonJung Cho

Clustering for Data Mining: A Data Recovery Approach, Second Edition
Boris Mirkin

Introduction to Machine Learning and Bioinformatics
Sushmita Mitra, Sujay Datta, Theodore Perkins, and George Michailidis

Introduction to Data Technologies
Paul Murrell

R Graphics
Paul Murrell

Correspondence Analysis and Data Coding with Java and R
Fionn Murtagh

Pattern Recognition Algorithms for Data Mining
Sankar K. Pal and Pabitra Mitra

Statistical Computing with R
Maria L. Rizzo

Statistical Learning and Data Science
Mireille Gettler Summa, Léon Bottou, Bernard Goldfarb, Fionn Murtagh, Catherine Pardoux, and Myriam Touati

Music Data Analysis: Foundations and Applications
Claus Weihs, Dietmar Jannach, Igor Vatolkin, and Günter Rudolph

Foundations of Statistical Algorithms: With References to R Packages
Claus Weihs, Olaf Mersmann, and Uwe Ligges

Chapman & Hall/CRC
Computer Science and Data Analysis Series

MUSIC DATA ANALYSIS
Foundations and Applications

edited by

Claus Weihs

Technical University of Dortmund, Germany

Dietmar Jannach

Technical University of Dortmund, Germany

Igor Vatolkin

Technical University of Dortmund, Germany

Günter Rudolph

Technical University of Dortmund, Germany

 **CRC Press**
Taylor & Francis Group
Boca Raton London New York

CRC Press is an imprint of the
Taylor & Francis Group, an **informa** business
A CHAPMAN & HALL BOOK

CRC Press
Taylor & Francis Group
6000 Broken Sound Parkway NW, Suite 300
Boca Raton, FL 33487-2742

First issued in paperback 2019

© 2017 by Taylor & Francis Group, LLC
CRC Press is an imprint of Taylor & Francis Group, an Informa business

No claim to original U.S. Government works

ISBN-13: 978-1-4987-1956-8 (hbk)
ISBN-13: 978-0-367-87281-6 (pbk)

Contents

Chapter 1

Introduction

CLAUS WEIHS, DIETMAR JANNACH, IGOR VATOLKIN, GÜNTER RUDOLPH
TU Dortmund, Germany

1.1 Background and Motivation

Whenever we listen to a piece of pre-recorded music today, it is, almost with certainty, a playback of a digital recording. This is not surprising, since music has been distributed in digital form since the 1980s on compact discs. Since then, we have observed major disruptions in the music sector. Today, with the advances in the context of media encoding formats, higher processing power even on small devices, and high-bandwidth Internet connectivity, many of us no longer have physical music collections anymore but carry our virtual collections with us on our smartphones.

But this digitization has not only changed the way music is distributed and how we consume it, many other applications became possible since music can be easily digitally processed by computers. Today, various online music platforms automatically generate personalized radio stations based on your favorite tracks or recommend new music that sounds similar to your favorites. Other online services help to identify a certain piece of music based on the hummed melody. Your mobile music player, finally, probably tries to automatically organize your music collection based on the musical similarity of the tracks.

Many of these applications are based on the results of a *music analysis process*. The goal of these analysis processes typically is to automatically extract characteristic *features* of the musical pieces. These features can, e.g., be used to find similar tracks since they include characteristics of tempo and key, the instruments that are played, or even the mood that is conveyed by a track.

This book introduces the reader to the foundations of such music analysis processes and sketches the most prominent types of applications that can be built on these analyses. Furthermore, it provides the reader with the background knowledge required throughout the paper, e.g., in terms of acoustics, music theory, signal processing, statistics, and machine learning.

1.2 Content, Target Audience, Prerequisites, Exercises, and Complementary Material

Content and Target Audience   This book is a university-level textbook and provides self-contained and interdisciplinary material for different target audiences. The primary audiences are university classes with a topic related to music data analysis, e.g., in the fields of computer science and statistics, but also in musicology and engineering.

The main features of this first comprehensive and self-contained book on music data analysis can be summarized as follows.

- The book covers both the foundations of music – including acoustics, physics and the human perception – as well as the basics of modern data analysis and machine learning techniques and the corresponding evaluation methodology.

- Based on these foundations, the book discusses various applications of music data analysis in depth including music recommendation, transcription and segmentation as well as instrument, chord, and tempo recognition.

- Finally, the book also covers implementation aspects of music data analysis systems including their architecture, user interface, as well as hardware-related issues.

Prerequisites, How to Read the Book, Exercises, and the Supporting Web Page   Basic mathematics and, for the exercises, basic programming skills – preferably in R or MATLAB – are the only recommended prerequisites. Obviously, being able to read a musical score is fundamental. Throughout the book, we will provide additional pointers to further literature.

The book is designed for readers with heterogeneous backgrounds. When you prefer to approach the field from the application perspective, you might probably start reading one of the corresponding chapters in the third part of the book. Pointers to the underlying terminology, methodology, and algorithmic approaches, which are described in the first two parts of the book, will be given within these application chapters. Some chapters furthermore include short technical "interludes" for the advanced reader. You might skip these details in case you are rather interested in a general understanding of the subject.

If you want to test your understanding of material presented in the book, you may want to try some exercises. The book itself does not include exercises. However, theoretical as well as practical exercises based on R and MATLAB will be provided at the book's *web site* http://sig-ma.de/music-data-analysis-book, which also includes example data sets partly needed for the exercises and errata.

Relation to Other Books   A number of books focusing on Music Data Analysis appeared in the last ten years, among them [1], [2], [5], [3], [4], and [6].

Almost all these books provide state-of-the-art research summaries containing comparably advanced material so that typically further literature has to be consulted when used in a lecture. In contrast to these works, our book aims to be more comprehensive in that it also covers the foundations of music and signal analysis and introduces the required basics in the fields of statistics and data mining. Furthermore, examples based on music data are provided for all basic chapters of this book. Nonetheless, the above-mentioned books can serve as valuable additional readings for advanced topics in music data analysis.

1.3 Book Overview

General Structure The book is structured in four parts.

I "Music and Audio": In this part we cover the basics of music in terms of the underlying physics, fundamental musical structures as well as the human perception of music. We then introduce the reader to the foundations of digital signal processing, the extraction of musical features from the audio signal and from other sources, and how to represent music in digital form.

II "Methods": This part is devoted to statistical and machine learning methods used for music data analysis. We discuss regression, unsupervised and supervised classification, feature processing and selection, as well as methods for the evaluation of the models that result from these methods. Moreover, optimization methods that form the basis of many advanced data analysis methods are introduced.

III "Applications": The third, and central part focuses on applications that can be based on automated music data analysis methods. The discussed applications for example include instrument, chord, and tempo recognition, the detection of emotions, music recommendation, automated composition, and the tool-supported organization of music collections.

IV "Implementation": In the last part of the book we focus on practical considerations when building certain types of music-related applications. The topics include a case study on architectural considerations, questions of the design of user interfaces for music applications, as well as considerations of how to implement parts of a music analysis system directly on hardware.

Parts I - III mainly comprise 200 pages each, part IV comprises around 50 pages.

1.4 Chapter Summaries

Part I: Music and Audio

Chapter 2 "The Musical Signal: Physically and Psychologically": The computerized processing of music requires some understanding of the physical and sensational aspects of musical tones. In this chapter, we discuss the key characteristics of a tone, which are its pitch, volume, timbre, and duration. For each of these "moments" we give a description based on concepts from the fields of physics, psychoacoustics and music.

Chapter 3 "Musical Structures and Their Perception": In this chapter, the basics of musical harmonies and polyphony – the basis of Western tonal music – are reviewed. Furthermore, the concepts of consonant and dissonant intervals, which form the basis of Western music aesthetics, are discussed and the rules of tone progression and the fundamentals of the craft of counterpoint are explained. Then, the basic concept of chords as combinations of several tones and their harmonic functions are presented along with an elementary theory of musical form. Finally, the notion of "Gestalt" is introduced to bridge the gap between music perception and music cognition.

Chapter 4 "Digital Filters and Spectral Analysis": Following the Introduction to the physics of sounds and the structures of music in Chapters 2 and 3, we focus in this chapter on methods for digital processing of music signals. One of the core tasks of music signal processing is the spectral decomposition of the signal in order to further analyze or process the signal in frequency subbands. We introduce and discuss several techniques based on filter banks and signal transforms such as the Fourier transform and the Constant-Q transform. The chapter concludes with a brief discussion of how the fundamental frequency of harmonic sounds can be automatically estimated.

Chapter 5 "Signal-Level Features": Many music-related applications require knowledge about certain characteristics of individual musical pieces. Just like a good disc jockey (DJ), an automated recommender system or virtual DJ could for example try to make sure that two consecutive pieces are not too different with respect to their rhythms, melodies, or even their instruments. The automatic identification of certain features of a musical piece is one of the core steps in music analysis. In this chapter, we review the typical features that are used today in research and practice, including features for timbre, pitch, harmony, tempo and rhythm, and describe how they can be extracted from a given audio signal.

Chapter 6 "Auditory Models": In some sense, music is simply organized sound waves that travel through the air to which humans are listening. This chapter covers the main foundations of human hearing and the most important models for the human reception of sound (auditory models). Based on these foundations, the chapter briefly discusses different algorithmic techniques, e.g., for pitch estimation, which are based on simulations of the human auditory process.

Chapter 7 "Digital Representation of Music": Today, music is mostly stored in digital form and we all have our MP3 files on our mobile phones and home entertainment systems. MP3 is, however, only one particular file format, which became popular because its digital representation is comparably compact on disc. In this chapter we introduce how the acoustic signal can in principle be transformed into a digital signal. We then discuss the various ways and techniques to digitally represent music. Moreover, we present different forms of storing musical scores and outline popular technical standards such as MIDI.

Chapter 8 "Music Data: Beyond the Signal Level": The music information retrieval (MIR) literature is historically focused on extracting musical features from the audio signal (Chapter 5). However, with the emergence of Social Music Platforms and the World Wide Web in general, various additional sources for obtaining the features or metadata for a certain musical track have become available, e.g., in the form of user-provided annotations (tags) or music databases. In this chapter we review a number of additional sources for music data that can be used alone or in combination with signal-level features for music data analysis applications.

Part II: Methods

Chapter 9 "Statistical Methods": Music data analysis is typically heavily based on statistical methods. This chapter introduces the main relevant terms and methods and gives examples for their usage for music data analysis problems. We provide the basics of probability theory, the concept of random variables and their distributions, introduce the concept of random vectors, discuss how to estimate the parameters of unknown distributions and how to test hypotheses on parameters, and finally present how to model relationships between variables.

Chapter 10 "Optimization": Many advanced methods in the field of music data analysis are solutions to mathematical optimization problems. In this chapter we review optimization methods that are often embedded in music data analysis approaches. We discuss single-objective as well as multi-objective optimization problems, discuss basic and heuristic local search methods, gradient methods, evolutionary algorithms as well as analytic methods to find the optimal solution for a given problem.

Chapter 11 "Unsupervised Learning": Classification is one of the main methods used for music data analysis tasks. In the unsupervised case, we may want to automatically build groups of tracks similar in a certain musical aspect from features derived from the audio signal. This problem of grouping objects is typically approached using so-called clustering methods. The chapter introduces the most typical clustering methods including k-means and Self-Organizing Maps. Moreover, Independent Component Analysis is introduced aiming at "splitting" observations into the underlying independent components.

Chapter 12 "Supervised Classification": In contrast to the unsupervised techniques discussed in the previous chapter, supervised classification techniques predict the class of a new object (e.g., the genre of a new track) based on its features and a model that was previously learned using examples for which the label (e.g., the genre) was known. The chapter will introduce the most common techniques from the literature, namely Naive Bayes and Linear Discriminant Analysis, nearest-neighbor prediction methods, decision trees, Support Vector Machines as well as ensemble methods that combine several classifiers.

Chapter 13 "Evaluation": Once a regression or classification model is learned and applied on new data examples, the question is how to evaluate whether one model or method works better than another one for a given task. This chapter introduces the readers into the typical evaluation methodology used in the research literature in the context of classification and regression tasks. The chapter focuses on offline evaluation scenarios where the goal is to predict a class label or function value for a selected subset of the data not used for learning the model. We discuss resampling procedures, common evaluation measures, as well as methods to analyze whether the eventually observed differences between different models are statistically significant.

Chapter 14 "Feature Processing": In Chapters 5 and 8, we discussed which kind of musical features can, in principle, be extracted from the audio signal. It may however be reasonable or required to further process these features, for example, because there are missing data, the data contains noise and outliers, or the amount of data to be processed is too huge to be processed efficiently. This chapter gives an overview of methods to further process a given set of features. We present techniques to modify or complete the existing feature information and discuss approaches which automatically construct additional features. Furthermore, techniques will be presented which help to reduce the computational complexity by only considering a smaller segment of a musical piece in the analysis.

Chapter 15 "Feature Selection": The number of features that can be extracted from the audio signal can be quite large. While it is in general good to know as much as possible about the musical pieces, having a large set of features can also lead to a number of problems including computational complexity and noise that is introduced through features that are not extracted with high accuracy or are not relevant for the given task. In this chapter we discuss algorithmic approaches to reduce the number of features to be taken into account in further analyses. The goal of such approaches typically is to select the subset of the features in a way that no or nearly no relevant information is lost.

Part III: Applications

Chapter 16 "Segmentation": The problem of segmentation is to automatically identify different parts of a digital music piece. On a fine-grained level, the goal can be to find parts as small as individual notes – this process is often called "onset detection" – where these notes can then, e.g., be the input to an automated transcription process. On the other hand, on a more coarse-grained level, the goal of segmentation can be to identify the individual musical parts of a track like the introduction, verses, bridges, and chorus. The chapter discusses techniques for onset detection as well as an approach for the identification of larger musical structures.

Chapter 17 "Transcription": Transcription is the process of deriving the musical score from the digital audio signal. In this chapter we discuss the challenges and different steps of a typical transcription process. These steps include the separation of the relevant part of the musical piece, e.g., the singing voice, the estimation of the fundamental frequency, the classification of notes and the estimation of their lengths, as well as the estimation of the key.

Chapter 18 "Instrument Recognition": Knowing which instruments are played in a musical piece can be a valuable input to different other music-related tasks including automated classification, music recommendation, playlist generation or mood detection. The chapter reviews the different variants of instrument recognition – e.g., for monophonic vs. polyphonic tracks, types of instrument taxonomies – and then discusses a typical processing chain where instrument recognition is considered as a supervised classification task.

Chapter 19 "Chord Recognition": Chords are sets of notes that are played nearly simultaneously. Automated chord recognition can have different purposes, in particular, the generation of so-called "lead sheets" for pop and jazz music where the harmony of a musical piece is simply represented as an ordered sequence of chord symbols (e.g., C Major, G minor, G7). This chapter will lead the reader step-by-step through a typical scheme for the automated recognition of such chord sequences. The main signal processing step, which is called chroma or pitch class profile extraction, will be discussed in detail and different knowledge-driven or data-driven approaches for the generation of chord sequences will be presented.

Chapter 20 "Tempo Recognition": Tempo or rhythm is one of the fundamental characteristics of musical pieces. Being able to automatically determine the tempo can be an important input or prerequisite for a number of other processes like transcription, segmentation, content analysis and recommendation, or automated audio synchronization. The chapter first introduces basic terms like rhythm, beat, tempo or meter and then presents the details of a typical processing workflow for tempo estimation. The main steps of this process include the extraction of a temporal sequence of relevant features from the audio signal, the estimation of periodicities in the signal, and the temporal positioning of the beats (beat tracking).

Chapter 21 "Emotions": Music is typically strongly connected to emotional processes. Music can evoke certain emotions in the listener and our moods on the other hand can influence which kind of music we like to hear. This chapter analyzes the various relationships between emotions and music. It first discusses the differences between emotions and moods, sketches what happens in our brains when we listen to music, and presents different theories of emotions and their connection to music. Then, different approaches for the automated recognition of emotions are discussed. The problem is often framed as a classification or regression task where the input features include, e.g., the melody, the played instruments, the dynamics or the lyrics of the tracks.

Chapter 22 "Similarity-Based Organization of Music Collections": The music collections that we can browse online on shopping platforms, consume through music services, or carry with us on our mobile devices can be huge and contain thousands of songs. To be able to find interesting tracks, such musical collections are typically organized in a number of different ways, e.g., by artist, genre or musical style. Since the manual classification of the tracks can be tedious, different approaches for the automated classification and organization of music collections have been proposed in the literature. This chapter focuses on a technique that automatically organizes the tracks of a collection based on a generic and adaptable concept of musical similarity. Furthermore, it presents techniques for the visualization of the elements of a music collection that can help the user to find similar tracks and visually explore a given collection.

Chapter 23 "Music Recommendation": The automated recommendation of music, e.g., on online music platforms, is probably one of the most visible applications of different music data analysis techniques discussed in this book. The chapter reviews the most common general techniques for building automated recommender systems – including collaborative filtering and content-based filtering – and then focuses on the particularities of the music recommendation and automated playlist generation problems. The chapter furthermore discusses the various challenges of evaluating music recommendation systems.

Chapter 24 "Automatic Composition": In contrast to many other chapters, which focus on the processing and analysis of a given musical piece, this chapter touches upon the topic of the computerized composition of music. The chapter first highlights how computers can support composers in their work and introduces the reader to the history of automated composition. Then, a number of different strategies like deterministic, stochastic as well as rule-based and grammar-based approaches are presented that can be used by a computer in the music generation process.

Part IV: Implementation

Chapter 25 "Implementation Architectures": Many of the music analysis processes presented throughout this book are computationally demanding. The automated extraction of features from the audio signal for many tracks or the training phase of complex classification models, for example, easily exceeds the computational resources of typical office computers or mobile devices. This chapter reviews different design alternatives for such systems using a case study of a music classification system. Furthermore, it discusses different possible architectures for a selected set of further applications.

Chapter 26 "User Interaction": Most music-related software applications or, more generally, music processing systems support direct interactions with their users, i.e., they accept different forms of "manual" inputs and provide outputs that are "consumed" by the user. The chapter reviews the various forms of input and outputs of systems designed for the generation, editing, or retrieval of digital music. It covers audio, visual, haptic input and output mechanisms as well as sensor input and actuator output modalities for a number of different music processing systems.

Chapter 27 "Hardware Architectures for Music Classification": For some music processing systems, e.g., for music classification, it can be advantageous to implement the solution directly on a chip, i.e., on hardware. This chapter discusses the challenges a system designer is confronted with when creating such a hardware-based music processing system. The specific problem lies in the fact that one single hardware architecture cannot satisfy the possibly conflicting goals of short computation times, low production costs, low power consumption, and programmability simultaneously. In the chapter we discuss the advantages and disadvantages of the different design alternatives.

1.5 Course Examples

The book is primarily designed for graduate-level courses at universities. Depending on the target audience (e.g., computer scientists, statisticians, or musicologists) and the corresponding background of the students, different course designs are possible.

Table 1.1 shows a course design mainly targeted at *machine learners*, *data analysts*, and *statisticians*, but also suited for musicologists and engineers. The whole course was held within one week at the TU Dortmund, Germany, with about 8 hours of classroom activities (lectures and exercises) per day.

Table 1.1: Example Course Design

Day 1	Music basics	Chapters 2 and 3
Day 2	Signal analysis basics	Chapters 4, 5, and 7
Day 3	Statistical methods	Chapters 9, 11, and 12
Day 4	Model Selection	Chapters 13, 14, and 15
Day 5	Applications	Chapters 16, 17, and 18

An alternative course design particularly for *musicologists* might include the basic Chapters 2, 3, 5, 7, and 8 as well as all chapters on applications (Chapters 16–24). The signal analytical and statistical methods that are needed to understand the application chapters might be briefly explained in passing.

A course design especially for *engineers* might include the basic Chapters 2–5 and 7, the Chapters 11–15 on methodology, the application Chapters 16–20 as well as Chapter 27 on hardware.

A course design targeted to *computer scientists* might include the basic Chapters 2–5 and 7, the application Chapters 17–19 and 22–24, and the hardware Chapters 26 and 27. The material of Chapters 9–15 on methodology should be interspersed when needed.

1.6 Authors and Editors

Authors
Many authors are involved in the book (see Table 1.2), having their background in different disciplines including computer science, engineering, music, and statistics. The core of this group of authors is formed by the *Special Interest Group on Music Data Analysis (SIGMA)*[1] of researchers mainly located in Dortmund und Bochum (Germany). The group has been working on music data analysis for almost a decade and several chapters in particular on applications are based on research works by members of the group. We particularly thank the authors who do not belong to the group, namely Geoffray Bonnin (LORIA, University of Lorraine, France), Johan Pauwels (Queen Mary University of London, England), Geoffroy Peeters (IRCAM, France), Sebastian Stober (University of Potsdam, Germany), and José R. Zapata (Universidad Pontificia Bolivariana, Colombia).

Editors
Claus Weihs is a full professor for Computational Statistics at the Department of Statistics at TU Dortmund, Germany. He received his Ph.D. in mathematics from the Department of Mathematics of the University of Trier, Germany. He has worked for more than nine years as a statistician in the chemical industry (CIBA-Geigy, Switzerland). His research interests include classification methods, engineering statistics,

---

[1] http://sig-ma.de/. Accessed 22 June 2016.

Table 1.2: Author list

Name	Affiliation	Email
Nadja Bauer	TU Dortmund	nadja.bauer@tu-dortmund.de
Holger Blume	Leibniz Universität Hannover	holger.blume@ims.uni-hannover.de
Geoffray Bonnin	LORIA, France	geoffray.bonnin@loria.fr
Martin Bottek	FH Südwestfalen	mbotteck@gmx.de
Martin Ebeling	TU Dortmund	martin.ebeling@tu-dortmund.de
Klaus Friedrichs	TU Dortmund	klaus2.friedrichs@tu-dortmund.de
Tobias Glasmachers	Ruhr-Universität Bochum	tobias.glasmachers@ini.rub.de
Maik Hester	TU Dortmund	mail@maikhester.net
Dietmar Jannach	TU Dortmund	dietmar.jannach@tu-dortmund.de
Sebastian Knoche	TU Dortmund	sebastian.knoche@tu-dortmund.de
Sebastian Krey	TU Dortmund	sebastian.krey@tu-dortmund.de
Bileam Kümper	TU Dortmund	bileam.kuemper@tu-dortmund.de
Uwe Ligges	TU Dortmund	uwe.ligges@tu-dortmund.de
Rainer Martin	Ruhr-Universität Bochum	rainer.martin@rub.de
Anil Nagathil	Ruhr-Universität Bochum	anil.nagathil@rub.de
Johan Pauwels	Queen Mary University of London, England	johan.pauwels@gmail.com
Geoffroy Peeters	IRCAM, France	geoffroy.peeters@ircam.fr
Günther Rötter	TU Dortmund	guenther.roetter@tu-dortmund.de
Günter Rudolph	TU Dortmund	guenter.rudolph@tu-dortmund.de
Ingo Schmädecke	Leibniz Universität Hannover	schmaedecke@ims.uni-hannover.de
Sebastian Stober	Universität Potsdam	sstober@uni-potsdam.de
Wolfgang Theimer	Volkswagen Infotainment GmbH	wolfgang.theimer@ieee.org
Igor Vatolkin	TU Dortmund	igor.vatolkin@tu-dortmund.de
Claus Weihs	TU Dortmund	claus.weihs@tu-dortmund.de
Kerstin Wintersohl	TU Dortmund	kerstin.wintersohl@tu-dortmund.de
José R. Zapata	Universidad Pontificia Bolivariana, Colombia	joser.zapata@upb.edu.co

statistical process control, statistical design of experiments, time series analysis, and, since 1999, statistics in music data analysis. He has co-authored more than 30 papers on the topic of the book.

Dietmar Jannach is a full professor of Computer Science at TU Dortmund, Germany. Before joining TU Dortmund he was an associate professor at University Klagenfurt, Austria. Dietmar Jannach's main research interest lies in the application of intelligent systems technology to practical problems, e.g., in the form of recommendation and product configuration systems.

Igor Vatolkin is postdoctoral researcher at the Department of Computer Science, TU Dortmund, where he received a diploma degree in Computer Science and Music as a secondary subject and a Ph.D. degree. His main research interests cover the optimization of music classification tasks with the help of computational intelligence techniques, in particular evolutionary multi-objective algorithms. He has co-authored more than 25 peer-reviewed papers on the topic of the book.

Günter Rudolph is professor of Computational Intelligence at the Department of Computer Science at TU Dortmund, Germany. Before joining TU Dortmund, he was with Informatics Center Dortmund (ICD), the Collaborative Research Center on Computational Intelligence (SFB 531), and Parsytec AG (Aachen). His research interests include music informatics, digital entertainment technologies, and the development and theoretical analysis of bio-inspired methods applied to difficult optimization problems encountered in engineering sciences, logistics, and economics.

Bibliography

[1] A. Klapuri and M. Davy, eds. *Signal Processing Methods for Music Transcription*. Springer, 2006.

[2] T. Li, M. Ogihara, and G. Tzanetakis, eds. *Music Data Mining*. Chapman & Hall/CRC Data Mining and Knowledge Discovery, 2011.

[3] M. Müller. *Information Retrieval for Music and Motion*. Springer, 2007.

[4] M. Müller. *Fundamentals of Music Processing: Audio, Analysis, Algorithms, Applications*. Springer, 2015.

[5] Z. W. Ras and A. Wieczorkowska, eds. *Advances in Music Information Retrieval*. Springer, 2010.

[6] J. Shen, J. Shepherd, B. Cui, and L. Liu, eds. *Intelligent Music Information Systems: Tools and Methodologies*. IGI Global, 2007.

MATLAB is a registered trademark of The MathWorks, Inc. For product information, please contact:

The MathWorks, Inc.
3 Apple Hill Drive
Natick, MA 01760-2098 USA
Tel: 508 647 7000
Fax: 508-647-7001
E-mail: info@mathworks.com
Web: www.mathworks.com

Part I

Music and Audio

Chapter 2

The Musical Signal: Physically and Psychologically

SEBASTIAN KNOCHE
Department of Physics, TU Dortmund, Germany

MARTIN EBELING
Institute of Music and Musicology, TU Dortmund, Germany

2.1 Introduction

In Ancient Greek, the *Muses* were the goddesses of the inspiration of literature, science and arts, and were considered the source of knowledge. The Latin word *musica* and eventually the word *music* derive from the Greek *mousike*, which means *the art of the Muses*. Although music is ubiquitous in all cultures, a commonly accepted definition of music has not yet been given. Instead, through the centuries a variety of definitions and classifications of music from different perspectives have been proposed [25]. All definitions of music agree about its medium: Music constitutes communication by artificially organized sound. For the purpose of communication, the structures of music must be comprehensible in connection with auditory perception and cognition.

The constitutive musical signals are tones. From a phenomenological point of view, a tone is a sensation. Different essential sensational moments of a tone can be discriminated such as pitch, loudness, duration, and timbre. Tones are distinguished by their tone names, which refer to the pitch. Two tones with different loudness values or different timbres but the same pitch are identified as the same tone merely played with different intensities and on different instruments. Thus, according to Carl Stumpf (1848–1934) the *tonal quality* of pitch is the crucial sensational moment of a tone, whereas the *tonal intensity* is its loudness [30].

According to the specific properties of the tonal sensation, music is organized in relation to the sensational moments pitch (tone name), intensity (dynamics), rhythm (onset and offset), and timbre (instrumentation). In what follows we will consider each single sensational moment of a tone by looking at the musical signal as a sound and as a sensation. This includes sound generation and sound propagation as well as the psychoacoustic preconditions of music perception and cognition.

2.2 The Tonal Quality: Pitch — the First Moment

2.2.1 Introduction

Pitch as the sensational quality of a tone is correlated with the frequency content of the stimulus. Perceiving two pitches, it is possible to decide whether they are equal or not. In case of inequality, it is also possible to determine which pitch is the higher one. Thus, it is possible to sort and linearly graduate tones according to their perceived tone height or pitch. As a result, the tonal quality as one dimension of tones is a dense sensational continuity of pitches [8]. This property of pitch perception is not at all trivial. Recall for example that a graduation of color perception is not possible. The organization of tonal systems is exclusively based upon pitch.

A number of different stimuli elicit a pitch sensation, e.g., pure tones (sine waves), harmonic complex tones, band-limited noise, amplitude modulated noise, click trains, and so on. All these stimuli are characterized by periodic time courses of air pressure variations propagating in a medium as waves from the sound source to the ear.

Let $x(t)$ be a function to describe the pressure variation of the medium. It is periodic with the period T, if $x(t+T) = x(t)$ for every time t. After each duration of T the waveform recurs again and again. The number of repetitions of the waveform in time is its frequency $f = 1/T$. The repetition rate corresponds to the perceived pitch. Next to a constant, the simplest periodic waveform is a sine wave which is often used as a theoretical idealization of a tone. A pure sine tone can only be produced by electronic devices. The tones of musical instruments are harmonic complex tones which consist of sine tones with harmonically related frequencies: All frequencies are integer multiples of a fundamental frequency. Normally, harmonic complex tones elicit only a single pitch percept, which corresponds to the fundamental frequency.

The identification of pitch with the frequency of a sine wave dates from the late 18th century, e.g., Daniel Bernoulli (1700–1782). Already Bard Taylor (1685-1731) and Leonard Euler (1707–1783) described vibrating strings by sums of sines and cosines. In 1822, Fourier presented his famous theorem to solve a problem on thermal conduction. This theorem demonstrates the expansion of periodic functions into trigonometric series, which are sums of sines and cosines. The starting point for the application of Fourier's theorem in acoustics was Georg Simon Ohm's (1843) definition of a tone as a sound containing a sine wave. Since Ohm's times, the frequencies of sine waves are used as measurable references of the pitch [21].

2.2.2 Pure and Complex Tones on a Vibrating String

As a very simple model of a musical instrument, we consider a one-dimensional string under a tensional force F_t. The following *Physics Interlude*[1] discusses the behavior of such a string. It is shown that a string of length L can vibrate in certain *eigenmodes*. The eigenmodes can be numbered with an index $n = 1, 2, 3, \ldots$ and they oscillate with frequencies which are integer multiples of the smallest frequency called *fundamental frequency* f_0 . Thus, the frequencies of the eigenmodes are $f_n = n \cdot f_0$. All eigenmodes can be played simultaneously on a single string (this is called superposition), which motivates Definition 2.3 of tones below.

Notation of the fundamental frequency: In acoustics, phonetics, and audio engineering it has become common practice to annotate the fundamental frequency by f_0 . But as $f_1 = 1 \cdot f_0 = f_0$, there are two correct ways to write the fundamental frequency: f_1 or f_0. In arithmetic expressions it is generally more reasonable to use the notation f_1 but if we explicitly refer to a fundamental frequency, we write f_0 (see [12, p. 82]).

A Physics Interlude

The string is oriented along the z-axis, and it can be displaced transversely. The displacement $u(z,t)$ is a function of the spatial coordinate z and time t, and its time evolution is governed by the one-dimensional wave equation [9].

Definition 2.1 (One-Dimensional Wave Equation). The *one-dimensional wave equation* is the linear partial differential equation

$$\ddot{u}(z,t) = c^2 u''(z,t) \tag{2.1}$$

that involves temporal and spatial derivatives of a function $u(z,t)$, denoted with a dot and a prime, respectively, so that $\ddot{u} = \partial^2 u / \partial t^2$ and $u'' = \partial^2 u / \partial z^2$. The constant c that appears in the wave equation is called the *phase velocity*.

For a string under tension, the phase velocity depends on the tensional force F_t and the line density ρ_l of the string (measured in kg/m) by $c = \sqrt{F_t/\rho_l}$. A solution $u(z,t)$ depends not only on the wave equation, but also on the initial conditions (i.e. initial configuration $u(z,t_0)$ at the starting time t_0, initial velocity $\dot{u}(z,t_0)$), and boundary conditions, which must be specified.

For infinitely long strings there is a very general class of solutions, the *D'Alembert solutions*.

Theorem 2.1 (D'Alembert Solution of the Wave equation). The functions $u_+(z,t) = f(z+ct)$ and $u_-(z,t) = f(z-ct)$ with any twice differentiable function f satisfy the wave Equation (2.1). They are called the *D'Alembert solutions* of the wave equation.

---

[1]The Physics Interludes in this chapter go deeper into the physical mechanisms underlying music, and may be skipped by readers who are satisfied with a purely phenomenological description of the musical signal.

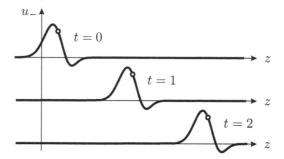

Figure 2.1: D'Alembert solution $u_-(z,t) = f(z-t)$ (phase velocity $c = 1$), plotted at three different times. For $t = 0$, we just see the plot of $f(z)$. A point on the curve is identified by its function value $f(z_0)$. At a later time t_1, this function value is found at a different position $z_1 = z_0 + t_1$ since $u_-(z_1,t_1) = f(z_0 + t_1 - t_1) = f(z_0)$.

They represent traveling waves of shape $f(z)$, which move in negative z-direction (u_+) or positive direction (u_-); see Figure 2.1 for an example. Because the wave propagates into a direction perpendicular to the movement of the string elements (which move up and down), this kind of wave is called a *transverse wave*.

Since the wave equation is linear, a superposition of solutions also satisfies the equation. Accordingly, the general solution of the wave equation is $u(z,t) = f(z - ct) + g(z + ct)$ consisting of a function f traveling to the right and a function g traveling to the left. The functions f and g must be determined from the given initial conditions.

The above considerations are important for propagating waves, like sound waves in air (whose detailed properties we will study in Section 2.3). However, vibrating strings, like in a piano or violin, are of finite length. The ends of the string are clamped, which imposes boundary conditions and necessitates a different analysis.

Let us now assume that the string is clamped at $z = 0$ and $z = L$, which imposes the boundary conditions $u(0,t) = u(L,t) = 0$ for all times t. The D'Alembert solutions generally fail to satisfy these boundary conditions. Another solution technique, called *separation*, is used instead. We make an ansatz that splits spatial from temporal variables, $u(z,t) = Z(z)T(t)$. Inserting it into Equation (2.1) yields

$$Z(z)\ddot{T}(t) = c^2 Z''(z)T(t) \quad \Leftrightarrow \quad \frac{1}{c^2}\frac{\ddot{T}(t)}{T(t)} = \frac{Z''(z)}{Z(z)}. \tag{2.2}$$

Note that the left-hand side of the last form only depends on the independent variable t, and the right-hand side only on z, but the equation has to be satisfied for all z and t. This is only possible when both sides of the equation are actually independent of z and t, i.e. equal to some constant $-k^2$ (this form of the constant is chosen so that the results can be written in a simple form).

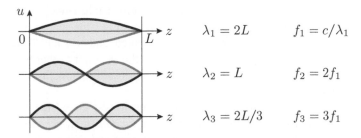

Figure 2.2: First three eigenmodes of a vibrating string. The spatial structures oscillate up and down with frequencies f_n. The first mode oscillates with the fundamental frequency $f_1 (= 1 \cdot f_0)$; higher modes oscillate with multiples of the fundamental frequency $f_n = n f_1 = (n f_0)$.

So the spatial differential equation reads $Z''(z) = -k^2 Z(z)$, which is the familiar differential equation of a harmonic oscillator and has sine and cosine functions as solutions. For our boundary conditions $X(0) = X(L) = 0$, only sine functions with a root at $z = L$ are appropriate, $Z(z) \sim \sin(k_n z)$ with $k_n = n\pi/L$ and $n \in \mathbb{Z}$. Thus the constant k is quantized, i.e. may only assume discrete values k_n.

The time-dependent differential equation $\ddot{T}(t) = -c^2 k_n^2 T(t)$ is of the same structure as the spatial differential equation, and its general solution is $T(t) \sim \sin(\omega_n t + \varphi_n)$ with $\omega_n = ck_n$ and an integration constant φ_n representing the initial phase at time $t = 0$.

To obtain the final solution, the results of the spatial and temporal parts must be inserted back into the ansatz. In summary, there are infinitely many solutions, indexed by $n = 1, 2, 3, \ldots$, which are called the eigenmodes of a clamped string:

Definition 2.2 (Eigenmodes of a Clamped String). The *eigenmodes of a clamped string* are the solutions

$$u_n(z,t) = A_n \sin(k_n z) \sin(\omega_n t + \varphi_n) \quad \text{with} \quad k_n = n\pi/L \quad \text{and} \quad \omega_n = ck_n \tag{2.3}$$

of the wave equation, with coefficients A_n and φ_n which must be determined from the initial conditions.

The *wave number* k_n and *angular frequency* ω_n describe the spatial and temporal dilatation of the sine functions. The wave number is related to the *wave length* λ, which measures the spatial distance between two oscillation maxima, by $\lambda = 2\pi/k$. Other typical parameters of the temporal structure are the frequency f, which measures the number of oscillations per second, and the period T, which measures the duration of one oscillation. They are related to the angular frequency by $f = \omega/2\pi$ and $T = 1/f$.

The relation $\omega = ck$ with which the angular frequency was introduced is called the *dispersion relation* and can be equivalently expressed as $f = c/\lambda$.

Although it looks as simple as the previous relations, it has much more physical meaning. Basically, λ and k describe the same thing – the spatial stretching of the sine wave – and are therefore obviously related, and ω, f and T all describe how fast the oscillations are. The dispersion relation, on the other hand, relates the spatial to the temporal characteristics, and describes how fast a wave with given wavelength will oscillate.

In Figure 2.2, the spatial structure of the lowest three eigenmodes of a vibrating string, called the *first harmonic* or *fundamental*, *second harmonic*, and *third harmonic* are sketched. It oscillates up and down with proceeding time, and thus represents *standing waves*. The boundary conditions enforce the wavelength $\lambda_n = 2\pi/k_n = 2L/n$ to be quantized, so that two nodes of the sine fall exactly on the string boundaries. The frequencies of the harmonics are also indicated in Figure 2.2. They are integer multiples of the fundamental frequency (concerning the notation of the fundamental frequency, see remark at the beginning of this section)

$$(1 \cdot f_0) = f_1 = \frac{c}{\lambda_1} = \frac{1}{2L}\sqrt{\frac{F_t}{\rho_l}}, \tag{2.4}$$

which depends on the force F_t acting on the string, its line density ρ_l and its length L. The fundamental frequency can be tuned by adjusting these characteristics of the system, corresponding to different notes being played on the string.

The general solution for the clamped string is a linear superposition of the eigenmodes:

Theorem 2.2 (General Solution for the Clamped String). The linear superposition of the eigenmodes with arbitrary coefficients A_n and φ_n,

$$u(z,t) = \sum_{n=1}^{\infty} A_n \sin(k_n z) \sin(\omega_n t + \varphi_n), \tag{2.5}$$

is the general solution of the wave equation with boundary conditions $u(0,t) = u(L,t) = 0$.

The coefficients of the superposition can be deduced from the initial conditions with a Fourier analysis (see Section 2.2.6). Again, a uniquely defined solution requires the initial shape $u(z,0)$ and initial velocity $\dot{u}(z,0)$ to be given, see [1] for a worked out example.

End of the Physics Interlude

The vibrating string generates sound waves that propagate through the air, and eventually arrive at the listener's ear. We will discuss the physics behind sound propagation below in Section 2.3; for now we are satisfied with the result that a time signal $x(t)$ arrives at the ear and is perceived. This signal has the same time dependence as the deflection Equation (2.5) of the vibrating string. Based on this signal, we define two types of tones, pure and complex ones. The tones of musical are typically complex tones.

Definition 2.3 (Pure and Complex Tones, Partials, Harmonics, Overtone series). A *pure tone* or *sine tone* is a signal that consists of only one frequency f,

$$p(t) = A\sin(2\pi f t + \varphi). \qquad (2.6)$$

It is completely described by its frequency f (or angular frequency $\omega = 2\pi f$), its amplitude A and its starting phase φ. On the other hand, a *complex tone* is a sum of arbitrary sine tones, called *partials* $p_n(t) = A_n\sin(2\pi f_n t + \varphi_n)$:

$$x(t) = \sum_{n=1}^{N} A_n\sin(2\pi f_n t + \varphi_n). \qquad (2.7)$$

If the frequencies of the partials are integer multiples of a fundamental frequency f_0: $f_n = nf_0$ (note that $f_1 = f_0$, see remark concerning the notation of the fundamental at the beginning of this section), $p_n(t)$ is a *harmonic partial tone* and $x(t)$ is a *harmonic complex tone*.

The series of tones with harmonic frequencies $\tilde{f}_n = (n+1)f_0$ with $n = 1, 2, \ldots$ is called an *overtone series*. Note, that the n-th overtone is the $(n+1)$-th partial: $\tilde{f}n = f_{n+1}$.

A complex tone is periodic if and only if it is a harmonic complex tone. According to our discussion above, this is the case for tones played on a monochord (which is the experimental realization of a vibrating string). The tones of musical instruments and of the singing voice are periodic complex tones or synonymously harmonic complex tones.

Periodic complex tones elicit a single clear pitch percept equal to the pitch of a sine tone with the frequency of the fundamental f_0 (of slightly lower). The strength of the individual partial and their phase shifts have (almost) no influence on the pitch percept. On the other hand, the strengths of the partials are crucial for timbre perception as described below. Under certain conditions, some subjects are able to identify single partials of a complex tone up to the sixth or seventh overtone [20].

A phenomenon widely discussed in psychoacoustics is the *pitch of the residue*. The pitch of the fundamental is heard even if the fundamental frequency component is eliminated from a harmonic complex tone. Because the *residue* of the remaining higher partials contains frequencies which are integer multiples of this missing fundamental frequency: $f_n = n \cdot f_0$ corresponding to the periods $T_n = 1/f_n$, the lowest common multiple of these periods is the period of the fundamental $T_0 = 1/f_0$ (note that $T_1 = 1/f_1 = 1/f_0 = T_0$; concerning the notation of the fundamental see the beginning of this section). Thus, although the frequency of the missing fundamental is not contained in the tone itself, the period of the fundamental is still the lowest period of the complex tone without fundamental thus eliciting the residue pitch [26]. Further, low partials can be eliminated from the complex tone without changing the pitch percept. Only the timbre of the tone changes. For some subjects, three (sometimes even two) adjacent harmonic partials are sufficient for the perception of a residue pitch. This indicates that a periodicity detection mechanism in the auditory system is responsible for pitch perception [16].

2.2.3   Intervals and Musical Tone Height

In musical systems, tones are sorted linearly according to their perceived tone height or pitch. Perceptionally, *intervals* are distances between tones, acoustically they are the relations of the corresponding frequencies (or equivalently periods) of the interval tones. It is remarkable that in tone perception and thus also in musical systems it is possible to join intervals. Together with the above-mentioned graduation of pitches in the dimension of the tonal quality (see Section 2.2.1), tones and intervals form an algebraic structure in perception: in tone perception we involuntarily calculate with the sensations of tones and intervals as if we were calculating with numbers [30, 8]. Consider two intervals. The first perceived interval Δ_1 is the perceptional distance between the pitch τ_1 and the pitch τ_2 and the second perceived interval Δ_2 is the perceptional distance between pitch τ_2 to the pitch τ_3, and we hear the successive intervals Δ_1 and Δ_2 constituting a melodic line:

$$\Delta_1 = \tau_2 - \tau_1, \quad \Delta_2 = \tau_3 - \tau_2. \tag{2.8}$$

If a melody moves from τ_1 to τ_2 and then to τ_3, both intervals are joined together and the melody has moved over the interval $\Delta_3 = \tau_3 - \tau_1$. Perceptually, the intervals that are tonal distances are added. Normally, we are still aware of the first tone and can hear that the melody has moved over a distance of the Interval Δ_3:

$$\Delta_1 + \Delta_2 = \tau_2 - \tau_1 + \tau_3 - \tau_2 = \tau_3 - \tau_1 = \Delta_3. \tag{2.9}$$

To give a simple example from music perception, let τ_1 be tone c' and let τ_2 be tone g'. These tones form the interval of a pure fifth: $\Delta_1 = pure\ fifth$, and we may note *pure fifth* = g' − c'. If τ_3 is the tone c'', the tones τ_3 and τ_2 (g') have a distance of a pure fourth: $\Delta_2 = pure\ fourth$, and we note *pure fourth* = c'' − g'. The tones c'' and c' are a pure octave apart, so that $\Delta_3 = pure\ octave$ and we note *pure octave* = c'' − c'. It is well known that a pure fifth and a pure fourth joined together lead to a pure octave, which can easily be verified by hearing or singing or on any instrument, for example on a monocord. According to Equation (2.9) we calculate

$$\Delta_1 + \Delta_2 = pure\ fifth + pure\ fourth = \text{g' − c' + c'' − g'} = \text{c'' − c'} = \Delta_3. \tag{2.10}$$

The summation of perceived intervals in hearing can be compared to the vibration ratios derived from the lengths of vibrating strings (tensions kept constant) as was done in ancient times. If l_i is the length of a string of the tone τ_i, the corresponding *vibration ratio* of these intervals are

$$s_1 = \frac{l_1}{l_2}, \quad s_2 = \frac{l_2}{l_3}, \quad s_3 = \frac{l_1}{l_3}. \tag{2.11}$$

Note that the vibration ratio s_1 corresponds to the interval Δ_1, the vibration ratio s_2 corresponds to the interval Δ_2, and the vibration ration s_3 corresponds to the interval Δ_3. Obviously, joining together the intervals Δ_1 and Δ_2 corresponds to the multiplication of the vibration ratios:

$$s_1 \cdot s_2 = \frac{l_1}{l_2} \cdot \frac{l_2}{l_3} = \frac{l_1}{l_3} = s_3. \tag{2.12}$$

Comparing Equation (2.9) to Equation (2.12) reveals that the addition of interval sensations $(\Delta_1 + \Delta_2)$ corresponds to the multiplication of vibration ratios $(s_1 \cdot s_2)$. The interval sensations can be regarded as an additive algebraic structure (see above), whereas the corresponding vibration ratios form a multiplicative algebraic structure. If L is a function that maps the vibration ratios $\{s\}$ onto the interval sensations $\{\Delta\}$, then L has to obey the logarithmic rule $L(s_1 \cdot s_2) = L(s_1) + L(s_2) = \Delta_1 + \Delta_2$. Logarithmic functions are the only functions to obey this rule. Thus there is a logarithmic relation between the multiplicative vibration ratios and the additive interval sensations. In principle, the basis of the logarithm can be chosen arbitrarily. If L is chosen to be the logarithm to the basis 2, the octave (vibration ratio 2:1) is mapped onto the number 1, as $\log_2(2:1) = 1$. Because of the importance of the octave in music theory, it is quite advisable to use the logarithm to the basis 2 for the construction of the psychoacoustic function L. Thus the physical definition of Tone Height as a Numerical Quantity, cp. Equation (2.13), is in line with perception.

As the overtone series has been regarded as a natural phenomenon since ancient times, intervals between two partials are called *natural* or *pure intervals*. The interval that is found between the fundamental and the first tone of the overtone series (that is, between the first and second partial) is called an *octave*. In the frequency region relevant for music, the octave corresponds to a vibration ration of about 2:1. Restricted to this frequency region, one may define:

Definition 2.4 (Octave). An *octave* is the interval between two tones whose frequencies are in a ratio of $f_2/f_1 = 2$. A tone is said to be one octave higher than another tone if it has the double frequency; and it is one octave lower if it has half the frequency.

The same ratio $f_2/f_1 = 2$ can be found between the fourth and second tone of the overtone series, the eighth and fourth, and so on. Each of these tone pairs forms the interval of one octave.

A tone that is one octave above another tone has a much higher pitch. Nevertheless, both tones are sensed to be quite similar. This phenomenon is called *octave identification*. In almost all tonal systems, pitches of tones with frequency ratios of 2^k (with integer k) are regarded as the same tone and are given the same tone name. The octave identification is crucial for almost all music systems. Some peculiarities of the perception of octaves are discussed in below in Section 2.2.5.

Music makes use of a discrete selection from the continuity of pitches to provide the tonal material for melodies. To construct this discrete selection, an octave is divided into smaller intervals. This can be achieved in many different ways, which lead to different tonal systems. In ancient times, some pure intervals of the overtone series were used to construct all musical notes. The *Pythagorean system* is constructed on the basis of the pure fifth and the pure octave, whereas *mean tone temperaments* use pure thirds, pure fifths, and pure octaves to get purely tuned triads. However, as a result only some chords sound more or less pure, whereas some others become unbearably mistuned. Equidistant tonal systems divide the octave into equal intervals. For mathematical reasons, except for the octave, the intervals of an equidistant tonal system cannot be pure intervals. Here we present the currently established tuning system of Western music, the *equal tempered system*, which is based on an equidistant division of the octave into 12 semitones.

To develop the mathematics behind the musical pitches, we start with some basic definitions that allow us to calculate with tones. A tone τ is, in principle, a sensation – and not a number to calculate with. The variable τ associated with a tone could be identified, for example, with the tone name. The frequency $f(\tau)$ of a pure tone or the fundamental of a complex tone ($f_0 = f(\tau)$), on the other hand, is a numerical quantity, given in the unit Hz. Based on the frequency, we define our numerical scale for the musical tone height by:

Definition 2.5 (Tone Height as a Numerical Quantity). A tone τ with frequency $f(\tau)$ has a *tone height*

$$H(\tau) = 12\log_2(f(\tau)/F) = 12(\log_2(f(\tau)) - \log_2(F)), \tag{2.13}$$

where F is the frequency of a reference tone, which in principle could arbitrarily be defined, but should be chosen according to the musical context.

Note that the tone height is a perceptional distance from a reference tone with frequency F. The definition thus resembles the widespread relative pitch: the sensation of tone height depends on a reference pitch. In contrast, the absolute pitch is the quite rare ability to name tones without any reference.

Let us calculate the tone heights of a harmonic complex tone. In this case, it is reasonable to take the fundamental frequency as reference: $F = f_0$. Take a monochord that is tuned to a fundamental frequency $f_0 = 65.406\,\text{Hz}$, which corresponds to the tone C (the other tone names will be defined below) and calculate the tone heights of all the harmonic partials h_1, $h_2 \ldots$, now regarded as tonal sensations. We already know that the frequency f_n of the n-th harmonic partial h_n is $f_n = f(h_n) = n \cdot f_0$ (see Definition 2.3). From Equation (2.13) we obtain the tone height $H(h_n) = 12\log_2(n)$ of the *n*-th harmonic . A plot of the function H is shown in Figure 2.3. This figure also shows the musical notation of the tones h_n and again demonstrates the logarithmic relation between frequency ratios and interval sensations. Note, that the musical notation of the overtone series approximates the shape of the logarithm function plotted above.

Our definition of the numerical tone height is closely related to the one used by the MIDI standard, cp. Section 7.2.3; it is merely shifted by a constant value. The tone C with $F = 65.406\,\text{Hz}$ is assigned to the MIDI number $M = 36$, and tones with other frequencies f are assigned the numbers $M(\tau) = 36 + 12\log_2(f(\tau)/F)$. With the logarithmic identity $\log x + \log y = \log(x \cdot y)$ this can be equivalently expressed as $M(\tau) = 12\log_2(f(\tau)/F')$ with $F' = F/8 = 8.176\,\text{Hz}$.

Definition 2.6 (Intervals as Numerical Quantities, Cent Scale). Intervals are differ-

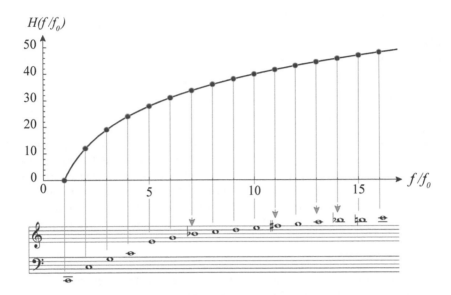

Figure 2.3: Plot of the sensational tone height $H(h_n)$ of the overtone series h_n of the tone C. On the abscissa, the ratio of the frequency to the fundamental frequency f_0 is shown. The numbers on the abscissa are the numbers n of the partials which are equal to f_n/f_0. The dots mark the tone heights of the partials. Note, that the n-th overtone is the $(n+1)$-th partial. At the bottom, the musical notation of the overtones is shown, where arrows indicate overtones that are a bit lower than the corresponding tones of the equal temperament.

ences in tone height between two tones. The interval between the tones τ_1 and τ_2 is assigned the numerical value

$$I(\tau_1, \tau_2) = H(\tau_2) - H(\tau_1) = 12\log_2(s(\tau_1, \tau_2)), \tag{2.14}$$

where $s(\tau_1, \tau_2) = f(\tau_2)/f(\tau_1)$ is the *vibration ratio* of the interval. A finer scale to measure intervals is the *cent scale*, often used in electronic tuners, on which the hundred-fold value of the above definition is given, i.e. $I(\tau_1, \tau_2) = 1200\log_2(s(\tau_1, \tau_2))$ *cent*.

Note that a semitone step comprises 100 cents; an octave corresponds to 1200 cents. In the preceding definition of intervals, the logarithmic identity $\log x - \log y = \log(x/y)$ was used. The reference frequency F cancels out, so that the interval between two tones only depends on the vibration ratio, that is, the ratio of their frequencies. An octave, for example, is assigned to the numerical value $I_O = 12$ in our system, because it has a vibration ratio $s_O = 2$ and because $\log_2(2) = 1$.

In our aim to divide the octave into twelve equidistant intervals, the logarithmic relationship between frequencies and tone heights must be taken into account. On the tone height scale, the octave $I_O = 12$ shall be divided into twelve equal intervals

I_H, such that $I_O = \sum_{n=1}^{12} I_H = 12 \cdot I_H$, which simply means that the interval we are
looking for has a numerical value of $I_H = 1$. However, on the frequency scale we are
looking for a vibration ratio s_H which leads to the vibration ratio $s_O = 2$ of an octave
when multiplied twelve times: $s_O = \prod_{n=1}^{12} s_H = s_H^{12}$, which leads to $s_H = \sqrt[12]{2}$. This
explains the following definition of a semitone step.[2]

Definition 2.7 (Semitone Step). A *semitone step* is an interval $I_H = 1$, so that twelve
semitone steps add up to one octave. Its corresponding vibration ratio is $s_H = \sqrt[12]{2}$.

Raising a tone τ_1 by a semitone step produces a tone τ_2 with the frequency
$f(\tau_2) = s_H \cdot f(\tau_1)$, and vice versa, lowering a tone by a semitone step produces
a tone with the frequency $f(\tau_2) = f(\tau_1)/s_H$. Intervals smaller than the semitone
step are not used in traditional Western music. The frequencies for all pitches of
equal temperament can be calculated by ascending and descending in semitone steps
from a reference tone, which is traditionally the musical note a' with a frequency of
$f(\text{a'}) = 440$ Hz and leads us to the following definition.

Definition 2.8 (Equal Temperament). The *equal temperament* is a tonal system comprised of all tones with a tonal distance of an integer number of semitones from the
reference pitch a' with a frequency of 440 Hz. The frequency of a tone that is $|k| \in \mathbb{Z}$
semitone steps apart from a' is

$$f = s_H^k \cdot f(\text{a'}) = 2^{k/12} \cdot 440 \text{ Hz}, \tag{2.15}$$

where $k > 0$ corresponds to a tone that is higher than a', and $k < 0$ to a tone that is
lower than a'.

2.2.4 Musical Notation and Naming of Pitches and Intervals

In music theory and practice, pitches are referred to by their names, and not by
their frequencies or numerical tone heights as defined by Equation (2.13). The usual
designation of a pitch consists of a letter, possibly an alteration sign, and an octave
indication. Because of the octave identification, the tone names are repeated in each
octave, and thus it is sufficient to explain them for one octave as shown in Figure 2.4.
The untransposed diatonic scale consists of the notes c, d, e, f, g, a, b, c. In German,
the b is replaced by h. These tones are not spaced equally: There are semitone
steps from e to f and from b to c, but between the other adjacent notes there is an
interval of two semitone steps or a whole tone step. Accordingly, there are additional
tones within this octave, namely those in between the whole tone steps, for example
between c and d.

These additional notes are called *chromatic* notes and are deduced from the diatonic notes by *alteration*. A *flat* sign (symbol: ♭) or a *sharp* sign (symbol: ♯) before
one of the notes of the diatonic scale changes its pitch: A flat lowers the original
pitch by a semitone and a sharp raises it by a semitone. The alteration of a note is

---

[2]This definition holds for the equal temperament; other tuning systems do not divide the octave
equidistantly.

Figure 2.4: Musical notation and naming of the diatonic notes (black). The brackets [and ⟨ indicate whole tone and semitone steps, respectively, between the diatonic notes. The *chromatic* notes between the diatonic notes are printed in gray.

revoked by the *natural* sign (symbol: ♮). Each of the diatonic notes can be raised by a sharp or lowered by a flat. Note that the same pitch can be represented by various notes. The note between c and d for example can be noted as c-sharp or alternatively as d-flat. Both c-sharp and d-flat have the same pitch. The notes e and f are separated by only one semitone step, so that there is no note in between them: e-sharp has the same pitch as f, and f-flat the same as e. The same holds for b and c. In the strict sense, these so-called *enharmonic changes* without pitch changes are only possible in the equal temperament. In other tuning systems, an enharmonic change slightly changes the pitch, e.g., the note f-sharp would be slightly higher than g-flat in pure tuning.

Finally, the octave indication completes the tone name. The octave presented in Figure 2.4 is marked with a prime – it is the one-line octave. Higher octaves are denoted with an increasing number of primes; as indicated in the figure, c" introduces the two-line octave, and so on. Lower octaves are marked as follows. Directly below the one-line octave, there is the small octave, indicated with lower-case letters for the tone names without any prime, for example c. Even below, there is the great octave, indicated by capital letters, like C. The contra octave is indicated by underlining a capital letter (like C̲) or by a comma following a capital letter (like C,) and this notation is continued by adding more underlines or commas, respectively, for even deeper octaves.

The *English notation*, also called the *scientific notation* of pitch, uses the same note names but numbers the octaves from the bottom up starting from C̲ with 32.703 Hz. Thus the tones from C̲ to H̲ are the first octave, denoted by C1, D1, E1 etc., the tones from C with 64.406 Hz to H are the second octave, denoted C2, D2, E2, etc. and so on. The one-line octave is the fourth octave and C4 corresponds to c' and has a frequency of 262.626 Hz. The concert pitch a' with 440 Hz is the tone A4 in the scientific system.

The *MIDI standard* assigns numbers to all notes from the bottom up starting with the even inaudible note A-1 with 8.176 Hz as midi-number 0. The notes are chromatically numbered all the way through so that the tone c' or C4 respectively with 262.626 Hz has the midi-number 60, the concert pitch of 440 Hz has the midi-number 69.

As we are now able to count the number n of semitone steps between any given note and the reference pitch a', we can calculate the frequencies of all tones of the equal temperament according to Equation (2.15). Table 2.1 summarizes all tones of

Table 2.1: Notes of the One-Line Octave and the Corresponding Frequencies According to the Equal Temperament

diatonic scale	chromatic scale	frequency [Hz]
c'	c'	261.626
	c' sharp = d' flat	277.183
d'	d'	293.665
	d' sharp = e' flat	311.127
e'	e'	329.628
f'	f'	349.228
	f' sharp = g' flat	369.994
g'	g'	391.995
	g' sharp = a' flat	415.305
a'	a'	440.000
	a' sharp = b' flat	466.164
b'	b'	493.883
c"	c"	523.251

the one-line octave. The tones in all other octaves can be obtained by doubling or halving these frequencies to ascend or descend by one octave.

Intervals are also referred to by names rather than numerical values as defined in Equation (2.14). Originally, the tonal distances were determined by counting the notes of the diatonic scale. Thus the intervals are named after ordinal numbers: prime, second, third, fourth, and so on. The distances between all chromatic tones can be determined by counting the contained whole tone and semitone steps on the basis of the diatonic scale.

Because of their strong tonal fusion and for historical reasons, the intervals prime, fourth, fifth, and octave are attributed as pure intervals. If these intervals are augmented by a semitone they are called augmented intervals, and if they are diminished by a half tone they are called diminished intervals. The other intervals exist in two forms: as minor and major intervals differing by a semitone. Diminishing a minor interval by a semitone results in a diminished interval. Augmenting a major interval by a semitone leads to an augmented interval.

The intervals of the chromatic scale starting from c' are summarized in Table 2.2. Along with the numerical values I of Definition 2.6, the vibration ratios s are given, which can be obtained as $s = 2^{I/12}$. Except for the prime and octave, vibration ratios are not rational numbers, which is an artefact of the equal temperament. Other attempts to construct tonal systems, such as the *pure intonation*, try to ascribe vibration ratios of simple rational numbers to the consonant intervals of fifth, fourth and major third. From the overtone series, see Figure 2.3, one obtains the vibration ratios $3 : 2 = 1.5$ for the pure fifth, $4 : 3 = 1.\overline{3}$ for the pure fourth and $5 : 4 = 1.25$ for the major third, which are not perfectly matched by the equally temperament. It was an assumption of speculative music theory, that simple vibration ratios are re-

Table 2.2: Intervals of the Chromatic Scale and their Numerical Values *I* and Vibration Ratios *s* According to Equation (2.14) in the Equal Temperament

interval	example	$I(\tau_1, \tau_2)$	$s = f(\tau_2)/f(\tau_1)$
pure prime	c' - c'	0	1
minor second	c' - d' flat	1	1.059
major second	c' - d'	2	1.122
minor third	c' - e' flat	3	1.189
major third	c' - e'	4	1.260
pure fourth	c' - f'	5	1.335
augmented fourth	c' - f' sharp	6	1.414
pure fifth	c' - g'	7	1.498
minor sixth	c' - a' flat	8	1.587
major sixth	c' - a'	9	1.682
minor seventh	c' - b' flat	10	1.782
major seventh	c' - b'	11	1.888
pure octave	c' - c"	12	2

sponsible for the consonance of two tones being played simultaneously. The smaller the numbers of the ratio the more consonant is the interval. An underlying reason might be that the period of the resulting superposition signal is quite short, because the periods of the individual tones have a relatively small common multiple. However, as a consequence of the irrational vibration ratios, the superposition of two equally tempered tone signals is not periodic at all, unless the two tones are in octave distance. Nevertheless, the consonant intervals of fifth, fourth and major third have the same character in the equal temperament; see Section 3.5.1 below. It needs a musically trained ear and certain circumstances to distinguish the irrational vibration ratios of the equal temperament from the ideal rational ones, and a demand of high standards to be bothered by these differences. These flaws of the equal temperament have therefore been accepted, and they are outweighed by the benefits concerning the possibility to modulate through all scales. On the other hand, ensembles of early music favor historical tuning systems, which noticeably change the sound and character of early music. Furthermore, these tuning systems are more convenient for historical instruments, i.e. old organs and keyboard instruments.

2.2.5 The Mel Scale

Definitions 2.5 and 2.6 of the numerical scales for tone height and intervals are based on the octave, which is the most consonant interval, besides the pure prime. Tones with a tone height difference of one octave sound very harmonic and if played simultaneously, they fuse to a perceptional entity because of the simple vibration ratio of 2 : 1. The mechanism behind this phenomenon is the periodicity detection mechanism of our auditory system. However, this period detection mechanism only works for frequencies that are not too high; roughly speaking, not higher than about 2000 Hz [16].

In contrast to the physical octave with a vibration ratio of 2 : 1, listeners prefer a slightly greater vibration ratio for the octave. Due to refractory effects, the interspike-intervals of the neural coding of intervals in the auditory system deviate from the integer ratios of the stimuli [18].

The perceptional scales of *melodic octaves* are equal to the scale of harmonic octaves up to a frequency of 500 Hz. But for higher frequencies the perception of melodic octaves deviates considerably from the vibration ratio of 2 : 1 as described in what follows.

If a listener is presented a pure tone a' with $f(a') = 440$ Hz and is asked to adjust the frequency of a second tone so that it has half the pitch of a', he will find a tone with about 220 Hz, at least when he is musically trained [34]. At higher frequencies, however, the frequency ratio of two tones must be much larger to elicit the sensation of half pitch (or double pitch). If the half pitch of a tone with a frequency of 8 kHz is to be determined, subjects on average find a tone around 1.8 kHz, and not around 4 Hz.

Systematic pitch halving and pitch doubling experiments with pure tones were used to define the psychoacoustic quantity *ratio-pitch* with its unity mel, so that a tone with the double pitch of another has the double ratio-pitch, i.e. the double mel value. From different experimental procedures, several mel scales with different reference frequencies have been determined. The mel scale of Stevens and Volkman [29] refers to the frequency of 1000 Hz corresponding to 1000 mel. The mel scale defined by Zwicker et al. refers to the *bark scale*, which is a scale of the width of auditory filters [34]. Its reference frequency is 125 Hz, which corresponds to 125 mel. Here we adopt the approximation of O'Shaugnessy [22] given in the following definition.

Definition 2.9 (Ratio-Pitch and Mel Scale). The psychoacoustic quantity *ratio-pitch* f_{mel} with its unity mel is defined for a pure tone with a frequency f by

$$f_{mel}(f) = \log_{10}\left(1 + \frac{f}{700\,\text{Hz}}\right) \cdot 2595\ \text{mel.} \tag{2.16}$$

This formula approximates the averaged results of psychoacoustic experiments. Its inversion is

$$f(f_{mel}) = \left(10^{f_{mel}/2595\,\text{mel}} - 1\right) \cdot 700\ \text{Hz.} \tag{2.17}$$

From the given formulas, we can calculate the frequencies of tones that have half the pitch of a given tone. For example, a tone with $f = 8$ kHz has a ratio-pitch of $f_{mel} = 2840$ mel. A tone that is perceived to have half the pitch must therefore have a ratio pitch of $f_{mel}/2 = 1420$ mel, which corresponds to a frequency of $f(1420\ \text{mel}) = 1768$ Hz.

Figure 2.5 shows a plot of the ratio pitch as a function of the frequency, $f_{mel}(f)$. A frequency of 1000 Hz corresponds to a ratio pitch of 1000 mel, which can be seen in the figure where the function $f_{mel}(f)$ crosses the identity. For small frequencies,

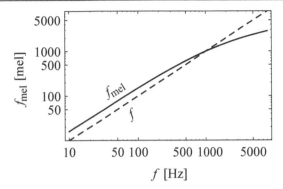

Figure 2.5: Double logarithmic plot of the mel scale $f_{mel}(f)$ as defined by Equation (2.16). The continuous line is a plot of the ratio pitch f_{mel} as a function of the frequency, and the dashed line is a plot of the identity function, i.e. a plot of the frequency as a function of itself.

the ratio pitch is approximately proportional to the frequency. This can also be seen directly from Equation (2.16) by expanding the logarithm in a Taylor series around $f = 0$, where we obtain $f_{mel} \approx 1.61f$ for $f \ll 700$ Hz. For larger frequencies, the slope of the function $f_{mel}(f)$ gets smaller and smaller, indicating that larger and larger vibration ratios are needed to produce tones with half (or double) pitch.

The mel scale is mainly used in psychoacoustics and finds applications in speech recognition, and recently also in music data analysis, see [33] for an overview. In the theory of Western music as developed in Sections 2.2.3 and 2.2.4, however, the mel scale is of little importance. On the other hand, the mel scale corresponds to physiological data: There is a linear relationship between the mel scale and the number of abutting haircells of the basilar membrane: 10 mel correspond to 15 abutting haircells, and the total of 2400 mel corresponds to the total of 3600 haircells. Furthermore, the mel scale reflects the width of auditory filters, which are described by the bark scale (see [34, p. 162]).

2.2.6   Fourier Transform

Analyzing a musical tone, one might be interested in deducing its frequency content from its waveform $x(t)$. The Fourier transformation is a mighty tool for the frequency analysis of vibrations and sounds. The theory of complex tones, the analysis of the overtones of a complex tone, and the distribution of sound energy are based on the famous theorem which Jean Baptiste Joseph Fourier (1789–1854) published in 1822 and which Georg Simon Ohm introduced into acoustics in 1843. So, we give a brief introduction to the mathematical background.

Definition 2.10 (Trigonometric Series). Let a_n and b_n be arbitrary constants. The finite or infinite ($N = \infty$) series

$$S_N = \frac{a_0}{2} + \sum_{n=1}^{N} \left(a_n \cdot \cos(n\omega_0 t) + b_n \cdot \sin(n\omega_0 t)\right) \qquad (2.18)$$

is called a *trigonometric series*.

There are equivalent notations for S_N, which can be obtained by writing the trigonometric functions as complex exponential functions and simplifying the expression. This way we can obtain

$$S_N = \frac{a_0}{2} + \sum_{n=1}^{N} C_n \cdot \sin(n\omega_0 t + \phi_n) = \frac{a_0}{2} + \sum_{n=1}^{N} C_n \cdot \cos(n\omega_0 t - \psi_n), \qquad (2.19)$$

with the amplitudes $C_n = \sqrt{a_n^2 + b_n^2}$ and phase shifts $\phi_n = \arg(a - ib)$ and $\psi_n = \arg(b + ia)$. The argument function $\arg z$ gives the angle of the complex number $z = x + iy$, and can be calculated from the real and imaginary parts as

$$\arg(x + iy) = \begin{cases} \arctan(y/x) & \text{for } x > 0 \\ \arctan(y/x) + \pi & \text{for } x < 0, y \geq 0 \\ \arctan(y/x) - \pi & \text{for } x < 0, y < 0 \\ \pi/2 & \text{for } x = 0, y > 0 \\ -\pi/2 & \text{for } x = 0, y < 0 \end{cases} \qquad (2.20)$$

or with the atan2-function provided by many calculators and programming languages.

Applying Euler's formula $e^{i\phi} = \cos(\phi) + i\sin(\phi)$ the trigonometric series can be written in complex form

$$S_N = \sum_{n=-N}^{N} c_n \cdot e^{in\omega_0 t} \qquad (2.21)$$

with the coefficients

$$c_0 = a_0/2 \quad \text{and} \quad c_n = \begin{cases} (a_n - ib_n)/2 & \text{for } n > 0 \\ (a_n + ib_n)/2 & \text{for } n < 0 \end{cases}. \qquad (2.22)$$

Note that in the complex notation, the summation runs from $-N$ to N; to recover the original Equation (2.18), which runs from 1 to N, the summands n and $-n$ must be collected.

It can be shown that if the series of coefficients a_n and b_n converge absolutely, i.e. if $\sum_{n=1}^{\infty} |a_n| < \infty$ and $\sum_{n=1}^{\infty} |b_n| < \infty$, the trigonometric series converges uniformly, thus defining a function $f(t) = \lim_{N \to \infty} S_N$ with period $T = 2\pi/\omega_0$.

From another viewpoint, we can take an arbitrary periodic function $f(t)$ with period T (which satisfies some regularity conditions presented below) and expand it into a trigonometric series. For such a *Fourier series expansion*, the a_n and b_n of the trigonometric series must be calculated from the function f with the *Euler–Fourier formulas* presented in the following definition.

Definition 2.11 (Fourier Coefficients). The *Fourier coefficients* of a function $f(t)$ are defined as

$$a_n = \frac{2}{T} \int_{-\frac{T}{2}}^{\frac{T}{2}} f(t)\cos(n\omega_0 t)dt \quad \text{and} \quad b_n = \frac{2}{T} \int_{-\frac{T}{2}}^{\frac{T}{2}} f(t)\sin(n\omega_0 t)dt. \quad (2.23)$$

Equivalently, when the complex form Equation (2.21) of the trigonometric series is used, the Fourier coefficients are defined as

$$c_n = \frac{1}{2T} \int_{-\frac{T}{2}}^{\frac{T}{2}} f(xt)e^{-in\omega_0 t}dt. \quad (2.24)$$

Dirichlet's theorem describes the regularity conditions that an arbitrary periodic function has to meet to have a Fourier series expansion, and it assures the convergence of the series [1].

Theorem 2.3 (Theorem of Dirichlet). If the single-valued, periodic function $f(t)$

- has a finite number of maxima and minima,
- has a finite number of discontinuities
- and if its absolute is integrable over one period, i.e. $\int_0^T |f(t)|dt$ is finite,

then the trigonometric series from Equation (2.18) with coefficients given by Equation (2.23) converges to $f(t)$ at all points where $f(t)$ is continuous. At points u where $f(t)$ is discontinuous, the series converges to the mean value $(f(u+) + f(u-))/2$ of the left and right limits of $f(t)$.

Most periodic functions that could be of interest to us will satisfy the Dirichlet conditions. Even functions with discontinuities, like square or sawtooth waves, can be expanded into Fourier series. At the jumps of these functions, the Fourier series converges to the "midpoint" of the jump.

The Fourier coefficients a_n, and b_n are positive or negative real numbers. In a *spectrum*, the values of a_0, a_n, and b_n are plotted against the frequency $n\omega_0$ or against the number n of harmonics. Equivalently, we can also plot the coefficients C_n and the phases of the cosine and sine expansion, respectively. In many cases, the phase shifts of the single partials are of no interest and only the coefficients C_n are plotted against the frequency to show the spectrum of a periodic function.

A Fourier series can be constructed for periodic functions only. However, non-periodic functions can be analyzed in a similar way using the *Fourier transform*. It is, in a sense, the limit of the Fourier series for periods $T \to \infty$. In this limit, the angular frequencies $n\omega_0 = n2\pi/T$ contained in the series come closer and closer, so that for $T = \infty$ a whole continuum of frequencies is involved. Then, the summation must be replaced by an integration. This motivates the following definition.

Definition 2.12 (Fourier Transform). The *Fourier transform* $F(\omega)$ of a function $f(t)$ is defined as

$$F(\omega) = \int_{-\infty}^{\infty} f(t)e^{-i\omega t}dt. \quad (2.25)$$

As $F(\omega)$ is complex, it is the sum of a real and a complex component and has a polar coordinate representation

$$F(\omega) = R(\omega) + iX(\omega) = A(\omega)e^{i\phi(\omega)}. \tag{2.26}$$

Definition 2.13 (Fourier Spectrum, Energy Spectrum, Phase Angle). The function $A(\omega) = |F(\omega)|$ is the *Fourier spectrum* of $f(t)$ and its square $A^2(\omega)$ is the *energy spectrum* of $f(t)$. The function $\phi(\omega) = \arg F(\omega)$ is the *phase angle*.

The Fourier transform $F(\omega)$ measures the magnitude and phase with which an oscillation of the angular frequency ω is contained in the function $f(t)$; very similar to the coefficients c_n of the complex Fourier series. The superposition of all these oscillations then reproduces the original function $f(t)$. As we are dealing with a continuum of oscillations, this superposition is written as an integral,

$$f(t) = \frac{1}{2\pi} \int_{-\infty}^{\infty} F(\omega)e^{i\omega t} dt. \tag{2.27}$$

This equation is the analogue of Equation (2.21), where the Fourier transform $F(\omega)$ plays the part of the Fourier coefficients c_n. For Equation (2.27) to hold, the function $f(t)$ must satisfy the Dirichlet conditions on any finite interval, and it must be absolutely integrable over the whole range $t \in (-\infty, \infty)$. In comparison with Equation (2.25), we see that the inverse transform, Equation (2.27), has the same structure as the Fourier transform, but with a factor of $1/2\pi$ and a sign change in the exponential.

2.2.7 *Correlation Analysis*

Correlation analysis is of special importance for the theory of hearing. In fact, the periodicity detection mechanism of the human auditory system is based on auto-correlation (for frequencies relevant for speech and music perception), and thus the correlation analysis is largely responsible for our perception of pitch. In the auditory system, a tone is coded by a periodic neural pulse train with a period equal to the inverse of the frequency of the tone. Neurally, tone height is thus coded and analyzed in the time domain but not in the frequency domain. The sophisticated mathematical method of Fourier transform is an analysis in the frequency domain which cannot be performed by "tiny" neurons. But an analysis in the time domain has been found by Gerald Langner ([16] in the midbrain (*colliculus inferior*) of the auditory system: It is a periodicity detection mechanism based on neuronal delay circuits and coincidence neurons. In essence, this periodicity detection mechanism represents an autocorrelation analysis. Thus, pitch perception is probably based on a neuronal autocorrelation (cp. Section 6.4). By autocorrelation, a signal is projected onto delayed versions of itself. The autocorrelation is a measure of similarity between the original signal and its delayed versions, and it can be applied to test for periodicity. The theorem of Wiener–Khintchine, which Norbert Wiener proved in 1931, grants that a periodicity detection by autocorrelation is equivalent to a frequency analysis by Fourier transform: It states that the Fourier transform of the autocorrelation function is the energy spectral density (see [12, p. 334]). As a remarkable property, the autocorrelation annihilates phase shifts. This is in line with pitch perception: Spatial changes of the sound source result in runtime differences of the sound wave which are mathematically represented by phase shifts. But the pitch percept is not at all altered by spatial changes of the sound source.

The nomenclature to classify signals $f(t)$ is adopted from physics. One defines that $|f(t)|^2$ has the meaning of the (instantaneous) *power* of the signal, and analogous to mechanics, power is defined as *energy per time*. This motivates the following definitions.

Definition 2.14 (Average Power and Total Energy). The *total energy E_f* and *average power P_f* of a signal $f(t)$ are defined by the integrals

$$P_f = \lim_{D \to \infty} \frac{1}{2D} \int_{-D}^{D} |f(t)|^2 dt \quad \text{and} \quad E_f = \int_{-\infty}^{\infty} |f(t)|^2 dt. \qquad (2.28)$$

When analyzing a signal $f(t)$, one must distinguish between two types of signals. *Energy signals* are signals with finite energy E_f, that is, $f(t)$ is square integrable. On the other hand, *power signals* are signals whose total energy E_f is infinite, but whose average power P_f is finite. The distinction between these two signals is necessary to avoid infinities (or to avoid that everything is zero), which is achieved by choosing suitable normalizations in the definitions of the autocorrelation functions.

Definition 2.15 (Autocorrelation Functions). The autocorrelation functions for energy signals and power signals are defined as

$$a(\tau) = \int_{-\infty}^{\infty} f^*(t)f(t+\tau)dt \quad \text{and} \quad a(\tau) = \lim_{D \to \infty} \frac{1}{2D} \int_{-D}^{D} f^*(t)f(t+\tau)dt, \quad (2.29)$$

respectively (the asterisk * denotes the complex conjugate).

The autocorrelation $a(\tau)$ measures how similar a signal $f(t)$ is to a time-shifted version $f(t+\tau)$ of itself. Of course, the maximal similarity is reached when the functions $f(t)$ and $f(t+\tau)$ are identical, which is at least the case for $\tau = 0$. Our definitions of the autocorrelation functions reflect this property: It can be shown that $|a(\tau)| \leq |a(0)|$ for arbitrary τ. The maximum values $a(0)$ are equal to the average power, $P_f = a(0)$ for power signals, or equal to the total energy, $E_f = a(0)$ for energy signals, respectively, which can be easily seen from the definitions.

Let us discuss a signal $f(t)$ which has a period of T, so that $f(t+T) = f(t)$. Furthermore, it shall be bound, which should be the case for any signal we want to analyze in music data analysis. Unless the signal is zero everywhere, it is not square integrable because its periodic form extends to infinity. Therefore, it is a power signal. From $f(t+nT) = f(t)$ (with integer n), it follows for the autocorrelation at $\tau = nT$ that

$$a(nT) = \lim_{D \to \infty} \frac{1}{2D} \int_{-D}^{D} f^*(t)f(t+nT)dt = \lim_{D \to \infty} \frac{1}{2D} \int_{-D}^{D} f^*(t)f(t)dt = P_f. \quad (2.30)$$

Thus, the autocorrelation itself is periodic, and has relative maxima equal to the average power P_f for all delays equal to the periods $\tau = nT$. One hint for calculating the average power or the autocorrelation of a periodic function in practice: It is sufficient to evaluate the averages $\frac{1}{2D}\int_{-D}^{D}\ldots$ over one period, i.e. with $D = T/2$, instead of forming the limit $D \to \infty$. Due to the periodicity, the average over one period is the same as the average over the whole time axis.

2.2.8 Fluctuating Pitch and Frequency Modulation

Stable sounds are rare in music and sound unnatural. As tones are man-made, sound fluctuations belong to the sound of musical instruments and determine the richness of their sound. Musicians strive for a vivid expressiveness of sustained notes and add fluctuations to them. Periodic fluctuations of the frequency, also called *frequency modulation*, are discussed in what follows, and periodic fluctuation of the amplitude, also called *amplitude modulation*, is discussed in the context of volume (see Section 2.3.4). The *vibrato* is a combined frequency and amplitude modulation to enhance the expressiveness of tones. Although the vibrato was an ornamentation in early music, it became an omnipresent instrumental technique for string instruments and the Italian art of *belcanto*.

Besides a vivid sound, a periodic frequency modulation adds extra partials to the original tones, thus enriching the timbre of the sound. In the following, the essence of frequency modulation is explained for a sine wave with a frequency varying in time. Consider a signal $x(t) = A\sin(\theta(t))$ with an instantaneous phase $\theta(t)$. For pure tones, $\theta(t) = \omega t$, and the angular frequency is the time derivative of the instantaneous phase: $\omega = \dot{\theta}(t)$. Conversely, when a time-dependent angular frequency $\omega(t)$ is given, the instantaneous phase can be obtained by integration, $\theta(t) = \int \omega(t)dt$.

Let us construct a signal with an angular frequency that deviates sinusoidally from a center frequency ω_c,

$$\omega(t) = \omega_c + \Delta\omega\cos(\omega_m t), \tag{2.31}$$

where $\Delta\omega$ is the maximum deviation from the center frequency. Integrating this equation gives the instantaneous phase (the integration constant is set to zero)

$$\theta(t) = \omega_c t + \beta\sin(\omega_m t) \quad \text{with} \quad \beta := \frac{\Delta\omega}{\omega_m}. \tag{2.32}$$

Here, the *modulation index* β has been defined. Thus, the sinusoidally frequency modulated (SFM) sinusoidal signal reads

$$x(t) = A\sin(\omega_c t + \beta\sin(\omega_m t + \phi)). \tag{2.33}$$

Its spectrum can be revealed by applying trigonometric identities, Fourier series expansions, and again, trigonometric identities. We skip this rather long calculation here and quote the result from [12],

$$x(t) = J_0(\beta)\sin(\omega_c t) + \sum_{k=1}^{\infty} J_k(\beta)\Big[\sin([\omega_c + k\omega_m]t) + (-1)^k \sin([\omega_c - k\omega_m]t)\Big]. \tag{2.34}$$

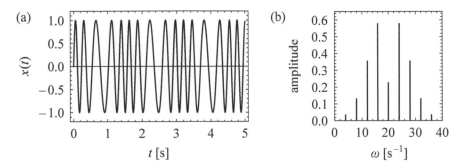

Figure 2.6: (a) Signal and (b) spectrum of a frequency modulated sine with carrier frequency $\omega_c = 20\,\mathrm{s}^{-1}$, a modulation frequency $\omega_m = 4\,\mathrm{s}^{-1}$, and a modulation index of $\beta = 2$.

The amplitudes $J_k(\beta)$ occurring here are the *Bessel functions of the first kind* [1, 12]. From the representation in Equation (2.34) of the signal, its frequency content can be immediately read off: There is a carrier with frequency ω_c and an infinite number of sidebands with frequencies $\omega_c \pm k\omega_m$.

Figure 2.6 shows the signal and its spectrum. The signal looks like a sine function that is periodically stretched and compressed, as a result of the periodically modulated angular frequency, Equation (2.31). In the spectrum, the absolute values of the amplitudes are plotted; the factor $(-1)^k$ occurring in Equation (2.34) turns some of the left sidebands negative, which is equal to a phase shift of half a period.

2.2.9 Simultaneous Pitches

The *superposition principle* says that simultaneous vibrations add up without any distortion. For example, consider a signal that is the sum of two sine tones. If the superposition principle would apply to hearing, two simultaneous pitches should be perceived without any additional effects. But on the contrary, the perception of simultaneous tones may elicit additional sensations. *Tonal fusion* is the perceptual phenomenon that consonant intervals are perceived as unities, whereas the interval tones are still present and do not mix. It was subject to extensive empirical investigations by Carl Stumpf [30], and was discussed by ancient Greek philosophers in the context of the octave identification. Since Hermann von Helmholtz (1862) investigated *roughness*, discovered *summation tones*, and described *combination tones*, these effects have extensively been examined in psychoacoustics [13]. Partly explanations of these phenomena are given in the field of acoustics (some of them are described below) partly properties of the auditory system are responsible for them.

We first discuss the phenomenon of *beats*, which occurs when two pure tones with different frequencies are superposed. Consider a signal consisting of two sine tones with equal amplitude,

$$x(t) = \sin(\omega_1 t) + \sin(\omega_2 t). \tag{2.35}$$

(a)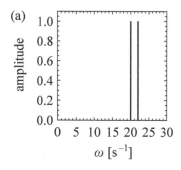

(b)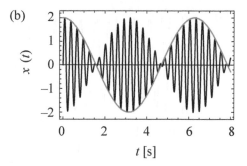

Figure 2.7: (a) Spectrum and (b) waveform $x(t)$ of a signal that is the sum of two sines with equal amplitude, angular frequencies of $\omega_1 = 20\,\mathrm{s}^{-1}$ and $\omega_2 = 22\,\mathrm{s}^{-1}$. The envelope $E(t)$, plotted in gray, fluctuates with the beating frequency $\omega_b = 1\,\mathrm{s}^{-1}$.

The trigonometric identity $\sin\alpha + \sin\beta = 2\sin([\alpha+\beta]/2)\cos([\alpha-\beta]/2)$, see [3], can be used to rewrite this equation as

$$x(t) = 2\sin(\omega_m t)\cos(\omega_s t) \quad \text{with} \quad \omega_m = \frac{\omega_1+\omega_2}{2}, \quad \omega_s = \frac{\omega_1-\omega_2}{2}. \quad (2.36)$$

The first factor fluctuates rapidly with the mean frequency ω_m, whereas the second factor oscillates very slowly with a low frequency ω_s, which is half of the frequency difference. The slow oscillation thus constitutes an envelope $E(t) = 2\cos(\omega_s)$ which is perceived as a periodic loudness fluctuation with the slow frequency ω_s. In this respect, beats remind us of an amplitude modulated sine. Note that due to the different kind of envelopes, the beating signal of two sines has two spectral components, whereas an amplitude modulate sine has three spectral components as discussed below in Section 2.3.4. Figure 2.7 shows the spectrum and waveform of a beating sinusoids, and the beating envelope.

The phenomenon of *roughness*, extensively investigated by Helmholtz [13], is closely related to beats. If the frequencies of both tones are very different, two simultaneous pitches and fast beats are perceived. If the frequency difference is small, only one pitch is heard, which corresponds to the algebraic mean of the original frequencies. In between is a region of frequency differences where *roughness* occurs: The perceived beats sound ugly and rough and the pitch percept is unclear. Helmholtz thought that a frequency difference of about $\Delta f = 33$ Hz, corresponding to $\omega_s = 2\pi\Delta f/2 \approx 104\,\mathrm{s}^{-1}$, elicits the highest degree of roughness. He developed a consonance theory based on roughness, which in essence is still en vogue in Anglo-American music theory, although it cannot sufficiently explain the phenomenon of consonance and dissonance, which is based on tonal fusion (see Section 3.5.1).

The discussion of beats was based on the assumption that the signals of two tones can be simply added to give the resulting signal. Now we will discuss what happens to simultaneous tones when they are processed by a non-linear transmission system, and we will find that those systems produce additional tones called *combination tones* as a result of a signal distortion. Non-linear transmission systems may be the body or the sound board of an instrument, a loudspeaker, or the human hearing system. Again, it was Helmholtz who studied the combination tones of the ear [13]. It must be mentioned that the auditory system itself may produce additional tones that are emitted into the ear and can even be record by microphone probes in the ear canal. These sounds are called *oto-acoustic emissions* [34, p. 35].

A transmission system converts an input signal $x(t)$ into an output signal $y(t)$. This can be virtually anything, for example, a loudspeaker converting an input voltage $x(t)$ into a pressure variation $y(t)$ in the air in front of the speaker membrane; another example would be the human ear where a pressure variation $x(t)$ that arrives at the outer ear must be transferred to the inner ear where it arrives as a signal $y(t)$. The transmission can be described by a continuous transfer function T so that $y(t) = T(x(t))$. As each continuous transmission function can be approximated by a polynomial function, some general properties of the transfer can be read from the output of a polynomial function $P(x) = \sum_{n=0}^{\infty} p_n x^n$.

For simplicity, let us consider a polynomial of degree 2. The output signal can be obtained as

$$y(t) = P[x(t)] = p_1 x(t) + p_2 x(t)^2 \qquad (2.37)$$

from the input signal, with coefficients p_1 and p_2. At first, we consider a pure tone $x(t) = cos(\omega t)$ to be transferred. The result is a signal

$$y(t) = p_1 \cos(\omega t) + p_2 \cos^2(\omega t) = p_1 \cos(\omega t) + \frac{p_2}{2}(1 + \cos(2\omega t)), \qquad (2.38)$$

which contains the octave of the pure tone. Responsible for the generation of the octave is the quadratic term in Equation (2.37).

Now consider a signal that is the sum of two pure tones, $x(t) = \cos(\omega_1 t) + \cos(\omega_2 t)$. In this case, the quadratic term in Equation (2.37) is proportional to

$$x(t)^2 = \cos(\omega_1 t)^2 + 2\cos(\omega_1 t)\cos(\omega_2 t) + \cos(\omega_2 t)^2 \qquad (2.39)$$

$$= 1 + \frac{1}{2}\cos(2\omega_1 t) + \frac{1}{2}\cos(2\omega_2 t) + \cos([\omega_1 - \omega_2]t) + \cos([\omega_1 + \omega_2]t).$$

The output signal thus contains, in addition to the linearly transmitted signal, two overtones of frequencies $2\omega_1$ and $2\omega_2$, one difference tone ($\omega_1 - \omega_2$), and one summation tone ($\omega_1 + \omega_2$). Similar calculations for higher-order polynomials, i.e. when P contains terms $\sim x^3$, x^4 etc., show that a cluster of summation tones and difference tones, altogether called combination tones, are produced by the distortion of the transmission system. The distortion products have frequencies ω_p of the form

$$\omega_p = \pm k_1 \omega_1 + k_2 \omega_2 \qquad (2.40)$$

where k_1 and k_2 are positive integers or zero. See [12, p. 613–617], for details.

2.2.10 *Other Sounds with and without Pitch Percepts*

Most musical instruments produce sounds with clear pitches. But there are several percussion instruments producing sounds without a pitch. A sharp beat onto a percussion instrument may have the characteristic of a click. Mathematically, a click corresponds to a *delta pulse*, which Fourier transforms to the constant 1 in the frequency domain. Thus, a click contains all frequencies and no pitch can be sensed. All drums are tuned to a certain frequency region, which gives them an individual timbre, but their inharmonic spectra do not elicit a fundamental pitch. The vibrations and spectra of church bells are highly complex, and often several pitches can be heard next to the strike tone. Some percussion sounds are noise-like. A vague pitch sensation can be evoked in case of sounds similar to band-limited noises. In the following, we discuss what kind of signals these sounds are.

In the auditory system, pitch is detected by a neuronal periodicity analysis [16]. All stimuli with a somehow periodic time course can elicit a pitch sensation. Fourier analysis shows that these stimuli are also somehow centered in frequency. Waveforms may vary from completely aperiodic over quasi-periodic to completely periodic sounds. Sounds may have a continuous frequency spectrum or they can be centered more or less in frequency. As a consequence there are sounds without any pitch or sounds with only a vague pitch percept. Pitch ambiguities may also occur. Complex tones are preferred as musical tones because of their clear and unambiguous pitch. As we will see later, the timbre of a sound is mainly determined by its spectral content. Thus, even sounds without any pitch may have a distinct tonal color.

As an example of sounds without pitch percept, we discuss noise. Roughly speaking, a noise is a randomly fluctuating signal. A precise definition in continuous time is mathematically quite elaborate, and we will keep the discussion on an abstract level.

Noises can be classified according to their *power spectral density*. The power spectral density $N_x(\omega)$ of a (power) signal $x(t)$ is the Fourier transform of its autocorrelation function (cf. Theorem of Wiener–Khintchine). Its meaning is that $N_x(2\pi f)df$ is the amount of power contributed by the signal components with frequencies between f and $f + df$ to the average power of the signal. Integrating this density over all frequencies thus gives the average power,

$$P_x = \int_{-\infty}^{\infty} N_x(2\pi f)df. \tag{2.41}$$

The power spectrum of white noise is independent of frequency $N_x(\omega) = N_0$. In practice, a noise has lower and upper cut-off frequencies f_l and f_o. The average power of such a (band-limited) white noise is therefore $P_x = N_0 \cdot (f_u - f_l)$.

White noise does not produce a pitch percept but has a diffuse timbre of its own. The attribute *white* does not describe a color sensation evoked by the noise but is a metaphor with respect to light: The color white contains all visible frequencies. Analogously, a continuum of arbitrarily many frequencies contribute to the white noise.

High-pass and low-pass noises with a steep cut-off frequency elicit a pitch sensation closely corresponding to the cut-off-frequency. Band-pass noises produce an ambiguous pitch. The pitch sensation corresponds either to the upper or to the lower cut-off frequency. Narrow-band noises elicit only one pitch sensation corresponding to the center frequency [34].

A further example of sounds with unclear pitch percepts is a complex tone with inharmonic partials. If there are only slight inharmonicities in an otherwise harmonic spectrum, the perception of a fundamental pitch is not entirely abolished. Instead, pitch-shift effects may be observed or single inharmonic partials may even be heard out [20]. But complex tones with completely inharmonic partials elicit sound sensations without a clear pitch. Examples of this sensation are the sounds of drums produced by vibrating two-dimensional membranes [9]. The vibration pattern of a circular membrane is described by Bessel functions and has a series of inharmonic partials, with vibration ratios f_n/f_1 of 1.59, 2.13, 2.29, 2.65, 2.91, ... This inharmonicity is the reason that membrane instruments generally produce pitch-less sounds. A remarkable exception is the sound of the timpani which has a pitch: The air in the kettle under the membrane imposes harmonic vibration ratios of the partials onto the membrane [9].

2.3 Volume — the Second Moment

2.3.1 Introduction

Loudness is the hearing sensation referring to the moment of intensity. Primarily, loudness corresponds to the sound intensity of the stimulus, but it also depends on the spectral content of the sound signal. Sound intensity as well as the spectral content are physical quantities that can be measured. By contrast, the sensational moment of loudness is a psychological quantity and can only be determined by the human listener.

If two audible stimuli are presented, the listener can immediately decide whether they are equal in loudness or whether there is a level difference. Estimations about the equality of loudness values are used in psychoacoustics to derive the *phon scale*. Estimations about loudness differences are used to derive the *sone scale*. Both psychoacoustical scales are measures of the human loudness sensation.

Before introducing the psychoacoustical loudness scales, we will first investigate the physics behind sound waves in air in the following section. This will lead to a physical description of the "magnitude" of a sound wave in air, measured by the *sound pressure level*. Once the physical basis is developed, psychoacoustical scales can be referred to the physical quantities as we will see in Section 2.3.3.

2.3.2 The Physical Basis: Sound Waves in Air

Sound waves propagate through the air surrounding us. The state of the air is characterized by spatial and temporal varying fields of pressure $p(r,t)$, density $\rho(r,t)$ and velocity $v(r,t)$, where $r = (x,y,z)^T$ is the position vector in the Cartesian space. When the air is in equilibrium, without any flow or sound, these fields are spatially and temporally constant with values $p(r,t) = p_0$, $\rho(r,t) = \rho_0$ and $v(r,t) = 0$. Numerical values for these constants and further properties of air are summarized in Table 2.3.

Small deviations of these fields from their equilibrium values, $p(r,t) = p_0 + p^*(r,t)$, $\rho(r,t) = \rho_0 + \rho^*(r,t)$ and $v(r,t)$, constitute sound. The amplitudes of the acoustic fields p^*, ρ^* and v characterize the "strength" or volume of the sound.

Table 2.3: Some Properties of Dry Air for Usual Ambient Conditions (20°C, Standard Pressure)

absolute temperature	T_0	293.15	K	(kelvin)
density	ρ_0	1.204	kg/m³	(kilogram per meter³)
pressure	p_0	1013.25	hPa	(hectopascal)
molar volume	V_m	24.055	L	(liter)
average molar mass	M	28.962	g/mol	(Gram per mole)

In the following Physics Interlude, the three-dimensional wave equation for sound waves is discussed and the speed of sound in air is determined. A nice calculation shows that the displacement amplitude of the air particles, which is caused by a sound at the hearing threshold, is only about 10^{-11} m, a tenth of an atom diameter.

A Physics Interlude

When the deviations p^*, ρ^* and v are small, their time evolution is governed by the following wave equation [15]. It is formulated in terms of an abstract quantity, the velocity potential, from which p^*, ρ^* and v can be calculated.

Definition 2.16 (Three-Dimensional Wave Equation, Velocity Potential). The equation

$$\frac{\partial^2 \phi}{\partial t^2} - c^2 \left(\frac{\partial^2 \phi}{\partial x^2} + \frac{\partial^2 \phi}{\partial y^2} + \frac{\partial^2 \phi}{\partial z^2} \right) = 0 \tag{2.42}$$

is called a *three-dimensional wave equation* for the *velocity potential* ϕ, from which the fields of interest can be derived by [15]

$$v = \left(\frac{\partial \phi}{\partial x}, \frac{\partial \phi}{\partial y}, \frac{\partial \phi}{\partial z} \right)^T, \qquad p^* = -\rho_0 \frac{\partial \phi}{\partial t}, \qquad \rho^* = \frac{1}{c^2} p^*. \tag{2.43}$$

In the wave equation, c is called the *phase velocity*, and it depends on the adiabatic bulk modulus K of the air (which is a constant and will be calculated below) and the air density ρ_0 by $c = \sqrt{K/\rho_0}$.

Together with appropriate boundary and initial conditions, Equations (2.42) and (2.43) entirely describe the dynamics of sound waves in air. The use of the velocity potential as the fundamental quantity has the advantage that all physical quantities can be calculated straightforwardly from it. However, the velocity potential has no particularly evident meaning; it is rather an auxiliary quantity. Sometimes we may be interested only in one of the physical fields v, p^* or ρ^*. For such cases, we should note that wave equations for these fields can be derived from Equation (2.42), which have exactly the same form as Equation (2.42) but with the respective field in place of ϕ. Thus, all quantities v, p^* and ρ^* satisfy the same wave equation. The solutions, however, will be different because of deviating boundary and initial conditions for the different fields.

The wave equation is quite general and holds for any compressible fluid or gas with negligible viscosity. The material properties particularly influence the speed of sound $c = \sqrt{K/\rho_0}$. For air, the speed of sound can be estimated quite accurately by basic thermodynamics of gases. To determine the bulk modulus K we consider an air volume V with mass m, which is compressed while its mass is conserved. By definition, $K = \rho \, dp/d\rho = -V \, dp/dV$ where the differential was transformed according to the relation $\rho = m/V$. To evaluate the derivative dp/dV, we must specify the appropriate thermodynamic process. The simplest suggestion might be an isothermal process; however, sound waves oscillate typically so fast that the heat exchange between compressed and decompressed regions is too slow to balance temperature differences. It is much more reasonable to assume no heat exchange at all between compressed and decompressed regions, i.e. an adiabatic compression. For an adiabatic process, thermodynamic teaches $pV^\gamma = \text{constant}$, where γ is the adiabatic index. For air, whose main constituents are diatomic gases, $\gamma = 7/5$. Thus, $dp/dV = -\gamma p/V$ for adiabatic processes, and the adiabatic bulk modulus at equilibrium is $K = \gamma p_0$. For the speed of sound we obtain $c = \sqrt{\gamma p_0 / \rho_0}$. In this formulation, however, ρ_0 itself depends on the pressure p_0. It is a good idea to resolve this dependency by defining the density $\rho_0 = M/V_m$ via the molar mass M and molar volume V_m. Furthermore, the temperature and pressure dependence of the molar volume is given by the ideal gas equation for one mole, $p_0 V_m = R T_0$ with the universal gas constant $R = 8.31446 \, \text{J/mol K}$. Thus, our final result for the speed of sound in air reads

$$c = \sqrt{\frac{\gamma R T_0}{M}} \approx 343 \, \text{m/s} \tag{2.44}$$

(see Table 2.3 for the numerical values). This table corresponds quite well to the measurements. Note that the final result shows that the speed of sound is independent of the ambient pressure, but proportional to the square root of the absolute temperature. Hence, sound propagates faster in warm air than in cold air. The dependence on the average molar mass M of the air has also practical significance: Moist air has a lower average molar mass than dry air, because the weight of a molecule of water is lower than that of nitrogen or oxygen (the two main constituents of air). Therefore, sound propagates faster in moist air than in dry air. Both effects play an important part in intonation problems of various instruments.

The d'Alembert solutions which we encountered in Section 2.2.2 can be easily transferred to three dimensions, and they are useful to describe propagating sound waves. Here we consider the special case of sinusoidal plane waves given by the velocity potential

$$\phi(\boldsymbol{r},t) = \hat{\phi} \sin(\boldsymbol{k} \cdot \boldsymbol{r} - \omega t) \tag{2.45}$$

with a certain amplitude $\hat{\phi}$, wave vector $\boldsymbol{k}$ and angular frequency ω. The wave vector is the three-dimensional generalization of the wave number, and can be

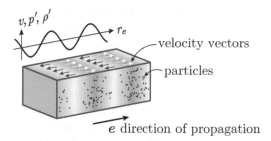

Figure 2.8: Section of a plane wave. With proceeding time, the spatial structure will be shifted in direction e with velocity c, but keep its shape. The phase velocity c, with which the whole structure moves, has to be well distinguished from the velocity field $v(r,t)$ (see indicated vectors), which describes the local velocities of air particles.

written as $k = ke$ with a unit vector e and norm $k = |k|$. It is easily seen that Equation (2.45) satisfies the wave Equation (2.42) if $\omega = ck$, so we obtain the same dispersion relation as for the one-dimensional wave equation.

The name "plane wave" comes from the observation that the spatial variations only occur along the directions e; in the directions perpendicular to e, the velocity potential is constant, thus defining planar *wave fronts* of constant ϕ (see Figure 2.8). That becomes clear by writing r in the orthonormal basis e, i, j via $r = r_e\, e + r_i\, i + r_j\, j$. Then the velocity potential is independent of the side components r_i and r_j since it reads $\phi = \hat{\phi}\sin(kr_e - \omega t)$.

According to Equations (2.43), the physically relevant fields are given by

$$v(r,t) = \hat{v}\cos(k \cdot r - \omega t), \quad p^*(r,t) = \hat{p}\cos(k \cdot r - \omega t) \quad \text{and}$$
$$\rho^*(r,t) = \hat{\rho}\cos(k \cdot r - \omega t) \tag{2.46}$$

with amplitudes $\hat{v} = k\hat{\phi}$, $\hat{p} = \rho_0\,\omega\,\hat{\phi}$ and $\rho = \rho_0\,\omega\,\hat{\phi}/c^2$. They all have the same spatial and temporal characteristics and differ only in amplitude. An example is sketched in Figure 2.8. Because the velocity v of air particles is oriented along the propagation direction e (which is parallel to k), sound waves belong to the class of *longitudinal waves*, in contrast to the transverse waves on a vibrating string.

A very nice calculation shows with which orders of magnitude we are dealing. The auditory threshold for a pure 1-kHz tone is $\hat{p} = 28\,\mu$Pa. The corresponding particle velocity is $\hat{v} = k\hat{\phi} = k\hat{p}/\rho_0\omega = \hat{p}/\rho_0 c = 6.8 \cdot 10^{-8}\,$m/s, so it is actually very slow compared to the propagation velocity c. The displacement of the air particles out of their equilibrium position is $\xi(t) = \int v(t)\,dt = -(\hat{v}/\omega)\sin(\omega t)$. Thus, the displacement amplitude is only

$$\hat{\xi} = \frac{\hat{v}}{\omega} = \frac{\hat{p}}{2\pi f\, \rho_0\, c} = 1 \cdot 10^{-11}\,\text{m}, \tag{2.47}$$

a tiny distance of approximately a tenth of an atom diameter which our ear can just detect!

For monochromatic plane waves, the amplitude $\hat{\phi}$, or alternatively the more seizable amplitudes $\hat{v}$, $\hat{p}$ and $\hat{\rho}$ of the physical fields, completely characterize the "strength" of the sound wave. However, when we hear sounds in daily life, we *never* hear such monochromatic waves.

End of the Physics Interlude

A general approach to measure the magnitude of a sound wave at a given position r_0 in space, and average over a certain time period D, uses a time-averaged value of the sound pressure $p^*(t)$ (we drop the spatial argument r_0 in the following discussion). Since $p^*(t)$ oscillates around 0 (not necessarily harmonically), the simple time average vanishes. In such cases, the *root mean square* value

$$p_{\text{RMS}} = \sqrt{(1/D) \int_0^D p^{*2}(t)\, dt} \tag{2.48}$$

suggests itself. Within the range of perceivable sounds, this value varies over many orders of magnitude, therefore it is advantageous to use a logarithmic scale. This leads us to the following definition.

Definition 2.17 (Sound Pressure Level (SPL)). The *sound pressure level*, abbreviated SPL, is defined as

$$L_p = 10\log\left(\frac{p_{\text{RMS}}^2}{p_{\text{ref}}^2}\right) \text{dB} \tag{2.49}$$

with a reference value $p_{\text{ref}} = 20\,\mu\text{Pa}$.

The reference value p_{ref} was believed to be the auditory threshold (for a pure sound of 1 kHz) at the time the sound pressure level was first defined, and the definition assures that a tone with $p_{\text{RMS}} = p_{\text{ref}}$ corresponds to a SPL of $L_p = 0$. Actually, the value is slightly wrong, as more recent psychoacoustic experiments showed and we will discuss later. Since the acoustic pressure is technically quite easy to measure using a microphone, it is perfectly suitable for defining a usable scale. In addition to this practical advantage, the acoustic pressure is the relevant quantity on the stimulus level for the hearing sensation of loudness. The relation between sound pressure level and loudness sensation is described below.

For pure tones, where $p(t) = \hat{p}\sin(\omega t)$, the root mean square value is simply obtained as $p_{\text{RMS}} = \hat{p}/\sqrt{2}$. Note that the integration time to calculate the RMS must be much longer than a single period. In the case of an incoherent superposition of pure tones, the squares of the individual root mean square pressures add up to the total:

$$p_{\text{RMS}}^2 = \frac{1}{D}\int_0^D \left(\sum_i \hat{p}_i \sin(\omega_i t)\right)^2 dt = \sum_i \frac{1}{D}\int_0^D \hat{p}_i^2 \sin^2(\omega_i t)\, dt \tag{2.50}$$

$$= \sum_i p_{\text{RMS},i}^2. \tag{2.51}$$

In the second step we use that mixed integrals $(1/D) \int \sin(\omega_i t) \sin(\omega_j t) \, dt$ are approximately zero for integration times much larger than the period, since $\omega_i \neq \omega_j$ in incoherent superpositions.

There is another important measure of the magnitude of sound, according to the following definition.

Definition 2.18 (Sound Intensity and Sound Intensity Level (SIL)). The *sound intensity I* of a sound wave is defined as

$$\boldsymbol{I}(t) = p^*(t) \, \boldsymbol{v}(t) \tag{2.52}$$

and measures the acoustic energy per area and time (unit W/m^2) transported by the wave. Its associated *sound intensity level L_I*, abbreviated SIL, is defined via its root mean square value,

$$L_I = 10 \log \left(\frac{I_{\text{RMS}}}{I_{\text{ref}}} \right) \text{dB} \quad \text{with} \quad I_{\text{ref}} = 10^{-12} \, W/m^2. \tag{2.53}$$

Again, the reference value L_{ref} was believed to be the auditory threshold. For propagating plane waves, the sound intensity level is identical to the sound pressure level: From Equation (2.46) we see that $v(t) = p^*(t)/\rho_0 c$, and consequently $I = p^{*2}/\rho_0 c$. The calculated reference value $I_{\text{ref}} = p_{\text{ref}}^2/\rho_0 c = 0.97 \cdot 10^{-12} \, W/m^2 \approx 10^{-12} \, W/m^2$ coincides, apart from rounding errors, with the definition. So we have

$$L_I = 10 \log \left(\frac{p_{\text{RMS}}^2/\rho_0 c}{p_{\text{ref}}/\rho_0 c} \right) = L_p, \tag{2.54}$$

but only in the case of propagating plane waves. Yet it is a good approximation when the sound comes from one direction and the listener is sufficiently far away from the sound sources. In other cases, L_I can only be determined by separately measuring $p^*(t)$ and $v(t)$, which is technically more complex.

2.3.3   Scales for the Subjective Perception of the Volume

Since the nineteenth century, the investigation of loudness perception has raised the question of how changes of the sound pressure or sound intensity, respectively, of the stimulus varies the loudness sensation. Which are the thresholds and the just noticeable intensity differences of loudness perception? Those questions are even of epistemological importance. Here, we cannot give details and refer to the literature, e.g., [12, p. 61-64]. Remarkable is the logarithmic relation between stimulus level and loudness sensation. Recall that there is also a logarithmic relation between vibration ratio and interval perception (see Section 2.2.3).

In acoustics the intensities of two sound stimuli are compared by considering the ratio of both intensities. Thus, a ratio of sound intensities is sensed as a difference of loudness values. This relation reminds us of a logarithmic function that has the property $\log(a/b) = \log(a) - \log(b)$ which is an immediate consequence of the logarithmic rule. And indeed, Weber–Fechner's law says that the difference of two loudness sensations $\Psi_1(I_1)$ and $\Psi_2(I_2)$ elicited by the Intensities I_1 and I_2 is logarithmically related to the ratio of the intensities.

$$\Delta L = \Psi(I_2) - \Psi(I_1) = k \cdot \log_{10}\left(\frac{I_2}{I_1}\right) \quad k = const. \tag{2.55}$$

However, Fechner made some assumptions which contradict experimental observations. Since Stevens, a power law is applied for a more realistic description of the relation between loudness and sound intensity (see [12, p. 64]).

Listeners are able to ascribe an equal loudness to sine tones of different frequencies. Successively presenting sine tones covering the whole audible frequency range from about 20 Hz up to almost 20,000 Hz and adjusting the sound pressure level of each tone to produce the same loudness sensation leads to curves of equal loudness in dependence of frequency (see Figure 2.9). To quantify this common loudness the unit *phon* is established in the following way.

Definition 2.19 (Loudness Level). For sine tones of $f = 1000$ Hz, the *loudness level* L_Φ, measured in phon, is equal to the sound pressure level L_p in dB,

$$L_\Phi = \frac{L_p}{1\,\text{dB}} \cdot 1\,\text{phon} \quad \text{for 1000-Hz sine tones.} \tag{2.56}$$

Hence, a 1000-Hz sine tone of $L_p = 20$ dB sound pressure level has a loudness level of $L_\Phi = 20$ phon (note that when inserting the given SPL into Equation (2.56), the unit dB cancels out and the unit phon is left). For sine tones of some other frequency, the corresponding loudness level is defined indirectly as the value of the SPL (in dB) of the 1000-Hz tone, which is perceived as equally loud. In [6], the SPLs of equally loud tones of different frequencies are defined by a diagram (and table of values) similar to Figure 2.9, which displays curves of constant loudness level in the f-L_p-plane.

Though isophonic sine tones, i.e. tones with equal loudness level, are perceived as equally loud they may have quite different SPLs, depending on their frequencies. It turns out that sine tones of very low or very high frequencies must have much higher intensities than sine tones in the midst of the audible frequency range to produce the same loudness level. The ear is most sensitive to frequencies between 2000 Hz and 5000 Hz.

For example, the reference sine tone of $f = 1000$ Hz and $L_\Phi = 30$ phon has an SPL of $L_p = 30$ dB, whereas a sine tone of 40 Hz and 30 phon must have an SPL of 65 dB to achieve the same loudness, and a sine tone with 5000 Hz and 30 phon has an SPL of only 23 dB as can be seen from the figure.

An isophone of particular importance is the *threshold in quiet* or *hearing threshold*, where the limit of loudness sensation is reached. It corresponds to the equal-loudness contour of 3 phon (not 0 phon, because the reference value in the SPL scale had been defined slightly falsely). People with average hearing capabilities fail to hear sounds below this threshold.

The definition of the loudness level L_Φ measured in phon lets us identify tones of different frequencies that all appear to be equally loud to the human ear. But it

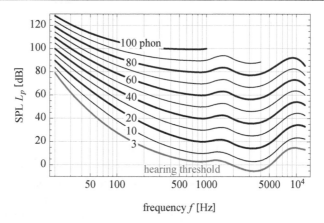

Figure 2.9: Curves of constant loudness level $L_\Phi = 3, 10, \ldots 100$ phon, so-called *iso-phones*, in the f-L_p-plane, after [6].

is not suitable to measure loudness differences as perceived by the human ear. If two sounds are sensed to be different in their loudness values, listeners can make estimations about this difference: They are able to judge how much louder or softer a sound is heard. Subjects are even able to estimate a doubling or halving of the loudness. Doubling and halving procedures were used to derive the sone scale. It turns out that a tone of 20 phon, for example, is not perceived as twice as loud as a 10-phon tone. Thus, another psychoacoustic scale has been defined to compensate for this.

Definition 2.20 (Loudness and Sone Scale). The *sone scale*, as the scale for the psychoacoustic quantity of *loudness* Ψ, is defined in the following way. A tone of loudness level $L_\Phi = 40$ phon is defined to have a loudness of $\Psi = 1$ sone, which serves as a reference loudness. If a tone is perceived as twice as loud as this reference loudness, it is assigned the loudness of 2 sone, and generally speaking a tone which appears n times as loud as the reference tone is assigned the loudness of n sone. The relation between loudness level L_Φ in phon and loudness Ψ in sone is given by [5]

$$\Psi = 2^{(L_\Phi - 40\,\text{phon})/10\,\text{phon}}\,\text{sone} \quad \text{if } L_\Phi \geq 40\,\text{phon} \tag{2.57}$$

for tones that are louder than the reference tone, and approximately

$$\Psi \approx \left[(L_\Phi/40\,\text{phon})^{1/0.35} - 0.0005\right]\,\text{sone} \quad \text{if } L_\Phi < 40\,\text{phon} \tag{2.58}$$

for tones that are softer than the reference tone.

So the loudness Ψ is proportional to the magnitude of the sound sensation of human listeners. Equations (2.57) and (2.58) were obtained by empirical studies on people with average hearing capabilities.

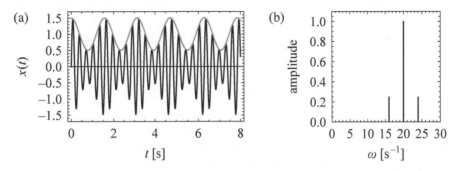

Figure 2.10: (a) Amplitude modulated signal with carrier angular frequency $\omega_c = 20\,\text{s}^{-1}$, envelope angular frequency $\omega_m = 4\,\text{s}^{-1}$, modulation depth $m = 1/2$ and initial phase shift $\phi = 0$. The envelope function $E(t)$ is plotted in gray. (b) Spectrum of the modulated signal.

2.3.4 Amplitude Modulation

Expressive musical tones rarely have a steady loudness: Playing a tone with a vibrato, for example, the musician modulates its frequency (see Section 2.2.8) as well as its amplitude. Preferentially the modulation rate is about 4 Hz. Like beats (see Section 2.2.9), amplitude modulations with a modulation rate of about 30 Hz evoke the sensation of roughness. The theoretical description of the amplitude modulations comes from radio engineering.

Definition 2.21 (Amplitude Modulated Signal, Carrier, and Envelope). An *amplitude modulated signal*

$$x(t) = E(t)c(t) = [1 + s(t)]c(t) \tag{2.59}$$

is the product of a fast oscillating *carrier* $c(t)$ and an *envelope* $E(t) = 1 + s(t)$ containing the signal $s(t)$ (which, in radio engineering, is to be broadcast).

If the carrier is a sine with the *carrier angular frequency* ω_c and the modulation is periodic with the angular frequency ω_m, the amplitude modulated signal becomes $x(t) = [1 + m\cos(\omega_m t + \phi)]\sin(\omega_c t)$. This signal is called a *sinusoidally amplitude modulated sinusoid* or SAM. The factor m describes the *modulation depth*. It is often expressed in percentage, and ϕ specifies the initial phase of the signal. For an example, see Figure 2.10 (a).

Applying trigonometric identities shows that $x(t)$ is the sum of three sine functions with frequencies ω_c, $\omega_c - \omega_m$, and $\omega_c + \omega_m$,

$$x(t) = \sin(\omega_c t) + \frac{m}{2}\sin([\omega_c - \omega_m]t - \phi) + \frac{m}{2}\sin([\omega_c + \omega_m]t + \phi). \tag{2.60}$$

The components with the angular frequencies $\omega_c \pm \omega_m$ are called *sidebands* and have an amplitude of $m/2$. Figure 2.10 (b) shows the spectrum of the amplitude modulated sine of Equation (2.60). Thus, an amplitude modulated sine is a complex tone consisting of the three equidistant partials and is very likely to elicit a residue pitch corresponding to the modulation frequency which is the frequency of the fundamental.

2.4 Timbre — the Third Moment

Besides the tonal quality of pitch and the tonal intensity perceived as loudness, the third important moment of a tone is its timbre. Although it seems to be evident what the notions *timbre* or *tonal color* perceptionally describe, it is difficult to give an unambiguous definition. For a historical overview of the historical evolution of the notions of timbre and "Klangfarbe," see [21].

Two tones may sound different even if they have the same pitch, loudness and duration – compare, for example, a note played on a piano to the same note played on a trumpet. All the perceived differences of these two tones are summarized under the notion of timbre: We say that the tone of the piano has a different timbre than the tone of the trumpet. However, tones played on the same instrument, but with different pitches, durations or loudness values, can also be interpreted as having different timbres – there is no constant "piano timbre" or "trumpet timbre" over the whole range of the instrument.

Tones of different heights are commonly described by metaphors of light, weight or volume and the like. High tones are said to be bright, brilliant, alert, light and tiny. Low tones on the other hand sound dark and dull, calm, heavy, thick and plump. These associations depend on the given context and cannot always be generalized over-individually. It can be shown that even a sine tone has a tonal color depending on its frequency region [14]. Accordingly, all frequency-centered sounds, such as bandpass noise not only elicit a somewhat uncertain pitch sensation, but also a timbre sensation. Depending on the cut-off frequencies, low-pass noises tend to sound dull and dark whereas high-pass noises might sound bright or even shrill.

The timbre of a tone is largely determined by its spectral content (that is, the intensity of the different partials plus additional noise components) and the temporal course of the tone, especially around its onset. To quantify the aspects of timbre mathematically, one needs a tool that analyzes a musical signal and gives its spectrum as a function of time.

From Section 2.2.6 we know that a Fourier transform permits the calculation of the spectrum of a sound, which shows the distribution of energy over all frequencies. However, the Fourier transform fails to monitor the temporal change of the spectrum: Any time information is integrated out, because the Fourier transform integrates over time from minus infinity to plus infinity, at least theoretically. Thus, temporal changes of a sound signal remain unveiled. But as sound signals, especially in music and speech, tend to change rapidly over time, it is desirable to determine their instantaneous frequency content and their temporal evolution. Quick sound fluctuations are decisive in instrument recognition and influence the perceived timbre.

To account for rapid sound fluctuations, the signal is resolved into tiny time-frequency atoms, which undergo a Gabor or wavelet transform. These transformations are similar to the Fourier transform, but they integrate the musical signal only over a short time interval, thus calculating the spectral content within this interval. However, the local precision of a time-frequency atom is limited by the uncertainty principle [17]. For the calculated time-dependent spectrum this means that we either have a high temporal resolution together with a low frequency resolution, or a low temporal resolution together with precisely determined frequencies – but both together, a high temporal resolution and a high frequency resolution, is impossible.

Note that in the following theory description, we will always work with angular frequencies ω because it is much more convenient than using frequencies f and spares a lot of factors 2π in the equations. As usual in the signal processing literature, the word "frequency" is often used synonymous with "angular frequency." Therefore, care must be taken: the real physical frequencies can be obtained from the angular frequencies via $f = \omega/2\pi$. Whenever the unit Hz is used, a real (and not an angular) frequency is indicated; angular frequencies are given in s^{-1}. Although both Hz and s^{-1} are of the same dimensions, this distinction proves useful to remove the confusion introduced by the theorists.

2.4.1 Uncertainty Principle

Let $f(t)$ be a time function with norm $\int_{-\infty}^{\infty} |f(t)|^2 dt = 1$, and let $F(\omega)$ denote the Fourier transform of $f(t)$.

Definition 2.22 (Time and Frequency Location and Spread). The *time location u* and the *duration* σ_t of $f(t)$ are defined by

$$u = \int_{-\infty}^{\infty} t|f(t)|^2 dt \quad \text{and} \quad \sigma_t^2 = \int_{-\infty}^{\infty} (t-u)^2 |f(t)|^2 dt, \tag{2.61}$$

respectively. Similarly,

$$\xi = \frac{1}{2\pi} \int_{-\infty}^{\infty} \omega|F(\omega)|^2 d\omega \quad \text{and} \quad \sigma_\omega^2 = \frac{1}{2\pi} \int_{-\infty}^{\infty} (\omega-\xi)^2 |F(\omega)|^2 d\omega \tag{2.62}$$

define the *(angular) frequency location* ξ and *(angular) frequency spread* σ_ω.

Theorem 2.4 (Uncertainty Principle). The *uncertainty principle* states that if $\lim_{t\to\pm\infty} \sqrt{t}\, f(t) = 0$, then

$$\sigma_t \sigma_\omega \geq \frac{1}{2}. \tag{2.63}$$

The uncertainty principle [17] can be illustrated by a box in the time-frequency plane centered at (u,ξ) with the time duration σ_t and the frequency spread σ_ω. It covers an area of $\sigma_t \sigma_\omega \geq 1/2$. Such a box is called a *Heisenberg Box*, see Figure 2.11. In case of equality, $f(t)$ is a Gaussian function:

$$f(t) = \sqrt{\frac{\alpha}{\pi}} \exp(-\alpha t^2). \tag{2.64}$$

angular frequency ω

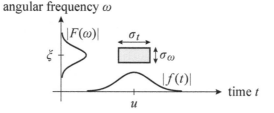

Figure 2.11: Heisenberg box in the time-frequency plane.

2.4.2 Gabor Transform and Spectrogram

The *Gabor transform* or *windowed Fourier transform* uses symmetric window functions $g(t) = g(-t)$ with norm $\int_{-\infty}^{\infty} |g(t)|^2 dt = 1$ to pick only limited portions from the musical signal and determine their frequency contents. Frequently used window functions are listed in Table 2.4. Note that in the table, the window functions are renormalized to $g(0) = 1$ and that $\int_{-\infty}^{\infty} |g(t)|^2 dt \neq 1$ [17]. Outside the interval $t \in [-1/2, 1/2]$ the functions are set to zero. On the right, the Gaussian window function is plotted. Note that all functions are bell shaped.

The signal $x(t)$ to be analyzed is multiplied with time-shifted versions of the time window and afterwards a Fourier transform is applied to the product, which leads us to the following definition.

Definition 2.23 (Time-Frequency Atom, Gabor Transform, and Spectrogram). The *Gabor transform* of a function $x(t)$ is the complex amplitude $WX \in \mathbb{C}$

$$WX(u, \xi) = \int_{-\infty}^{\infty} x(t) g_{u,\xi}^*(t) dt, \tag{2.65}$$

where the function $g_{u,\xi}^*(t) = g(t - u)e^{i\xi t}$ is called the *time-frequency atom*. The Gabor transform characterizes the strength and phase with which the frequency ξ is contained in the signal at about time u. Its energy density, i.e. the square of its absolute value, is called the *spectrogram*

$$P_{WX}(u, \xi) = |WX(u, \xi)|^2. \tag{2.66}$$

Table 2.4: Frequently Used Windows for the Gabor Transform

Name	$g(t)$	
Rectangle	1 for $|t| < 0.5$ else 0	
Hamming	$0.54 + 0.46\cos(2\pi t)$	
Gaussian	$\exp(-18t^2)$	
Hanning	$\cos^2(\pi t)$	
Blackman	$0.42 + 0.5\cos(2\pi t) + 0.08\cos(4\pi t)$	

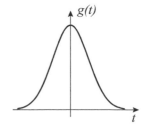

The time-frequency atom $g_{u,\xi}$ has a time location u and frequency location of approximately ξ, and its Fourier transform reads $G_{u,\xi}(\omega) = G(\omega - \xi)e^{-iu(\omega - \xi)}$, where G is the Fourier transform of the window function g. The duration and frequency spread of this time-frequency atom do not depend on the instantaneous time u or frequency ξ but only on the window function g, as

$$\sigma_t^2 = \int_{-\infty}^{\infty} (t-u)^2 |g(t-u)|^2 dt = \int_{-\infty}^{\infty} t^2 |g(t)|^2 dt \qquad (2.67)$$

$$\sigma_\omega^2 = \frac{1}{2\pi} \int_{-\infty}^{\infty} (\omega - \xi)^2 |G_{u,\xi}(\omega)|^2 d\omega = \frac{1}{2\pi} \int_{-\infty}^{\infty} \omega^2 |G(\omega)|^2 dt. \qquad (2.68)$$

Therefore, all time-frequency atoms $g_{u,\xi}$ have the same duration and frequency spreads and thus all their Heisenberg Boxes, which are centered at points (u,ξ) in the time-frequency plane, are of the same aspect ratio and of the same area $\sigma_t \sigma_\omega > 1/2$; see Figure 2.19. Loosely speaking, a Heisenberg Box at the point (u,ξ) indicates the region of the time-frequency plane, which influences the value of the spectrogram $P_{wx}(u,\xi)$. Two values $P_{wx}(u_1,\xi_1)$ and $P_{wx}(u_2,\xi_2)$ are independent of each other, only if the Heisenberg Boxes around the two sampling points (u_1,ξ_1) and (u_2,ξ_2) do not overlap. Hence, the time-frequency resolution of the Gabor transform is limited by the size of the Heisenberg Boxes, which in turn underlies the uncertainty principle, Equation (2.63).

2.4.3 Application of the Gabor Transform

As explained before, musical signals change over time, and thus the static Fourier transform is not appropriate to analyze music. The Gabor transform is one of the most up-to-date methods to analyze "a class of signals called music" [7]. Its *spectrogram* shows the evolution of the signal in time. It is an ideal tool for the analysis of recorded music [32] and reveals the spectral content over small time intervals as well as the evolution of the fine structures relevant for timbre perception.

A window function is stepwise shifted along the time axes by small equidistant intervals Δt. This results in a sequence of shifted window functions $\{g(t - m\Delta t)\}_{m \in M}$. Multiplying the signal $x(t)$ with the elements of this series results in the sequence $\{x(t)g(t - m\Delta t)\}_{m \in M}$. Further, each product $x(t)g(t - m\Delta t)$ is Fourier transformed. According to Equation (2.65) the result is a series of Gabor transforms $\{WX(m\Delta t, \omega_0)\}_{m \in M}$. Applying a discrete series of frequencies with appropriate frequency steps, a complete superimposition of the relevant time-frequency plane with Heisenberg Boxes is achieved. For each of the boxes, a Gabor transform has been performed, Equation (2.65), from which according to Equation (2.66) a spectrogram is derived. Thereby, the distribution of the energy density of the signal over the whole time-frequency plane is calculated and can graphically be represented [32]. Note the similarity between the time-frequency plane and a musical score.

Figure 2.12 shows two spectrograms, that is, plots of the energy density, Equation (2.66), as a function of time u and frequency $f = \xi/2\pi$. Both represent the same melody consisting of four quarter notes e, f, g, a. The short tune was played softly as well on a piano as on a trumpet. Both recordings underwent a Gabor transform

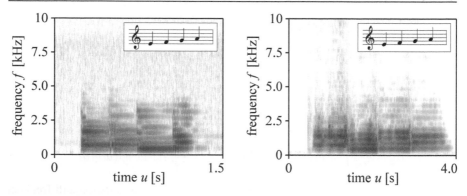

Figure 2.12: Spectrograms of a short tune (see inset) played on a piano (left) and on a trumpet (right).

performed with the software package *Praat* [2]. Energy concentrations, i.e. large values of P_{wx}, are represented by dark gray colors. The spectrograms reveal that the piano tones have higher strong partials than the trumpet tones. The onsets of the piano tones are quicker and sharper than the onsets of the trumpet tones. On the other hand, the onsets of the piano tones are very noisy. Recall that the Fourier transform of a click contains all frequencies. Correspondingly, the long vertical lines at the beginning of each tone of the piano spectrogram represent the stroke of the piano hammer. On the other hand, the initial white stripes of the trumpet spectrogram demonstrate that the trumpet player cared about soft onsets.

2.4.4   Formants, Vowels, and Characteristic Timbres of Voices and Instruments

The notion of the *formant*, coined by the German physiologist Ludimar Hermann (1838–1914), denotes the concentration of sound energy in certain frequency regions. As the timbre sensation is foremost determined by the spectral content of the sound, the positions and strengths of formants determine the characteristics of the timbre.

Formants arise from oscillations and subsequent frequency filtering. A primary sound is produced by an oscillator. For example, such an oscillator may be the larynx or a vibrating string or plate. The spectrum of this primary sound has a certain characteristic spectrum which is altered by a filter. This filter is a resonator such as the corpus of a musical instrument or the vocal tract of a speaker or singer. Normally, oscillator and resonator are coupled in a feedback loop, which gives the player of an instrument the opportunity to control the sound production and especially the timbre of the sound. Mathematically, a filter function is applied to a Fourier spectrum of the primary sound (see Figure 2.13).

Formants do not change their frequency position and frequency spread when the fundamental frequency, and thus the frequency spectrum, changes. Thereby it is granted that an instrument keeps its characteristic timbre over a wide range of pitches. The same is true for changes in volume. As most instruments have several formants in different frequency regions, their timbre may change with the tone height. These

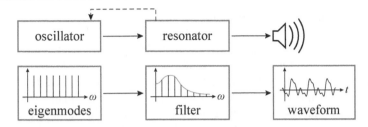

Figure 2.13: Scheme of the formant filtering of a music instrument.

frequency regions of different tonal color are the so-called registers of the instrument. In 1929 the German physicist Erich Schumann (1898–1985) demonstrated the influence of the stability of formants on the changing parameters of a sound such as pitch and volume [19]. Some examples of formant center frequencies are given in Table 2.5.

The formants of string instruments do not only depend on the strike and position of the bow, but vary widely from instrument to instrument and depend on the form of the corpus, the volume of air in the corpus, the wood, and the coating. Nevertheless, averaged values of the formant frequencies can be given (see Table 2.6).

If an instrument player changes the register, e.g., by over-blowing his wind instrument, or if a violinist changes the sound of his violin by altering the stroke of the bowing, the formants also change. Some instruments have very stable formants,

Table 2.5: Center Frequencies of the Formants of Woodwind Instruments (left) and Brass Instruments (right) [4]

instrument	f_1 [Hz]	f_2 [Hz]	instrument	f_1 [Hz]	f_2 [Hz]
flute	810	-	french horn	340	750
oboe	1400	2960	trumpet	1200	2200
cor anglais	950	1350	trombone	500	1500
clarinet	1180	2700	bass trombone	370	720
bassoon	440	1180	tuba	230	400
double bassoon	250	450			

Table 2.6: Center Frequencies of the Formants of String Instruments [4]

instrument	f_1 [Hz]	f_2 [Hz]
violin	400	1000
viola	230/350	600/1600
violoncello	250/400	600/900
contrabass	70-250	400

e.g., brass instruments, whereas other instruments show a great flexibility of their formants, e.g., string instruments or the human voice.

Speech recognition is based on the human ability to form and to discern different vowels. Each vowel has characteristic formants of its own. Generally, four vowel formants are distinguished. The first and second formants f_1 and f_2 are important for vowel recognition (see Table 2.7) whereas the higher formants are individually different and determine the timbre of the speaker's voice. In the table, the frequencies of the main formants are bold-faced. On the right, there is a spectrogram of the five vowels.

A formant of special interest is the *singer's formant*. This physical designation describes a special energy concentration in the human voice which is a result of the highly artificial old Italian *belcanto* singing technique. The voice of an educated opera singer produces a strong formant in a frequency region of about 2000 Hz and 3000 Hz, which give the voice its sonority and noble timbre. In contrast, an opera orchestra produces much less sound energy in this frequency region as demonstrated in Figure 2.14. Recall that the human ear is highly sensitive for those frequencies as can be read from the isophones shown in Figure 2.9. As a result, a voice with a well-developed singer's formant is heard out against a whole orchestra although the singer is not at all able to produce as much sound energy as the orchestra [31].

2.4.5 Transients

Two successive notes in music or two vowels in speech are joined by transitory sound components called *transients*. In contrast to tones or vowels, transients are not stationary and thus do not have a clear pitch. Instead, transients are characterized by a quickly changing frequency content and noisy sound components. The consonants of speech are transients with great importance for recognition. In music, the onsets of tones are transient sounds decisive for instrument recognition. Next to the spectrum of the stationary tones, the transients determine the timbre of the instruments. The short transient oscillation from tone onset to the stationary sound shows a quiet individual evolution of the spectral components. As an example, the first 50 mil-

Table 2.7: Center Frequencies of the First and Second Formants of the Five Vowels [4]

vowel	f_1 [Hz]	f_2 [Hz]
u	**320**	800
o	**500**	1000
a	**1000**	1400
e	500	**2300**
i	320	**3200**

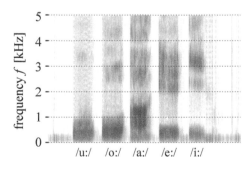

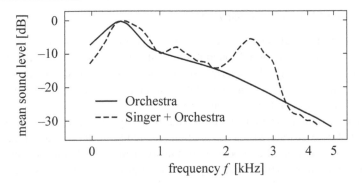

Figure 2.14: The singer's formant. Between 2000 Hz and 3000 Hz, a professional opera singer produces a sound energy concentration much higher than that of an orchestra [31].

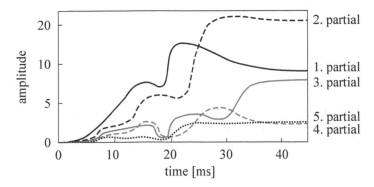

Figure 2.15: Transient oscillations (50 milliseconds) of the first five partials of a saxophone [28].

liseconds from onset to the stationary sound of a saxophone are shown in Figure 2.15 [28]. Other instruments have a different evolution of their spectral components in the onset phase, which may last even longer than 100 ms.

For most instruments, the onset is more intense than the following stationary part of the sound. In sound synthesis applications, the intensity evolution of tone onset is simulated by an *ADSR* envelope. The acronym *ADSR* stands for the four phases of the envelope (see Figure 2.16): An intense *attack* is followed by a quick *decay* to the designated sustain level of the sound. The long *sustain* phase is determined by the finishing *release* phase, during which the sound intensity is turned off. Depending on the different time evolutions of onsets, a number of variations of the *ADSR*-envelope are applied.

2.4.6   Sound Fluctuations and Timbre

Sound fluctuations affect the timbre of many instrumental sounds. Because of their expressive nature, intentional sound fluctuations like a *vibrato* are of aesthetic importance. The term vibrato characterizes sound fluctuations of sustained tones. A vibrato in music can be regarded as a combination of an amplitude (see Section 2.3.4) and a frequency modulation (see Section 2.2.8).

The tone of an instrument contains a harmonic spectrum. If this tone is frequency modulated by a vibrato with a modulation frequency of ω_m and a maximum deflection from ω_c of $\Delta\omega$, all components of the spectrum are also frequency modulated with the same modulation frequency and maximum deflection. As the formants of the instrument are stable, the different partials of the tone may enter or leave the resonant regions of the formants. Thus, the individual partial not only fluctuates in frequency but also in loudness: Entering a formant region amplifies the partial while leaving it attenuates the partial. As the timbre of a complex tone is determined by the strength of its partials, it becomes plausible that also the timbre of a tone fluctuates during a vibrato.

2.4.7   Physical Model for the Timbre of Wind Instruments

In the preceding sections, we learned a lot about how to characterize the timbre of instruments. However, one question remains to answer: What are the physical mechanisms behind the generation of timbre? Clearly, this question cannot be answered in general within this book, but we will try to shed a little light on the sound generation in brass instruments and see where the overtone spectrum contained in the tone comes into play.

The following Physics Interlude contains a minimal example for the different timbres of wind instruments with cylindrical and conical shape. Solutions of the wave equation in form of standing waves have different overtone spectra in these geometries: The cylinder has overtones with odd multiples of the fundamental frequency, whereas the cone has a full harmonic overtone spectrum.

A Physics Interlude

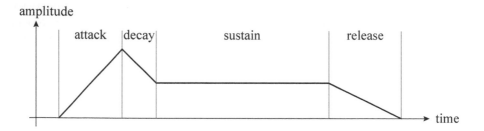

Figure 2.16: Schematic drawing of an *ADSR* envelope.

In Section 2.2.2 we saw that a clamped string can vibrate in different modes – the fundamental and the overtones – which gives rise to the generation of complex tones on that string. A similar concept can be applied to wind instruments, where the air inside the instrument vibrates itself in different modes.

Two kinds of boundary conditions for the air column inside the instrument can be specified:

- Closed boundary (rigid wall): Since particles cannot move into a rigid wall, the normal velocity must vanish, e.g., $\mathbf{v} \cdot \mathbf{e}_z = 0$ if the wall is in the x-y-plane. This implies for the velocity potential, see Definition 2.16, that for all times $\partial \phi / \partial z = 0$ on this boundary.
- Open boundary (the "outlet" of the instrument): For the idealized case that the instrument doesn't irradiate any sound, i.e. the waves are entirely reflected back into the instrument, the acoustic pressure vanishes in the ambient atmosphere in front of the boundary, $p^* = 0$. In terms of the velocity potential, this condition reads $\partial \phi / \partial t = 0$ on the open boundary.

Obviously, an instrument without any sound irradiation at its outlet is entirely useless; but this is the only simplification which gives a manageable boundary condition in the framework of eigenmodes.

Let us discuss two special instrument geometries that are relevant for wind instruments. One of the simplest geometries in use is the cylinder, found for example in flutes, clarinets, and mostly anything with a "pipe" in its name like organ pipes, bagpipes or panpipes. The full analysis of the eigenmodes of a cylinder requires a coordinate transformation of the wave equation and the use of Bessel functions. However, we will be content with finding the most obvious, and yet most relevant, modes.

We denote the direction of the cylinder axis as z, so that the circular cross section lies in the x-y-plane (see Figure 2.17, left). Due to the boundary conditions, the velocity field must be parallel to the z-axis at the cylinder walls. In the simplest case, it is parallel to the z-axis and constant throughout the whole circular cross section. In this plane wave assumption, we only need to solve the wave equation in the z direction for the velocity field. As boundary conditions, we take a closed boundary at $z = 0$, because the opening is blocked by a reed or the player's lips, and an open boundary at $z = L$, so our task is to solve

$$\ddot{\phi}(z,t) - c^2 \phi''(z,t) = 0 \quad \text{with} \quad \phi'(0,t) = 0, \quad \dot{\phi}(L,t) = 0, \qquad (2.69)$$

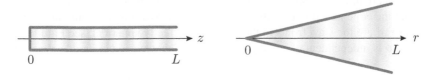

Figure 2.17: Coordinates of a cylinder with plane wavefronts (left) and a cone with spherical wavefronts (right).

where a prime denotes differentiation with respect to z.

The solution proceeds analogous to Section 2.2.2; it is the same differential equation as for a vibrating string, but with different boundary conditions. The ansatz $\phi(z,t) = Z(z)\,T(t)$ leads to separation

$$\frac{\ddot{T}}{T} = -c^2 k^2 \qquad \text{and} \qquad \frac{Z''}{Z} = -k^2 \qquad (2.70)$$

$$\Rightarrow \qquad T(t) \sim \sin(\omega t + \varphi) \qquad \text{and} \qquad Z(z) \sim \sin(kz + \psi) \qquad (2.71)$$

where, as usual, the angular frequency is defined by the dispersion relation $\omega = ck$. The boundary condition at $z = 0$ enforces $\psi = \pi/2$, which is equivalent to replacing the sine by a cosine, $Z(z) \sim \cos(kz)$. Then, the other boundary conditions imply a constraint on the wave vector: From the condition $\cos(kL) = 0$, we obtain $kL = (n - 1/2)\pi$. This leads to quantized wave numbers, wavelengths and frequencies

$$k_n = (2n-1)\pi/2L, \quad \lambda_n = \frac{4L}{2n-1} \quad \text{and} \quad f_n = \frac{c}{4L}(2n-1) \qquad (2.72)$$

respectively, with $n = 1, 2, 3, \ldots$. The whole solution for a single eigenmode then reads

$$\phi_n(\boldsymbol{r}, t) = A_n \cos(k_n z)\,\sin(\omega_n t + \phi_n) \qquad (2.73)$$

and the general solution is a superposition of the eigenmodes with the respective amplitudes and the phase shift ϕ_n as coefficients. The behavior of the pressure field, which can be obtained by Equation (2.43) from the velocity potential, is illustrated in Figure 2.18. We note that pressure and velocity are off-phase in time and space: At points where the pressure oscillates most, the particle velocity is always zero; and at times where the pressure wave is at its full deflection, the velocity field is zero in the whole cylinder.

The musically relevant part of this calculation are the frequencies from Equations (2.72) of the harmonics, which will be present in a complex tone and determine its timbre. We have a fundamental of $f_0 = c/4L$, and harmonics with *odd multiples* of this frequency, $3f_0, 5f_0, 7f_0$, and so on (concerning the notation of the fundamental frequency see the beginning of Section 2.2.2).

Next, we want to consider the eigenmodes of a cone (see Figure 2.17, right). A cone is a section of a sphere, and the wave equation with the boundary conditions of a conical instrument can be solved best when it is expressed in spherical coordinates (r, θ, φ). The cone's tip shall be placed at $r = 0$, so that the cone axis points in the radial direction. We consider only the simplest case where the wavefronts in the cone are perfectly spherical, see Figure 2.17, and ϕ is independent of the angles θ and φ. The wave equation then reads [15]

$$\frac{\partial^2}{\partial t^2}\phi(r,t) - c^2 \frac{1}{r^2} \frac{\partial}{\partial r}\left(r^2 \frac{\partial}{\partial r}\phi(r,t) \right) = 0. \qquad (2.74)$$

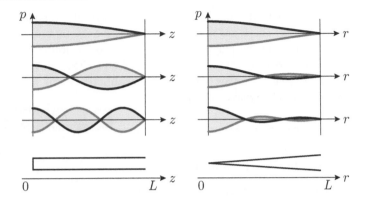

Figure 2.18: Eigenmodes in a cylinder (left) and cone (right) with one end open. For the cylinder, plane waves are assumed, and an appropriate scaled cosine function with a maximum at $z = 0$ and root at $z = L$ describes the spatial pressure variation. In the case of the cone, spherical waves are assumed. The pressure variation is described by an appropriately scaled cardinal sine, again with a maximum at $r = 0$ and root at $r = L$. Despite their similarity (besides the decay in amplitude for the cone), these different modes produce very different harmonics – odd ones for the cylinder, and the full spectrum for the cone.

Inserting the separation ansatz $\phi(r,t) = \frac{R(r)}{r} T(t)$ gives

$$\frac{\ddot{T}}{T} = -c^2 k^2 \qquad \text{and} \qquad \frac{R''}{R} = -k^2 \qquad (2.75)$$

$$\Rightarrow \qquad T \sim \sin(\omega t + \alpha) \qquad \text{and} \qquad R \sim \sin(kr + \beta). \qquad (2.76)$$

We have a closed boundary condition at the tip, $\phi(0,t) = 0$ and an open boundary at the other end, $\dot{\phi}(L,t) = 0$. The first one implies $\beta = 0$, which is also the correct choice to obtain a finite value for $\phi(0,t)$ since $\sin(kr)/r$ converges to 1 when r approaches zero. The second boundary condition quantizes the wave number, and thus the wavelength and frequency, as

$$k_n = n\pi/L, \quad \lambda_n = \frac{2L}{n} \quad \text{and} \quad f_n = \frac{c}{2L}n \qquad (2.77)$$

with $n = 1, 2, 3, \ldots$ and so on. To summarize, an eigenmode of the cone is given by

$$\phi_n(r,t) = A_n \frac{\sin(kr)}{r} \sin(\omega t + \alpha_n). \qquad (2.78)$$

For an illustration, see Figure 2.18. Note that the ansatz used in this calculation has a very special form. There are other possible solutions involving spherical Bessel functions (and spherical harmonics for modes which are not homogeneous in the angles).

However, the modes calculated above are musically very relevant, since they represent a complete spectrum of harmonics with the fundamental frequency $f_0 = c/2L$. These are not only important for the timbre of wind instruments, but also for the playing practice. On brass instruments, different harmonics can be played without using any valve or slide. In fact, the first brass instruments had no valves or slides, and the different harmonics were the only tones that could be played on them.

End of the Physics Interlude

2.5   Duration — the Fourth Moment

2.5.1   *Integration Times and Temporal Resolvability*

Due to physiological and psychological processing times, a sound is not perceived at the same moment as the sound pressure wave hits the ear. It takes up to about 250 milliseconds from the onset to a thorough perception of a tone or harmony. Three integration times are distinguished for auditory processing [24]:

- First integration time up to 10 ms: It takes the ear about 10 ms to establish the auditory filters necessary for frequency filtering. Thus, in the first 10 ms a spectral analysis of sounds is impossible for the hearing system. Partials remain unresolved in the first milliseconds after the onset of a complex tone. Only changes of timbre may be perceived due to amplitude fluctuations.

   As a result of the first integration time, attack and decay times shorter than 10 ms seconds elicit clicks. The attack time necessary for avoiding clicks is strongly related to frequency and loudness. As a rule of thumb one can state: The higher the frequency, the shorter the possible attack time for sound onsets without eliciting clicks, and the smaller the loudness, the shorter the possible attack time for sound onsets without eliciting clicks. To achieve smooth sound onsets and offsets without clicks it is advisable to ramp sound for the first and last 10 to 50 ms. Instead of ramps, sinusoidally shaped envelopes are also frequently used.

- Second integration time up to 50 ms: As soon as the auditory filter bank has been established at about 10 ms after sound onset, an auditory spectral analysis becomes possible. Between 10 ms to about 50 ms after sound onset, the time evolution of single partials becomes detectable. Thus, the second integration time describes a time span at which the detection of tone height and timbre is being established. As a rule of thumb one can state again: The higher the frequency, the shorter the integration time for beginning tone height and timbre sensation.

- Third integration time up to 250 ms: At the end of the third integration time, at 250 ms after sound onset, a distinct sensation of tone height and tonal timbre can be observed due to a thorough auditory analysis of the whole spectral content and all harmonic partials. From 250 ms on, quasi-periodic sounds are perceived as being totally periodic. In the time span up to 250 ms, quick changes of the sound lead to an unclear and diffused tonal perception and a noisy and smeared timbre.

2.5.2 Time Structure in Music: Rhythm and Measure

Rhythm and measure are the musical means to organize the musical course of time. They are marked by the regulated succession of strong and weak notes that imprint an apparent structure onto the flow of time. Rhythmic patterns in music have reference to regularly recurrent pulses that determine the perceived tempo of the music. In a piece of music, normally several layers of regular pulse patterns from slow to fast are simultaneously intertwined. Every pulse pattern has a certain repetition rate of beats ranging from less than 1 Hz to not more than 20 Hz. Every pulse is a release of sound energy at a certain moment in time. A time-frequency analysis of music in the low frequency range below 20 Hz is appropriate to analyze the rhythms of music. This can be performed by a wavelet transform [27, 32].

2.5.3 Wavelets and Scalograms

Similar to the Gabor transform, the *wavelet transform* projects a signal $x(t)$ onto a time window called a *wavelet*. But instead of applying a Fourier transform, the signal power in the time-frequency atom of the wavelet is determined.

Definition 2.24 (Wavelet). The wavelet $\psi(t)$ is a window function with

$$\int_{-\infty}^{\infty} \psi(t)dt = 0, \qquad \int_{-\infty}^{\infty} |\psi(t)|^2 dt = 1, \qquad (2.79)$$

that is, a function with zero average and unit norm that is centered in the neighborhood of $t = 0$.

Let $\Psi(\omega)$ be its Fourier transform. Its frequency location is given by

$$\eta = \frac{1}{2\pi} \int_{-\infty}^{\infty} \omega |\Psi(\omega)|^2 d\omega. \qquad (2.80)$$

The wavelet $\psi(t)$ can be scaled by a real factor $1/s$ to alter its width,

$$\psi_s(t) = \frac{1}{\sqrt{s}} \psi(t/s), \qquad (2.81)$$

where the prefactor $1/\sqrt{s}$ ensures that the scaled version ψ_s is still normalized. From Fourier theory it follows that the Fourier transform $\Psi_s(\omega)$ of $\psi_s(t)$ is also rescaled, $\Psi_s(\omega) = \sqrt{s}\Psi(s\omega)$. Applying this relation, it becomes clear that the frequency location η_s of $\Psi_s(\omega)$ is equal to the scaled frequency location of $\Psi(\omega)$, that is, $\eta_s = \eta/s$.

In addition to this rescaling, which achieves a shift of the frequency location, the wavelet can be time-shifted to any time location u,

$$\psi_{u,s}(t) = \frac{1}{\sqrt{s}} \psi\left(\frac{t-u}{s}\right). \qquad (2.82)$$

From the time-shift theorem it follows for the Fourier transform of $\psi_{u,s}(t)$ that $\Psi_{u,s}(\omega) = \Psi_s(\omega) e^{-iu\omega}$. Thus, the frequency location of $\Psi_{u,s}(\omega)$ is equal to the frequency location of $\Psi_s(\omega)$, which is η/s as shown above; the time shift does not influence the frequency properties of the wavelet.

The duration $\sigma_{t,s}$ and frequency spread $\sigma_{\omega,s}$ of the shifted and scaled wavelet $\psi_{u,s}(t)$ are also only affected by the scaling parameter s. A quick calculation shows that they are related to the duration σ_t and frequency spread σ_ω of the original wavelet $\psi(t)$ by $\sigma_{t,s} = s\sigma_t$ and $\sigma_{\omega,s} = \sigma_\omega/s$. When the scale s decreases, the width (or duration) of $\psi_{u,s}(t)$ is reduced; and at the same time, its frequency location η_s and frequency spread $\sigma_{\omega,s}$ increase. Thus, the Heisenberg box of the wavelet is shifted upwards to higher frequencies in the time-frequency plane, and changes its aspect ratio to be more slender; see Figure 2.19 (b).

Definition 2.25 (Wavelet Transform and Scalogram). The *wavelet transform* of a function $x(t)$ is the complex amplitude $WX \in \mathbb{C}$

$$WX(u,s) = \int_{-\infty}^{\infty} x(t)\psi_{u,s}^*(t)dt = \int_{-\infty}^{\infty} x(t)\frac{1}{\sqrt{s}}\psi^*\left(\frac{t-u}{s}\right)dt. \qquad (2.83)$$

Analogous to the spectrogram of the Gabor transform, the energy density of the wavelet transform is called the *scalogram* and is defined as

$$P_{WX}(u,s) = |WX(u,s)|^2. \qquad (2.84)$$

According to the Heisenberg Boxes of the shifted and scaled wavelets, the high-frequency components of a musical signal are analyzed with a higher time resolution than the low-frequency components with a wavelet transform. The smaller the value of the scaling factor s, the more the energy of the wavelet $\psi_{u,s}(t)$ is confined to a smaller time interval $\sigma_{t,s}$, but the higher the frequency location η_s is and the greater the frequency spread $\sigma_{\omega,s}$. This is in contrast to the Gabor transform, whose Heisenberg Boxes have the same aspect ratio everywhere on the time-frequency plane; see Figure 2.19 for a comparison.

Plotting a scalogram, a common choice for the scale parameter is $s = 2^{-k/M}$ with $k \in \{0,1,\ldots,I\cdot M\}$. The positive integer I is called the *number of octaves* and the integer M is the *number of voices* per octave. Choosing $M = 12$ would correspond to the twelve notes of the chromatic scale [17].

In the following example a wavelet transform is applied to a series of clicks. The number of octaves is $I = 8$ and the window function is a *Mexican Hat*

$$\psi(t) = \frac{2}{\pi^{1/4}\sqrt{3\sigma}}\left(1 - t^2/\sigma^2\right)\exp\left(-t^2/2\sigma^2\right) \qquad (2.85)$$

as wavelet, which is the normalized second derivative of a Gaussian function. This function is real and symmetric; thus its Fourier transform $\Psi(\omega)$ is also real and symmetric. That means that the frequency location $\eta = 0$ vanishes, so that this wavelet transform cannot detect the frequency of a signal. However, it can detect irregularities like jumps and kinks in a signal [17], and is therefore suitable for the analysis of click series or rhythmic patterns. Figure 2.20 (a) shows an equidistant click sequence with a frequency of about 4.8 Hz and its wavelet transform. Bright regions of

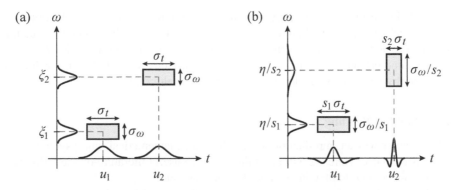

Figure 2.19: Heisenberg Boxes in the time-frequency plane for (a) the Gabor transform and (b) the wavelet transform. The small functions on the axes symbolize the window or wavelet functions. In case of the Gabor transform, the time and frequency resolution is constant throughout the time-frequency plane, whereas in case of the wavelet transform, the time resolution increases for higher frequencies.

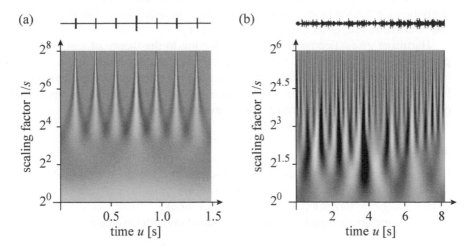

Figure 2.20: (a) A signal of seven equidistant clicks undergoes a wavelet transform. The top line shows the signal, the bottom figure the scalogram. (b) A percussion clip of the song "Buenos Aires" is analyzed by a wavelet transform. The rhythmic patterns on the different frequency levels are analyzed and can be read from the scalogram. For the analyzes, the software *FAWAVE*, written by the mathematician James S. Walker, was applied.

the scalogram indicate high energy concentration. As discussed before on the level of Heisenberg Boxes, the scalogram clearly shows a more precise time resolution for higher factors $1/s$. The limits of time resolution can be observed directly: Only at large enough values of $1/s$, the brighter regions are clearly separated.

Detailed insights into the rhythmic structure of a piece of music can be obtained from a scalogram. Figure 2.20 (b) shows an analysis of the highly complex structure of a percussion clip from the song "Buenos Aires" [32]. On the level of lower values of $1/s$, superordinate rhythmic patterns are detected; on the level of higher values of $1/s$, the rhythmic fine structure is reflected by the scalogram.

2.6 Further Reading

This chapter was an introduction to the physical, mathematical, and psychological aspects of music. Here we will point the reader to useful literature on the individual topics that come into play in this interdisciplinary field.

First of all, our description requires the mathematical knowledge of typical undergraduate math courses. Reference [1] is a mathematical textbook that covers all prerequisites, for example complex numbers, Fourier series and differential equations.

The physical background of sound waves in air was presented along the lines of [15], where additional information and the derivation of the equations can be found. The basics of acoustics are also contained in [9], where also the physics of musical instruments is treated in great detail. A thorough introduction to musical acoustics can be found in [10] or its German translation [11].

Hartmann gives a comprehensive introduction to the theory of auditory signal processing and psychoacoustics including the mathematical modeling in this field [12]. The various mathematical tools of signal analysis such as the different kinds of Fourier and Wavelet transforms are discussed in [17]. Examples for the application of Gabor and Wavelet transforms to music are demonstrated in [32].

2.7 Exercises

Exercises, theoretical as well as practical based on the software packages R [23] and MATLAB®, will be provided at the book's *web site*
`http://sig-ma.de/music-data-analysis-book`,
which also includes example data sets partly needed for the exercises.

Bibliography

[1] M. L. Boas. *Mathematical Methods in the Physical Sciences*. Wiley, 2006.

[2] P. Boersma and D. Weenink. Praat: Doing phonetics by computer. `www.praat.org`.

[3] I. N. Bronshtein, K. A. Semendyayev, G. Musiol, and H. Mühlig. *Handbook of Mathematics*. Springer, 2007.

[4] M. Dickreiter. *Der Klang der Musikinstrumente*. TR-Verlagsunion, 1977.

[5] DIN 45630. *Physical and Subjective Magnitudes of Sound*. Beuth, 1971.

[6] DIN ISO 226. *Acoustics: Normal Equal-Loudness-Level Contours*. Beuth, 2003.

[7] M. Dörfler. *Gabor Analysis for a Class of Signals Called Music.* Diss. University of Vienna, 2002.

[8] M. Ebeling. Die Ordnungsstrukturen der Töne. In W. G. Schmidt, ed., *Faszinosum Klang. Anthropologie – Medialität – kulturelle Praxis*, pp. 11–27. De Gruyter, 2014.

[9] N. Fletcher and T. Rossing. *The Physics of Musical Instruments.* Kluwer Academic Publishers, 1998.

[10] D. E. Hall. *Musical Acoustics.* Brooks/Cole, 2001.

[11] D. E. Hall. *Musikalische Akustik.* Schott/Mainz, 2008.

[12] W. M. Hartmann. *Signals, Sound and Sensation.* Springer, 2000.

[13] H. v. Helmholtz. *Die Lehre von den Tonempfindungen als physiologische Grundlage der Theorie der Musik.* Olm, 1862 / 1983.

[14] W. Köhler. *Akustische Untersuchungen.* Zeitschrift für Psychologie. Barth, 1909.

[15] L. Landau and E. Lifšic. *Fluid Dynamics.* Course of Theoretical Physics. Butterworth-Heinemann, 1995.

[16] G. Langner. Die zeitliche Verarbeitung periodischer Signale im Hörsystem: Neuronale Repräsentation von Tonhöhe, Klang und Harmonizität. *Zeitschrift für Audiologie*, 2007.

[17] S. Mallat. *A Wavelet Tour of Signal Processing.* Elsevier, 2009.

[18] M. McKinney and B. Delgutte. A possible neurophysiological basis of the octave enlargement effect. *J Acoust Soc Am.*, 106(5):2679–2692, 1999.

[19] P.-H. Mertens. *Die Schumannschen Klangfarbengesetze und ihre Bedeutung für die Übertragung von Sprache und Musik.* Bochinsky, 1975.

[20] B. Moore and K. Ohgushi. Audibility of partials in inharmonic complex tones. *J. Acoust. Soc. Am. 93*, 1993.

[21] D. Muzzulini. *Genealogie der Klangfarbe.* Peter Lang, 2006.

[22] D. O'Shaughnessy. *Speech Communications: Human and Machine.* Addison-Wesley, 1987.

[23] R Core Team. *R: A Language and Environment for Statistical Computing.* R Foundation for Statistical Computing, Vienna, Austria, 2014.

[24] C. Reuter. *Die auditive Diskrimination von Orchesterinstrumenten.* Peter Lang, 1996.

[25] A. Riethmüller and H. Hüschen. Musik. In L. Finscher, ed., *Die Musik in Geschichte und Gegenwart*, volume 6. Bärenreiter, 2006.

[26] J. F. Schouten, R. J. Ritsma, and B. Lopes Cardozo. Pitch of the residue. *J. Acoust. Soc. Am. 34*, 1962.

[27] H. Smith, L. M. & Honing. Time-frequency representation of musical rhythm by continuous wavelets. *J. of Mathematics & Music 2/2*, 2008.

[28] W. Stauder. *Einführung in die Akustik.* Florian Noetzel Verlag, 1990.

[29] S. Stevens and J. Volkman. The relation of pitch to frequency. *J. Acoust. Soc. Am. 34*, 1940.

[30] C. Stumpf. *Tonpsychologie*. S. Hirzel, 1883 / 1890.

[31] J. Sundberg. *The Science of the Singing Voice*. Northern Illinois University Press, 1987.

[32] J. S. Walker. *A Primer on Wavelets and Their Scientific Application*. Chapman & Hall/CRC, 2008.

[33] C. Weihs, U. Ligges, F. Mörchen, and D. Müllersiefen. Classification in music research. *Advances in Data Analysis and Classification*, 1(3):255–291, 2007.

[34] E. Zwicker and H. Fastl. *Psychoacoustics: Facts and Models*. Springer, 1999.

Chapter 3

Musical Structures and Their Perception

MARTIN EBELING

Institute of Music and Musicology, TU Dortmund, Germany

3.1 Introduction

The sensation of tone and the auditory perception of time patterns are the foundations of all music. In the previous chapter we started from the sensation of tone and investigated the relation of the most prominent moments of the tonal sensation, which are pitch, loudness, duration, and timbre, to the properties of the tonal stimulus. In the following we demonstrate how the elementary components of music, that is to say tones and time patterns, form musical structures and discuss the psychological foundations of their perception. We use the notion of *Gestalt* coined by v. Ehrenfels and the rules of *Gestalt perception* of Max Wertheimer to reflect the emergence of musical meaning. For this purpose, we consider some essential elements of Western music theory in the light of music perception and cognition. It must be pointed out, that the grasping of musical structures is foremost implicitly learned by mere exposure to music and does not require knowledge of the musical system and of music theory. The perception of *Gestalt in time* ("Zeitgestalten") is still an open question. Musical structures are objects of musical thinking, which is essentially different from rational thinking. But to grasp musical structures is likewise a powerful source of emotions. Musical thinking and feeling are intertwined to evoke the aesthetic effects of music, or as Ludwig v. Beethoven took it: *Music is a higher offering than all wisdom and philosophy* ("Musik ist höhere Offenbarung als alle Weisheit und Philosophie").

3.2 Scales and Keys

3.2.1 Clefs

The notation system of Western music uses a five-line staff. Pitch is shown by the position on the staff. At the beginning of each staff, a *clef* indicates a reference pitch. On the basis of the concert pitch of a' (scientific: A4) with 440 Hz, the *f-clef* indicates the small f (scientific: F3) with 174.614 Hz, the *c-clef* denotes the one-lined c' (scientific: C4) with 261.626 Hz, and the *g-clef* or *treble clef* marks the one-lined g' (scientific: G4) with 391.995 Hz (concerning the note name, refer to Section 2.2.4). Note that the clefs are a pure fifth apart. Theoretically, every clef can be positioned on each of the five lines of a staff. But in practice, only few of them have been used. These are shown in Figure 3.1.

Figure 3.1: The clefs used in Western music notation.

The *treble clef* and the *bass clef* are the commonly used clefs. The *alto* or *viola clef* is regularly used for notation of viola music. The *tenor clef* is often used in instrumental music, e.g., for the notation of higher parts of the violoncello or bassoon. All c-clefs were in use up to the end of the 19th century, especially in scores of choir music. All clefs can be found in older music scores.

3.2.2 Diatonic and Chromatic Scales

Generally, music uses a finite number of discrete tones from the continuum of pitch. Because of the octave identification, a small number of tones (5–7) within the range of an octave is selected. To form a *musical scale*, this set of tones is ordered by pitch in an ascending sequence which is analogously repeated in the other octaves, preserving the interval structure. The Western music is based on the diatonic major and minor scales. As mentioned, the diatonic scale is a subset of seven notes from the twelve tones of the chromatic scale (see Section 2.2.4). The original (untransposed) diatonic scale consists of the notes: *c, d, e, f, g, a, b* (in German: *h*). It is comprised of two *tetrachords* with the same interval structure: two whole tone steps and a semitone step follow each other. The first tetrachord consists of the notes *c d e f* and the second of the notes *g a h c*. Both tetrachords are separated by a whole tone step (see Figure 3.2). Note that each tetrachord comprises a pure fourth. Furthermore, the lowest notes of both tetrachords are a pure fifth apart. Both tetrachords together fill up a pure octave. The concept of two tetrachords within the interval of a pure octave stems from ancient Greek music theory and belongs to the heritage of Western music.

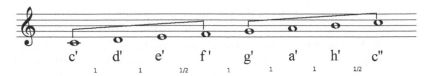

Figure 3.2: The notes of the diatonic C major scale. The number 1 beneath the staff marks a whole tone step whereas 1/2 indicates a semitone step. Each bracket spans a tetrachord.

Table 3.1: The Church Modes

keynote	scale
d	Dorian mode
e	Phrygian mode
f	Lydian mode
g	Mixolydian mode

The *chromatic* notes are deduced from the diatonic notes by *alterations*. A *flat*-sign (German: B) (symbol: ♭) or a *sharp* sign (German: Kreuz) (symbol: ♯) as an *accidental* before one of the seven notes of the chromatic scale changes its pitch: a flat lowers the original pitch by a semitone and a sharp raises it by a semitone. A *double flat* sign (German: Doppel-B) (symbol: ♭♭) lowers the original pitch by a whole tone and a *double sharp* (German: Doppelkreuz) (symbol: x) raises it by a whole tone. The alteration of a note is revoked by the *natural* sign (symbol: ♮).

Each of the seven notes of the diatonic scale can be raised by a sharp or lowered by a flat. As a consequence, there are two ways to notate a chromatic note. For example, the note *C sharp* has the same pitch as the note *D flat*, the note *D sharp* has the same pitch as the note *E flat*, etc. In the strict sense, this so-called *enharmonic change* without pitch change is only possible in the equal temperament. In other tuning systems, an enharmonic change slightly changes the pitch, e.g., in pure intonation the note *F sharp* is slightly higher than the note *G flat* (for a detailed discussion of tuning systems see [19]).

The diatonic major scale of Figure 3.2 has the keynote *c* as starting point. The *natural minor scale* has the same notes as the diatonic major scale but its keynote is the tone *a*, which is a minor third below the keynote *c* of the major scale. Medieval music and the music of the Renaissance and early Baroque commonly used the *ecclesiastical modes* or *church modes*, which are also based on the diatonic scale but differ in their keynotes.

Only the degrees II, III, IV, and V of the diatonic scale serve as keynotes in the medieval system of church modes (see Table 3.1). In order to construct a closed system of church scales in which every degree of the diatonic scale can be a keynote of a scale, later theorists in the time of the Renaissance invented the *Aeolian scale*, which is just the natural minor scale with keynote *a* (degree VI of the diatonic scale). They further constructed the *Ionic scale* with keynote *c* (degree I of the diatonic scale), which is equal to the major diatonic scale of the modern music theory. They also invented the *Locrian mode* with keynote b (degree VII of the diatonic scale), which however has no significance in music. Church modes are still important in modern music theory: in jazz harmony, every chord is identified with a certain church mode.

A scale (like any melody) can be *transposed*: it is moved upward or downward in pitch so that the keynote changes. Of course, each of the twelve tones of the chromatic scale can be the key of a diatonic scale. To preserve the diatonic structure of the scale under transposition, key signatures are added to the staff. If the diatonic scale is moved upwards by the interval of a pure fifth, a sharp must be added to the seventh degree of the transposed scale to make it a leading note. If the diatonic scale is moved downwards by the interval of a pure fifth, a flat must be added to the fourth step of the transposed scale to preserve the interval of a pure fourth between the new keynote and the fourth degree of the transposed scale.

Starting from the key *c* and successively ascending by the interval of a pure fifth leads to the ascending circle of fifth with the keys *c, g, d, a, e, h, f sharp, c sharp* (see Figure 3.3). Each upward transposition by a pure fifth makes it necessary to add a further sharp ($\sharp$). Successive downward transposition by the interval of a pure fifth yields to the descending circle of fifth with the keys *c, f, b flat, e flat, a flat, d flat, g flat,* and *c flat* (see Figure 3.4). A further flat ($\flat$) must be added to the new key on proceeding to it by a downward transposition of a pure fifth. To each major key corresponds a minor key with the same number of flats or sharps. Its keynote is equal to the sixth degree (VI) of the corresponding major scale. The seventh degrees of the minor scales are commonly altered by a semitone upwards to turn them into leading tones.

3.2.3   *Other Scales*

Pentatonic scales consist of five tones instead of seven tones and are probably the oldest scales in music. There are pentatonic scales with semitones called *hemitonic pentatonic*, as in classical Japanese or Indonesian music, and the pentatonic scale without semitone steps called *anhemitonic pentatonic* (see Figure 3.5). Pentatonic scales seem to be elementary to music from all over the world. In historical Chinese music theory, the anhemitonic pentatonic scale is derived from the chord of five pure fifths (see Figure 3.5 (a)): transposing its tones into the same octave gives the pentatonic scale (see Figure 3.5 (b)). Pentatonic scales can be found in most folk music, e.g., old European tunes and children's songs are often pentatonic. Pentatonic scales can be embedded into the diatonic scale so that Western music can be enriched with the exotic atmosphere of pentatonic tunes as Giacomo Puccini (1858–1924) demonstrated in his famous operas *Madama Butterfly* (Japanese hemitonic pentatonic) and *Turandot* (Chinese anhemitonic pentatonic).

But there are also scales completely different from the diatonic scale. The Javanese *slendro* is a tonal system which divides the octave into seven almost equally

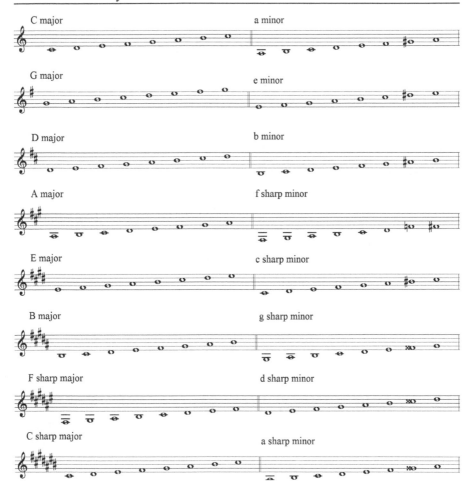

Figure 3.3: The circle of fifths: upward direction with seven major and corresponding minor keys starting from the diatonic C major scale.

spaces tones. It is obvious that none of these tones can be equal to any of the twelve tones of the equidistant chromatic scale. The Javanese *Pelog* is another system using five tones from a set of seven tones within an octave not equally spaced. None of these seven tones is equal to the tones of the Western chromatic scale. The blue notes of *blues* are another example of tones that are not contained in the chromatic scale. The *blues third* lies between the minor and the major third. The other *blue notes* are the *blues fifth* the *blues seventh*. On a keyboard, instead of the *blue notes*, the minor third and the minor seventh and the diminished fifth are played. Different blues scales are used and the most prominent are shown in Figure 3.6. Note the minor seventh, which represents the *blues seventh*. The bottom scale has a minor and a major third. This ambiguity reflects the *blues third*.

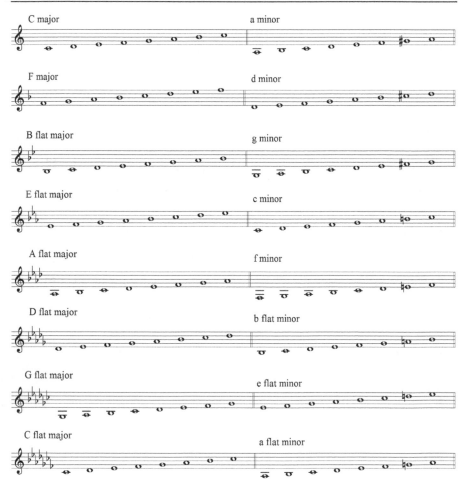

Figure 3.4: The circle of fifths: downward direction with seven major and corresponding minor keys starting from the diatonic C major scale.

3.3 Gestalt and Auditory Scene Analysis

Elementary sensory processes bear perceptional entities which form the perceptional scene. These sensory phenomenons were the focus of investigations by early psychologists in the nineteenth century, e.g., Carl Stumpf (1848–1936). Their epistemological background was the philosophy of *the whole and its parts*. The decisive matter of interest is not the analysis of all parts, but the determination of the relations between all parts and between the parts and the whole. According to Stumpf, the *Gestalt* is the paragon of these relations and therewith an abstraction from a lot of *Komplexes* with same relations between their parts and between the parts and the whole, thus bearing the same *Gestalt* [26, p. 229–240]. For example, a major triad is a *Gestalt* whereas the F major triad and the C major triad are two different

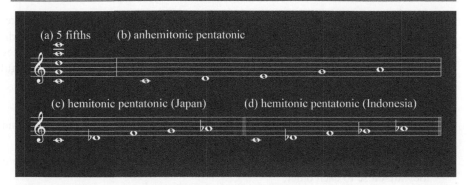

Figure 3.5: Pentatonic scales: (a) the tones of the chord of five pure fifths have the same note names as the tones of the anhemitonic pentatonic scale (b); classical Japanese music uses the hemitonic pentatonic scale shown in (c), which is different from the hemitonic pentatonic scale as used in Indonesian music.

Figure 3.6: Two blues scales on c.

Komplexes of the major triad. Symmetry operations preserve relations and thus parts of a *Gestalt*, and have always been a topic of aesthetics. The elementary symmetry operations are translation, reflection, rotation, and dilatation. The notion of *Gestalt* was coined by Christian von Ehrenfels (1859–1932), who referred to the example of a melody and pointed out that its *Gestalt* is more than the sum of its parts (notes) and that it is transposable [28]. In music, a *transposition* is a translation of pitch: the same melody can start from another tone and is played in another key. A canon consists of a melody and of one or more time-shifted versions of the same melody in other parts. This is a translation in time. A rhythm can be played with augmented or diminished note values. These are dilatations in time. Reflections and rotations are also applied in music, e.g., in counterpoint or in twelve-tone technique (*Krebs, Inversion*), but may not always be auditorily evident to the listener.

We cannot but conceptualize our perceptions as *Gestalten*, which become the content of higher cognitive functions and which may be one source of meaning. *Gestalt* emerges from the form-generating capacities of our sensory systems. Max Wertheimer (1880–1943) postulated *Gestalt principles* that apply not only to vision but also to music, although vision and hearing are quite different in many other respects [29]. These principles may, for example, explain why and under which conditions a series of tones is perceived as connected, thereby forming a melody, or they may explain why certain simultaneous tones are heard as a musically meaningful chord perceived as an entity.

Gestalt Principles

1. *Figure-ground articulation* – two components are perceived: a figure and a ground. Example: A soloist accompanied by an orchestra.

2. *Proximity principle* – elements tend to be perceived as aggregated into a group if they are close to each other.
   Example: A good melody prefers small tone steps and avoids great jumps, which would destroy the continuous flow of the music.

3. *Common fate principle* – elements tend to be perceived as grouped together if they move together.
   Example: If all parts of a piece of music move with the same rhythm, a succession of harmonies is heard instead of single independent voices.

4. *Similarity principle* – elements tend to be grouped if they are similar to each other. Example: Tones played by one instrument are heard as connected because they have a quite similar timbre.

5. *Continuity principle* – oriented units or groups tend to be integrated into perceptual wholes if they are aligned with each other.
   Example: The oriented notes of a scale are perceived as a single upward-moving figure.

6. *Closure principle* – elements tend to be grouped together, if they are parts of a closed figure.
   Example: The notes of an arpeggio are heard as a harmony, which is the closed figure in this case, e.g., a triad or seventh chord.

7. *Good Gestalt principle* – elements tend to be grouped together if they are part of a pattern that is a good Gestalt, which means that it is as simple, orderly, balanced, coherent, etc., as possible.
   Example: A two-part piece of music; each part is perceived as a closed and good Gestalt, so that the tones of two voices are perceptually segregated.

8. *Past experience principle* – elements tend to be grouped together if they appeared quite often together in the past experience of the subject.
   Example: Certain chord successions such as cadences have so often been heard that the chords are perceived as an entity. Any deviation from the chord scheme surprises and irritates the listener as in case of an interrupted cadence.

Integration and Segregation and Auditory Scene Analysis   The reign of the *Gestalt principles* in the sense of sight and the visual perception of static pictures is obvious and was the main topic of the Gestalt psychologists. On the other hand, the perception and cognition of temporal forms (German: *Zeitgestalten*) are not yet completely clear. Different levels and durations of memory, short-term memory as well as long-term memory and psychological grouping processes resuming simultaneous and successive percepts into entities seem to be important. Albert Bregman [2] picked up the concept of the Gestalt principles and combined it with the investigation of grouping processes in hearing. He investigated the conditions of *integration* and *segregation* in auditory perception and made a distinction between *primitive segregation* determined by unlearned constraints and *schema-based segregation*, which is based on learned constraints. Unlearned constraints are imposed on perception by, for example, the structure of the sensory systems, neuro-physiological processes, and psycho-physiological pre-conditions of perception [12], which on the whole are the result of a long evolutionary process. Or as Bregman takes it: "To me, evolution seems more plausible than learning as a mechanism for acquiring at least a general capacity to segregate sounds. Additional learning-based mechanisms could then refine the ability of the perceiver in more specific environments" [2, p. 40].

3.4 Musical Textures from Monophony to Polyphony

In music theory, the textures of music are classified according to their complexity [17]. Unaccompanied melodies are elementary in all musical cultures. This simplest texture in music is called *monophony* or *monody*. Successive tones as single musical events are integrated into a melodic line.

The term *monody* also describes music that consists of a single melodic line accompanied by instruments. On the one hand, the melody and the accompaniment are perceived as different layers of the musical texture. On the other hand, the melody consists of successive tones integrated into a line, whereas the accompaniment consists of successive chords that integrate into a coherent structure of harmonies supporting the melodic line.

If a melody is simultaneously played by several instruments, the melody can independently be varied by the musicians. *Heterophony* describes a musical texture that is characterized by a melody simultaneously presented together with variations of the melody. It is a widespread feature of non-Western music but is also known in Western music (e.g., Anton Bruckner).

Homophony refers to musical textures with two or more parts moving together in harmony. In homophonic vocal music the texts of all parts are identical and move together. Homophonic music tends to be integrated into a single stream of harmonies.

In contrast to *homophony*, the term *polyphony* describes a musical texture of several parts that move independently. On the one hand, all parts are conceived as components of the same piece of music. Harmony ensures the integration of the parts. On the other hand, the melodic lines of all parts move independently and are perceived as segregated. The highly artificial interplay of integration and segregation was condensed into the craft of counterpoint [9].

3.5 Polyphony and Harmony

Polyphony and *harmony* are the core of Western tonal music. A polyphonic composition consists of two or more clearly distinguishable free parts which are perceived as independent from each another. Nevertheless, perceptually, all parts should fit together well. To this aim, a collection of rules should be attended which form the basis of the craft of *counterpoint*. Rules of counterpoint were collected in textbooks, i.e. in the seminal compendium *Gradus ad Parnassum* written 1725 by the German composer Johann Joseph Fux (1660–1741), a textbook that has been studied by generations of composers [10]. We briefly discuss the elementary rules of a two-part counterpoint as they provide insight into the psychological preconditions of music perception and cognition. A thorough introduction to the craft of counterpoint is given by Lemacher and Schroeder [16]. The following discussion of the rules of counterpoint is widely based on this book. De la Motte [4] describes the development of counterpoint in Western music. Many textbooks on harmony cover the theory of chords, chord progressions, cadences, and modulations. The word *counterpoint* stems from the Latin *punctus contra punctum*, which describes the technique of how to put a suitable note (*punctus*) against the notes of a given tune called *cantus firmus*. It describes the simplest form of a two-part counterpoint. The concept of *harmony* is a (logical and historical) consequence of the rules that form the craft of counterpoint. The theory of harmony describes different types of chords and their significance in music and gives rules for the chord progression.

3.5.1   Dichotomy of Consonant and Dissonant Intervals

Counterpoint and harmony are based on the fundamental concept of consonance and dissonance, which is a strict dichotomy of the musical intervals. Within the range of an octave the perfect consonances are the pure intervals prime, octave, fifth, and fourth. The imperfect consonances are the major and minor thirds and sixths. The dissonant intervals are the seconds and sevenths. Perceptionally, the pure fourth is a consonant, but in strict counterpoint and classical harmony it is regarded as a dissonant interval and must be resolved (see Figure 3.10 (a)). However, other composition techniques regard the pure fourth as a consonant interval (see Figure 3.7).

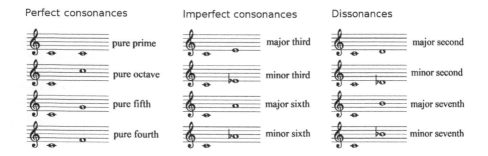

Figure 3.7: Intervals. The pure or perfect consonants are prime (unison), octave, fifth and fourth (left), the imperfect consonances are thirds and sixths (middle), the dissonant intervals are seconds and sevenths (right).

By alterations, which are chromatic changes of one or both interval tones, all intervals can be augmented or diminished. Disregarding their perceptional qualities, all augmented and diminished intervals are counted as dissonances in counterpoint and must be resolved.

The dichotomy of musical intervals simplifies music theory, but perceptionally, each interval has its specific degree of consonance. Since ancient times the phenomenon of consonance has been debated and especially the quest for the cause of consonance led to different explanations.

The octave phenomenon is of special importance in psychoacoustics and in music theory. Remarkably, in most tonal systems tones an octave apart are regarded as the same note. This is due to the intense sensation of *tonal fusion* of the two tones of a simultaneous octave. The term tonal fusion describes the sensation of a unity when listening to two simultaneous tones an octave apart. The phenomenon of tonal fusion has already been discussed by the ancient Greek philosophers. Other consonant intervals also show a more or less pronounced tonal fusion. The degree of tonal fusion is directly correlated to the degree of consonance of the interval. The German psychologist and philosopher Carl Stumpf (1848–1936) was the first to investigate this phenomenon systematically in extensive hearing experiments [25]. He concluded that there must be a physiological cause of tonal fusion in the brain. Indeed, tonal fusion and the sensation of consonance has neurophysiological reasons. Licklider [18] had already proposed a neuronal autocorrelation mechanism for pitch detection. This idea is based on the theorem of Wiener–Khintchine which states that the Fourier transform of the power spectrum of a signal is equal to the autocorrelation function of the signal [11]. Thus, a spectral analysis by Fourier transform is equivalent to an autocorrelation analysis (see Section 2.2.7). Unfortunately, Licklider's model is physiologically infeasible. But a neuronal periodicity detection mechanism for pitch and timbre perception in the auditory systems has been found in neural nodes of the brain stem (nucleus cochlearis) and the mid-brain (inferior colliculus) [14, 15]. Essentially, it performs an autocorrelation analysis and is based on a bank of neuronal circuits. Each circuit adds a specific delay to the signal. The neural codes of the original signal and the delayed signal are projected onto a coincidence neuron of the circuit. If the specific delay is equal to a period of the signal, the coincidence neuron fires a pulse thus indicating that the specific delay of the circuit is equal to a period of the signal. Physiological data suggest a time window of $\varepsilon = 0.8$ ms for coincidence detection. In the auditory system, a tone is represented by a periodic train of neuronal pulses. Its period is the same as the period of the tone [30]. In the case of musical intervals, the periodicities of both interval tones is well preserved in the auditory nerve [27]. Thus, in the model, an interval is represented by two pulse trains $x_1(t)$ and $x_2(t)$ with periods p_1 and p_2 which are related by the vibration ratio s of the interval tones: $p_2 = s \cdot p_1$. The width of the pulses $I_\varepsilon(t)$ is adjusted to the width $\varepsilon = 0.8$ ms of the time window for coincidence detection.

$$x_1(t) := \sum_n I_\varepsilon(t - n \cdot p_1) \tag{3.1}$$

$$x_2(t) := \sum_n I_\varepsilon(t - n \cdot p_2) = \sum_n I_\varepsilon(t - n \cdot s \cdot p_1) \tag{3.2}$$

Now, the interval with the vibration ratio s is represented by the sum of both pulse trains representing the interval tones. The autocorrelation function of this sum is calculated to simulate the neuronal periodicity detection mechanism applied to neuronal representation of the interval. For arbitrary vibration ratios s, an autocorrelation function $a(\tau, s)$ of the corresponding pulse trains can be calculated over the range of all audible periods from about 0 ms, corresponding to 20,000 Hz, to $D = 50$ ms corresponding to 20 Hz.

$$a(\tau, s) := \int_0^D (x_1(t) + x_2(t))(x_1(t + \tau) + x_2(t + \tau))dt \tag{3.3}$$

The General Coincidence Function $\Gamma(s)$ as defined by Ebeling [6, 7] integrates over the squared autocorrelation function for arbitrary vibration ratios s, thus calculating the power of the autocorrelation function for every possible vibration ratio s and the corresponding interval.

$$\Gamma(s) := \int_0^D a(\tau, s)^2 d\tau \tag{3.4}$$

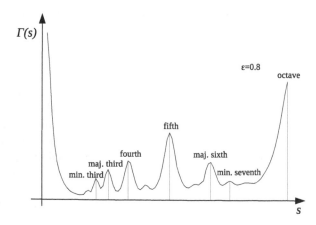

Figure 3.8: The Generalized Coincidence Function [6, 7]; the vibration ratio of the two interval tones for arbitrary intervals within the range of an octave are shown on the abscissa.

The graph of the General Coincidence Function predicts high firing rates for consonant intervals and low firing rates for dissonant intervals. It shows the same qualitative course as the "Curve der Verschmelzungsstufen" which Carl Stumpf determined from extensive hearing experiments [25]. The predictions can experimentally be confirmed [1].

3.5.2 Consonant and Dissonant Intervals and Tone Progression

The phenomenon of tonal fusion is the reason for an elementary rule of tonal progression: successive parallel octaves and fifths between two parts as shown in Figure 3.9 are prohibited as these strongly fusing intervals would void the independence of both parts. The perceptual reason is the segregation of parts.

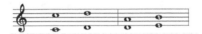

Figure 3.9: Prohibited parallels of pure octaves and pure fifths.

Counterpoint strives for a balance in the interplay of consonance and dissonance. As consonant intervals are sensed as unities, they support the integration of the parts whereas dissonant intervals segregate the parts. The perceived tension of dissonant intervals evokes the interest of the listener [9]. On sustained times, all consonant intervals are allowed. The perceptual reason is the integration of the parts. Dissonant intervals on sustained times must be introduced by foregoing consonant intervals and are resolved into consonant intervals normally by stepwise downward movement of one or both parts. The perceptual reason is the segregation of parts by the sensation of dissonance [2]. The dissonant tone is introduced to soothe the harshness of the dissonances. The downward movement corresponds to the remission of tension which is conjoined with room associations.

Resolution of Dissonances   Two examples of a dissonant second are shown in Figures 3.10 (a) and (b). The octave distribution of the intervals is of no theoretical importance in counterpoint, as is demonstrated by the example on the right of Figure 3.10 (b): the dissonant ninth, which is a second plus an octave, is regarded as a dissonant second. The second is resolved by stepwise downward motion of the lower part.

Normally, a dissonant seventh is resolved into a sixth by stepwise downward movement of the upper part (see Figure 3.10 (c)). But if the upper note is a leading tone, the upper part moves upwards. In this case the seventh is resolved into a perfect octave as shown in Figure 3.10 (d). In strict counterpoint, the perfect fourth is regarded as a dissonant interval. To resolve it into a perfect fifth, the lower part moves stepwise down. If the upper part goes stepwise down it is resolved into a third (see Figures 3.10 (e)–(g)). The resolution of the diminished fifth is only possible in the lower part. It is resolved into a sixth (see Figure 3.10 (h)).

Changing Notes and Passing Notes   On unsustained times, in addition to consonant intervals, all dissonant intervals may be used as passing or changing notes. The perceptual reason is that on unsustained times, dissonant intervals are perceived as

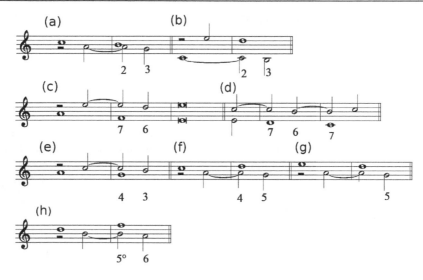

Figure 3.10: Resolution of introduced dissonances in the two-part counterpoint.

incidental deviations from the overall consonant structure quickly passing by. Examples of changing notes are given in Figure 3.11. Dissonant changing notes are only allowed in stepwise motion.

Figure 3.11: Examples of changing notes.

Passing notes may be consonant or dissonant. In Figure 3.12 dissonant intervals are indicated by numbers.

Figure 3.12: Examples of passing notes.

3.5.3 Elementary Counterpoint

Consider the tune in the Dorian mode of Figure 3.13 from "Gradus ad Parnassum" by Johann Joseph Fux (1725) [10]. This simple tune serves as a *cantus firmus*. Counterpoints shall be composed against it.

Figure 3.13: Dorian *cantus firmus* (c.f.) by Fux.

Two-Part Counterpoint   There are five species for a two-part counterpoint:

1. note against note (1:1),
2. two notes against one (2:1),
3. three or more notes against one (3:1, 4:1, 6:1),
4. suspensions (notes offset against each other (1/2:1), and
5. florid counterpoint.

The fourth species involves dissonant intervals and their resolution.

In a tone-against-tone two-part counterpoint (1:1), only consonant intervals are suitable. To underline the independence of both parts, contrary movement is preferred. A counterpoint in the upper part against the *cantus firmus* in the lower part is shown in Figure 3.14.

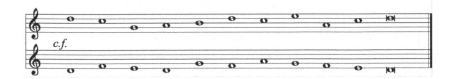

Figure 3.14: Dorian *cantus firmus* (c.f.) by Fux in the lower part and a 1:1 counterpoint in the upper part.

A few elementary rules have to be observed when composing a 1:1 two-part counterpoint [16] which are of psychological interest as they are based on elementary auditory perception and grouping processes [9, 2]. The most essential rules are as follows:

1. Only consonant intervals are allowed. The perceptual reason is the integration of both parts.
2. The first and the last intervals should be perfect primes, fifths, or octaves. The perceptual reason is that the first and the last tones of both parts fuse to entities supporting the integration of both parts.
3. Stepwise motion should be predominant. The perceptual reason is the integration of the tones of each voice into a coherent melodic line.
4. Contrary motion of both parts should be preferred. The perceptual reason is the independent movement of both parts supporting their segregation.

5. In every part, successive skips in the same direction should be avoided. The perceptual reason is that the integration of the tones into a melodic line would be disturbed by two successive skips.

6. The interval of the tenth should not be exceeded between both parts. The perceptual reason is that a separation of both parts by intervals greater than a tenth would disturb their integration.

The second and third species follow the easy rule that on sustained notes only consonant intervals are allowed, whereas consonant- and stepwise-reached dissonant intervals–passing and changing notes–may be used on all unsustained notes as demonstrated in Figure 3.15. Again, the perceptual reason is that consonant intervals support the integration of the parts, dissonant intervals lead to their segregation.

Figure 3.15: Dorian *cantus firmus* (c.f.) by Fux in the upper part and a 1:2 counterpoint in the lower part. The dissonant intervals are indicated by numbers.

The fourth species instructively demonstrates the usage of dissonant intervals (see Figure 3.16). Series of introduced dissonant and consonant intervals as their resolution evoke the sensation of continuously alternating tension and relaxation, which is one of the most important sources of the emotional effects in music.

Figure 3.16: Dorian *cantus firmus* (c.f.) by Fux in the lower part and a 1/2:1 counterpoint (bindings) in the upper part.

The fifth species is no longer bounded to strict rhythmical relations but still observes the rules of the first four species. Arbitrary rhythms are used to achieve a more lively expression. In vocal polyphony, the rhythms of the parts should follow and support the rhythm of the text. As an example of a florid two-part counterpoint, Figure 3.17 presents an original composition of the German composer and music theorist Michael Prätorius (1571–1621). The *cantus firmus* in the upper part is the Protestant chorale "Jesus Christus unser Heiland" against which Prätorius composed a counterpoint in the lower part. Observe the independent movement of two parts concerning their texts as well as the music.

Michael Prätorius
(1571 - 1621)

Figure 3.17: Michael Prätorius, "Jesus Christus unser Heiland" from "Musae Sioniae," 1610.

Three- and Four-Part Counterpoint and Harmony In a similar manner as demonstrated with a two-part counterpoint, several simultaneous parts can be composed against a given *cantus firmus*. All rules of the two-part counterpoint are also valid for a counterpoint with three or more parts.

Counterpoints with three (or more) parts form the bridge from counterpoint to harmony. The three-part counterpoint of the first species (1:1:1) is of special interest as it introduces triads. As for the first species of the two-part counterpoint (1:1), only consonant intervals between the parts are allowed. An additional rule for counterpoint with more than two parts says that the pure fourth between the upper parts is regarded as a consonance and may be used without hesitation. But between the two lowest parts a pure fourth must be avoided. It can easily be checked that according to this rule only major and minor triads and their first inversions are possible. The first inversion of a triad is called a *sixth chord* as the frame interval of the triad inversion is a sixth. The second inversion of the triad is a six-four chord, which has to be avoided as the interval between the two lower parts is a fourth. Figure 3.18 shows a C major and an a minor triad and their inversions in *closed position* (left) as well as in *open position* (right). Beneath the notation system, digits label the interval structure as conventional with figured basses: all intervals are related to the lowest or bass part.

3.5.4 Chords

Generally, *chords* are two or more simultaneous pitches. A synonym of a chord of two pitches is an *interval*, a *triad* is a chord of three pitches. The chords of the Western music system consist of stacked thirds upon the *root*, which is the lowest and fundamental tone of the chord. The name of the root denotes the chord. Chord

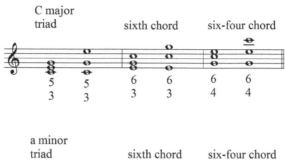

Figure 3.18: C major and a minor triads and their inversions. In strict counterpoint the six-four chord is regarded as a dissonant chord.

inversions contain the same notes as the original chord but their order is inverted so that a note other than the root is in the lowest part.

Triads   There are four kinds of triads. Starting from the root,

- the *major triad* is composed of a major third followed by a minor third (Figure 3.19 (1));
- the *minor triad* is composed of a minor third followed by a major third (Figure 3.19 (2));
- the *diminished triad* is composed of two minor thirds (Figure 3.19 (3)) and
- the *augmented triad* is composed of two major thirds (Figure 3.19 (4)).

Figure 3.19: Major, minor, diminished, and augmented triads.

The triads of Figure 3.19 are in *root position*, which means that the root of the triad is in the lowest part. Each triad has two inversions. The lowest note of the first inversion is the third of the triad. The first inversion is called a *sixth chord* as the interval between the lower part and the root is a sixth. In a *figured bass* the sixth chord is denoted by the numeral 6. The lowest note of the second inversion of a triad is the fifth. It is called a *six-four chord* because it contains a fourth and a sixth as intervals. In a *figured bass* the six-four chord is denoted as $\frac{6}{4}$ (see Figure 3.18).

A triad can be built up on every degree of the diatonic scale. Consider the diatonic major scale (see Figure 3.20 top):

- major triads are on degrees I, IV, and V;
- minor triads are on degrees II, III, and VI and
- a diminished triad is on degree VII.

Thus, three different kinds of triads are in the major diatonic scale in contrast to four kinds of triads in the diatonic scale of the harmonic minor (note the altered leading note; see Figure 3.20 bottom):

- minor triads are on degrees I and IV;
- major triads are on degrees V and VI;
- diminished triads are on degrees II and VII and
- an augmented triad is on degree III.

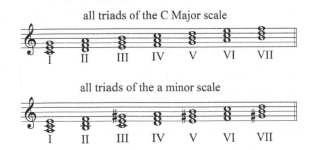

Figure 3.20: Diatonic major and minor scales with triads.

Note that only the major triad and the minor triad, but neither the diminished nor the augmented triad, appear on degree I. This coincides with the rule of music theory that only a major or minor triad can finish a piece of music. The diminished fifth of the diminished triad is a dissonant interval.

Seventh Chords   On each of the four triads, another minor or major third can be stacked up to get a seventh chord. Seven types of seventh chords are used in Western music theory (see Figure 3.21):

- Two seventh chords are derived from the major triad: (1) the major seventh chord and (2) the dominant seventh chord.
- Two seventh chords are derived from the minor triad: (3) the minor major seventh chord and (4) the minor seventh chord.
- Two seventh chords are derived from the diminished triad: (5) the half diminished seventh chord and (6) the diminished seventh chord.
- One seventh chord is derived from the augmented triad: (7) the augmented major seventh chord.

Figure 3.21: All seven kinds of seventh chords.

Consider the seventh chords with the notes of the major diatonic scale as roots (see Figure 3.22 top):

- major seventh chords are on degrees I and IV;
- a dominant seventh chord is on degree V;
- minor seventh chords are degrees II, III, and VI and
- a half diminished seventh chord is on degree VII.

Only four of the seven possible seventh chords can be built up in the major mode, but all seven seventh chords occur in the harmonic minor mode (note the altered leading note; compare Figure 3.22 bottom). The minor mode has a greater harmonic variety than the major mode:

- a major seventh chord is on degree VI;
- a dominant seventh chord is on degree V;
- a minor major seventh chord is on degree I;
- a minor seventh chords is on degree IV;
- a half diminished seventh chord is on degree II;
- a diminished seventh chord is on degree VII and
- an augmented seventh chord is on degree III.

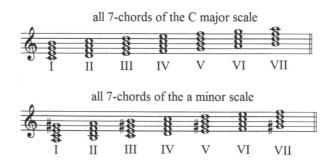

Figure 3.22: Diatonic major and minor scales with seven chords.

Each seventh chord has three inversions. The first inversion of a seventh chord is a six-fifth chord ($^6_5$-chord, third in the lowest part), the second inversion is a quarter-third chord ($^4_3$-chord, fifth in the lowest part), and the third inversion is a second chord (2-chord, seventh in the lowest part).

Note, that all tones of a chord, triads as well as seventh chords, can be altered. Altered tones are additional leading notes and must be resolved following the direction of alteration.

Further Chords   By stacking up more than three thirds, further chords can be formed. A ninth chord or even eleventh and thirteenth chords are upward extensions of seventh chords that encompass the intervals of a ninth (=octave plus second), an eleventh (=octave plus fourth), or even a thirteenth (=octave plus sixth). In modern music, all kinds of chords with arbitrary interval structures are conceivable. In most atonal music, the dichotomy of consonant and dissonant intervals is abolished. As a consequence, the requirement to resolve dissonant intervals is dropped. Unresolved dissonant intervals of a chord are not perceived as much as disturbances of the sound but as an individual timbre. A succession of chords with unresolved dissonant intervals evokes the effect of a timbre melody. The timbral richness of impressionistic music, e.g., of Claude Debussy, is mostly evoked by successions of unresolved dissonant chords (see [3]). In jazz and some kinds of popular music, seventh chords without resolution are ubiquitous and are one source of the characteristic jazz sound. Obviously, the seventh is treated as a consonant interval. Note, that the seventh is a weakly fusing interval (see Figure 3.8).

Chord Notation in Jazz and Popular Music   In jazz and pop music, a shorthand notation of tones and harmonies has been developed to facilitate notation and to organize group improvisation. A letter indicates the tone on which the chord is build up. The notes of the diatonic scale are C-D-E-F-G-A-B (German: H). Alterations are indicated by sharps and flats. For example, C♯ denotes a c sharp and an e flat is written as E♭. Though one always has to be aware of individual variations, some rules of chord notation a generally observed.

- Capitals label major triads. For example, D denotes the d-major triad, B♭ indicates the b flat major triad. Nothing is said about the inversion of the actual triad.
- A minus sign or the small letter m added to the capital indicate a minor triad. For example, A− or Am are the symbols for the a minor triad. F♯− or F♯m denote an f sharp minor triad.
- Added tones are indicated by index numbers corresponding to the interval between the fundamental note and the added tone. The *Berklee system* uses unambiguous prefixes to indicate whether this interval is pure (no prefix), minor (−), diminished (♭), major (M), or augmented (♯). Other usual prefixes and their meanings in the *Berklee system* are listed in Table 3.2 (see: [13, p. 11]).
- The index number 7 without any prefix always denotes a minor 7. Thus G^7 is the minor seventh chord on the note g. To denote a major seventh chord, a variety of prefixes are common: MAJ^7, Maj^7, maj^7, M^7, j^7, Δ^7 etc.

Tonal Functions   In a musical context the degrees of a scale are ascribed certain musical functions. The key of a scale is the *tonic* and the fifth degree of a scale is the *dominant*. One step below the dominant is the fourth degree, which is the *subdominant* (see Figure 3.23 (a)).

Rearranging the scale from the fourth degree, an octave lower to the fifth degree

Table 3.2: The Berklee System of Chord Notation

Symbols	Berklee system
♭9 / 9-	-2
9	M2
♯	♯2
sus4 / 11	4
♯11	♯4
♭13	-6
13 / 6	M6
° 7 / dim. 7	♭7
7	-7
Maj^7 etc.	M7

shows a constellation of the scale with the dominant a fifth above the tonic and the subdominant a fifth beneath the tonic. Note that the fifth is the most consonant interval besides the prime and the octave. The tonic is, so to speak, framed by the subdominant and the dominant which are harmonically closely related to the tonic by the strong consonant of a fifth (see Figure 3.23 (b)).

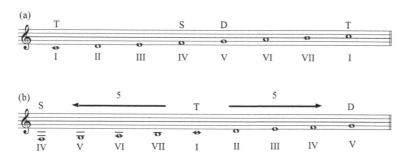

Figure 3.23: (a) Diatonic scale with the functions *tonic* T, *dominant* D, and *subdominant* S. (b) In the rearranged diatonic scale from degree IV to degree V an octave above, the subdominant and dominant are both a fifth apart from the tonic in the center.

Jumping a fifth up or down, a voice can change between the three functions. If this voice is the bass part, the degrees of this functions can be the roots of triads. And as the fifth is harmonically stable, a two times falling fifth first from the tonic to the subdominant and then from the dominant to the tonic stabilizes the tonic, which is the key (see Figure 3.24 (a)). By replacing the jump of a ninth by a second, the classical formula of the bass part of a complete cadence with a twice falling fifth is obtained (see Figure 3.24 (b)). Some theorists claim that these falling fifths are described by the name *cadence*: *cadere* means *to fall* in Latin.

Figure 3.24: The bass formular of the classical cadence.

Cadence and Harmonic Functions A clear structure facilitates the comprehension of music. To group a piece of music in its time evolution, cadences are the most effective harmonic mean to indicate intersections or the end of a piece of music. In the long history of music, different kinds of cadences were used. Since the early baroque era the classical cadence has become common to most Western music.

So-called *dominant chords* are erected on the fifth degree, *subdominant chords* are erected on the fourth, and *tonic chords* are build up on the first degree. A finalizing chord succession from the fourth to the first degree (IV-I) in the bass (from the subdominant to the tonic) is called a *plagal cadence* (see Figure 3.25 (a)). A finalizing chord succession from the fifth to the first degree (V-I) in the bass (from the dominant to the tonic) is called an *authentic cadence* (see Figure 3.25 (b)).

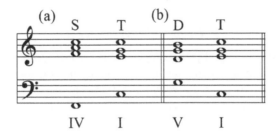

Figure 3.25: (a) Plagal cadence IV-I or S-T, (b) authentic cadence V-I or D-T.

The formula of Figure 3.24 (b) in the bass with the degrees I-IV-V-I represents the harmonic succession *tonic-subdominant-dominant-tonic*. It is the basic harmonic pattern of Western music. Figure 3.26 shows the cadence in the *fifth position* (left: the fifth of the bass note is in the upper part of the first chord), the *octave position* (middle: the octave of the bass note is in the upper part of the first chord), and the *third position* (right: the third of the bass note is in the upper part of the first chord):

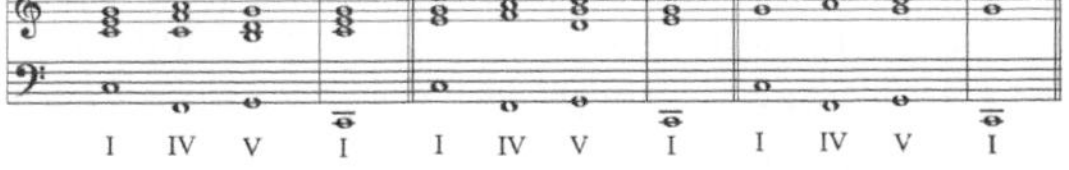

Figure 3.26: Cadence, left: fifth position, middle: octave position, right: third position.

The triads of the tonic, the dominant, and the subdominant can be substituted by other triads that have two tones in common with the original triad. These substituting triads are called *mediant chords*. Instead of triads, seventh chords may also be used. Note that the finalizing tonic chord must be a consonant triad. The minor seventh chord on the fifth degree is also called the *dominant seventh chord*. It is the most prominent seventh chord and strongly quests for the tonic as resolution. Especially the triad of the subdominant has a lot a substitutions that enrich the harmonic repertoire of music.

The pattern of the *jazz cadence* differs from the classical cadence. Its bass voice consists of the following sequence of degrees: I-VI-II-V-I, so that this chord progression is also called *sixteen - twenty-five*. Except for the first chord, it can be regarded as a succession of three falling pure fifths. As seventh chords are the standard chords of jazz, a basic jazz cadence may be:

Figure 3.27: Jazz cadence.

In the harmony theory of jazz, every chord is regarded as part of a church mode. Further notes from this church scale may optionally be play together with the chord and these additional tones are called *options*. For example, consider a minor seventh chord on the root c. It may be regarded as part of a Dorian scale if it is the chord of the II degree in b-flat major, or it is part of a Phrygian scale, if it is the chord of the III degree in a-flat major, or finally, it may be part of an Aeolian scale, if it is the chord of the VI degree in e-flat major. In Figure 3.28 the optional notes are indicated: in case of the Dorian scale (Figure 3.28 (a)), the options 9 and 11 (d or f) are possible, the 13 (a) may be used instead of the seventh, in case of the Phrygian scale (Figure 3.28 (b)) the option 11 (f) may be used, in case of the Aeolian scale (Figure 3.28 (c)) the options 9 and 11 (d and g) may be added. Some notes are crossed out as they are inappropriate as options although they belong to the particular modal scale (for reasoning see [13, p. 19].

As each chord of the harmonic repertoire of jazz is identified with a certain modal scale when used in a piece of music, the number of options that may be added to the chord is quite reduced. When improvising in a group, the musicians are aware of the chord successions. The elaborated system of chord and modal scale identifications grants that no inappropriate and disturbing tones are added by improvisation.

Certain chord successions are characteristic of a musical style. For example, the blues is a musical style of Afro-American music that had a great influence on other musical styles such as jazz (*blues jazz, boogie woogie*), pop music, and rock (*blues rock*). The original *blues scheme* consists of an easy chord succession within twelve bars. Let Latin numbers represent the degrees of the diatonic scale. The twelve bars

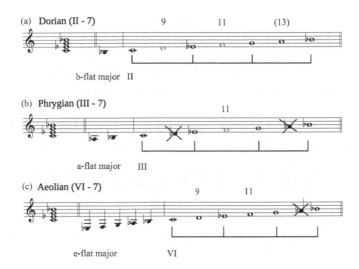

Figure 3.28: The minor seventh chord as part of three different modal scales. The identification of a chord with a certain modal scale depends on the musical context. The grey notes are possible options; theoretically possible but inappropriate options are crossed out.

of the *blues scheme* are now given by a pattern of three times four bars as shown in Figure 3.29 (a). In case of a *quick change*, the second bar has a chord on degree IV instead of the chord of degree I. Figure 3.29 (b) shows the chords of this pattern with a *quick change* applied to C major.

(a)

| I |I(IV)| I | I |

|IV| IV | I | I |

| V | IV | I | I |

(b)

|C7|F7| C | C |

|F7|F7| C | C |

|G7|F7| C | C |

Figure 3.29: The figure shows the original *blues scheme* (a) and its realization with a *quick change* in C major (b).

This chord pattern can be found in the American folk song "Blackwater Blues". Note that all chords are based on the *blues scale* as presented in Section 3.2.3. Thus the minor seventh, which reflects the original *blues seventh*, can be used on every degree. In bar nine, a clash of the minor third of the melody and the major third of the chord reminds us of the *blues third*. Of course, besides the original *blues scheme* there exist a number of variations of this chord pattern.

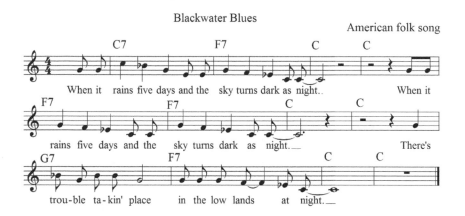

Figure 3.30: The American folk song "Blackwater Blues" is based on a blues scheme with a *quick change* in bar two.

3.5.5 Modulations

The term *modulation* refers to the process of changing from one key to another. In case of a *melodic modulation* a single and possibly unaccompanied melodic line audibly changes to a new key. In a *harmonic modulation* a certain chord mediating between both keys functions as a *means of modulation*. Different kinds of harmonic modulation are distinguished depending on the means of modulation.

In a *diatonic* or *common-chord modulation*, the means of modulation is a chord shared by both keys. For example, a modulation from C-major to B-flat-major can be mediated by an F-major triad, which is on the fourth degree of C-major (subdominant) and on the fifth degree (dominant) of B-sharp major (see Figure 3.31).

Figure 3.31: Diatonic modulation from C major to B flat major. The F major six-chord of the first bar is the modulation means. The F major triad is on the fourth degree of C major and on the fifth degree of B flat major.

In a *chromatic modulation*, a chord of the original key is changed into a chord of the destination key by alteration (see Figure 3.32).

In an *enharmonic change*, a chord of the original key is interpreted as a chord of the destination key. At least one tone name is changed without any change of pitch. For example, the dominant seventh chord of the original key sounds like the

Figure 3.32: Chromatic modulation from G major to e minor. By alteration, the G major triad of the first bar is chromatically changed into an augmented triad leading to e minor.

augmented six-fifth chord of the minor key a semitone under the original key (see Figure 3.33).

Figure 3.33: Enharmonic modulation from F major to e minor. The b flat of the $\frac{6}{5}$-chord of the second bar is read as a sharp in the following augmented $\frac{6}{5}$-chord of e minor.

3.6 Time Structures of Music

3.6.1 *Note Values*

A *note value* determines the relative duration of a tone. The note values of monodic songs follow the rhythm of the speech so that the durations of the single tones are somewhat arbitrary. But to allow for the simultaneity of all voices of polyphonic music, the durations of single tones must be defined exactly. The composers of early polyphony had to invent a notation system that involves a measure of tone durations. Around 1250, Franco von Köln invented different symbols for different durations. In the end, a notation system for note duration was developed that is based on halving the note values of longer notes to derive short ones. The most common note values and their relations are shown in Figure 3.34. A *whole note* is equal to

- two *half notes,*
- four *quarter notes,*
- eight *eighth notes,*
- 16 *sixteenth notes,* and
- 32 *thirty-second-notes.*

whole note

half note

quarter note

eighth note

sixteenth note

thirty-second-note

Figure 3.34: Relations of note values.

Analogously, the durations of the rests must be measurable. The different signs of the rests and the corresponding note values are shown in Figure 3.35.

Figure 3.35: Note values and rests.

A dot behind a note or rest prolongs its duration by half of its original value. Let N be the note value, then $N\cdot = (1+1/2)N$. Figure 3.36 shows dotted note values.

Figure 3.36: Dotted note values.

Two dots behind a note or rest prolong its duration by half and a quarter of its original value. Let N be the note value, then $N\cdot\cdot = (1+1/2+1/4)N$.

Instead of halving a note value, irregular divisions are also applied. A division by three results in a *triplet*, a division by five leads to a *quintuplet*, a division by seven creates a *septuplet* (see Figure 3.37).

Figure 3.37: Triplets, quintuplets, and a septuplet and their durations.

3.6.2 Measure

The term *measure* originally refers to the metrical foots of ancient Greek poetry which describe patterns of long and short syllables. Thus, the *measure* is an ordering principle. Applied to music, metrical foot refers to either patterns of tone durations or to patterns of the accentuation of notes (ordering of heavy/stressed and light/unstressed notes). Basic metrical foots with relevance in music theory are listed in Table 3.3. They are binary (*trochee, iambus, spondee*) or ternary (*dactyl, anapaest, tribrach*). Figure 3.38 shows these metric foots as patterns of tone durations.

Table 3.3: Ancient Metrical Foots

metrical foot	durations	accentuation	symbol
trochee	long - short	heavy - light	−U
iambus	short - long	light - heavy	U−
dactyl	long - short - short	heavy - light - light	−UU
anapaest	short - short -long	light - ligth - heavy	UU−
spondee	long - long	heavy - heavy	− −
tribrach	short - short - short	light - light - light	UUU

Each of these time patterns encompass a time frame of the psychological present, which is a short time interval conceived by the individual as the present moment [21] and which may merely last less than a second but does not last more than three or four seconds. The repetitions of these metric patterns evoke the sensation of coherence and support the integration over time. As an example of a continued trochee, the famous "Marcia funebre" from Beethoven's piano sonata *Pathétique* demonstrate that the ancient metric foots are still an appropriate resource to describe the elementary time structure of music (see Figure 3.39).

3.6.3 Meter

In music, the continuous flow of time is perceptually discretized by regular (equidistant) *beats*. The *meter* is a schematic ordering of stressed and unstressed (heavy and light) beats. Mostly, the meter is indicated by the numbers of a fraction behind the clef. The denominator represents the note value of the beats to be counted and the numerator specifies the number of beats in one bar of the meter. Simple meters are distinguished from compound meters. Simple meters have only one stress on the first

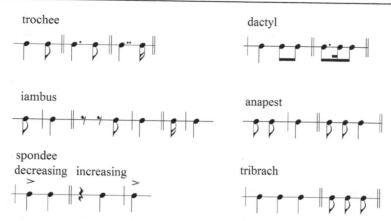

Figure 3.38: Binary (left) and ternary metric foots (right).

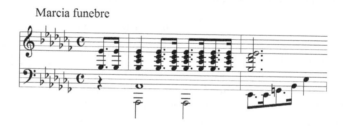

Figure 3.39: Trochee: Beethoven, *Marcia funebre* from the piano sonata *Pathétique*, op. 13.

beat, the other beats are unstressed. Compound meters have their main stress on the first beat and secondary stresses on other beats. Simple meters are all meters with two or three beats per bar, for example: two four times $\frac{2}{4}$, two-eight times $\frac{2}{8}$, and three four times $\frac{2}{3}$, three-eight times $\frac{3}{8}$. Compound meters either have equal parts (2+2 or 3+3) or unequal parts (2+3 or 3+2). Next to the main stress on the first beat, the first beat of the other part (or parts) has a secondary and somewhat lighter stress. Examples of compound meters are four-four times $\frac{4}{4} = \frac{2}{4} + \frac{2}{4}$, six-four times $\frac{6}{4} = \frac{3}{4} + \frac{3}{4}$, five-four times $\frac{5}{4} = \frac{2}{4} + \frac{3}{4}$ or $\frac{5}{4} = \frac{3}{4} + \frac{3}{4}$, six-eight times $\frac{6}{8} = \frac{3}{8} + \frac{3}{8}$, seven-eighth times $\frac{7}{8} = \frac{3}{8} + \frac{4}{8}$ or $\frac{7}{8} = \frac{4}{8} + \frac{3}{8}$, nine-eight times $\frac{9}{8} = \frac{3}{8} + \frac{3}{8} + \frac{3}{8}$, twelve-eight times $\frac{12}{8} = \frac{3}{8} + \frac{3}{8} + \frac{3}{8} + \frac{3}{8}$. The term *alla breve* describes meters in which half notes are counted as beats: one-two meter $\frac{1}{2}$, two-two meter $\frac{2}{2}$, three-two meter $\frac{3}{2}$, four-two meter $\frac{4}{2}$.

Generally, in Western music the first beat of a bar has the strongest stress. Thus, the regular bar lines at the beginning of each bar indicate this scheme of stressed and unstressed beats. Note that there are exceptions to this rule: for example, a four-four time in jazz music is often stressed on the second and fourth beats instead of the first and third beats.

For simple meters, binary and ternary patterns of beats can be distinguished according to whether the beats or pulses are organized in two or three. That means either a stressed beat is followed by one unstressed beat or a stressed beat is followed by two unstressed beats. In music theory the quarter note is the standard value. Thus there are two basic simple meters: the two-four and the three-four.

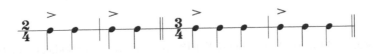

Figure 3.40: The two-four meter (left) and the three-four meter (right) are the basic simple meters.

Each beat may be subdivided by twos, threes, or fours. Occasionally, even unitary patterns with subdivisions occur, especially if the music has a fast pulse. For example, a Vienna waltz is an unitary pattern subdivided by three.

In medieval times, ternary patterns of beats were associated with the trinity of God and thus named *tempus perfectus*. Especially in sacred music, the *tempus perfectus* symbolized heavenly spheres and the perfection of God. As this divine perfection was graphically symbolized by a circle, the *tempus perfectus* was also indicated by a circle. On the other hand, the sinfulness of the human live on earth and the imperfection of manhood was characterized by binary patterns of beats called *tempus imperfectus* and symbolized by an imperfect circle (semicircle or three-quarter circle like the letter C). This example demonstrates how religious sense or ideas of world view can be associated with elementary musical structures imposing meaning on the music. Even today, an imperfect circle is used to indicate the four-four time. An example is given in Figure 3.39.

The regular four-four time is a compound meter with two parts. But there is also an alla-breve four-four meter which has four beats but only a single stress on the first beat whereas all other beats are unstressed. Thus the alla-breve four-four time is a simple and unitary meter with four subdivisions. As only every fourth beat is stressed, the alla-breve meter has a wafting expression, especially in combination with a slow tempo. A prominent example is Mozart's "Ave verum" which is signed *Adagio alla breve* and which shows the *alla breve sign* which looks like a struck-through C: it is equal to the C-sign for the four-four time but with a Roman number I to symbolize *unity* and to indicate that the meter is simple (see Figure 3.41).

3.6.4 Rhythm

Rhythm stems from Greek and means *the floating*. Originally it described the constant alterations of tension and relaxation. Rhythm freely combines metric units and is a quantifying ordering principle of tone durations and accentuations. It is a super-order of measure and meter. Rhythm can be independent of any scheme so that tensions and relaxation freely alternate. Thus, measure and meter are not presuppo-

Figure 3.41: Mozart: Ave verum K.V. 618, the *alla breve* sign indicates that only the first beat has to be stressed.

sitions of rhythm. This becomes obvious with chants in which the music follows the text as in the Gregorian chants of the medieval Roman church.

   If music has an underlying meter, rhythm is integrated into the metric scheme so that the stresses of the bars and the stresses of the rhythm pattern coincide. But the independence of a rhythm may result in a segregation of the bar scheme and the stresses of the rhythm, which leads to syncopation. A *syncope* originally means *beating together* and it results from slurring a stressed note to the prevenient unstressed one, which induces a shift of the stress onto the actually unstressed beat (see Figure 3.42). A syncope induces deviations from the accentuation scheme of the entrenched meter. Its effect may be surprising, yields diversion, and arouses interest. Syncopes support stream segregation especially if they occur only in one part.

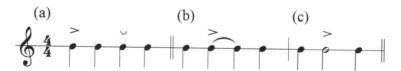

Figure 3.42: Syncope: (a) regular four-four meter, (b) the slur shifts the stress of the third beat to the second beat, (c) notation of resulting syncope with the same sound as in (b).

   Certain syncopation schemes are distinctive of certain music styles and associated dance styles. Figure 3.43 shows the rhythmic patterns of two Latin-American dances, a *rumba* and a *bossa nova*. Note the syncopations of the bass voices and the alternating rhythms of the chords, which are characteristics of these dances.

3.7   Elementary Theory of Form

Theory of musical form   The experience of a work of art like a piece of music has two aspects: its content and its form or *Gestalt*. The relations between the single parts and the whole define the *Gestalt*, elicit the aesthetic emotions and cogitations, and determine the architecture and tectonic of music. The theory of musical form

Figure 3.43: Syncopes in dance music: the vividness of syncopated rhythms are a characteristic of Latin-American dances; rhythmic patterns (a) of a *rumba*, and (b) of a *bossa nova*.

analyzes how the ordered elements of rhythm, melody, and harmony constitute a musical *Gestalt* which is always experienced as a whole or entity with distinguishable parts. Elementary *Gestalten*-like motifs form musical *Gestalten* of higher order like musical themes or melodies which themselves are parts of *Gestalten* at an even higher level like a whole composition or a movement of a sonata for example. Thus, the scope of the theory of musical form are the structural principles of music that evoke the experience of musical meaning and educe extra-musical associations, which both belong to the content of a composition.

Motif The *motif* is the smallest musical entity. Sometimes, a motif is the musical nucleus from which the whole musical structure of a piece of music evolves. It is characterized by a succession of certain pitches and bears an individual rhythmic content of one metric unit. A motif encompasses a time frame of a psychological present and represents an elementary musical *Gestalt*. Translating its pitches transposes the motif. A *progression* is a repeated translation of the same interval. A dilatation of the note values of a motif is called *augmentation* if all note values are proportionally elongated. The term *diminution* describes the proportional shortening of all note values of a motif. Figure 3.44 (a) shows a simple motif. Possible variations are demonstrated in Figure 3.44.

- A motif can be *transposed*, which is a translation operation on pitch. A series of transpositions is called a *sequence*. Figure 3.44 (b) shows a *diatonic sequence*.
- A motif can be *augmented*, which is a dilatation (elongation) of note values (Figure 3.44 (c)).
- A motif can be *diminished*, which is a shortening of note values (Figure 3.44 (d)).
- A motif can be augmented or diminished in its interval structure (Figures 3.44 (e) and (f)).

- A motif can be augmented and/or diminished with respect to its note values as well as in its interval structure (Figures 3.44 (g)–(j)).
- A motif and its variations can be inverted (Figures 3.44 (k)–(t)).

Figure 3.44: Sequence and rhythmic augmentation and diminution of the motif (a).

Melody   A *melody* (*tune*) is a succession of tones in a voice that are perceived as an entity. Psychologically, a melody is a *tonal Gestalt* with the following virtues:

- A melody consists of successive discrete pitches from a musical scale.
- It is perceptionally autonomous.
- A melody is perceived as a complete entity.
- All possible subdivisions of a melody are sensed to be incomplete.
- Each melody has a characteristic individual *rhythmic Gestalt*.
- A melody has an individual tempo. Too strong deviations from this tempo corrupt the melody.

In other words, a melody should be comprehensible and make sense in the musical context and the particular musical style. Vocal melodies should be singable. The melody of songs are often identified with a certain text. Figure 3.46 is an example of the melody of a medieval love song, Figure 3.47 shows the simple melody of a German folk song.

Musical Theme   A *musical theme* consists of one or more motifs forming a musical entity from which greater musical structures can be developed. Its characteristic *Gestalt* provides the musical expression that elicits the emotional effects of the music and may engross a whole piece of music. Themes formed by only one motif have an evolutionary tendency as the motif is resumed in several variations. Figure 3.45

gives an example of this type of theme from *Die Meistersinger von Nürnberg* by R. Wagner.

Figure 3.45: R. Wagner: *Vorspiel* of the opera *Die Meistersinger von Nürnberg*: a sequence of phrases form the theme of the pseudo-fugue.

Closed Forms and Sequential Forms   Closed forms comprise a sequence of musical phrases that are normally two or four measures long. Their separations are marked by cadences. The ending cadence is a *full cadence* on the tonic (or first degree). The prototype of a closed musical form is the *period*, which is a binary form. Normally it encompasses eight bars and consists of two phrases, each four measures long. The antecendent phrase ends on a weak imperfect cadence mostly on the fifth degree, whereas the second phrase ends on an authentic cadence on the first degree, which is also the tonic determined by the key tone.

Gestalt psychologists claim that symmetries are a means to improve *Gestalt* perception and thus symmetries are of aesthetic importance. As music is time dependent, a repetition (which is a translation in time) is the symmetry which is most effective for *Gestalt* perception in hearing as the German musicologist Hugo Riemann (1849–1919) pointed out [22]. Repetitions and varied repetitions are the core of elementary as well as higher-level structures in music. Compare Figure 3.44 for examples of variations of a motif.

Closed periods can be composed to form simple higher-level structures. The single periods of this form may be repetitions or variations of one another or they may be completely independent. A simple example of a form composed of independent periods is a chain of periods. Those chains already emerge from improvised singing or playing of a group of musicians. With capitals denoting the single periods, this form reads as A-B-C-D-....

The simplest binary song form consists of only one Period A that is repeated with only slight variations or an altered final cadence: A-A or A-A'.

Of course, the second period (now denoted B) can also be contrasting and completely different: A-B. Repeating part A results in a *bar form*, which was widespread in the medieval art of *minnesong*: A-A-B (see Figure 3.46).

Another elementary example is the ternary song form, which consists of two different phrases A and B and a repetition of A or its variations A', A". The resulting

Figure 3.46:* The medieval love song "Kum geselle min."

forms are: A-A'-A'', A-B-A, or A-B-A'. The German Volkslied "Alle Vögel sind schon da" has the form A-B-A, cp. Figure 3.47).

Figure 3.47: The German spring song "Alle Vögel sind schon da" clearly has a ternary A-B-A form.

Numerous other combinations of phrases are conceivable and can be found in sequential forms of music. The next example shows John Lennon's "Yellow Submarine" (see Figure 3.48). The song has two parts. Part A is repeated three times by its variations A1, A2, A3. The second part B, the *chorus*, is repeated once without any alterations.

All sequential forms follow the same elementary principles:

- The basic building blocks are closed phrases.
- The phrases are structured by cadences.
- Repetitions or phrases lead to symmetry and *Gestalt* perception.
- Variations of phrases elicit the listeners' concern.

Highly developed symmetric forms are ubiquitous in music and lead to the sense of order and stability. Listeners take delight in repetitions and variations. On the other hand, asymmetric structures deviate from expected symmetries and may express dynamic, rapidness, or disturbance as can be observed in the overture of Mozart's opera *La nozze di Figaro* (Figure 3.49).

Higher-level structures may also be organized according to the same structural principles, which results in a hierarchic organization of form. An easy example is the *great ternary song form.* Its higher-level structure can be denoted as A-B-A. The

Figure 3.48: The Beatles song "Yellow Submarine" has the simple sequential form A-A1-A2-A3-B-B.

Figure 3.49: The opening theme of the overture of Mozart's *La nozze di Figaro* has a complete asymmetric structure.

third part is a repetition of the first one so that part A frames the contrasting part B. Parts A and B are also ternary with parts a, b, and parts c, d respectively, so that A has the structure a-b-a and B has the subdivision c-d-c. The whole formal organization is as follows:

    A   B   A
   a-b-a   c-d-c   a-b-a

The different kinds of the *rondo*-form also illustrate the principles of sequential forms. Originally the rondo (Latin: rondellus, French: rondeau) was a chain of independent songs or a series of successive dances with the formal structure: A-B-C-D- ... . Repeating the first part as a so-called *ritornelle* leads to the formal structure

A-B-A-C-A-D- ... . The classical rondo combines the sequential organization with a balance of the harmonic structure. Denoting the key of the tonic by the letter *T* and the key of the dominant as *D*, the structure of the classical rondo can be written as follows:

A	B	A	C		A	B	A
T	D	T	parallel minor key		T	T	T
			or parallel major key				

Developmental Forms Whereas *sequential forms* with their interplay of repetition, variation, and contrast evoke a variety of musical ideas, *developmental forms* are restricted to a musical base material which may be a single theme or even a single motif. A whole piece of music can be developed by variations and transformations of this base material. The most important developmental forms are the *fugue* and the *sonata*. The *fugue* is regarded as the culmination of polyphonic composition and counterpoint. Fugues of the baroque era are to be regarded as unsurpassed master-pieces of this composition technique, which is based on the recurrent *imitation* of a theme in all parts of the composition and on different pitches. The composition starts with one voice to introduce the *theme*. The other voices, one after the other, imitate the theme on other tonal degrees. While one voice imitates the theme, the other voices play independent melodic lines composed against the subject according to the craft of counterpoint. After the theme has been played in all parts, one *exposition* of the fugue has been played. Three or more *expositions* form a fugue composition. Between the expositions, modulating interplays with sequences are inserted. Figure 3.50 shows the first exposition of a three-part "Fugue in C major" by J.S. Bach.

First of all, the theme determines the character of the composition. The counter-point may support or contrast the effect of the theme. The technique of *fugue* can be realized in different higher-order forms as demonstrated in J.S. Bach's *Die Kunst der Fuge*.

The baroque *suite* is a succession of historical dances of the Renaissance and Baroque. At the end of the 18th century the *classical sonata* as the most prominent musical form of the 19th century evolved from the suite. Regarded from the stand-point of musical form, a *Symphony* is a *classical sonata* composed for an orchestra. It consists of four movements: the first movement, *Allegro*, is contrasted by a second slow movement. As a relict of the *Suite*, the third movement is a *Minuet*, or later in the historical course, a fast *Scherzo*. The last fast movement is often a *rondo*. Joseph Haydn is renowned as the inventor of the *sonata*. The term sonata-form refers to the first movement of the whole sonata. The first part is called the *exposition*, which presents and develops the musical material: a robust or virile first theme is followed by a second theme in a singing stile. In case of a major key, the second theme is on the key of the fifth degree (*dominant key*); in case of a minor key, the second theme is in the parallel major key. The second part is the *development*: modulation, splitting of the themes, variations of their motifs, and other compositional means are applied for dramatic effects and to produce musical tension. The third part is a *recapitulation* of the *exposition* and returns to the original key for both themes. Often a final part called the *coda* finishes the movement.

Figure 3.50: J.S. Bach: first exposition of a "Fugue in C major."

3.8 Further Reading

For all issues of music, the two great encyclopaedias of music are the primary sources to be consulted:

- *Musik in Geschichte und Gegenwart MGG* [8]
- *Grooves Dictionary of Music* [23]

As an introduction to different aspects of music psychology, the article collection *The Psychology of Music* by Diana Deutsch is recommended [5]. Concerning the perception of *Gestalt* in music, one should refer to *Erkenntnislehre* by Carl Stumpf [26] and *The Auditory Scene Analysis* by Alber Bregman. Still highly recommendable is Arnold Schönberg's textbook *Harmonielehre* (1911) [24] and its English translation *Theory of Harmony*, which are both available in several reprints. Those who want to study counterpoint from the ground up are advised to read the textbook by Lemacher / Schoeder [16], which partly follows Fux's *gradus ad parnassum*. A different approach that involves the music from the baroque era to the 19th century is the textbook by Walter Piston [20]. A *theory of form* with a richness of examples is provide by Lemacher / Schroeder [17].

Bibliography

[1] G. M. Bidelman and A. Krishnan. Neural correlates of consonance, dissonance, and the hierarchy of musical pitch in the human brainstem. *The Journal of Neuroscience*, 29(42):13165–13171, 2009.

[2] A. S. Bregmann. *Auditory Scene Analysis*. MIT Press, 1990.

[3] D. de la Motte. *Harmonielehre*. dtv / Bärenreiter Kassel, 1976.

[4] D. de la Motte. *Kontrapunkt*. Bärenreiter, 2010.

[5] D. Deutsch, ed. *The Psychology of Music*. Academic Press, New York, 1982.

[6] M. Ebeling. *Verschmelzung und neuronale Autokorrelation als Grundlage einer Konsonanztheorie*. Peter Lang Verlag, 2007.

[7] M. Ebeling. Neuronal periodicity detection as a basis for the perception of consonance: A mathematical model of tonal fusion. *The Journal of the Acoustical Society of America*, 124(4):2320–2329, 2008.

[8] L. Finscher, ed. *Die Musik in Geschichte und Gegenwart (MGG)*. Bärenreiter, 2nd, revised edition, 2003.

[9] C. W. Fox. Modern counterpoint: A phenomenological approach. *Notes: Quarterly Journal of the Music Library Association*, 6(1):46–57, 1948.

[10] J. Fux. *Gradus ad Parnassum oder Anführung zur regelmäßigen musicalischen Composition. (Nachdr. d. Ausg. Leipzig 1742)*. Olms, 1742 / 2004.

[11] W. M. Hartmann. *Signal, Sounds, and Sensation*. Springer, 2000.

[12] W. M. Hartmann and D. Johnson. Stream segregation and peripheral channeling. *Music Perception*, 9(2):155–183, 1991.

[13] A. Jungblut. *Jazz Harmonielehre*. Schott, 1981.

[14] G. Langner. Evidence for neuronal periodicity detection in the auditory system of the guinea fowl: Implications for pitch analysis in the time domain. *Experimental Brain Research*, 52(3):333–355, 1983.

[15] G. Langner. Die zeitliche Verarbeitung periodischer Signale im Hörsystem: Neuronale Repräsentation von Tonhöhe, Klang und Harmonizität. *Zeitschrift für Audiologie*, 46(1):8–21, 2007.

[16] H. Lemacher and H. Schroeder. *Kontrapunkt*. Schott, 1978.

[17] H. Lemacher and H. Schroeder. *Formenlehre der Musik*. Gerig, 1979.

[18] J. C. R. Licklider. A duplex theory of pitch perception. *Experimenta VII/4*, 1951.

[19] M. Lindley and R. Turner-Smith. *Mathematical Models of Musical Scales.* Verlag für systematische Musikwissenschaft GmbH, 1993.

[20] W. Piston. *Counterpoint.* W. W. Norton & Company, New York, 1947.

[21] E. Poeppel. *Grenzen des Bewusstseins: Über Wirklichkeit und Welterfahrung.* Stuttgart : Deutsche Verlags-Anstalt, 1985.

[22] H. Riemann. *Wie hören wir Musik?* Max Hesse's Verlag, 1888.

[23] S. Sadie. *The Grove Dictionary of Music and Musicians.* Oxford University Press, 2001.

[24] A. Schönberg. *Harmonielehre.* Universal Edition, 1922.

[25] C. Stumpf. *Tonpsychologie*, volume 1. S. Hirzel, 1883 / 1890.

[26] C. Stumpf. *Erkenntnislehre.* Barth, reprint: Pabst Science Publishers 2011, 1939 / 1940.

[27] M. J. Tramo, P. A. Cariani, B. Delgutte, and L. D. Braida. Neurobiological foundations for the theory of harmony in western tonal music. In R. J. Zatorre and I. Peretz, eds., *The Biological Foundations of Music*, volume 930, pp. 92–116. New York Academy of Sciences, 2001.

[28] C. von Ehrenfels. Über Gestaltqualitäten. *Vierteljahrsschrift für wissenschaftliche Philosophie*, 14:249–292, 1890.

[29] M. Wertheimer. Über Gestalttheorie. *Philosophische Zeitschrift für Forschung und Aussprache 1*, 1925.

[30] W. Yost. *Fundamentals of Hearing: An Introduction.* Academic Press, 2000.

Chapter 4

Digital Filters and Spectral Analysis

RAINER MARTIN, ANIL NAGATHIL
Institute of Communication Acoustics, Ruhr-Universität Bochum, Germany

4.1 Introduction

In this chapter we will review fundamental concepts and methods of digital signal processing with emphasis on aspects which are important for music signal analysis. We thus lay the foundations for chapters dealing with feature extraction, feature selection, and feature processing (Chapters 5, 15, 14). The chapter starts with the definition of continuous, discrete, and digital signals and a brief review of linear time-invariant systems. We explain the design and implementation of digital filters with finite impulse response (FIR) and infinite impulse response (IIR). These filters are frequently used in filter banks for spectral analysis of audio signals. Besides filter banks, we present transformations for spectral analysis such as the discrete Fourier transformation (DFT), the constant-Q transformation (CQT), and the cepstrum. The chapter concludes with a brief introduction to fundamental frequency estimation.

4.2 Continuous-Time, Discrete-Time, and Digital Signals

The machine-based analysis of music most often utilizes audio signals which represent the acoustic waveform. We therefore start this chapter with definitions of basic types of signals.

Definition 4.1 (Continuous-time Signal). A one-dimensional *continuous-time* signal $x(t)$ is a function which may be real-valued $x : \mathbb{R} \to \mathbb{R}$ or complex-valued $x : \mathbb{R} \to \mathbb{C}$. A signal that is continuous in both time and amplitude is often called an *analog* signal.

An example of a real-valued signal is the sound pressure as picked up by a microphone in a music recording. A complex-valued signal might be obtained after a band-pass filtering operation or as the output of a frequency domain transformation.

Definition 4.2 (Discrete-time Signal). A *discrete-time* signal $x[k]$ is a sequence of real or complex-valued numbers, i.e. $x : \mathbb{Z} \to \mathbb{R}$ or $x : \mathbb{Z} \to \mathbb{C}$.

111

In many cases a discrete-time signal $x[k]$ is the result of sampling a continuous-time signal $x(t)$ using a sampling period of $T_s = 1/f_s$, where f_s is the *sampling frequency*. Then, the discrete-time signal $x[k]$ represents the analog signal at the sampling instances, e.g., $x[k] = x(kT_s)$, $k \in \mathbb{Z}$. k is called the sampling or time index. According to the Nyquist–Shannon sampling theorem, the reconstruction of a bandlimited analog signal from its sampled representation is possible if the sampling rate f_s is more than twice as large as the bandwidth B of the signal. $f_s/2$ is also known as the *Nyquist frequency*.

In commercially available recordings, music signals are typically sampled with a minimum rate of $f_s = 44.1$ kHz such as on the audio compact disc (CD). Many recording devices support other sampling rates as well, e.g., $f_s = 32$ kHz, $f_s = 48$ kHz, or $f_s = 96$ kHz as introduced in the legacy *Digital Audio Tape* (DAT) system. The sampling rate of $f_s = 48$ kHz corresponds to a maximum audio bandwidth of $B = 24,000$ Hz which is sufficient to cover the frequency range of the human auditory system.

Definition 4.3 (Digital Signal). In principle, the discrete-time signal is continuous in amplitude. By contrast, a *digital signal* $x_q[k]$ is both discrete in time and in amplitude. The quantized amplitude is denoted by x_q.

Figure 4.1 depicts examples of an analog, a discrete-time, and a digital signal. In most cases the discrete amplitudes are the result of a quantization process: in Figure 4.1 only four quantization levels are used, corresponding to 2 bits per sample. However, to render quantization effects inaudible significantly more quantization levels are required. Therefore, music signals are quantized with about 16–24 bits per sample corresponding to 65,536–16,777,216 quantization levels.

In summary, the conversion from an analog to a digital signal (*A/D conversion*) comprises sampling and quantization steps. In the context of this chapter, however, sampling and quantization errors will not be discussed in greater detail.

The inverse digital-to-analog conversion (*D/A conversion*) is commonly used in music players to convert digital data streams into analog audio signals. Note that the reconstruction of the original analog signal from a sampled and quantized signal cannot be perfect as quantization errors are irreversible.

4.3 Discrete-Time Systems

Music signals are processed almost exclusively on digital hardware (also see Chapter 27) using discrete-time processing approaches. Typical examples are filters, filter banks, signal transformations, or feature extraction modules. When processing music signals on digital hardware, quantization effects might play an important role. However, in the scope of this chapter, we assume that all signals are represented with a sufficient number of bits in all stages of the processing chain. Then, quantization errors can be neglected and we deal with discrete-time systems as outlined below.

Definition 4.4 (Discrete-time System). A discrete-time system $T[\bullet]$ maps one (or more) input signal onto one (or more) output signal. In the basic case of a *single-*

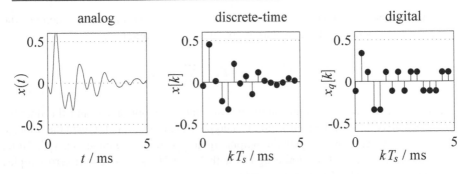

Figure 4.1: Analog, discrete time, and digital signals. t, k, and T_s denote, respectively, the continuous time variable, the discrete time index, and the sampling period.

input / single-output (SISO) system we have

$$y[k] = T[x[k]], \tag{4.1}$$

where $T[\bullet]$ is a mapping from signal $x[k]$ to signal $y[k]$.

The most important class of discrete-time systems are *linear time-invariant* systems.

Definition 4.5 (Linear and Time-Invariant (LTI) System). A system is called *linear* if the superposition of two input signals $x_1[k]$ and $x_2[k]$ results in the same superposition of the corresponding two output signals. Therefore, for $y_1[k] = T[x_1[k]]$ and $y_2[k] = T[x_2[k]]$ we have for a linear system and any $a, b \in \mathbb{C}$

$$ay_1[k] + by_2[k] = T[ax_1[k] + bx_2[k]]. \tag{4.2}$$

A system is *time-invariant* if any temporal shift k_0 of the input signal $x[k]$ results in the corresponding shift of the output signal $y[k] = T[x[k]]$, i.e.

$$y[k - k_0] = T[x[k - k_0]]. \tag{4.3}$$

An LTI system may be characterized by its *impulse response* $h[k]$ and its *initial conditions*. Frequently, we assume that initial conditions are zero, i.e., the system is at rest before a signal is applied. For zero initial conditions the impulse response is obtained by submitting a unit impulse[1] $\delta[k]$ to the system input; see Figure 4.2. Using linearity and time-invariance we obtain for any input signal $x[k]$

$$y[k] = T[x[k]] = T\left[\sum_{l=-\infty}^{\infty} x[l]\delta[k-l]\right] = \sum_{l=-\infty}^{\infty} x[l]T[\delta[k-l]] \tag{4.4}$$

$$= \sum_{l=-\infty}^{\infty} x[l]h[k-l] = x[k] * h[k],$$

---

[1]The unit impulse is defined as a sequence $\delta[k]$ with $\delta[0] = 1$ and $\delta[k] = 0$ for all $k \neq 0$.

where the above summation is known as the *discrete convolution* and $*$ denotes the convolution operator. The convolution operation is commutative,

$$y[k] = \sum_{l=-\infty}^{\infty} x[l]h[k-l] = \sum_{l=-\infty}^{\infty} h[l]x[k-l], \tag{4.5}$$

which leads to the interesting interpretation that the input signal and the impulse response may be interchanged without changing the output signal. Given the impulse response $h[k]$ we may compute the output signal $y[k]$ to any input signal $x[k]$. Note, that the convolution of two finite signals with N and M successive non-zero samples has at most $N + M - 1$ non-zero samples.

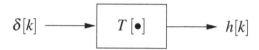

Figure 4.2: Generation of the impulse response of an LTI system.

Definition 4.6 (Frequency Response). The *frequency response* $H(e^{i\Omega})$ of an LTI system is given by the *discrete-time Fourier transform* (DTFT) of its impulse response, and, vice versa, the impulse response may be computed via an inverse DTFT of the frequency response. Thus we have[2]

$$H(e^{i\Omega}) = \sum_{k=-\infty}^{\infty} h[k]e^{-i\Omega k} \quad \text{and} \quad h[k] = \frac{1}{2\pi} \int_{-\pi}^{\pi} H(e^{i\Omega})e^{i\Omega k}d\Omega. \tag{4.6}$$

Computing the DTFT of Equation (4.4) we find that $Y(e^{i\Omega}) = H(e^{i\Omega})X(e^{i\Omega})$, where $X(e^{i\Omega})$ and $Y(e^{i\Omega})$ denote the DTFT of the input and the output signals. Therefore, the convolution of the input signal and the impulse response of an LTI system leads to a multiplication of their respective frequency responses.

The frequency response is a complex-valued quantity which is often displayed in terms of its *magnitude response* $|H(e^{i\Omega})|$ and its *phase response* $\phi(\Omega)$ such that $H(e^{i\Omega}) = |H(e^{i\Omega})|e^{i\phi(\Omega)}$. Since the magnitude response often covers a large dynamic range, it is common to denote it in decibels (dB) as $A(\Omega) = 20\log_{10}\left(|H(e^{i\Omega})|\right)$. Furthermore, it is common to display the frequency response on a scale normalized to the sampling frequency, i.e. $\Omega = 2\pi f/f_s$. Then, the Nyquist frequency corresponds to $\Omega = \pi$.

Definition 4.7 (Group Delay of an LTI System). The *group delay* $\tau_g(\Omega)$ of an LTI

---

[2]To remain consistent with the widely used z-transform notation, we write the frequency response as a function of $e^{i\Omega}$; see [19] for further explanations.

system characterizes the delay of the output signal w.r.t. the input signal. In general, this delay depends on frequency and is defined as

$$\tau_g(\Omega) = -\frac{d\phi(\Omega)}{d\Omega}. \tag{4.7}$$

An important special case concerns systems that have a linear phase response. A system with a linear phase response has a constant group delay. It will delay the input signal uniformly across frequency and thus will not introduce dispersion. For music signal processing such behavior is beneficial, for instance, for the reproduction of sharp onsets.

Example 4.1 (Sampling rate conversion). *To reduce the computational effort of music analysis tasks we may lower the sampling rate from $f_s = 48$ kHz to $f'_s = 16$ kHz. Prior to this decimation step the bandwidth of the audio signal must be limited to frequencies below $f'_s/2 = 8$ kHz. The impulse, amplitude, and phase responses of a suitable low-pass filter are shown in Figure 4.3. This filter has an impulse response of 181 non-zero coefficients and a linear phase response. The attenuation of frequency components outside the desired frequency band (stopband attenuation) is about 80 dB. Its design is discussed in Section 4.3.2.*

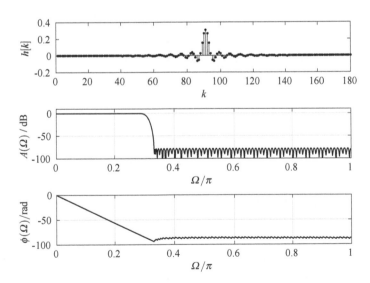

Figure 4.3: Impulse response $h[k]$ (top), amplitude response $A(\Omega) = 20\log_{10}\left(|H(e^{i\Omega})|\right)$ (middle), and phase response $\phi(\Omega)$ (bottom) of a low-pass filter for bandwidth reduction to one third of the original bandwidth.

4.3.1 Parametric LTI Systems

In digital music signal processing, discrete-time systems are widely used to emphasize or to filter out specific signal components. In their most basic form, these filters
are recursive or non-recursive LTI systems defined by a small set of parameters. In
the general case of a causal parametric LTI system, input samples with indices $l \leq k$
and output samples with indices $l < k$ contribute to an output sample $y[k]$ as

$$
\begin{aligned}
y[k] = {}& b_0 x[k] + b_1 x[k-1] + \cdots + b_N x[k-N] \\
& + a_1 y[k-1] + a_2 y[k-2] + \cdots + a_M y[k-M].
\end{aligned}
\tag{4.8}
$$

This *difference equation* is visualized in the *block diagram* of Figure 4.4. In the case

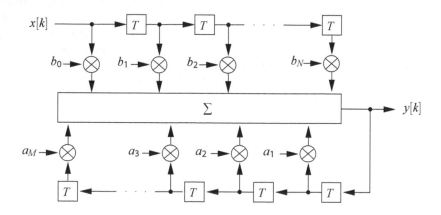

Figure 4.4: A block diagram of a causal and linear parametric system.

of constant coefficients a_μ and b_ν we obtain a causal LTI system. The maximum of
N and M, $\max(N,M)$, is called the *order* of the system. Because of the recursive
part, the impulse response is of infinite length in general.

Definition 4.8 (Infinite Impulse Response (IIR) System). A system with an infinite
number of non-zero coefficients in its impulse response is called an *infinite impulse
response* (IIR) system.

Applying the DTFT defined in Equation (4.6) to the left-hand and the right-hand
sides of Equation (4.8) the frequency response $H(e^{i\Omega})$ of the parametric LTI system
is derived as

$$
\begin{aligned}
H(e^{i\Omega}) = {}& \frac{Y(e^{i\Omega})}{X(e^{i\Omega})} = \frac{\sum_{\nu=0}^{N} b_\nu e^{-i\nu\Omega}}{1 - \sum_{\mu=1}^{M} a_\mu e^{-i\mu\Omega}} \\
= {}& \frac{b_0 + b_1 e^{-i\Omega} + \cdots + b_N e^{-iN\Omega}}{1 - a_1 e^{-i\Omega} - a_2 e^{-i2\Omega} \cdots - a_M e^{-iM\Omega}}.
\end{aligned}
\tag{4.9}
$$

Thus, the frequency response and its inverse DTFT, the impulse response $h[k]$, depend on the coefficients a_μ and b_ν only.

Example 4.2 (First-Order Recursive System). *First-order recursive systems are widely used in (music) signal processing for the purpose of smoothing fluctuating signals and for the generation of stochastic autoregressive processes. While the statistical view on models of stochastic time series is treated in depth in Section 9.8.2, we here explain the first-order recursion in terms of a digital filter. Using Equation (4.8) and the stability condition $|a_1| < 1$ we find for the difference equation and the frequency response of a first-order recursive system*

$$y[k] = b_0 x[k] + a_1 y[k-1] \Leftrightarrow H(e^{i\Omega}) = \frac{b_0}{1 - a_1 e^{-i\Omega}}, \tag{4.10}$$

and for its magnitude response

$$\left|H(e^{i\Omega})\right| = \frac{|b_0|}{\sqrt{1 + a_1^2 - 2a_1 \cos(\Omega)}}. \tag{4.11}$$

Obviously, b_0 controls the overall gain of the filter. In order to normalize the overall response on its maximum, it may be set to $b_0 = 1 - |a_1|$. The frequency characteristic of the filter is determined by the coefficient a_1: When this coefficient is positive we achieve a low-pass filter which attenuates high-frequency components and thus smooths the input signal. When this coefficient is negative a high-pass filter results, which leads to a relative emphasis of high frequencies up to the Nyquist frequency $\Omega = \pi$. An example for $a_1 = \pm 0.8$ is shown in Figure 4.5.

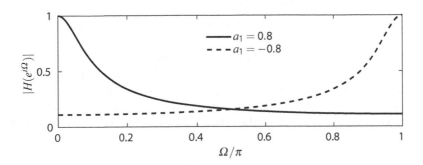

Figure 4.5: Frequency responses of first-order recursive systems with coefficient $a_1 = 0.8$ (solid line, low-pass filter) and $a_1 = -0.8$ (dashed line, high-pass filter).

When we set $a_\mu = 0$ for all μ we eliminate the feedback path and obtain the block diagram in Figure 4.6. Then, the frequency response simplifies to

$$H(e^{i\Omega}) = \frac{Y(e^{i\Omega})}{X(e^{i\Omega})} = b_0 + b_1 e^{-i\Omega} + \cdots + b_N e^{-iN\Omega} = \sum_{\nu=0}^{N} b_\nu e^{-i\nu\Omega}. \tag{4.12}$$

Since there is no feedback, the impulse response of this system has a finite number of non-zero coefficients.

Definition 4.9 (Finite-Impulse Response (FIR) System). A system with a finite number of non-zero coefficients in its impulse response is called a *finite-impulse response* (FIR) system.

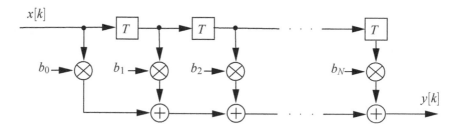

Figure 4.6: A block diagram of a causal and linear parametric system with finite impulse response (FIR).

The filter discussed in Example 4.1 is a FIR filter and could be implemented using the block diagram in Figure 4.6. Based on this implementation, the impulse response is given by $h[k] = b_k \ \forall k \in [0, 1, \ldots, N]$, where N is the order of the filter. FIR filters may also be used to compute a *moving average* of the input signal and are thus useful for the online computation of audio features.

4.3.2 Digital Filters and Filter Design

In general, the design of linear and time-invariant digital filters reduces to finding coefficients a_μ and b_ν such that a prescribed (ideal) frequency response $H_T(e^{i\Omega})$ is approximated. Ideal frequency responses are shown for basic low-pass, high-pass, and band-pass filters in Figure 4.7 as a function of the normalized frequency Ω. In music analysis tasks, digital filters are used, for instance, for the computation of audio features. Recursive filters may also serve as models for acoustic resonators such as the human vocal tract or as used in musical instruments. The specification of the

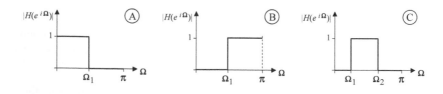

Figure 4.7: Ideal magnitude responses of a low-pass filter (A), a high-pass filter (B), and a band-pass filter (C). Ω_1 and Ω_2 denote cut-off frequencies.

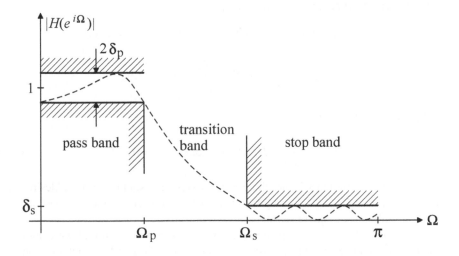

Figure 4.8: Specification of a low-pass filter with tolerances (after [18]).

desired frequency response $H_T(e^{i\Omega})$ includes the frequency ranges where the input signal should be passed or attenuated by the filter, also known as the passband(s) and stopband(s), respectively. As the transitions between these bands cannot be abrupt, the width of corresponding transition band(s) is also part of the specification. Furthermore, tolerance intervals which, for instance, specify the maximum permissible ripple in the passband(s) and the stopband(s) of the filter, are necessary. Figure 4.8 plots a tolerance specification for a discrete low-pass filter which entails a passband, a transition band, and a stopband with their respective parameters. In most applications it will be desirable to make the maximum passband distortion δ_p, the width of the transition band $\Delta\Omega = \Omega_s - \Omega_p$, and the stopband ripple δ_s as small as possible while satisfying a constraint on the filter order and hence on the group delay and the computational complexity.

Digital filters may be designed as FIR or IIR filters. FIR filters are often preferred as they are non-recursive and therefore always stable. Furthermore, they can be designed to have a linear phase response avoiding the detrimental effects of phase distortions and non-uniform group delays.

Popular design methods for FIR filters are based on the *modified Fourier series approximation* or the *Chebychev approximation*. In both cases we strive to approximate an ideal frequency response $H_T(e^{i\Omega})$ by the FIR filter response given in Equation (4.12).

4.3.2.1 Modified Fourier Approximation

The Fourier series approximation minimizes the mean-square error

$$J = \int_{-\pi}^{\pi} \left| H_T(e^{i\Omega}) - \sum_{v=0}^{N} b_v e^{-iv\Omega} \right|^2 d\Omega \tag{4.13}$$

and provides the solution

$$b_v = \frac{1}{2\pi} \int_{-\pi}^{\pi} H_T(e^{i\Omega}) e^{i\Omega v} d\Omega, \quad v = 0\dots N. \tag{4.14}$$

Thus, the coefficients b_v are the Fourier series representation of the desired frequency response $H_T(e^{i\Omega})$. For a given filter order N this constitutes the best approximation in the mean-square sense.

When the filter coefficients are (even or odd) symmetric around the center bin, the filter has a linear phase response. Then, for a filter of order N we obtain a constant group delay of $\tau_g(\Omega) = N/2$.

Example 4.3 (Fourier Approximation). *Figure 4.9 depicts a linear-phase approximation of the ideal low-pass filter for two values of N. Here the filter coefficients are arranged in a non-causal fashion, i.e., symmetric around the time index k = 0. Clearly, the filter of higher order achieves a smaller transition interval between the passband and the stopband. Note, however, that the maximum ripple in the passband and the stopband is not improved when the filter order is increased.*

This design may be modified by multiplying the impulse responses in Figure 4.9 by a tapered *window* function $w[k]$, thus achieving a higher stopband attenuation and less passband ripple. For a Hamming window (as defined in Equation (4.24)) the resulting filter coefficients and the corresponding frequency response are shown in Figure 4.10. We now observe significantly less ripple in the passband and the stopband but a wider transition band.

The filter length and the choice of the window clearly depends on the desired stopband attenuation and the width of the transition region. For the widely used *Kaiser* window (see Equation (4.26)) the following design rule for the filter order has been established [12, 19]

$$N \approx \frac{A_s - 7.95}{2.2855\Delta\Omega}, \tag{4.15}$$

where $A_s = -20\log_{10}(\delta_s)$ specifies the desired stopband attenuation and $\Delta\Omega = 2\pi\frac{\Delta f}{f_s}$ is the normalized transition width. The shape parameter α of the Kaiser window controls its bandwidth and is found by

$$\alpha = \begin{cases} 0 & , A_s < 21 \\ 0.5842(A_s - 21)^{0.4} + 0.07886(A_s - 21) & , 21 \leq A_s \leq 50 \\ 0.1102(A_s - 8.7) & , A_s > 50. \end{cases} \tag{4.16}$$

4.3.2.2 Chebychev Approximation

The second widely used method is based on the Chebychev approximation. The Chebychev approximation minimizes the maximum approximation error (or the L_∞ norm) and is implemented via the Remez algorithm [22], which often yields lower filter orders than the modified Fourier approximation when a maximum ripple in the passband and the stopband is prescribed. The Chebychev approximation leads to an *equiripple* error. In the context of filter design, this procedure is also known as the *Parks–McClellan* algorithm [20]. The Parks–McClellan algorithm may be also used to design high-pass, band-pass, differentiators, and Hilbert transform filters [19].

Example 4.4 (Chebychev Approximation). *An example of an equiripple low-pass filter design of order 40 is shown in Figure 4.11. The equiripple property is clearly observed in the passband and in the stopband.*

For the filter design via the Parks–McClellan algorithm, a rule-of-thumb relation has been established to estimate the required filter order [19]

$$N \approx \frac{-10\log_{10}(\delta_s\delta_p) - 13}{2.324\Delta\Omega}, \tag{4.17}$$

where $\Delta\Omega$ is the width of the transition band and δ_s and δ_p are the ripples in the stopband and passband, respectively.

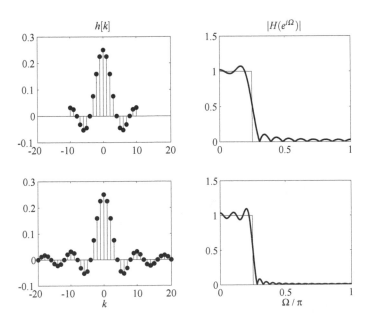

Figure 4.9: Fourier approximation of the ideal low-pass filter response. Left: impulse response $h[k]$, right: magnitude response $|H(e^{i\Omega})|$. Top: $N = 20$, Bottom: $N = 40$.

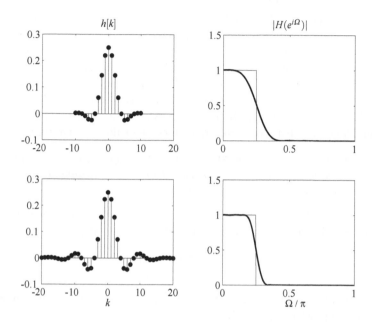

Figure 4.10: Modified Fourier approximation of an ideal low-pass filter using a Hamming window. Left: impulse response $h[k]$, right: magnitude response $|H(e^{i\Omega})|$. Top: $N = 20$, Bottom: $N = 40$.

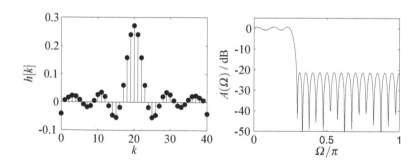

Figure 4.11: Filter design based on the Chebychev approximation and the Parks-McClellan (Remez) algorithm. Left: impulse response $h[k]$, right: magnitude response $A(\Omega) = 20\log_{10}\left(|H(e^{i\Omega})|\right)$. The filter order is 40 and the width of the transition band is $\Delta\Omega = 0.04\pi$.

4.4 Spectral Analysis Using the Discrete Fourier Transform

The Fourier spectrum of any sequence $x[k]$ of finite or infinite length may be computed by means of the DTFT as defined in Equation (4.6). However, the DTFT results in a spectrum which is a function of the continuous frequency variable Ω and is thus not directly suited for numeric computations. Furthermore, many signals, and especially music signals, change their temporal and spectral structure rapidly over time. For these reasons, the Fourier analysis should be confined to short, quasi-stationary signal segments. This is accomplished using the discrete Fourier transform (DFT) which we introduce in the following section.

4.4.1 The Discrete Fourier Transform

Definition 4.10 (Discrete Fourier Transform). The discrete Fourier transform (DFT) of a sequence $x[k]$, $k = 0, \ldots, M-1$, is defined as

$$X[\mu] = \sum_{k=0}^{M-1} x[k] e^{-i\frac{2\pi\mu k}{M}}, \quad \mu = 0, \ldots, M-1, \tag{4.18}$$

and the inverse relationship (IDFT) by

$$x[k] = \frac{1}{M} \sum_{\mu=0}^{M-1} X[\mu] e^{i\frac{2\pi\mu k}{M}}, \quad k = 0, \ldots, M-1. \tag{4.19}$$

On the normalized frequency axis $\Omega = 2\pi f/f_s$ the *frequency bins* of the DFT are spaced by $2\pi/M$. Thus, when the signal samples $x[k]$ are generated by sampling a continuous-time signal $x(t)$ with sampling rate f_s, the center frequencies of the DFT bins are located at $\Omega_\mu = 2\pi\mu/M$ or $f_\mu = \mu f_s/M$, for $\mu = 0, \ldots, M-1$. Note that center frequencies of the DFT bins $f_\mu = \mu f_s/M$ are sometimes called *Fourier frequencies* (see, e.g., right before Definition 9.45). The frequency bin at $\mu = 0$, i.e. $X[0]$, is M times the DC value or mean of the sequence $x[k]$, $k = 0, \ldots, M-1$. When the DFT length M is even, the frequency bin at $\mu = M/2$ is known as the *Nyquist* bin. In what follows we review some properties of the DFT.

Theorem 4.1 (*Linearity*). The DFT is a linear transformation. Hence, the DFT of $ax[k] + by[k]$ yields $aX[\mu] + bY[\mu]$.

Theorem 4.2 (*Periodicity*). The DFT provides a periodic continuation (with period M) of the input sequence and the sequence of spectral coefficients, i.e.,

$$X[\mu] = \widetilde{X}[\mu + rM] = \sum_{k=0}^{M-1} x[k] e^{-i\frac{2\pi(rM+\mu)k}{M}}, \quad \text{for } \mu = 0, \ldots, M-1 \quad \text{and } r \in \mathbb{Z},$$

$$\tag{4.20}$$

and

$$x[k] = \widetilde{x}[k + rM] = \frac{1}{M} \sum_{\mu=0}^{M-1} X[\mu] e^{i\frac{2\pi\mu(rM+k)}{M}}, \quad \text{for } k = 0, \ldots, M-1 \quad \text{and } r \in \mathbb{Z},$$

$$\tag{4.21}$$

where $\widetilde{X}[\mu]$ and $\widetilde{x}[k]$ are the periodically continued sequences.

Theorem 4.3 (*Symmetry*). When the sequence $x[k]$, $k = 0,\ldots,M-1$, is real-valued, the sequence of DFT coefficients is conjugate-symmetric, i.e. when M is even we have $X[\mu] = X^*[M-\mu]$, $\mu = 1,\ldots,M/2-1$.

Theorem 4.4 (*Cyclic convolution of two sequences*). The cyclic convolution

$$x[k] \overset{M}{\circledast} y[k] := \sum_{\ell=0}^{M-1} x[\ell]\,y[(k-\ell)_{\mathrm{mod}M}]$$

of two time domain sequences $x[k]$ and $y[k]$ corresponds to a multiplication $X[\mu]Y[\mu]$ of the respective DFT sequences $X[\mu]$ and $Y[\mu]$. $(k)_{\mathrm{mod}M}$ denotes the modulo operator.

Theorem 4.5 (*Multiplication of two sequences*). The multiplication $x[k]\,y[k]$ of two sequences $x[k]$ and $y[k]$ corresponds to the cyclic convolution

$$\frac{1}{M}X[\ell] \overset{M}{\circledast} Y[\ell] = \frac{1}{M}\sum_{\ell=0}^{M-1} X[\ell]\,Y[(\mu-\ell)_{\mathrm{mod}M}]$$

of the corresponding DFT sequences.

Theorem 4.6 (*Parseval's theorem*). For two sequences $x[k]$ and $y[k]$ the following correspondence holds:

$$\sum_{k=0}^{M-1} x[k]y^*[k] = \frac{1}{M}\sum_{\mu=0}^{M-1} X[\mu]Y^*[\mu]\,, \tag{4.22}$$

and for the special case $x[k] = y[k]$ we have

$$\sum_{k=0}^{M-1} |x[k]|^2 = \frac{1}{M}\sum_{\mu=0}^{M-1} |X[\mu]|^2\,. \tag{4.23}$$

Whenever we select a length-M sequence of the signal $x[k]$ prior to computing the DFT, we may describe this in terms of applying a *window* function $w[k]$ of length M to the original longer signal. Thus, the DFT coefficients $X[\mu]$ are equal to the DTFT $X_w(e^{i\Omega})$ of the windowed sequence $w[k]x[k]$ at the discrete frequencies $\Omega_\mu = \frac{2\pi\mu}{M}$.

The implementation of the DFT makes use of fast algorithms known as the *fast Fourier transform* (FFT). The FFT algorithm achieves its efficacy by segmenting the input sequence (or the output sequence) into shorter sequences in several steps. Then, DFTs of the shortest resulting sequences are computed and recombined in several stages to form the overall result. The most popular versions of the FFT algorithms use a DFT length M being equal to a power of two. This allows a repeated split into shorter sequences until, in the last stage, only DFTs of length two are required. The FFT algorithm then recombines these length-2 DFTs in $\log_2(M) - 1$ stages following a regular pattern.

Example 4.5 (Window functions). *In Figure 4.12 we illustrate the effect of applying a rectangular window w[k] to a sinusoidal signal x[k] = sin(Ωk) of infinite length prior to computing the DFT. The plots on the left side show the sinusoidal signal for two different signal frequencies Ω while the plots on the right side show the corresponding DTFT (dashed line) and DFT magnitude spectra. We find that the DFT coefficients result from a sampled version of the spectrum $X_w(e^{i\Omega}) = X(e^{i\Omega}) * W(e^{i\Omega})$, where $W(e^{i\Omega})$ is the DTFT of the window function w[k]. The effect of convolving the spectrum of an infinite sinusoidal signal with the spectrum of the window function is clearly visible in the DTFT. For the plots in the upper graphs, the length of the DFT is equal to an integer multiple of the period of the sinusoidal signal, a condition that is not met in the plots of the lower row. Hence, the DFT spectra are quite different. While the sampling of the DTFT in the upper graph results in two distinct peaks in the DFT (as it would be expected for a sinusoidal signal), the lower graph shows the* typical *spectral leakage as the DTFT is now sampled on its side lobes. This spectral leakage will obfuscate the signal spectrum and should be minimized.*

The spectral leakage may be reduced by using a *tapered* window function, however, at the cost of a reduced spectral resolution. Well-known window functions are the Hamming, the Hann, the Blackman, and the Kaiser window. Some widely used

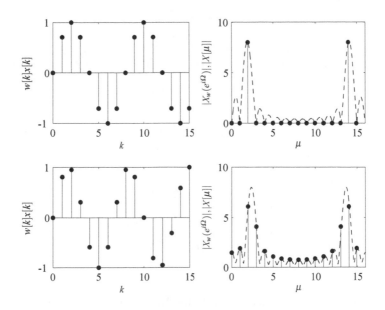

Figure 4.12: DFT analysis of a sinusoidal signal multiplied with a rectangular window. The DFT length is $M = 16$. Upper plots: signal and magnitude spectrum for $\Omega = 3/16$. Lower plots: signal and magnitude spectrum for $\Omega = 10/48$.

windows may be written in a parametric form as

$$w[k] = a - (1-a)\cos\left(k\frac{2\pi}{M-1}\right), \quad k = 0\ldots M-1,\qquad(4.24)$$

where the shape parameter a is set to

- $a = 1$ for the rectangular (boxcar) window,
- $a = 0.54$ for the Hamming window,
- $a = 0.5$ for the Hann window.

The frequency response of the window function, Equation (4.24), is given by

$$W(e^{i\Omega}) = e^{-i\frac{M-1}{2}\Omega}\left[a\frac{\sin\left(M\Omega/2\right)}{\sin\left(\Omega/2\right)} + \frac{1-a}{2}\times\right.\qquad(4.25)$$
$$\left.\left(\frac{\sin\left(M\left(\Omega-2\pi/(M-1)\right)/2\right)}{\sin\left(\left(\Omega-2\pi/(M-1)\right)/2\right)} + \frac{\sin\left(M\left(\Omega+2\pi/(M-1)\right)/2\right)}{\sin\left(\left(\Omega+2\pi/(M-1)\right)/2\right)}\right)\right].$$

The rectangular, the Hamming, the Blackman window and their corresponding magnitude responses are shown in Figure 4.13. The Kaiser window is specified as

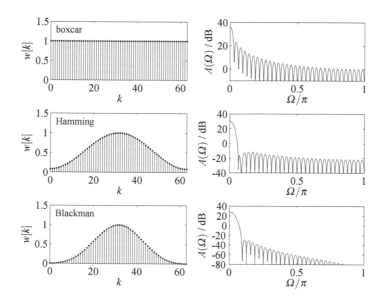

Figure 4.13: The rectangular (boxcar), the Hamming, the Blackman window and their respective magnitude responses $A(\Omega) = 20\log_{10}\left(|W(e^{i\Omega})|\right)$.

$$w_{Kaiser}[k] = \begin{cases} I_0\left(\alpha\sqrt{1-\left(\frac{k-(M-1)/2}{(M-1)/2}\right)^2}\right)\Big/I_0(\alpha) & 0 \le k \le M-1 \\ 0 & \text{otherwise}, \end{cases}\qquad(4.26)$$

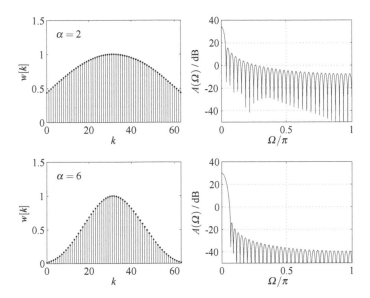

Figure 4.14: Kaiser windows of length $M = 64$ and shape parameters $\alpha = 2$ (top) and $\alpha = 6$ (bottom) and their magnitude response $A(\Omega) = 20\log_{10}\left(|W(e^{i\Omega})|\right)$.

where $I_0(\cdot)$ denotes the zero-order modified Bessel function of the first kind and α is a *shape parameter* which controls the bandwidth of its frequency response.

The effects of a tapered window are shown in Figure 4.15, where the same sinusoidal signals as in Example 4.5 and a Hamming window $w[k]$ are used. Clearly, the amplitudes of the spectral side lobes are now significantly reduced and the amount of leakage in the DFT spectra depends much less on the signal frequency. The main lobe, however, is wider indicating a loss of spectral resolution.

4.4.2 Frequency Resolution and Zero Padding

The DFT length M also determines the number of discrete bins in the frequency domain. These bins are spaced on the normalized frequency axis according to

$$\Delta\Omega = \frac{2\pi}{M}. \tag{4.27}$$

However, the spacing between frequency bins must not be confused with the *frequency resolution* of the DFT. Frequency resolution may be defined via the capability of the DFT to resolve closely spaced sinusoids. In general, the frequency resolution depends on the number of elements of the input sequence $x[k]$ and the window function chosen. The maximum resolution is obtained for the rectangular window for which the 3-dB bandwidth is less than but close to

$$\Delta\Omega_{3dB} \approx \frac{4\pi}{M}. \tag{4.28}$$

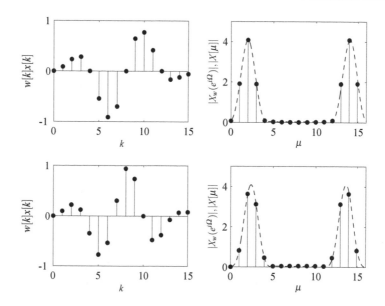

Figure 4.15: DFT analysis of a sinusoidal signal multiplied with a Hamming window. The DFT length is $M = 16$. Upper plots: signal and magnitude spectrum for $\Omega = 3/16$. Lower plots: signal and magnitude spectrum for $\Omega = 10/48$.

The 3-dB bandwidth is defined as the frequency interval for which the magnitude response of a band-pass filter is not more than 3 dB below its maximum value. Typically, the maximum response is achieved for the center frequency of a frequency bin. For the windows specified in Equation (4.24) the 3-dB bandwidth $\Delta\Omega_{3dB}$ relative to $4\pi/M$ is shown in Figure 4.16 as a function of parameter a.

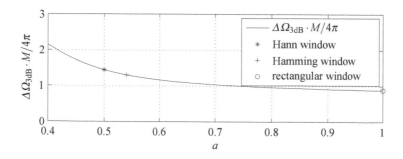

Figure 4.16: 3-dB bandwidth of the window function from Equation (4.24) relative to $4\pi/M$ as a function of the window design parameter a.

A technique known as *zero-padding* extends the time-domain sequence $x[k]$, $k = 0 \ldots M-1$ with zero samples prior to the computation of the DFT. Accordingly, zero-padding increases the number of frequency bins in the DFT domain and yields an interpolation of the DFT of the original sequence. For finite-length signals, the DFT and zero-padding allows us to compute the DTFT for any number of frequency bins. Zero-padding, however, is not suitable to increase the frequency resolution.

4.4.3 Short-Time Spectral Analysis

When we apply the DFT to successive signal segments of length M, we may write this as

$$X[\lambda,\mu] = \sum_{k=0}^{M-1} w[k]x[k+\lambda R]e^{-i\frac{2\pi\mu k}{M}}, \quad \mu = 0,\ldots,M-1, \qquad (4.29)$$

where R is the shift between successive segments and λ the index to these segments. This constitutes a *sliding-window* short-term Fourier analysis.

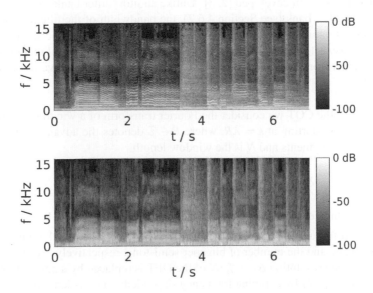

Figure 4.17: Narrowband (top) and wideband (bottom) spectrogram of a music signal (pop song). In the top plot the harmonics of the singing voice are clearly resolved while in the lower plot the formants are more prominently displayed.

The log-magnitude spectra $20\log_{10}(|X[\lambda,\mu]|)$ of successive signal segments are then plotted on a dB scale to obtain the *spectrogram* of an audio signal.

Example 4.6 (Narrowband and Wideband Spectrogram). *In Figure 4.17 we depict two versions, a narrowband spectrogram using a Hann window* $w[k]$ *of length 512, and a wideband spectrogram using a Hann window* $w[k]$ *of length 128. The sampling rate is* $f_s = 44100\,Hz$ *and the DFT length is* $M = 1024$. *While in the upper spectrogram the harmonics of the singing voice are resolved and are visible as horizontal lines in the spectrogram, the lower spectrogram displays less spectral resolution but gives a better indication of formant frequencies. The formants are the resonances of the vocal tract and are clearly visible in terms of several broad frequency bands below 5 kHz.*

4.5   The Constant-Q Transform

The DFT with its equispaced frequency bins is not well suited to represent musical signals over a large range of frequencies. To resolve closely spaced harmonics at low frequencies, a substantial length of the transform would be required which would in turn not be efficient at high frequencies. Because of the geometric progression of musical notes (see Chapter 3), a lower resolution is sufficient at high frequencies. Therefore, to cope with the specific structure of musical signals the *constant-Q transform* (CQT) has been developed [2, 3]. Unlike auditory filter banks, for instance the gammatone filter bank (see Section 4.6.2), the bandwidth of the filters is not bound to a constant value at low frequencies. Similar to the gammatone filter bank, the CQT has no exact inverse transformation, however, recently several methods for the approximate reconstruction of a signal from its CQT spectrum have been proposed [25, 17].

To derive the CQT we consider the Fourier transform of a windowed signal segment $x[k + \lambda R]$ starting at $k = \lambda R$, where $R \in \mathbb{Z}$ denotes the advance between successive signal segments and N is the window length,

$$X_{\text{stft}}[\lambda, \mu] = \sum_{k=0}^{N-1} x[k + \lambda R]\, w[k] \exp\left(-i\frac{2\pi f_\mu k}{f_s}\right), \qquad (4.30)$$

and replace the equispaced subband center frequencies $f_\mu = f_s \frac{\mu}{N}$ by the geometrically spaced frequencies $f_\mu = f_{\min} 2^{\frac{\mu}{12b}}$, where $f_{\min}$ and b denote the minimal analysis frequency and the number of bins per semi-tone, respectively. Further, the uniform frequency resolution $\Delta f = f_s/N$ of the DFT is replaced by a non-uniform resolution $\Delta f_\mu = f_s/N_\mu$ by applying frequency-dependent window lengths N_μ such that a constant quality factor $Q = f_\mu/\Delta f_\mu = f_\mu/(f_{\mu+1} - f_\mu) = 1/(2^{\frac{1}{12b}} - 1)$ is achieved. For $b = 1$ the quality factor is adjusted to a scale of 12 semi-tones and for $b > 1$ we obtain more than 12 frequency bins per octave and thus a higher frequency resolution.

Definition 4.11 (Constant-Q Transform).   For a frequency grid $f_\mu = f_{\min} 2^{\frac{\mu}{12b}}$, $\mu \in \mathbb{N}$, and a quality factor $Q = f_\mu/(f_{\mu+1} - f_\mu) = 1/(2^{\frac{1}{12b}} - 1)$ the CQT is defined as

$$X_{\text{cqt}}[\lambda, \mu] = \frac{1}{N_\mu} \sum_{k=0}^{N_\mu-1} x[k + \lambda R]\, w_\mu[k] \exp\left(-i\frac{2\pi Q k}{N_\mu}\right), \qquad (4.31)$$

where b denotes the number of frequency bins per semi-tone. The length of the window functions $w_\mu[k]$ depends on the frequency band index μ and is given by

$$N_\mu = Q\frac{f_s}{f_\mu} = \frac{f_s}{f_{\mu+1}-f_\mu}.$$

Example 4.7 (Comparison of DFT and CQT). *Figure 4.18 shows an example of a sliding window DFT (left) vs. a sliding window CQT (right) of a sustained multi-tone mixture with a spacing of four semitones sampled at $f_s = 16$ kHz and with $b = 2$. Clearly, the resolution of the DFT is not sufficient at low frequencies.*

Example 4.8 (Short-time spectral analysis using the CQT). *Figure 4.19 depicts an example where a music signal was analyzed using the CQT. The base line and the singing voice are clearly resolved.*

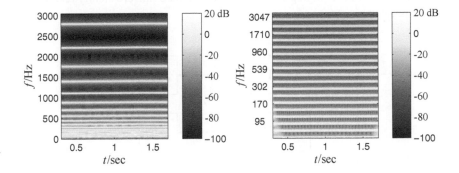

Figure 4.18: Comparison of spectral resolution of the sliding-window DFT (left) and the sliding-window CQT (right) for a sustained mixture of musical notes with a spacing of four semitones.

4.6 Filter Banks for Short-Time Spectral Analysis

As an alternative to using the DFT or the CQT we may use a bank of filters to decompose the music signal into multiple frequency bands. The general block diagram of a filter bank for signal analysis is shown in Figure 4.20, where for each frequency band we employ a discrete filter followed by a decimation step which reduces the sampling rate by a factor of R. When these band-pass filters have a reasonably high stopband attenuation, the bandwidth in each band may be reduced in accordance with the sampling theorem.

The center frequencies $f_c(i)$ of these frequency bands might be equispaced over frequency or non-uniformly distributed, as indicated in Figure 4.21. Furthermore, the bandwidth of these filters might be uniform across the filters, or might vary. Typically, the center frequencies and the bandwidths of the subband filters are designed such that there is no gap in between these filters.

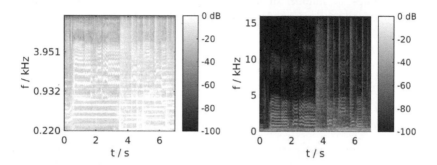

Figure 4.19: CQT spectrum of a pop song (left) and narrowband spectrogram (right). The harmonics of the singing voice as well as the bass line are clearly resolved in the CQT spectrum.

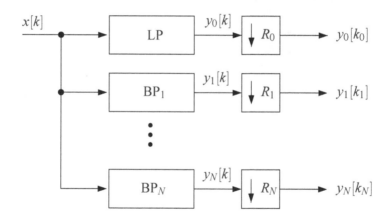

Figure 4.20: Filter bank for spectral analysis composed of a low-pass (LP) filter, several band-pass filters (BP_i), and decimators for sampling rate reduction by R_i.

4.6.1   Uniform Filter Banks

Uniform filter banks are widely used as their constituent filters can be implemented efficiently using a common *prototype* low-pass filter $h[k]$ and the discrete Fourier transform [29]. In fact, it can be shown that a *sliding window* short-time Fourier analysis is equivalent to a filter bank, where the window function corresponds to the impulse response of the prototype low-pass filter. Furthermore, by increasing the length of the prototype filter impulse response beyond the number of frequency bands (and the DFT length), filter banks with high stopband attenuation and hence with excellent channel separation may be designed. Also in this case, efficient im-

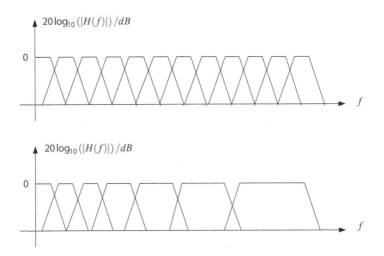

Figure 4.21: Magnitude response of subband filters with uniform (top) and non-uniform (bottom) filter bandwidths.

plementations via a *poly-phase* decomposition and the DFT are possible. As an example, we briefly discuss complex-modulated uniform filter banks.

We consider a filter bank that decomposes a broadband signal $x[k]$ into M narrowband subband signals $\bar{x}_\mu[k]$, with $\mu = 0,\dots,M-1$. The normalized center frequencies of these subbands are denoted by Ω_μ.

Since the frequency response $H_\mu^{BP}(e^{i\Omega}) = H\left(e^{i(\Omega-\Omega_\mu)}\right)$ of the μ-th subband filter may be written as a shifted version of a prototype response $H(e^{i\Omega})$, we find for the corresponding impulse response $h_\mu^{BP}[k] = h[k]e^{i\Omega_\mu k}$. Therefore, the impulse responses $h_\mu^{BP}[k]$ as well as the subband signals are complex-valued. Often, the narrowband signals at the output of any individual filter are modulated into the low-pass band (base-band)

$$\bar{x}_\mu[k] = e^{-i\Omega_\mu k} \sum_{\ell=-\infty}^{\infty} x[\ell]h_\mu^{BP}[k-\ell] = e^{-i\Omega_\mu k} \sum_{\ell=-\infty}^{\infty} x[\ell]h[k-\ell]e^{i\Omega_\mu(k-\ell)} \qquad (4.32)$$

$$= \sum_{\ell=-\infty}^{\infty} x[\ell]h[k-\ell]e^{-i\Omega_\mu \ell}. \qquad (4.33)$$

The prototype filter impulse response $h[k] = h_0^{BP}[k]$ may now be designed for a desired number of subbands, for a desired bandwidth and stopband attenuation. Furthermore, we like to achieve an overall perfect response (unity response)

$$h_A[k] = \sum_{\mu=0}^{M-1} h_\mu^{BP}[k] = \delta[k-k_0], \qquad (4.34)$$

where k_0 is the overall group delay (latency) of the filter bank. Thus, for a uniform frequency spacing $\Omega_\mu = \frac{2\pi}{M}\mu$ we have

$$h_A[k] = \sum_{\mu=0}^{M-1} h_\mu^{BP}[k] = \sum_{\mu=0}^{M-1} h[k]e^{i\frac{2\pi}{M}\mu k} = h[k]Mp^{(M)}[k], \qquad (4.35)$$

where $p^{(M)}[k] = \frac{1}{M}\sum_{\mu=0}^{M-1} e^{i\frac{2\pi}{M}\mu k}$ is different from zero only for $k = \lambda M$ with $\lambda \in \mathbb{Z}$. The condition in Equation (4.34) can thus be fulfilled if

$$h[\lambda M] = \begin{cases} 1/M & , \lambda M = k_0 \\ 0 & , \lambda M \neq k_0. \end{cases} \qquad (4.36)$$

All other samples of $h[k]$ may be used to optimize the channel bandwidth and the stopband attenuation. Note that the discrete tapered sinc function[3]

$$h[k] = w[k]\mathrm{sinc}[\pi(k-k_0)/M]/M, \qquad (4.37)$$

where $w[k]$ denotes an arbitrary window function, satisfies the above constraint. Thus, the frequency response of the individual band-pass filters can be controlled via the shape and the length of the window function $w[k]$; see Section 4.4.1.

Example 4.9 (Uniform Filter Bank). *Figure 4.22 depicts an example for* $M = 8$ *channels of which 5 channels are located between 0 and the Nyquist frequency. For the design of the prototype low-pass filter, a Hann window of length 81 was used.*

When the length of the prototype impulse response $h[k]$ equals the DFT length M the above complex-modulated filter bank with uniform frequency spacing corresponds to a *sliding window* DFT analysis system. With $\Omega_\mu = \frac{2\pi}{M}\mu$, the substitution $u = \ell - k$, and

$$h[-u] = \begin{cases} w[u] & u = 0, \ldots, M-1 \\ 0 & \text{otherwise} \end{cases} \qquad (4.38)$$

we rewrite Equation (4.33) to yield

$$X[k,\mu] = \bar{x}_\mu[k] = e^{-i\frac{2\pi}{M}\mu k} \sum_{u=0}^{M-1} x[k+u]w[u]e^{-i\frac{2\pi}{M}\mu u}. \qquad (4.39)$$

For a critical decimation of the subband signals, i.e., $R = M$ and $k = \lambda M$, we have $e^{-i\Omega_\mu k} = e^{-i\frac{2\pi}{M}\mu\lambda M} = 1$. Then, the right-hand side corresponds to a (sliding-window) DFT of length M.

---

[3]The cardinal sine function is defined here as $\mathrm{sinc}[x] = \sin[x]/x$ for $x \neq 0$ and $\mathrm{sinc}[x] = 1$ for $x = 0$.

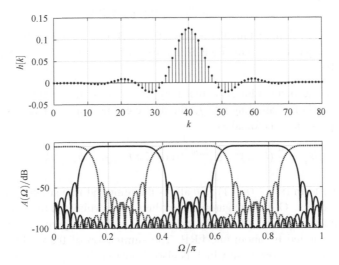

Figure 4.22: Impulse response of prototype low-pass filter (top) and magnitude response of a complex-modulated uniform filter bank (bottom). The center frequencies are spaced by $\pi/4$. For clarity, the line style toggles between dashed and solid lines. These frequency responses add to a constant value of one.

4.6.2 Nonuniform Filter Banks

Non-uniform filter banks are often used when the frequency decomposition of the human auditory system is to be taken into account. Popular approaches are based on the mel frequency scale (see Chapters 2 and 5) and the gammatone filter bank. Furthermore, tree-structured filter banks, which are also related to the *wavelet packet* decomposition [14, 6], are of interest. As an example we consider the gammatone filter bank.

4.6.2.1 Gammatone Filter Bank

The *gammatone* filters model the frequency decomposition of the human auditory system. Their impulse response is composed of a sinusoidal tone and a gamma distribution.

Definition 4.12 (Gammatone Filter Bank). The impulse response of a gammatone filter is defined as

$$g(t) = at^{n-1}e^{-2\pi bt}\cos(2\pi f_c t + \phi) \quad \forall t > 0. \tag{4.40}$$

Here, a is a gain factor, n is the filter order, f_c the center frequency, b the bandwidth, and ϕ the phase of the cosine signal. The bandwidth of a filter at center frequency f_c

is given by [8]

$$b = 24.7 \cdot (4.37 \cdot f_c/1000 + 1)BW_C, \qquad (4.41)$$

which is the equivalent rectangular bandwidth (ERB) of a human auditory filter centered at the frequency f_c. This ERB is multiplied by a bandwidth correction factor $BW_C = 1.019$.

Note that the bandwidth is bounded by $24.7BW_C$ as the center frequency approaches zero. For a sampling rate f_s, the discrete-time impulse response of one of these band-pass filters is found as

$$g[k] = \frac{a}{f_s}(k/f_s)^{n-1}e^{-2\pi bk/f_s}\cos\left(2\pi f_c k/f_s + \phi\right) \quad \forall k \in \mathbb{N} \qquad (4.42)$$

and the phase term $\phi = -(n-1)f_c/b$ aligns the temporal fine structure of the filter impulse responses. Often, the gammatone filters are normalized to provide a maximum amplitude response of 0 dB. An example is shown in Figure 4.23 for 21 bands which are separated by about one ERB. The summation of these filters results in an almost flat overall response which is also shown in this figure. Several authors have developed code for the efficient implementation of the gammatone filter bank in terms of parametric recursive LTI systems which approximate the above impulse response [26]. The gammatone filter bank has no exact inverse, but approximations are available [11].

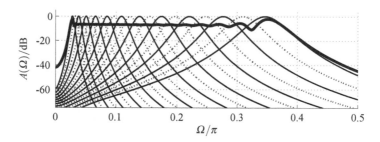

Figure 4.23: Magnitude response of a gammatone filter bank with center frequencies in the range of 442–5544 Hz. The center frequencies are spaced by the corresponding ERB. The sampling frequency is 32 kHz. For clarity, the line style toggles between dashed and solid lines. The bold line indicates the sum of all subband responses.

4.7 The Cepstrum

The *cepstrum* is a versatile tool for the analysis of audio signals especially for signals which obey a source-filter model. In this model the observed signal is generated by filtering a (possibly stochastic) source signal with a filter that imposes a certain spectral shape onto this signal. This model is often used to describe speech signals but could also be used to characterize the spectrum of individual musical instruments or the singing voice. The cepstrum is also the basis for audio features such as *mel frequency cepstral coefficients (MFCC)*, see Chapter 5, and has been used in automatic instrument recognition tasks [5]. The cepstrum was introduced in [1], in which its name and related vocabulary were also coined: the word cepstrum is derived by reversing the order of the first four letters of the word spectrum. The cepstrum can be defined as a complex-valued or real-valued quantity. In what follows, we only consider the real cepstrum.

Definition 4.13 (Cepstrum). The *real cepstrum* of a discrete-time signal $x[k]$ is computed using the inverse DTFT as

$$c_x[q] = \frac{1}{2\pi} \int_{-\pi}^{\pi} \log\left(|X(e^{j\Omega})|\right) e^{j\Omega q} d\Omega,$$

where $X(e^{j\Omega})$ is the DTFT of signal $x[k]$.

The distinctive feature of the cepstrum is the logarithmic compression of the spectral amplitudes. Then, the cepstrum delivers a Fourier decomposition of the log-magnitude spectrum. Therefore, low-order cepstral coefficients describe the coarse structure of the log-magnitude spectrum (spectral envelope), while the high-order coefficients describe its fine structure. Numeric computation of the real cepstrum is accomplished using the discrete (inverse) Fourier transform or the discrete cosine transforms.

Example 4.10 (The cepstrum of a speech sound). *Figure 4.24 displays an example of a short voiced speech sound, its power spectrum, and the corresponding cepstrum. The first cepstral coefficients encode the spectral envelope: A strongly positive first cepstral coefficient indicates a decreasing spectral slope which is typical for voiced speech. A strongly negative first cepstral coefficient would indicate an increasing spectral slope. Furthermore, the peak between 40 and 50 indicates the fundamental frequency f_0 of the sound. Its index is computed according to $q_0 = \text{round}(f_s/f_0)$ where $f_s = 8\,kHz$ is the sampling frequency.*

The relation $q_0 = \text{round}(f_s/f_0)$ as used in the above example can be motivated as follows: When the DFT/IDFT is used to compute the real cepstrum,

$$c_x[q] = \frac{1}{M} \sum_{\mu=0}^{M-1} \log\left(|X_\mu|\right) e^{j\frac{2\pi\mu q}{M}},$$

we note that a constructive summation of periodic spectral components is achieved for values $\mu q_\ell = (\ell+1)M$ for $\ell = 0, 1, 2, 3, \ldots$, or in terms of frequencies

$$\frac{\mu f_s}{M} q_\ell = (\ell+1)\frac{M f_s}{M} = (\ell+1)f_s.$$

For $\ell = 0$ we obtain $f_0 q_0 = f_s$ and the corresponding cepstral bin as

$$q_0 = \text{round}\left(\frac{f_s}{f_0}\right).$$

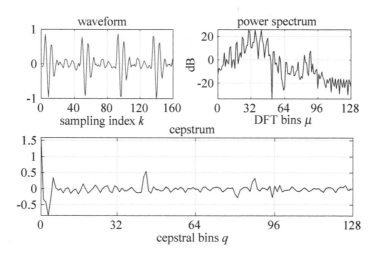

Figure 4.24: Sampled waveform, DFT power spectrum, and cepstrum of a short voiced sound. For clarity all signals are displayed as solid lines. The sampling rate is $f_s = 8$ kHz.

Multiples of f_0 are computed for $\ell = 1, 2, 3, \ldots$ and constitute the *rahmonics* $q_\ell = (\ell + 1)q_0$ in the cepstral domain. Again, the term rahmonics is constructed by inverting the order of the first three letters of the word harmonics [1]. The cepstrum is also useful for fundamental frequency estimation, especially for instruments with many equispaced spectral harmonics.

4.8 Fundamental Frequency Estimation

The fundamental frequency f_0 of a periodic signal is the inverse of the duration τ_0 of its shortest segment that, when repeated, will generate the periodic signal. The fundamental frequency is closely related to pitch (see Chapter 2) which denotes the corresponding perceived height of a tone. The estimation of f_0 is at the core of many speech [10] and music signal analysis tasks, such as melody tracking and extraction [24]. Strict periodicity may be observed in pieces of electronic music but most often, modulations of the temporal envelope and of the fundamental frequency will result in non-periodic signals. Therefore, for music as well as speech signals, we find approximately periodic (or *quasi-periodic*) structures only on short time intervals, e.g., on segments corresponding to one musical note or one voiced phone.

The accurate estimation of the fundamental frequency therefore entails the segmentation of the music signal into quasi-periodic segments, the extraction of the fundamental frequency from these segments, and a smoothing procedure to avoid sudden and non-plausible variations of the f_0-estimate due to estimation errors. Given quasi-periodic segments of monophonic music, the fundamental frequency can be estimated with some precision, however, for polyphonic music it becomes a challenging (and in general unsolved) task. Many methods were proposed, either working in the time domain, the spectral domain, the cepstral domain, or in several domains simultaneously [7].

A widely used f_0 estimation method is based on the short-time autocorrelation function

$$\varphi_{xx}[k,\tau] = \sum_{\ell=k-N+1}^{k} x[\ell]x[\ell+\tau], \quad \tau \in \mathbb{Z}, \tag{4.43}$$

computed over a segment of N signal samples. Here, k is the current time index and τ the correlation lag. The autocorrelation function is an even symmetric function and attains its global maximum for $\tau = 0$. For quasi-periodic segments the autocorrelation $\varphi_{xx}[k,\tau]$ will also exhibit strong local maxima at lags $\tau_i = i\tau_0, |i| = 1,2,\dots$ corresponding to multiples of the fundamental period $\tau_0 = 1/f_0$. Obviously, the lag τ_1 of the first of these peaks is an estimate of $1/f_0$, while the ratio of its amplitude and the amplitude of the global maximum at $\tau = 0$ is an indication of the periodicity or the *harmonic strength*. However, without prior knowledge of the range of admissible fundamental frequencies and/or further refinement this method is prone to errors. Depending on the variations of the signal envelope within the analysis segment of length N, the peak at τ_2 may be larger than the peak at τ_1 resulting in an f_0 estimate one octave below the true fundamental frequency.

A more versatile framework [4] exploits the shift invariance of periodic signal segments and minimizes the mean-squared difference

$$d[k,\tau] = \sum_{\ell=k-N+1}^{k} (x[\ell] - x[\ell+\tau])^2 \tag{4.44}$$

$$= \sum_{\ell=k-N+1}^{k} x[\ell]^2 + \sum_{\ell=k-N+1}^{k} x[\ell+\tau]^2 - 2 \sum_{\ell=k-N+1}^{k} x[\ell]x[\ell+\tau] \tag{4.45}$$

$$= \varphi_{xx}[k,0] + \varphi_{xx}[k+\tau,0] - 2\varphi_{xx}[k,\tau] \tag{4.46}$$

with respect to τ. Along with a normalization on the average of $d[k,\tau]$, this modification reduces errors which arise from variations of the signal power [4]. Another source of error is the search grid for τ_0 imposed by the sampling rate. To increase the resolution, an interpolation of the signal $x[\ell]$ or of the optimization objective $d[k,\tau]$ is often used. Other widely used methods employ the *average magnitude difference function* (AMDF) [23]

$$d_{amd}[k,\tau] = \sum_{\ell=k-N+1}^{k} |x[\ell] - x[\ell+\tau]|, \tag{4.47}$$

which exhibits small values when τ is equal to multiples of τ_0. Furthermore it avoids multiplications and is thus easy to implement.

Example 4.11 (*f_0-tracking*). *An example is shown in Figure 4.25 where the YIN algorithm [4] was used to compute a fundamental frequency estimate on a piece of monophonic music. In the spectrogram in Figure 4.25 the fundamental frequency corresponds to the frequency of the lowest of the equispaced harmonics.*

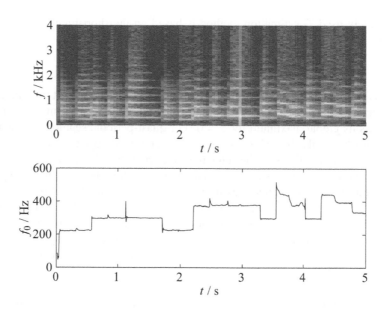

Figure 4.25: Spectrogram (top) and fundamental frequency estimate (bottom) of a music recording (French horn playing the first three bars of the solo part in the second movement of Mozart's *Concerto No. 1* (KV 412)). The fundamental frequency is estimated using the YIN algorithm [4].

4.9 Further Reading

This chapter provides an introduction to several basic signal processing techniques which are useful for music signal processing. To probe further, there are many excellent books on the general theory and applications of digital signal processing, most notably the books by Oppenheim and Schafer, [18, 19], Kammeyer [13], and Proakis [21]. These books provide an in-depth theoretical background and access to many signal processing techniques which could not be covered in this chapter, e.g., the design of recursive digital filters. Many more specialized books are available as well. For instance, filter and filter bank design is extensively discussed in the classic text of Vaidyanathan [28], and e.g., in [6] and [9]. There are many competing methods for the design of complex- or cosine-modulated filter banks, and transformation methods, such as the Lapped Transform [15].

Non-uniform filter banks may also be designed using frequency warping techniques. In this context the bilinear transformation has been used to approximate an auditory filter bank [27]. Furthermore, in many applications the filter bank for spectral analysis is followed by a signal modification step and a synthesis filter bank. Then, the design method has to take the overall response into account [29] and *perfect signal reconstruction* becomes a desirable design constraint. Prominent methods are explained in, e.g., [29] for uniform filter banks and, e.g., in [11] and [17], the latter two offering near perfect reconstruction for the gammatone filter bank and for the sliding-window CQT.

A more recent treatment of music signal processing is presented in the overview article [16] where the authors emphasize specific methods for music signal analysis and applications, such as onset detection, periodicity and tempo analysis, beat and fundamental frequency tracking, and musical instrument identification. Fundamental frequency estimation is also the basis of melody extraction algorithms, query-by-humming applications, and music transcription. An overview on melody extraction methods is provided in [24].

Bibliography

[1] B. Bogert, M. Healy, and J. Tukey. The quefrency alanysis of time series for echoes: Cepstrum, pseudo-autocovariance, cross-cepstrum and saphe cracking. In *Proc. of the Symposium on Time Series Analysis*, pp. 209–243, 1963.

[2] J. Brown. Calculation of a constant Q spectral transform. *J. Acoust. Soc. of America.*, 89:425–434, 1991.

[3] J. Brown and M. Puckette. An efficient algorithm for the calculation of a constant Q transform. *J. Acoust. Soc. of America*, 92(5):2698–2701, 1992.

[4] A. de Cheveigne and H. Kawahara. Yin, a fundamental frequency estimator for speech and music. *J. Acoust. Soc. of America.*, 111(4):1917 – 1930, 2001.

[5] A. Eronen and A. Klapuri. Musical instrument recognition using cepstral coefficients and temporal features. In *Proceedings of IEEE International Conference on Acoustics, Speech, and Signal Processing (ICASSP '00)*, volume 2, pp. II753–II756, 2000.

[6] N. Fliege. *Multirate Digital Signal Processing: Multirate Systems—Filter Banks—Wavelets*. Wiley, 1999.

[7] D. Gerhard. Pitch extraction and fundamental frequency: History and current techniques. Technical Report TR-CS 2003-06, Department of Computer Science, University of Regina, 2003.

[8] B. Glasberg and B. Moore. Derivation of auditory filter shapes from notched-noise data. *Hearing Research*, 47:103–108, 1990.

[9] H. Göckler and A. Groth. *Multiratensysteme: Abtastratenumsetzung und digitale Filterbänke*. J. Schlembach, 2004. (in German).

[10] W. Hess. *Pitch Determination of Speech Signals*. Springer, Berlin, 1983.

[11] V. Hohmann. Frequency analysis and synthesis using a gammatone filterbank. *Acta Acoustica united with Acoustica*, 88(3):433–442, 2002.

[12] J. Kaiser. Nonrecursive digital filter design using the I0-sinh window function. In *IEEE Symp. Circuits and Systems*, pp. 20–23, 1974.

[13] K. Kammeyer and K. Kroschel. *Digitale Signalverarbeitung: Filterung und Spektralanalyse mit MATLAB-Übungen*. B.G. Teubner, 5th edition, 2002. (in German).

[14] S. Mallat. *A Wavelet Tour of Signal Processing*. Elsevier Ltd, Oxford, 3rd edition, 2009.

[15] H. Malvar. *Signal Processing with Lapped Transforms*. Artech House, Boston, London, 1992.

[16] M. Müller, D. Ellis, A. Klapuri, and G. Richard. Signal processing for music analysis. *IEEE Journal on Selected Topics in Signal Processing*, 5(6):1088–1110, 2011.

[17] A. M. Nagathil and R. Martin. Optimal signal reconstruction from a Constant-Q Spectrum. In *Proc. IEEE Int. Conf. on Acoustics, Speech, and Signal Processing (ICASSP)*, pp. 349–352, 2012.

[18] A. Oppenheim and R. Schafer. *Digital Signal Processing*. Prentice Hall, 1975.

[19] A. Oppenheim and R. Schafer. *Discrete-Time Signal Processing*. Pearson New International Edition. Pearson Education, 2013.

[20] T. Parks and J. McClellan. Chebyshev approximation for nonrecursive digital filters with linear phase. *IEEE Transactions on Circuit Theory*, 19(2):189–194, Mar 1972.

[21] J. Proakis and D. Manolakis. *Digital Signal Processing*. Pearson, 4th edition, 2006.

[22] E. Remez. Sur un procédé convergent d'approximations successives pour déterminer les polynômes d'approximation. *Compt. Rend. Acad. Sci.*, 198:2063–2065, 1934.

[23] M. Ross, H. Shaffer, A. Cohen, R. Freudberg, and H. Manley. Average magnitude difference function pitch extractor. *IEEE Transactions on Acoustics, Speech, and Signal Processing*, 22(5):353–362, 1974.

[24] J. Salamon, E. Gómez, D. P. W. Ellis, and G. Richard. Melody extraction from polyphonic music signals: Approaches, applications, and challenges. *IEEE Signal Processing Magazine*, 31(2):118–134, 2014.

[25] C. Schörkhuber and A. Klapuri. Constant-Q Transform Toolbox for music processing. In *SMC Conference*, 2010. online: http://smcnetwork.org/node/1380.

[26] M. Slaney. An efficient implementation of the Patterson-Holdsworth auditory filter bank. Technical Report 34, Apple Technical Report, Apple Computer Library, Cupertino, 1993.

[27] J. Smith III and J. Abel. Bark and ERB bilinear transforms. *IEEE Transactions*

on Speech and Audio Processing, 7(6):697–708, 1999.

[28] P. Vaidyanathan. *Multirate Systems and Filter Banks*. Pearson, 2002.

[29] P. Vary and R. Martin. *Digital Speech Transmission: Enhancement, Coding and Error Concealment*. John Wiley & Sons, Chichester, 2006.

Chapter 5

Signal-Level Features

ANIL NAGATHIL, RAINER MARTIN
Institute of Communication Acoustics, Ruhr-Universität Bochum, Germany

5.1 Introduction

A musical signal carries a substantial amount of information that corresponds to its timbre, melody, or rhythm properties and that may be used to classify music, e.g., in terms of instrumentation, chord progression, or musical genre. However, apart from these high-level musical features it also contains a lot of additional information which is irrelevant for an analysis or a classification task, or even degrades the performance of the task. It is therefore necessary to extract relevant and discriminative features from the raw audio signal, which can be used either to identify properties of a music piece or to assign music to predefined classes.

In this chapter some of the most commonly used features are reviewed. These features are often referenced in the literature and have proven to be well suited for music-related classification tasks. A feature value is obtained by following a defined calculation rule which can be defined in the time, spectral, or cepstral domain depending on the musical property to be modeled. Often it is computed for short, possibly overlapping signal segments which cover approximately 20–30 ms, thereby resulting in a feature series which may then describe the temporal evolution of a specific aspect.

As a starting point, a raw audio signal $x[\kappa]$ is given, where κ denotes the discrete time index. The time interval between successive time indices is defined by the inverse sampling frequency $1/f_s$. This signal is segmented into L frames $x[\lambda,k]$ of length K

$$x[\lambda,k] = x[\lambda R + k], \qquad k \in \{0,1,\ldots,K-1\}, \tag{5.1}$$

where λ and R denote the frame index and frame shift, respectively. If necessary, a spectral transform such as the short-time Fourier transform (STFT) or the constant-Q transform (CQT), as introduced in Chapter 4, can be applied to the frames yielding a complex-valued spectral coefficient $X[\lambda,\mu] = |X[\lambda,\mu]| e^{i\phi[\lambda,\mu]}$, where μ denotes the discrete frequency index and $\phi[\lambda,\mu]$ is the phase. Often $X[\lambda,\mu]$ is computed using the discrete Fourier transform (DFT) of length K. Note that each frame is assumed to contain a quasi-stationary portion of the signal. Therefore, a meaningful analysis can be performed which eventually yields a set of short-time features. These features are supposed to highlight the most important signal characteristics with respect to a certain task and therefore are a compact representation of the signal itself.

The remainder of this chapter is organized as follows. In Section 5.2 timbre-related features are introduced, which are commonly used in applications such as instrument recognition. Section 5.3 presents features which describe harmony properties and characteristics of partial tones in music signals. Features used for the extraction of note onsets and the description of rhythmic properties are discussed in Section 5.4. The chapter is concluded with a reference to related literature in Section 5.5.

5.2 Timbre Features

Timbre is a multidimensional characteristic (see Section 2.4) and there exist a number of features which aim at representing different aspects of timbral texture. In this section we outline some of the most commonly used timbral features which are extracted either from the time domain, the spectral domain, or the cepstral domain.

5.2.1 Time-Domain Features

Definition 5.1 (Zero-Crossing Rate). There are only a few features which are extracted directly from the time domain representation, see Equation (5.1). One of them is the relative number of zero-crossings

$$t_{\text{zcr}}[\lambda] = \frac{1}{2(K-1)} \sum_{k=1}^{K-1} |\text{sgn}\,(x[\lambda,k]) - \text{sgn}\,(x[\lambda,k-1])|, \qquad (5.2)$$

where the signum function $\text{sgn}(\cdot)$ yields 1 for positive arguments and 0 for negative arguments. The zero-crossing rate is a rough measure of the noisiness and the high-frequency content of the signal.

Definition 5.2 (Low-Energy). Another measure based on the time domain representation of the signal is the low-energy feature. Unlike many other features the low-energy feature is calculated using the whole signal $x[\kappa]$. The *root mean square* (RMS) energy of each frame λ is evaluated and normalized on the RMS energy of $x[\kappa]$

$$e(\lambda) = \frac{\sqrt{\frac{1}{K} \sum_{k=0}^{K-1} x^2[\lambda,k]}}{\sqrt{\frac{1}{K_{\text{T}}} \sum_{\kappa=0}^{K_{\text{T}}-1} x^2[\kappa]}}, \qquad (5.3)$$

where K_{T} is the total number of samples. This ratio is less than 1 if the RMS energy of frame λ is lower than the total RMS energy. Otherwise the ratio is greater than or equal to 1. The actual low-energy feature is defined as the relative frequency of frames with less RMS energy than the RMS energy of the signal $x[\kappa]$. Hence,

$$t_{le} = 1 - \frac{\sum_{\lambda=1}^{L} \text{sgn}(e(\lambda) - 1)}{L}. \tag{5.4}$$

This measure accounts for the continuity of the audio signal level. A piano piece containing many portions of silence will have a large low-energy value while continuous signals such as orchestral sounds will have a small low-energy value.

5.2.2 Frequency-Domain Features

There are a number of features which are defined in the frequency domain and which characterize certain properties of the spectral shape within a signal frame. In what follows, the most prominent ones are explained. Note that these features model short-time properties of the spectrum and do not capture any temporal aspects of timbre. We assume that these features are computed based on the DFT of length M.

Definition 5.3 (Spectral Centroid). The spectral centroid determines the frequency bin around which the highest amount of spectral energy is concentrated. It is defined as

$$t_{\text{cent}}[\lambda] = \frac{\sum_{\mu=0}^{M/2} \mu |X[\lambda, \mu]|}{\sum_{\mu=0}^{M/2} |X[\lambda, \mu]|}, \tag{5.5}$$

which is the center of gravity of the magnitude spectrum. For symmetry reasons, the summations range from 0 to $M/2$ only. Lower values correspond to dull sounds, whereas higher values denote brighter sounds.

Example 5.1 (Spectral Centroid). *We consider the note c' played on a piano and an oboe, respectively. Their spectrograms are computed using a DFT of length $M = 512$ at the sampling frequency $f_s = 16,000$. The frame shift is set to $R = 256$. Then, we obtain the frame-based spectral centroid, Equation (5.5). Both, spectrograms and the temporal evolution of the spectral centroid, are shown in Figure 5.1. The spectrogram of the piano note (left) shows that the energy of the harmonics steadily decreases with increasing time. This effect is also revealed in the slightly negative trend of the spectral centroid shown by the continuous black line. At the same time the harmonics of the sustained oboe tone (right) behave relatively stably in time which is also demonstrated by the corresponding evolution of the spectral centroid. It is also worth noting that on average the spectral centroid attains higher values for the oboe tone (1552.5 ± 68.0 Hz) than for the piano tone (740.0 ± 111.8 Hz) which points towards a higher energy concentration in the harmonics of the oboe sound.*

The following three features are based on the estimation of spectral moments; see also Definition 9.7.

Definition 5.4 (Spectral Spread). A measure which characterizes the frequency range

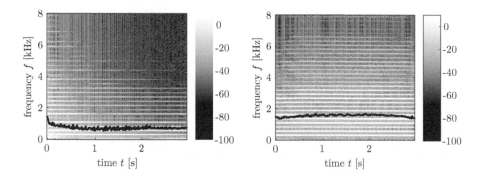

Figure 5.1: Temporal evolution of the spectral centroid (black line) of the note c' played on a piano (left) and an oboe (right) with their respective spectrograms.

of a sound around the spectral centroid is given by the spectral spread. It is defined as the normalized, second centered moment of the spectrum, i.e. the spectral variance. In order to make the spectral spread comparable in units with the magnitude spectrum, it is advisable to take the square root of the spectral variance which yields the standard deviation of the magnitude spectrum in a given frame

$$t_{\text{spread}}[\lambda] = \frac{\sqrt{\sum_{\mu=0}^{M/2} (\mu - t_{\text{cent}}[\lambda])^2 \, |X[\lambda,\mu]|}}{\sqrt{\sum_{\mu=0}^{M/2} |X[\lambda,\mu]|}}. \tag{5.6}$$

The spectral spread accounts for the sensation of timbral fullness or richness of a sound.

Definition 5.5 (Spectral Skewness). The ratio between the third centered moment and the spectral spread raised to the power of three is defined as the skewness. It can be obtained by

$$t_{\text{skew}}[\lambda] = \frac{\sum_{\mu=0}^{M/2} (\mu - t_{\text{cent}}[\lambda])^3 \, |X[\lambda,\mu]|}{\left(t_{\text{spread}}[\lambda]\right)^3 \sum_{\mu=0}^{M/2} |X[\lambda,\mu]|} \tag{5.7}$$

and describes the symmetry property of the spectral distribution in a frame. For negative values the distribution of spectral energy drops faster if frequencies exceed the spectral centroid while it develops a wider tail towards lower frequencies. For positive values the opposite behavior is observed. A value of zero indicates a symmetric distribution of spectral energy around the spectral centroid.

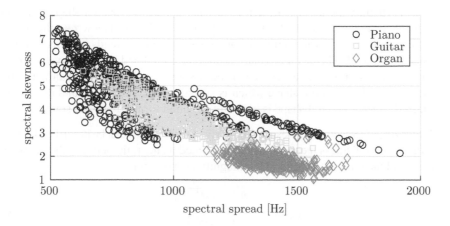

Figure 5.2: Scatter plot of spectral skewness values vs. spectral spread values for exemplary recordings of different instruments.

Example 5.2 (Spectral Spread and Skewness). *We consider excerpts of a piano piece (F. Chopin, "Waltz No. 9 a-flat Major Op. 69 No. 1"), a classical guitar piece (F. Tarrega, "Prelude in D minor, Oremus (lento)"), and an organ piece (J.S. Bach, "Toccata and Fugue in D minor, BWV 565") and compute their spectral spread and skewness values in a frame-wise fashion. The spectral analysis is performed using a DFT of length $M = 512$ at the sampling frequency $f_s = 16,000$. The frame shift is set to $R = 256$. In Figure 5.2 the spectral spread and skewness values are plotted against each other for each instrument. The scatter plot shows that piano and guitar sounds exhibit a lower spectral spread on average than an organ sound. However, we can also observe that in particular the piano sounds have a higher variation in terms of the spectral spread than the other two instruments. Further, it is worth noting that for all instruments the spectral skewness only attains positive values, which is a sign of a negative spectral tilt towards increasing frequencies. Obviously, in this feature space, organ sounds are well separated from piano and guitar sounds, respectively. Such an observation can be utilized for training supervised classifiers on instrument recognition (see Chapters 12, 18). A complete separation of piano sounds and guitar sounds, however, is not possible using only these features.*

Definition 5.6 (Spectral Kurtosis). Furthermore, the spectrum can be characterized in terms of its peakiness. This property can be expressed by means of the spectral kurtosis

$$t_{\text{kurt}}[\lambda] = \frac{\sum_{\mu=0}^{M/2} (\mu - t_{\text{cent}}[\lambda])^4 \, |X[\lambda,\mu]|}{(t_{\text{spread}}[\lambda])^4 \sum_{\mu=0}^{M/2} |X[\lambda,\mu]|} - 3. \tag{5.8}$$

In particular, the kurtosis describes to what extent the spectral shape resembles or differs from the shape of a Gaussian bell curve. For values below zero the spectral shape is subgaussian, which implies that the spectral energy tends towards a uniform distribution. Such a behavior typically occurs for wide-band sounds. A value of zero points towards an exact bell-curved spectral shape. Values larger than zero characterize a peaked spectral shape which is strongly concentrated around the spectral centroid. Such a spectral shape is typically obtained for narrow-band sounds.

Definition 5.7 (Spectral Flatness). Similarly, the peakiness can be described by means of the spectral flatness measure, which is defined as the ratio between the geometric mean and the arithmetic mean of the magnitude spectrum

$$t_{\text{flat}}[\lambda] = \frac{\sqrt[M/2+1]{\prod_{\mu=0}^{M/2} |X[\lambda,\mu]|}}{\frac{1}{M/2+1} \sum_{\mu=0}^{M/2} |X[\lambda,\mu]|}. \tag{5.9}$$

A higher spectral flatness value points towards a more uniform spectral distribution, whereas a lower value implies a peaked and sparse spectrum.

Definition 5.8 (Spectral Rolloff). The distribution of spectral energy to low and high frequencies can be characterized by the spectral rolloff feature. It is defined as the frequency index μ_{sr} below which 85% of the cumulated spectral magnitudes are concentrated. μ_{sr} fulfills the equation

$$\sum_{\mu=0}^{\mu_{\text{sr}}} |X[\lambda,\mu]| = 0.85 \sum_{\mu=0}^{M/2} |X[\lambda,\mu]|. \tag{5.10}$$

The lower the value of μ_{sr}, the more spectral energy is concentrated in low-frequency regions.

Definition 5.9 (Spectral Brightness). Instead of keeping the energy ratio fixed, it is also possible to choose a fixed cut-off frequency, e.g. $f_c = 1500$ Hz, above which the percentage share of cumulated spectral magnitudes is measured. Thereby, the amount of high-frequency energy can be quantified. The spectral brightness feature is hence defined as

$$t_{\text{bright}}[\lambda] = \frac{\sum_{\mu=\mu_c}^{M/2} |X[\lambda,\mu]|}{\sum_{\mu=0}^{M/2} |X[\lambda,\mu]|}, \tag{5.11}$$

where μ_c is the discrete frequency index corresponding to the cut-off frequency.

Definition 5.10 (Spectral Flux). The amount of spectral change between consecutive signal frames can be measured by means of the spectral flux, which is defined as the sum of the squared difference between the (normalized) magnitudes of successive short-time spectra over all frequency bins μ

$$t_{\text{flux}}[\lambda] = \sum_{\mu=0}^{M/2} (|X[\lambda,\mu]| - |X[\lambda-1,\mu]|)^2. \tag{5.12}$$

5.2.3   Mel Frequency Cepstral Coefficients

So far, we have introduced a number of one-dimensional features which describe different spectral characteristics of a music signal. A more elaborate set of features which models the spectral envelope is given by mel frequency cepstral coefficients (MFCCs) [2]. They are widely used in the field of automatic speech recognition (ASR), where they have been applied successfully in the feature extraction stage and thus have become a standard feature set. MFCCs entail a psychoacoustic representation of the spectral content and are therefore perceptionally motivated. The usefulness of MFCCs in areas other than ASR such as the discrimination between speech and music or musical genre classification was demonstrated, for instance, in [7], [8], or [20].

The relationship between the acoustic frequency of a stimulus and its perceived pitch is, in fact, non-linear. We recall that this relationship can be modeled by [18] (see Definition 2.9):

$$f_{\text{mel}} = 2595 \log_{10} \left(1 + \frac{f}{700\,\text{Hz}} \right), \tag{5.13}$$

where f denotes the frequency in Hertz and f_{mel} is the *mel* frequency, a pseudo unit which relates to the perception of pitch. For two given frequencies f_1 and f_2 the mel frequency allows a comparison of the perceived pitch. Moreover, it shows that a mere doubling of the acoustic frequency f in general does not result in a pitch which is perceived as being doubled. For instance, a tone at frequency $f_1 = 1000$ Hz is equivalent to $f_{1,\text{mel}} = 1000$. In order to perceive a tone with doubled pitch, i.e. $f_{2,\text{mel}} = 2000$, we have to change the frequency to $f_2 \approx 3500$ Hz. However, for frequencies below $f_1 = 1000$ Hz the relationship can be approximated by a linear function. Therefore, in this frequency range, which essentially covers the range of fundamental frequencies played by musical instruments, a doubled acoustic frequency approximately yields a doubled pitch perception.

The mel scale is then segmented into Q bands of constant width, so that the mel frequencies within each band can be aggregated. On the linear frequency scale this yields Q bands with non-uniform bandwidth which are closely related to the critical bands. The bands are formed using half-overlapping triangular weighting functions $\Delta_q(f)$, with $q \in \{1, 2, \ldots, Q\}$. In the implementation of Slaney [17] the number of mel bands is set to $Q = 40$, where the first 13 are spaced linearly, i.e. the center frequencies $f_{c,q}$ for 0 Hz $< f <$ 1000 Hz are equidistant. The center frequencies of the remaining 27 mel bands are arranged logarithmically. Hence,

$$f_{c,q} = \begin{cases} 133.33 + 66.67(q-1), & q \in \{1, 2, \ldots, 13\} \\ 1.07\, f_{c,q-1}, & q \geq 14. \end{cases} \tag{5.14}$$

The triangular weighting functions can then be obtained by

$$\Delta_q(f) = \begin{cases} \frac{f - f_{c,q-1}}{f_{c,q} - f_{c,q-1}}, & f_{c,q-1} \leq f \leq f_{c,q} \\ \frac{f_{c,q+1} - f}{f_{c,q+1} - f_{c,q}}, & f_{c,q} \leq f \leq f_{c,q+1} \\ 0, & \text{otherwise.} \end{cases} \tag{5.15}$$

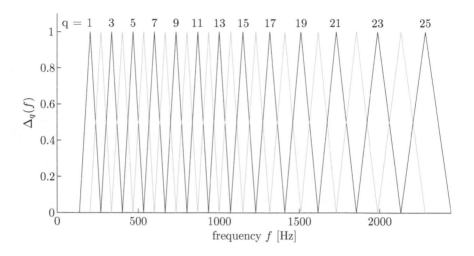

Figure 5.3: Weighting functions of the first 25 mel frequency band-pass filters.

In total, this results in a so-called mel filter bank. In Figure 5.3 the weighting functions are depicted for the first 25 filters.

For discrete frequency indices $\mu = Mf/f_s$ the weighting function of the q-th filter shall be denoted as $\Delta_q[\mu]$. Then, the short-time power spectrum for frame λ is weighted with $\Delta_q[\mu]$ for $q \in \{1, 2, \ldots, Q\}$. Subsequently, each filter output is summed up across all frequency indices to obtain the mel spectrum

$$\widetilde{X}[\lambda, q] = \sum_{\mu=0}^{M/2} |X[\lambda, \mu]|^2 \Delta_q[\mu]. \tag{5.16}$$

To account for the non-linear relationship between the sound pressure level and the perceived loudness, the logarithm of Equation (5.16) is evaluated. As a last step the mel spectrum $\widetilde{X}[\lambda, q]$ is decorrelated using the discrete cosine transform (DCT), yielding the MFCCs

$$\tilde{x}[\lambda, \xi] = v_\xi \sum_{q=0}^{Q-1} \log\left(\widetilde{X}[\lambda, q]\right) \cos\left(\frac{\pi q \xi}{Q}\right), \tag{5.17}$$

for $\xi \in \{0, 1, \ldots, Q-1\}$, with $v_0 = 1/\sqrt{M}$ and $v_\xi = \sqrt{2/M}$, for $\xi \in \{1, \ldots, Q-1\}$. This step decomposes the mel spectrum into cepstral coefficients which describe the spectral envelope and the spectral fine structure, respectively.

Example 5.3 (MFCCs).   We consider the note a' played by two different instruments, e.g. a cello and a piano, which differ considerably in terms of their timbral characteristics. The temporal evolution of the MFCCs for $\xi \in \{2, 3, \ldots, 13\}$ are shown in Figure 5.4. We observe that the piano exhibits higher values for $\xi \in \{2, 4, 5\}$. Furthermore, the MFCCs corresponding to the tone played by the cello have a more

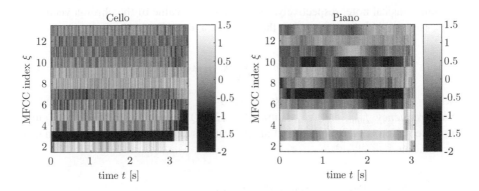

Figure 5.4: Temporal evolution of MFCCs $\tilde{x}[\lambda, \xi]$, $\xi \in \{2, 3, ..., 13\}$, for the note a'
played by a cello (left) and a piano (right).

rapidly fluctuating behavior which reveals a tremolo effect. Therefore, for describing timbre properties of an instrument or a music piece, it is useful to take temporal changes of short-term features into account.

5.3 Harmony Features

The simultaneous occurrence of multiple musical notes results in a harmony. This section introduces features which characterize properties of harmonies and hint towards applications for which these features are useful.

5.3.1 Chroma Features

The chromagram is a special time-frequency representation which achieves a frame-wise mapping of the spectral energy onto spectral bins which correspond to the twelve semi-tones of the chromatic scale; cp. Figure 2.4. Since this version of the chromagram accumulates the spectral energy across all octaves it is also denoted as the wrapped chromagram. Note that the unwrapped chromagram, i.e. the octave-wise variant, is provided by the sliding window CQT (see Section 4.5). Therefore, only the wrapped chromagram is considered henceforth.

The chroma vector, which constitutes one frame within the chromagram, can be computed based on an arbitrary spectral representation in which, however, even the lowest musical notes must be clearly resolved. This implies, that for a STFT-based chromagram the analysis window size of the STFT must be chosen appropriately. Alternatively, a non-uniform filter bank can be designed which decomposes a music signal into subbands which correspond to different musical notes [11] and facilitates a note-wise temporal representation of the spectral energy. Another straightforward approach is based on the CQT. Since it already provides an unwrapped chroma representation, the elements of the wrapped chroma vector can be obtained by performing an octave-wise summation over all magnitudes of CQT bins which correspond to a certain musical note, respectively. Hence, the p-th value of the chroma vector, with $p = \{0, 1, ..., 11\}$, is computed as the summation over O octaves

$$t_{\text{chroma}}[\lambda, p] = \sum_{o=1}^{O} |X_{\text{CQT}}[\lambda, p + 12o]|, \qquad (5.18)$$

where the first tone of the lowest octave $o = 1$ corresponds to the minimal analysis frequency f_{min}, and we assume $B = 12$ CQT bands per octave. The chroma representation is typically utilized in applications such as chord transcription or key estimation.

5.3.2   Chroma Energy Normalized Statistics

Differences in sound dynamics can lead to strongly fluctuating chroma values. In part, this can be compensated for by normalizing the chroma vector $t_{\text{chroma}}[\lambda, p]$ by its ℓ^1 norm

$$t_{\text{chroma,L1}}[\lambda, p] = \frac{t_{\text{chroma}}[\lambda, p]}{\sum_{p=1}^{12} t_{\text{chroma}}[\lambda, p]}. \qquad (5.19)$$

In order to make the normalized chroma values more robust against variations in tempo or articulation, the normalized chroma values can be quantized. To this end, in [11] the intuitive quantization function

$$t_{\text{chroma,Q}}[\lambda, p] = \begin{cases} 0, & \text{for} & 0 \leq & t_{\text{chroma,L1}}[\lambda, p] & < 0.05, \\ 1, & \text{for} & 0.05 \leq & t_{\text{chroma,L1}}[\lambda, p] & < 0.1, \\ 2, & \text{for} & 0.1 \leq & t_{\text{chroma,L1}}[\lambda, p] & < 0.2, \\ 3, & \text{for} & 0.2 \leq & t_{\text{chroma,L1}}[\lambda, p] & < 0.4, \\ 4, & \text{for} & 0.4 \leq & t_{\text{chroma,L1}}[\lambda, p] & < 1, \end{cases} \qquad (5.20)$$

is proposed which is applied to each normalized chroma value. In particular, chroma values which carry more than 40% of the energy are assigned to a maximal value of 4, whereas chroma values below a 5% threshold are mapped to zero in order to suppress noise.

A smoothed representation of the chroma time series can be obtained by convolving successive chroma values with a Hann window (see Section 4.5) which is another step towards a more robust chroma representation. In order to make the chroma representation usable for computationally inexpensive methods, decimating the feature rate is a recommended step. These steps lead towards the chroma energy normalized statistics (CENS).

Example 5.4 (Chromagram and CENS). *An example for the CQT-based chromagram of a classical piano piece (Chopin, "Grande valse brillante in e-flat major op. 18") and the corresponding CENS representation is provided in Figure 5.5. The CQT spectrogram is computed for a quarter semitone resolution, corresponding to a quality factor of $Q = 67.75$, and a minimal analysis frequency of $f_{\text{min}} = 261.63$ Hz*

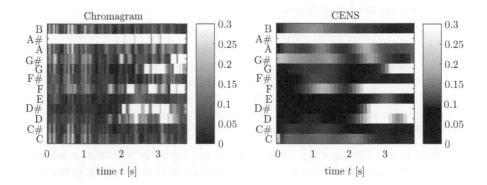

Figure 5.5: Chromagram (left) for an excerpt of a classical piano piece and the CENS representation of the same excerpt (right).

which corresponds to the note c'. The frame shift is set to $R = 32$ samples at the sampling frequency $f_s = 16$ kHz. This yields a feature rate of 500 Hz. For the temporally smoothed CENS features a 500-point Hann window and a decimation factor of 50 were used. This reduces the feature rate to 10 Hz, i.e. 10 features per second. Note that in this example the dominant A# is prominent.

5.3.3 Timbre-Invariant Chroma Features

Chroma features describe tonal aspects of a music signal and should therefore be independent of timbre. A particular chord progression that is played by two different instruments should ideally result in the same chroma representation. However, the spectral representation based on which the chromagram is obtained carries harmonics which are modulated by a spectral envelope. This envelope is characteristic of the timbre of a specific instrument. Therefore, the same chord progression played by two different instruments may result in two substantially different chroma representations.

In order to alleviate this effect, the spectrogram of a music signal can be whitened which means that the spectral envelope is flattened. This can be achieved, for example, by transforming the signal to the linear-frequency cepstral domain. Here, the lower cepstral coefficients which model the spectral envelope can be discarded by setting them to zero. An inverse transformation then yields a flattened spectrum which essentially only consists of the signal harmonics. Computing the chromagram based on this flattened spectral representation then results in a more timbre-invariant solution [11]. A more recent implementation which discards the timbre information of low MFCCs led to chroma DCT-reduced log pitch (CRP) representation [12].

5.3.4 Characteristics of Partials

Besides the analysis of chroma properties, it is worthwhile to analyze characteristics of partial tones (cp. Definition 2.3) in a music signal. In particular, it is of interest to know if partial tones exhibit a harmonic relationship or not, which has implications on the emotions a music piece can create in a listener (cf. Chapter 21). Therefore, in the following we will introduce features which account for properties of partial tones in music. The explanations are adapted from [22]. Let $A[\lambda, \hat{\mu}]$ be the amplitude of the $\hat{\mu}$-th partial tone in the λ-th frame, with $\hat{\mu} \in \{1, 2, \ldots, M\}$.

Definition 5.11 (Irregularity). A measure of the degree of variation in successive peaks of the short-time spectrum can then be obtained by the irregularity feature

$$t_{\text{irreg}}[\lambda] = \frac{\sum_{\hat{\mu}=1}^{M} (A[\lambda, \hat{\mu}] - A[\lambda, \hat{\mu}+1])}{\sum_{\hat{\mu}=1}^{M} (A[\lambda, \hat{\mu}])^2}, \tag{5.21}$$

which is obtained by accounting for the squared difference between amplitude values of adjacent partials.

Definition 5.12 (Inharmonicity). Another meaningful property of partial tones is their degree of harmonicity, which is perceived as the amount of consonance or dissonance in a music piece. Harmonic tones consist of a fundamental tone and overtones whose frequencies are integer multiples of the fundamental frequency f_0. Note that generally these frequency components are referred to as harmonics, where the first harmonic equals the fundamental tone, i.e. $f_1 = f_0$ (cf. Chapter 2.2.2). Given a set of extracted partial tones, a measure of inharmonicity can be obtained by computing the energy-weighted absolute deviation of the estimated partial tone frequencies $f_{\hat{\mu}}$ and the idealized harmonic frequencies $\hat{\mu} f_0$

$$t_{\text{inharm}}[\lambda] = \frac{2}{f_0} \frac{\sum_{\hat{\mu}=1}^{M} |f_{\hat{\mu}} - \hat{\mu} f_0| (A[\lambda, \hat{\mu}])^2}{\sum_{\hat{\mu}=1}^{M} (A[\lambda, \hat{\mu}])^2}, \tag{5.22}$$

which ranges from 0 (purely harmonic) to 1 (inharmonic).

Definition 5.13 (Tristimulus). Furthermore, it is insightful to measure the relative energy in subsets of partial tones compared to the total amount of tonal energy. For instance, the tristimulus feature quantifies the relative energy of partial tones by three parameters which measure the energy ratio of the first partial

$$t_{\text{trist1}}[\lambda] = \frac{(A[\lambda, 1])^2}{\sum_{\hat{\mu}=1}^{M} (A[\lambda, \hat{\mu}])^2}, \tag{5.23}$$

of the second, third, and fourth partial

$$t_{\text{trist2}}[\lambda] = \frac{\sum_{\hat{\mu} \in \{2,3,4\}}^{M} (A[\lambda, \hat{\mu}])^2}{\sum_{\hat{\mu}=1}^{M} (A[\lambda, \hat{\mu}])^2}, \tag{5.24}$$

and the remaining partials

$$t_{\text{trist3}}[\lambda] = \frac{\sum_{\hat{\mu}=5}^{M} (A[\lambda,\hat{\mu}])^2}{\sum_{\hat{\mu}=1}^{M} (A[\lambda,\hat{\mu}])^2},$$ (5.25)

respectively.

Definition 5.14 (Even-harm and Odd-harm). Similarly, the energy ratios of partials can be analyzed in terms of even-numbered and odd-numbers partial indices. Corresponding even-harmonic and odd-harmonic energy ratios can be defined as

$$t_{\text{even-harm}}[\lambda] = \sqrt{\frac{\sum_{\hat{\mu}=1}^{\lfloor M/2 \rfloor} (A[\lambda,2\hat{\mu}])^2}{\sum_{\hat{\mu}=1}^{M} (A[\lambda,\hat{\mu}])^2}}$$ (5.26)

and

$$t_{\text{odd-harm}}[\lambda] = \sqrt{\frac{\sum_{\hat{\mu}=1}^{\lfloor M/2+1 \rfloor} (A[\lambda,2\hat{\mu}-1])^2}{\sum_{\hat{\mu}=1}^{M} (A[\lambda,\hat{\mu}])^2}},$$ (5.27)

respectively.

5.4 Rhythmic Features

In this section we introduce features which are related to rhythmic properties of a music signal. Some of these features are defined as short-term features, as in the sections before, whereas other features are computed within larger time intervals in order to capture enough information for evaluating rhythmic patterns.

5.4.1 *Features for Onset Detection*

Onset detection is an important task in music information retrieval which is required, e.g., for the automatic segmentation of music signals or tempo estimation. In order to achieve a robust detection of note onsets, features are needed which are susceptible to sudden as well as slow spectral changes which occur for (pitched) percussive and non-percussive instruments, respectively. A feature which can be used for onset detection is the spectral flux, Equation (5.12). Other features that characterize local changes in spectral power are outlined in the following. An extensive introduction to onset detection methods based on these features is provided in Chapter 16.

Definition 5.15 (High-Frequency Content). Often a local energy increase which arises from a note onset can be observed more easily at higher frequencies since sharp onsets result in a broad spectrum. In order to accentuate the spectral content at higher frequencies the local spectral power can be weighted with a factor proportional to its frequency. Computing the mean across the linearly weighted spectral power yields the high-frequency content

$$t_{\text{hfc}}[\lambda] = \frac{1}{M/2+1} \sum_{\mu=0}^{M/2} \mu \, |X[\lambda,\mu]|^2.$$ (5.28)

This feature works well for detecting percussive onsets, but has weaknesses for other types of onsets [1].

Definition 5.16 (Phase Deviation). Besides defining features solely based on the magnitude spectrum, features related to the phase spectrum can also be taken into account. The phase spectrum contains additional details about the temporal structure of the signal. Considering the difference between two successive frames of the phase spectrum for a particular frequency bin yields an estimate of the instantaneous frequency

$$\phi'[\lambda,\mu] = \phi[\lambda,\mu] - \phi[\lambda-1,\mu]. \tag{5.29}$$

The difference in the instantaneous frequency between two successive frames

$$\phi''[\lambda,\mu] = \phi'[\lambda,\mu] - \phi'[\lambda-1,\mu] \tag{5.30}$$

then indicates a possible onset. A measure of the phase deviation is obtained by computing the averaged magnitude of Equation (5.30) yielding [3]

$$t_{\mathrm{pd}}[\lambda] = \frac{1}{M/2+1} \sum_{\mu=0}^{M/2} \left| \phi''[\lambda,\mu] \right|. \tag{5.31}$$

Since this feature is not robust against low-energy noise stemming from frequency bins which do not carry partials of a musical sounds, a normalized weighted phase deviation was proposed [3]

$$t_{\mathrm{nwpd}}[\lambda] = \frac{\sum_{\mu=0}^{M/2} |X[\lambda,\mu] \phi''[\lambda,\mu]|}{\sum_{\mu=0}^{M/2} |X[\lambda,\mu]|}, \tag{5.32}$$

which takes into account the strength of partial tones and consequently reduces the impact of irrelevant frequency bins. This feature shows a better performance for pitched non-percussive onsets than features based on the spectral amplitude [1].

Definition 5.17 (Complex Domain Features). Instead of a separate treatment of the magnitude spectrum and phase spectrum, a joint consideration of magnitude and phase is also possible. Assuming a constant amplitude and constant rate of phase change, a spectral estimate of the current frame λ based on the two previous frames can be obtained by

$$X_{\mathrm{T}}[\lambda,\mu] = |X[\lambda-1,\mu]| e^{i\left(\phi[\lambda-1,\mu]+\phi'[\lambda-1,\mu]\right)}. \tag{5.33}$$

By computing the sum of absolute differences between the spectrum and the target function $X_{\mathrm{T}}[\lambda,\mu]$ we arrive at a complex domain onset detection function which measures the deviation from a stationary signal behavior

$$t_{\mathrm{cd}}[\lambda] = \sum_{\mu=0}^{M/2} |X[\lambda,\mu] - X_{\mathrm{T}}[\lambda,\mu]|. \tag{5.34}$$

This feature treats increases and decreases in energy equally and therefore cannot distinguish between onsets and offsets. As we are only interested in onsets, the rectified sum of absolute deviations from the target function can be used instead of $t_{cd}[\lambda]$. This results in the rectified complex domain feature [3]

$$t_{rcd}[\lambda] = \sum_{\mu=0}^{M/2} t_{rcd,2}[\lambda, \mu] \tag{5.35}$$

with

$$t_{rcd,2}[\lambda, \mu] = \begin{cases} |X[\lambda, \mu] - X_T[\lambda, \mu]|, & \text{if } |X[\lambda, \mu]| \geq |X[\lambda - 1, \mu]| \\ 0, & \text{otherwise} \end{cases}. \tag{5.36}$$

The rectification ensures that only increases in spectral power which correspond to note onsets are considered while note offsets are discarded. This feature is well suited for detecting non-pitched percussive as well as pitched non-percussive onsets [1].

5.4.2 Phase-Domain Characteristics

In [10], two features were proposed which were successfully used for the discrimination between percussive and non-percussive music. In the following, we will consider the discrete-time signal $x[\lambda, k]$ as the input, but the transform can be applied to any time series. Before extracting these features, the phase domain vectors have to be computed.

Definition 5.18 (Phase Domain Transform).

$$\boldsymbol{p}_k = (x[\lambda, k], x[\lambda, k+d], x[\lambda, k+2d], ..., x[\lambda, k+(m-1)d])^T \tag{5.37}$$

is a phase vector of dimension m which contains a subsampled version of $x[\lambda, k]$ starting at time index k, where d determines the temporal spacing between successive values in $\boldsymbol{p}_k$. The phase vector describes the evolution of the discrete-time signal and highlights differences within the m sampled values. Successive phase vectors can be stored as a matrix $\boldsymbol{P} \in \mathbb{R}^{m \times (L-(m-1)d)}$, where the value $P_{k,z}$ corresponds to the z-th dimension of $\boldsymbol{p}_k$.

For the sake of simplicity, the two features are introduced for $m = 2$ and $d = 1$. The first one, the average length of differences between successive phase domain vectors, is calculated as

$$t_{AVG-L} = \frac{1}{L-2} \sum_{k=1}^{L-2} \|\boldsymbol{p}_{k+1} - \boldsymbol{p}_k\|. \tag{5.38}$$

The second feature, the average angle between the differences of successive phase domain vectors, is defined as:

$$t_{AVG-A} = \frac{1}{L-3} \sum_{k=1}^{L-3} \alpha\left(\boldsymbol{p}_{k+1} - \boldsymbol{p}_k, \boldsymbol{p}_{k+2} - \boldsymbol{p}_{k+1}\right), \tag{5.39}$$

where

$$\alpha \left(\boldsymbol{p}_{k+1} - \boldsymbol{p}_k, \boldsymbol{p}_{k+2} - \boldsymbol{p}_{k+1} \right) = \frac{\left\langle \boldsymbol{p}_{k+1} - \boldsymbol{p}_k, \boldsymbol{p}_{k+2} - \boldsymbol{p}_{k+1} \right\rangle}{\left\| \boldsymbol{p}_{k+1} - \boldsymbol{p}_k \right\| \cdot \left\| \boldsymbol{p}_{k+2} - \boldsymbol{p}_{k+1} \right\|}. \tag{5.40}$$

Note that in the general case, where an m-dimensional phase vector is used with a spacing step d, "$L-2$" should be replaced with "$L - (m-1)d - 1$" in Equation (5.38) and "$L-3$" with "$L - (m-1)d - 2$" in Equation (5.39).

Example 5.5 (Difference Vectors in Phase Domain). *Figure 5.6 illustrates the differences between successive phase vectors using $d = 1$ and $m = 2$. The horizontal axis corresponds to the first dimension $(p_{k+1,1} - p_{k,1})$, and the vertical to the second one $(p_{k+1,2} - p_{k,2})$. In Figure 5.6(a), 13 difference vectors are plotted for 15 original values of the function $\sin(x)$ evaluated at integer multiples of 0.5 radian, yielding $(0, 0.4794, 0.8415, \ldots)$. Hence, we obtain $\boldsymbol{p}_1 = (0, 0.4794)^T$, $\boldsymbol{p}_2 = (0.4794, 0.8415)^T$, etc. The periodic structure of the original series is visualized by an elliptic progression of the differences between phase vectors. Figure 5.6(b), shows 13 differences for a vector with 15 uniformly drawn random numbers between -1 and 1 $(-0.7162, -0.1565, 0.8315, \ldots)$. Figure 5.6(c) plots the difference vectors for one second of a rock song (AC/DC, "Back in Black"), and (d) for a classical piano piece (Chopin, "Mazurka in e-minor Op. 41 No. 2"). In both cases, the phase transform was applied to discrete-time signals samples at 44.1 kHz. We can observe a significantly broader distribution for the rock song, similar to examples in [10].*

5.4.3 Fluctuation Patterns

A desirable information about a music signal is its rhythmic periodicity which sheds light on the underlying musical genre. In [15] a method for extracting features that describe this musical aspect is proposed. In the following we outline a simplified version of the method.

Since rhythmic periodicity cannot be captured by means of short-term features, we have to consider longer segments of a music signal. The signal is therefore segmented into partitions with a duration of six seconds as this is long enough for humans to get an impression of the musical style [15]. The first two and last two partitions are then discarded in order to avoid fade-in and fade-out effects. For each of the remaining partitions the STFT $X[\lambda, \mu]$ is computed. In order to compensate for the frequency response of the outer ear, a model of the absolute threshold of hearing [19] is applied

$$\frac{L_{\mathrm{hs}}(f)}{\mathrm{dB}} = 3.64 \left(\frac{f}{1\mathrm{kHz}} \right)^{-0.8} - 6.5 e^{-0.6 \left(\frac{f}{1\mathrm{kHz}} - 3.3 \right)^2} + 10^{-3} \left(\frac{f}{1\mathrm{kHz}} \right)^4. \tag{5.41}$$

On a linear scale the hearing threshold is expressed by $L_{\mathrm{hs}}^{\mathrm{lin}} = 10^{L_{\mathrm{hs}}/20}$. Hence, the compensated power spectrum is obtained by

$$X_{\mathrm{comp}}[\lambda, \mu] = \left(10^{-L_{\mathrm{hs}}(f_\mu)/20} \right)^2 |X[\lambda, \mu]|^2, \tag{5.42}$$

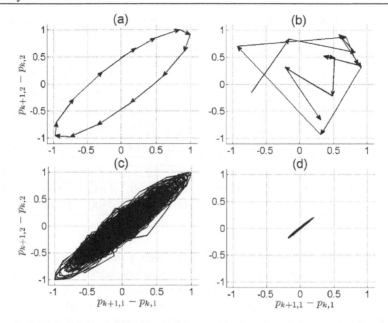

Figure 5.6: Examples of differences between successive phase vectors for the four input series. (a) sampled sine function; (b) a random sequence; (c) the time signal of a rock song (AC/DC); (d) the time signal of a classical piano piece (Chopin).

where $f_\mu = \frac{\mu f_s}{M}$ is the frequency corresponding to the μ-th STFT bin. Complying with the concept of critical bands, we then sum up the spectral power within each Bark band

$$X_{\text{Bark}}[\lambda, i] = \sum_{\mu_{1,i}}^{\mu_{u,i}} X_{\text{comp}}[\lambda, \mu], \qquad (5.43)$$

where i denotes the Bark band index and $\mu_{1,i}$ and $\mu_{u,i}$ are the STFT bins corresponding to the lower and upper frequency limits of the i-th Bark band as listed in Table 5.1. This results in a spectro-temporal representation with $I = 24$ Bark bands and L frames. In order to account for spectral masking, a spreading function

$$\frac{B(i)}{\text{dB}} = 15.81 + 7.5\,(i + 0.474) - 17.5\,(1 + (i + 0.474)^2)^{1/2} \qquad (5.44)$$

proposed in [16] is convolved with the Bark spectrum yielding

$$\hat{X}_{\text{Bark}}[\lambda, i] = X_{\text{Bark}}[\lambda, i] * 10^{B(i)/10}. \qquad (5.45)$$

To analyze the rhythmic periodicity, a DFT is computed across all L frames for each Bark band, respectively, resulting in

$$\widetilde{X}_{\text{mod}}[\nu, i] = \sum_{\lambda=0}^{L-1} \hat{X}_{\text{Bark}}[\lambda, i] e^{-i\frac{2\pi\lambda\nu}{L}}. \qquad (5.46)$$

Table 5.1: Lower Frequencies $f_{l,i}$ and Upper Frequencies $f_{u,i}$ of i-th Bark Band

i	1	2	3	4	5	6	7	8
$f_{l,i}/\mathrm{Hz}$	0	100	200	300	400	510	630	770
$f_{u,i}/\mathrm{Hz}$	100	200	300	400	510	630	770	920

i	9	10	11	12	13	14	15	16
$f_{l,i}/\mathrm{Hz}$	920	1080	1270	1480	1720	2000	2320	2700
$f_{u,i}/\mathrm{Hz}$	1080	1270	1480	1720	2000	2320	2700	3150

i	17	18	19	20	21	22	23	24
$f_{l,i}/\mathrm{Hz}$	3150	3700	4400	5300	6400	7700	9500	12000
$f_{u,i}/\mathrm{Hz}$	3700	4400	5300	6400	7700	9500	12000	15500

Here, $\widetilde{X}_{\mathrm{mod}}[\nu, i]$ is referred to as the Bark-frequency modulation spectrum and $\nu \in \{0, 1, \ldots, L/2 - 1\}$ denotes the modulation frequency index. Following this procedure a two-dimensional representation of dimension $I \times (L/2 + 1)$ is obtained for each partition of a music signal. In the last step, these modulation spectrograms can be temporally aggregated by computing the median values of the corresponding sequences which yields a single modulation spectrogram per music signal.

5.5 Further Reading

In this chapter we introduced a selection of basic and often used features. Many more features exist, and obviously, a complete overview of all available features is not possible. A different view on feature extraction is provided by e.g. [4] or [6]. In [11] more details about the variants of chroma features are explained.

Furthermore, a fundamental topic which has not been addressed in this chapter is the temporal aggregation of short-time features. An introduction to this topic will be given in Chapter 14. Feature aggregation by means of feature modulation analysis, feature autoregressive modeling (cp. Definition 9.40), and cepstral modulation analysis was also studied in [8], [9], and [14], respectively.

There are also many freely available tools for feature extraction. An extensive set of features can be extracted using, e.g., the MIR Toolbox [5], the AMUSE framework [21], the Auditory Toolbox [17], or the Chroma Toolbox [13].

Bibliography

[1] J. P. Bello, L. Daudet, S. Abdallah, C. Duxbury, M. Davies, and M. B. Sandler. A tutorial on onset detection in music signals. *IEEE Trans. Speech and Audio Processing*, 13(5):1035–1047, 2005.

[2] S. B. Davis and P. Mermelstein. Comparison of parametric representations for monosyllabic word recognition in continuously spoken sentences. *IEEE Trans. Acoustics, Speech, and Signal Processing*, 28(4):357–366, August 1980.

[3] S. Dixon. Onset detection revisited. In *Proc. Intern. Conf. Digital Audio Effects (DAFx)*, pp. 133–137. McGill University Montreal, 2006.

[4] H.-G. Kim, N. Moreau, and T. Sikora. *MPEG-7 Audio and Beyond: Audio Content Indexing and Retrieval.* John Wiley & Sons, 2006.

[5] O. Lartillot and P. Toiviainen. A MATLAB toolbox for musical feature extraction from audio. In *Proc. Intern. Conf. Digital Audio Effects (DAFx)*, pp. 237–244. Université Bordeaux, 2007.

[6] T. Li, M. Ogihara, and G. Tzanetakis, eds. *Music Data Mining.* CRC Press, 2011.

[7] B. Logan. Mel frequency cepstral coefficients for music modeling. In *Proc. Intern. Soc. Music Information Retrieval Conf. (ISMIR)*, 2000.

[8] M. F. McKinney and J. Breebaart. Features for audio and music classification. In *Proc. Intern. Soc. Music Information Retrieval Conf. (ISMIR)*, 2003.

[9] A. Meng, P. Ahrendt, J. Larsen, and L. K. Hansen. Temporal feature integration for music genre classification. *IEEE Trans. Audio, Speech, and Language Processing*, 15(5):1654–1664, July 2007.

[10] I. Mierswa and K. Morik. Automatic feature extraction for classifying audio data. *Machine Learning Journal*, 58(2-3):127–149, 2005.

[11] M. Müller. *Information Retrieval for Music and Motion.* Springer, 2007.

[12] M. Müller and S. Ewert. Towards timbre-invariant audio features for harmony-based music. *IEEE Transactions on Audio, Speech, and Language Processing*, 18(3):649–662, 2010.

[13] M. Müller and S. Ewert. Chroma toolbox: MATLAB implementations for extracting variants of chroma-based audio features. In *Proc. Intern. Soc. Music Information Retrieval Conf. (ISMIR)*, 2011.

[14] A. Nagathil, P. Göttel, and R. Martin. Hierarchical audio classification using cepstral modulation ratio regressions based on Legendre polynomials. In *Proc. IEEE Intern. Conf. on Acoustics, Speech and Signal Processing (ICASSP)*, pp. 2216–2219. IEEE Press, 2011.

[15] E. Pampalk, A. Rauber, and D. Merkl. Content-based organization and visualization of music archives. In *Proc. ACM Intern. Conf. on Multimedia*, pp. 570–579. ACM Press, 2002.

[16] M. Schroeder, B. Atal, and J. Hill. Optimizing digital speech coders by exploiting masking properties of the human ear. *J. Acoust. Soc. Am. (JASA)*, 66(6):1647–1652, 1979.

[17] M. Slaney. Auditory toolbox: A MATLAB toolbox for auditory modeling. Technical Report 45, Apple Computer, 1994.

[18] S. Stevens, J. Volkmann, and E. B. Newman. A Scale for the Measurement of the Psychological Magnitude Pitch. *J. Acoust. Soc. Am. (JASA)*, 8:185–190, January 1937.

[19] E. Terhardt. Calculating virtual pitch. *Hearing Research*, 1(2):155–182, 1979.

[20] G. Tzanetakis and P. Cook. Musical Genre Classification of Audio Signals. *IEEE Trans. Speech and Audio Processing*, 10(5):293–302, July 2002.

[21] I. Vatolkin, W. M. Theimer, and M. Botteck. AMUSE (Advanced MUSic Explorer): A multitool framework for music data analysis. In *Proc. Intern. Soc. Music Information Retrieval Conf. (ISMIR)*, pp. 33–38, 2010.

[22] Y.-H. Yang and H. H. Chen. *Music Emotion Recognition*. CRC Press, 2011.

Chapter 6

Auditory Models

KLAUS FRIEDRICHS, CLAUS WEIHS
Department of Statistics, TU Dortmund, Germany

6.1 Introduction

Auditory models are computer-based simulation models which emulate the human auditory process by mathematical formulas which transform acoustic signals into neural activity. Analyzing this activity instead of just the original signal might improve several tasks of music data analysis, especially tasks where human perception still outperforms computer-based estimations, e.g., several transcription tasks like onset detection and pitch estimation. For applying auditory models as a front-end for such tasks, some basic knowledge about the different stages of the auditory process is advisable in order to decide which one are most important and which stages can be ignored. Naturally, this depends on the actual application and might be important to simplify the auditory model in order to reduce the computation times.

The hearing process of humans consists of several stages located in the ear and the brain. In recent decades several computational models have been developed which can help to prove assumptions about the different stages of the hearing process by comparing psycho-physical experiments and animal observations to simulation outputs on the same acoustical data. While our understanding of these processes improves and the results of the models are getting more realistic, their application for automatic speech recognition and for several tasks in music research increases.

While some later stages of the hearing process taking place in several parts of the brain are more difficult to observe, the beginning of the process located directly in the ear is far more investigated. This stage is called the auditory periphery and models the transformation from acoustical pressure waves in the air to release events of the auditory nerve fibers as introduced in Section 6.2. Though several models of the auditory periphery have been proposed, in Section 6.3 we describe the popular Meddis model ([17]). In Section 6.4 technical models for pitch detection are compared to auditory models of the midbrain which try to simulate the pitch extraction process of humans. In Section 6.5 we give an overview of further reading. Additionally, music classification systems are described which process the output of auditory models. While the applications are explained in detail in other chapters, here, the main focus is a brief overview of the feature generating process.

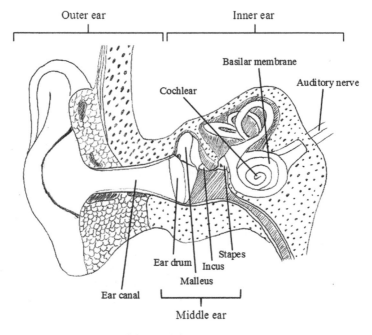

Figure 6.1: Model of the human ear.

6.2 Auditory Periphery

The auditory periphery consists of the outer ear, the middle ear and the inner ear (see Figure 6.1). The main task of the outer ear is collecting sound waves and directing them further into the ear. Additionally, it contributes to sound localization by a directional filtering where the degree of sound enhancement depends on its angle of incidence. At the back end of the outer ear the ear-drum vibrates. This vibration is transmitted to the stapes (bone) in the middle ear and then directed further to the cochlear in the inner ear. Inside the cochlear, the basilar membrane vibrates at specific locations depending on the stimulating frequencies (see Figure 6.2). On the basilar membrane, inner hair cells (IHC) are located which are activated by the velocity of the membrane and provoke spike emissions (neuronal activity) of the auditory nerve fibers. Additionally, outer hair cells (OHC) provide the human ability to enhance and reduce specific frequency regions, enabling tasks like speaker discrimination. The stimulus content defines the degree of vibration of specific regions on the basilar membrane and hence also the neural activity of the corresponding nerve fibers. High frequencies stimulate the membrane at its base and low frequencies at its apex (see Figure 6.2). For lower frequencies up to approximately 2 kHz – which includes all fundamental frequencies of musical tones (see Chapter 2) – neural activity occurs phase-locked with the stimulus. Phase-locking means that neural activity is periodic with peaks and tails where the period corresponds to periodicities of the stimulus. In later stages in the brain, this effect can be utilized to encode the frequency content of a stimulus by additionally analyzing the temporal structure besides the neural intensity of different fibers. The human auditory system consists of roughly 30,000 auditory nerve fibers, each responsible for the recognition of an individual but overlapping frequency range. In auditory models this is simplified and simulated by a much smaller quantity of filters.

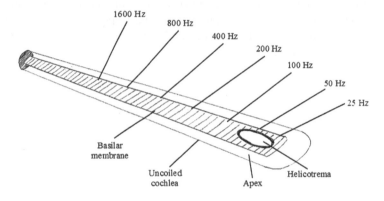

Figure 6.2: Basilar membrane.

From a signal processing view, a simplified model of the auditory periphery can be seen as a bank of linear filters (see Chapter 4) with a succeeding half-wave rectifier (only positive values pass). In this context linear means that a higher stimulus results in higher output levels of all simulated auditory nerve fibers independent of additional simultaneous noise. However, modern models of the auditory periphery are far more complex, modeling nonlinear and asynchrony properties which can more precisely explain psychoacoustic phenomena. One example is two-tone suppression: From two simultaneous tones, the louder one can mask the softer one even if they consist of entirely different frequencies. This means a drastic reduction of auditory nerve activity corresponding to the softer tone in contrast to the case where this tone is presented alone. This masking phenomenon violates the assumption of a linear model.

6.3 The Meddis Model of the Auditory Periphery

A popular model of the auditory periphery is the Meddis model [17]. It is a cascade of several consecutive modules, which emulate the spike firing process of multiple auditory nerve fibers. From a signal processing perspective it can be seen as a cascade of several filter banks with 41 channels in the standard setting getting a 41-dimensional vector of neural activity (firing probabilities) for each sampling moment as intermediate result. In a final step, several auditory nerve fibers are simulated based on the output of one channel by transforming release probabilities into binary

filled again. Transmitter release is only indirectly controlled by the electrical voltage $V(t)$. Actually, it controls the calcium stream into the cell and the calcium promotes the release of transmitter. Other auditory models make the simplifying assumption that the content of the transmitter reservoir into the synaptic cleft is proportional to the stimulus intensity. In the Meddis models, transmitter release is modeled more realistically using a cascade of transmitter reservoirs.

6.3.5   Auditory Nerve Activity

In the final step of the Meddis model, release probabilities are transformed into binary release events (spikes). This is a stochastic process where the probability of a spike is dependent on the release probability and the time since the last generated spike. For each channel, n auditory nerve fibers are simulated by repeating this process n times (typically $n \in [10, 100]$).

For many applications this final step can be skipped. For most music classification systems and also for automatic speech recognition, release probabilities are the adequate input. Skipping this step also saves lot of computation time. Nevertheless it is needed for biologically motivated applications which require binary release events as input.

6.4   Pitch Estimation Using Auditory Models

In what follows, pitch estimation is understood as a synonym for fundamental frequency (f_0) estimation. Apart from the frequency resolution which begins in the basilar membrane, it is assumed that temporal periodicities of spike emissions occurring in the auditory nerve fibers are responsible for pitch perception. Licklider proposed an autocorrelation analysis [13] which is nowadays widely accepted (see also Chapters 2 and 4) after Langner showed its physiological plausibility [10]. In this section, modern variants of autocorrelation analysis (cp. Section 9.8.2) based on the output of the auditory periphery are introduced. While for a long time it was controversial if autocorrelation can be achieved by the brain, in recent decades, neural models of the midbrain have been developed which are mathematically equivalent to an autocorrelation analysis. These models are described in the second part of this section.

6.4.1   Autocorrelation Models

One challenge of autocorrelation analysis applied to the output of a model of the auditory periphery is that there are several channels which have to be combined in some way. In [18] and [20], this is achieved by first computing the individual running autocorrelation function (ACF) for each channel and combining them by averaging over all channels (SACF). The running ACF of a channel k at time t and lag l is based on the spike probabilities, $p(t, k)$, and is recursively defined by

$$h(t,l,k) = p(t,k) \cdot p(t-l,k) \cdot \frac{\Delta t}{\tau(l)} + h(t-\Delta t, l, k) \cdot e^{\frac{-\Delta t}{\tau(l)}}, \tag{6.1}$$

where Δt is the sampling interval and $\tau(l)$ is a time constant (10 ms) which defines the length of the time period over which regularities are assessed.

The running SACF is defined by

$$s(t,l) = \frac{1}{N} \sum_{k=1}^{N} h(t,l,k), \tag{6.2}$$

where N is the number of channels used and $h(t,l,k)$ is the ACF at time t and lag l in channel k. The peaks of the SACF are indicators for the perceived pitch and a natural variant of monophonic pitch detection simply identifies the maximum peak for every time point t. The lag l achieving the maximum peak at a given time point t corresponds to the dominant periodicity, and hence its reciprocal is the estimated fundamental frequency. The model is successfully compared to several psychophysical phenomena like pitch detection with missing fundamental frequency.

After some recent psychophysical studies which indicate that the autocorrelation approach is inconsistent with human perception for some special stimuli, in [3] the autocorrelation approach is improved by a low-pass filter, resulting in the new function LP-SACF. Additionally, the time constant $\tau(l)$ is linked to the given lag and is set to $2l$ since it is assumed that for some specific stimuli the pitch characteristic can only be assessed over a longer period. The low-pass filter of LP-SACF is recursively defined as an exponentially decaying average,

$$P(t,l) = s(t,l) + P(t-\Delta t,l) \cdot e^{\frac{-\Delta t}{\lambda}}, \tag{6.3}$$

where λ is the time constant (120 ms) of the filter. Results indicate that the modified LP-SACF can overcome the limitations of the SACF.

6.4.2 Pitch Extraction in the Brain

While autocorrelation models can explain the pitch perception of humans very well, for a long time their physiological relevance was controversial since to many researchers this mathematical model appeared to be physiologically implausible. However, in the last decades pitch perception models of the brain have been developed which perform the mathematical function of autocorrelation by combining thousands of physiologically plausible neurons. The brainstem consists of several units, three of them are assumed to be relevant for pitch perception: the cochlear nucleus, the superior olivary complex and the inferior colliculus. Auditory nerve activities are further processed by the cochlear nucleus whose output is transferred to the superior olivary complex. Both the output of the cochlear nucleus and the output of the superior olivary complex are input to the inferior colliculus where the pitch extraction of harmonic tones is actually performed.

A simulation model for human pitch perception is proposed by Langner ([10], [5], [1] and [2]) which is based on several electro-physical experiments on different animal species. The basic idea of the model is the assumption that apart from the frequency decomposition in the cochlear, the temporal processing of spike events is also responsible for pitch extraction. This is achieved by a correlation analysis of spectral information and periodicity information which yields a spatial representation of spectro-temporal information. Cells in the inferior colliculus respond to specific frequencies and to specific modulations. In the model they react as coincidence detectors to the input of stellate and spindle neurons, which means they only react if both inputs are active. This is only the case if the delay time of the spindle neuron is equal to either the modulation period or to one of its integer multiples. Thereby, the coincidence neuron reacts as a comb filter (cp. Chapter 12).

A computer simulation of a similar model can be found in [21]. It describes the human perception of pitch changes and consists of four consecutive stages where stage 1 is the Meddis model of the auditory periphery described in Section 6.3. In [7], the model of Langner is applied as a basis to explain consonance perception in the context of tonal fusion. Therefore, autocorrelation functions of different intervals are analyzed and ranked by defining the so-called generalized coincidence function. The ranking by this function is equivalent to statements of music theorists, e.g., even slightly mis-tuned consonances which still sound consonant for humans achieve a high ranking.

6.5 Further Reading

This chapter gives only a brief overview over auditory models. For a comprehensive description of auditory models for the different stages of the hearing process, we refer the reader to [19]. Further insights in mathematical modeling of human pitch and harmony perception can be found in [11].

Another popular auditory model of the auditory periphery is the model of Zilany and Bruce [31]. In contrast to the Meddis model, in [31] the outer hair cells are explicitly modeled. This means, depending on the level input, that best frequencies and bandwidths are dynamically regulated. In the Meddis model, this effect is achieved implicitly by combining a linear and a nonlinear path which is a simpler but also more static approach.

Regarding auditory-based pitch estimation, in [8], an autocorrelation model is proposed which is more efficient in terms of computational complexity. The simple model consists of only two channels, one for low frequencies below 1 kHz and one for frequencies above 1 kHz. While the first channel is analyzed directly by autocorrelation only, the second channel is passed through an auditory model. Afterwards, the channels are combined by the SACF, defined in Equation (6.2). It is argued that the benefit of using an auditory model as a front end for pitch estimation is only reliable for higher frequencies since for lower frequencies the auditory system acts as a linear channel. In [9], an iterative approach for multipitch analysis is proposed. Instead of just picking the maximum peak of the SACF as in [18], the strength of a period candidate is calculated as a weighted sum of the amplitudes of its harmonic partials. Amplitudes of higher partials are more distorted by other tones and therefore they are weighted less than lower partials. The period with the highest strength is assessed as a pitr ch in the polyphonic signal. The next iteration starts with a cancelation process where harmonics and subharmonics of this pitch are suppressed. Subsequently, again the strength of each period candidate is calculated. This process is iterated until a specified number of pitches is achieved or until the maximum strength is not above a specified threshold.

In [12], the IPEM toolbox is proposed, a MATLAB® toolbox which constructs perception-based features for different applications. However, it relies on a somewhat outdated auditory model. Features based on an auditory model have been applied for several tasks of music classification which are separately described in the following paragraphs.

Onset Detection (see also Chapter 16)   In [4], onset detection is applied to the output of the Meddis model of the auditory periphery. The output of 40 simulated nerve fibers are first transformed into 40 individual onset detection functions and afterwards combined by using a specific quantile. Certainly, this is a simple approach leaving space for improvements. However, the results show that even by using this method the auditory model approach performs as well as the original approach on the acoustic waveform.

Transcription and Melody Detection (see also Chapter 17)   Mathematically, a melody is a sequence of notes where each note has a pitch, an onset, and an offset. In [26], the detection of the singing melody is performed using features based on an auditory model with a consecutive autocorrelation analysis. Here, again, the problem arises of how to choose the correct peak of the autocorrelation function. The naive variant consists of picking the maximum peak for each frame and merging successive frames with the same pitch to one tone. Better approaches additionally take the temporal development into account. For each frame the pitch candidates and their strengths with respect to the autocorrelation function are taken as features. As a second feature, for each pitch candidate the ratio of its strength to its strength in the predecessor frame is considered enabling the separation of consecutive tones with the same pitch. In [16], the ratio of each peak strength to the neighborhood average strength is computed as an additional feature. Furthermore, the zero-lag correlation of each channel is calculated to obtain a running estimate of the energy in each channel where local maxima might indicate onsets. Other examples for melody detection using an auditory model are described in [6], [15], and [23].

Instrument Recognition (see also Chapter 18)   In [30], instrument recognition is applied to features based on the output of the Meddis model. Under most circumstances these features lead to better results than the features based on the original waveform. Other promising approaches using an auditory model as the front end for instrument recognition are described in [22], and [29].

Genre Classification   Approaches for genre classification with an auditory model as the front end are described in [14] and [24].

Bibliography

[1] A. Bahmer and G. Langner. Oscillating neurons in the cochlear nucleus: I. Experimental basis of a simulation paradigm. *Biological Cybernetics*, 95(4):371–379, 2006.

[2] A. Bahmer and G. Langner. Oscillating neurons in the cochlear nucleus: II. Simulation results. *Biological Cybernetics*, 95(4):381–392, 2006.

[3] E. Balaguer-Ballester, S. L. Denham, and R. Meddis. A cascade autocorrelation model of pitch perception. *The Journal of the Acoustical Society of America*, 124(4):2186–2195, 2008.

[4] N. Bauer, K. Friedrichs, D. Kirchhoff, J. Schiffner, and C. Weihs. Tone onset detection using an auditory model. In M. Spiliopoulou, L. Schmidt-Thieme, and R. Janning, eds., *Data Analysis, Machine Learning and Knowledge Discovery*, pp. 315–324. Springer International Publishing, 2014.

[5] M. Borst, G. Langner, and G. Palm. A biologically motivated neural network for phase extraction from complex sounds. *Biological Cybernetics*, 90(2):98–104, 2004.

[6] L. P. Clarisse, J.-P. Martens, M. Lesaffre, B. De Baets, H. De Meyer, and M. Leman. An auditory model based transcriber of singing sequences. In *Proceedings of the 3rd International Conference on Music Information Retrieval (ISMIR)*, pp. 116–123. IRCAM, 2002.

[7] M. Ebeling. Neuronal periodicity detection as a basis for the perception of consonance: A mathematical model of tonal fusion. *The Journal of the Acoustical Society of America*, 124(4):2320–2329, 2008.

[8] M. Karjalainen and T. Tolonen. Multi-pitch and periodicity analysis model for sound separation and auditory scene analysis. In *IEEE International Conference on Acoustics, Speech, and Signal Processing*, volume 2, pp. 929–932. IEEE, 1999.

[9] A. Klapuri. Multipitch analysis of polyphonic music and speech signals using an auditory model. *IEEE Transactions on Audio, Speech, and Language Processing*, 16(2):255–266, 2008.

[10] G. Langner. Neuronal mechanisms for pitch analysis in the time domain. *Experimental Brain Research*, 44(4):450–454, 1981.

[11] G. Langner and C. Benson. *The Neural Code of Pitch and Harmony*. Cambridge University Press, 2015.

[12] M. Leman, M. Lesaffre, and K. Tanghe. Introduction to the IPEM toolbox for perception-based music analysis. *Mikropolyphonie – The Online Contemporary Music Journal*, 7, 2001.

[13] J. C. R. Licklider. A duplex theory of pitch perception. *Experientia, 7*, pp. 128–134, 1951.

[14] S. Lippens, J.-P. Martens, and T. De Mulder. A comparison of human and automatic musical genre classification. In *IEEE International Conference on Acoustics, Speech, and Signal Processing (ICASSP'04)*, volume 4, pp. 233–236. IEEE, 2004.

[15] M. Marolt. A connectionist approach to automatic transcription of polyphonic piano music. *IEEE Transactions on Multimedia*, 6(3):439–449, 2004.

[16] K. D. Martin. Automatic transcription of simple polyphonic music: Robust front end processing. *Massachusetts Institute of Technology Media Laboratory Perceptual Computing Section, Tech. Rep. 399*, 1996.

[17] R. Meddis. Auditory-nerve first-spike latency and auditory absolute threshold: A computer model. *The Journal of the Acoustical Society of America*, 119(1):406–417, 2006.

[18] R. Meddis and M. J. Hewitt. Virtual pitch and phase sensitivity of a computer model of the auditory periphery. I: Pitch identification. *The Journal of the Acoustical Society of America*, 89(6):2866–2882, 1991.

[19] R. Meddis, E. A. Lopez-Poveda, R. R. Fay, and A. N. Popper. *Computational Models of the Auditory System*. Springer, 2010.

[20] R. Meddis and L. O'Mard. A unitary model of pitch perception. *The Journal of the Acoustical Society of America*, 102(3):1811–1820, 1997.

[21] R. Meddis and L. P. OMard. Virtual pitch in a computational physiological model. *The Journal of the Acoustical Society of America*, 120(6):3861–3869, 2006.

[22] M. J. Newton and L. S. Smith. A neurally inspired musical instrument classification system based upon the sound onset. *The Journal of the Acoustical Society of America*, 131(6):4785–4798, 2012.

[23] R. P. Paiva, T. Mendes, and A. Cardoso. An auditory model based approach for melody detection in polyphonic musical recordings. In *Computer Music Modeling and Retrieval*, pp. 21–40. Springer, 2005.

[24] Y. Panagakis, C. Kotropoulos, and G. R. Arce. Non-negative multilinear principal component analysis of auditory temporal modulations for music genre classification. *IEEE Transactions on Audio, Speech, and Language Processing*, 18(3):576–588, 2010.

[25] M. R. Panda, W. Lecluyse, C. M. Tan, T. Jürgens, and R. Meddis. Hearing dummies: Individualized computer models of hearing impairment. *International Journal of Audiology*, 53(10):699–709, 2014.

[26] M. Ryynanen and A. Klapuri. Transcription of the singing melody in polyphonic music. In *Proceedings of the 7th International Conference on Music Information Retrieval (ISMIR)*, pp. 222–227. University of Victoria, 2006.

[27] M. Slaney et al. An efficient implementation of the Patterson–Holdsworth auditory filter bank. *Apple Computer, Perception Group, Tech. Rep*, 35, 1993.

[28] C. J. Sumner, L. P. O'Mard, E. A. Lopez-Poveda, and R. Meddis. A nonlinear filter-bank model of the guinea-pig cochlear nerve: Rate responses. *The Journal of the Acoustical Society of America*, 113(6):3264–3274, 2003.

[29] S. K. Tjoa and K. J. R. Liu. Musical instrument recognition using biologically inspired filtering of temporal dictionary atoms. In *Proceedings of the 11th International Society for Music Information Retrieval Conference (ISMIR)*, pp. 435–440. Utrecht University, 2010.

[30] K. Wintersohl. *Instrumenten Klassifikation mit Hilfe eines auditorischen Modells*. Bachelor Thesis, Department of Statistics, TU Dortmund University, 2014.

[31] M. S. A. Zilany and I. C. Bruce. Modeling auditory-nerve responses for high sound pressure levels in the normal and impaired auditory periphery. *The Journal of the Acoustical Society of America*, 120(3):1446–1466, 2006.

Chapter 7

Digital Representation of Music

GÜNTER RUDOLPH
Department of Computer Science, TU Dortmund, Germany

7.1 Introduction

The computer-aided generation, manipulation, and analysis of music requires a digital representation of music. Historically, music could be passed on only by singing or playing instruments from memory. This form of propagation inevitably leads to alterations in the original melodies over the years. The development of musical notations finally provided a method to record music in written form not only preventing uncontrolled changes in existing music but also enabling the reliable storage, reproduction, and dissemination of music over time and space.

Graphical notations were already used for Gregorian chants (about 800 AD) in the form of graphical elements called *neume* for representing the melodic shape – initially without, but later in a four-line staff notation. A five-line staff notation was created by Guido von Arezzo (about 1000 AD) which developed further to a standard notation basically valid since 1600 AD in Western music [3]. With the advent of electronic music it became necessary to develop new graphical notations; many of them may be considered as art work themselves [11]. In this chapter, however, we stick to the standard graphical notation of Western music as introduced in Chapter 3.

The processing of music with a computer requires mappings from the analog to the digital world and vice versa at different levels. Figure 7.1 sketches some conversion paths between different notational representations and file formats. Written sheet music may be mapped to a standardized file format by an appropriately educated person or a computer system that scans the sheet music from the paper, recognizes the music notation, and maps the information to a digital representation (*OMR: optical music recognition*) which can be stored in and re-read from files with some specified file format. Examples of such file formats (e.g. *MIDI* or *abc*) and the principles of OMR are presented in Section 7.2. Another path from the analog to the digital world is the sounding music performed by artists where the analog signals are recorded, mapped to digital signals by A/D-converters (cf. Chapter 4.2), and stored digitally in a format called PCM encoding. This field of business is dis-

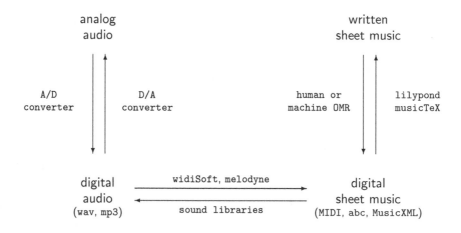

Figure 7.1: Conversion paths between musical notations and file formats.

cussed in Section 7.3. The transcription of digital audio data to digital sheet music
(cf. Chapter 17) is a complex task which is hardly possible without errors and loss in
precision. Nevertheless, there exist (mainly proprietary) software systems like `widi`
[15] or `melodyne` [8] that claim to accomplish such a conversion. A compilation
of those systems is given in Chapter 17 whereas Section 7.4 presents some tools for
rendering digital sheet music to written sheet music. Finally, Section 7.5 is devoted
to transforming digital music representation into analog sound. Digital sheet music
can be translated into digital audio data with the help of sound libraries and synthesizers, whereas music available as digital audio signals can be rendered to sound by
converting them to analog signals using D/A-converters (cf. Chapter 4.2).

7.2   From Sheet to File

7.2.1   Optical Music Recognition

Generating a digital representation of written sheet music by a human is a tedious
and error-prone task. Therefore, the idea to scan the sheets before recognizing the
notational structure and information automatically, similar to OCR (optical character
recognition) methods, appears to be obvious. Usage of OCR is ubiquitous in our
daily life and it works almost perfectly in recognizing letters, digits and punctuation.
But OCR is much simpler than OMR (optical music recognition) [1, 2, 10]. As a
consequence, current tools for recognizing sheet music can only serve to generate a
digital representation at some error rate—a human must verify the result and correct
the recognition errors afterward. Nevertheless, OMR tools may be a valuable assis-
tance. Details about the working principles of OMR can be found in Section 8.4.

7.2.2 abc *Music Notation*

The *abc* music notation has been designed to notate music in plain text format which is easily readable by humans and also amenable to digital processing. Originally, it was intended for folk and traditional Western-style tunes which can be written on one staff in standard classical notation, but its notational capabilities go far beyond simple single staff music.

The abc notation is registered as an Internet media (MIME) type and there are many software tools available that support the work with abc. Since 2009 the website http://abcnotation.com/ has collected tunes, software, tutorials and further information related to abc.

The format of an abc file is divided into a header and a body. The header contains information about the tune in general whereas the body encloses information about the notes. The structure of a header is presented below.

```
X:<int>           tune counter within file
T:<string>        title of the tune
M:<int>/<int>     time signature
L:1/<int>         default note length
C:<string>        name of composer(s)
Q:<int>           number of default note lengths per minute
K:<string>        key signature
```

The semantics of each header's entry is indicated by a capital letter at the beginning of the line followed by a colon. For example, the letter T indicates that the title of the tune is written after the colon, M specifies the meter by two integers separated by a slash, and L the default note length where 1/2 denotes a half, 1/4 a quarter and 1/8 a eighth note (and so forth).

Typically, the default note length is the note length appearing most frequently in the tune. For longer notes one has to put an integer factor after the note (e.g., C2 for double default length of note C), for shorter notes one puts a slash followed by the integer divisor after the note (e.g., C/2 for half the length). C/3 would be used for triplets.

Rests are represented by the lower-case letter z, and its length is defined by the default note length. Analogous to the notes, longer and shorter rests are expressed by a subsequent integer factor or a subsequent slash and an integer divisor.

The note pitches are specified by letters. The notation differs from usual octave designation systems (ODS) as summarized in Table 7.1. Accidentals are indicated by the symbols _ and ^. For example, _G means "G flat" and ^G means "G sharp". A tie is indicated by a hyphen after the first note.

Table 7.1: Notational Differences between the Helmholtz and abc Octave Designation Systems

ODS / octave	contra	great	small	one-line	two-line	three-line
Helmholtz	C_1 - B_1	C - B	c - b	c' - b'	c" - b"	c''' - b'''
abc ODS	C,,, - B,,,	C,, - B,,	C, - B,	C - B	c - b	c' - b'

Smoke on the water

Figure 7.2: Conversion of abc file of Example 7.1 with the tool abcm2ps.

The optional mark C: opens the field for the composer(s) and the entry beginning with Q: specifies how many notes with the default length are to be played per minute. Note lengths deviating from the default length may be used in the specification. For example, the expression Q:1/4=36 means that 36 quarter notes should be played per minute. Finally, the entry K: declares the key by using, for example, a "G" for *G major* and "Gm" for *g minor*. The body of the file follows directly after the header without any notification. Bars lines are indicated by a vertical line.

Example 7.1 (Smoke on the Water).
A version of the first four bars of the intro of Deep Purple's Smoke on the Water *from 1972 written in abc notation might look as follows:*

```
X:1
T:Smoke on the water
M:4/4
L:1/8
C:Ritchie Blackmore,Ian Gillian,Roger Glover,Jon Lord,Ian Paice
Q:1/4=120
K:Bb
G2 B2 c3 G | z B z _d c4 | G2 B2 c3 B | z G-G6 ||
```

The entry X:1 *indicates that the following data describe the first tune in the file. The title (T) of the tune is* Smoke on the Water. *The tune in four-four time (M) is specified with the eighth note as the default note length (L), composers are listed after the* C, *the speed (Q) is set to 120 quarter notes per minute (moderate rock tempo), and the key is B-flat major (K). The first four bars of the tune follow directly after the header. The score music shown in Figure 7.2 was generated with the tool* abcm2ps *directly from the abc listing of this example.*

7.2.3 Musical Instrument Digital Interface

MIDI (musical instrument digital interface) is basically a digital protocol that allows multiple hardware and software devices to communicate over a network [6]. The protocol is designated for the exchange of music-related messages about notes, lengths, pitches, tempo, and the like. These messages are sent through the network serially at a speed of 31,250 bits per second. The resulting data stream can be stored to a file for later reuse in a specific format that is known as *standard MIDI file (SMF)* format. The SMF format is different from the native MIDI protocol since it needs time-stamps for the realization of playback in a proper sequence.

A standard MIDI file typically has the extension `.mid` or `.smf`. It consists of a header and a body, which in turn may consist of several tracks. The header always has a length of 14 bytes whose structure is displayed in Table 7.2. The first four bytes indicate that this file is actually an SMF. The next four bytes contain the length of the header without the first 8 bytes; therefore the value is always 6. Three different track formats are possible: single track (0), multiple track (1), or multiple song (2). The code $\in \{0,1,2\}$ of the track format is stored in two bytes starting at offset 8. In the *multiple track* format, individual parts are saved on different tracks in the sequence whereas everything is merged onto a single track in *single track* format. The rarely used *multiple song* format stands for a series of tracks of type 0. The next two bytes contain the number of tracks. In the case of the single track format, the value is always 1. The last two bytes specify the meaning of the time stamps. If bit 15 is zero, then bits 0 to 14 encode the number of ticks per quarter note. If bit 15 is one, bits 14 through 8 correspond to an LTC time code [9] whereas bits 0 to 7 encode the resolution within a frame.

Table 7.2: Structure of the Header of a Standard MIDI File

offset	length	type	content	description
0	4	char4	"MThd"	tags file as a MIDI file
4	4	uint4	0x06	remaining length of header (= 6 bytes)
8	2	unit2		track format $\in \{0,1,2\}$
10	2	uint2		number of tracks in the body
12	2	uint2		unit of time for delta timing

A track consists of a track header and a sequence of track events (= the track body). The track header always has a size of 8 bytes. The first four bytes indicate the beginning of a track. The second four bytes contain the number of bytes in the track body. Table 7.3 provides a summary.

Table 7.3: Structure of the Track Header within an SMF

offset	length	type	content	description
0	4	char4	"MTrk"	tags file as a MIDI file
4	4	uint4		length of track body (= length of track - 8)

A track event within the track body consists of *delta time* (i.e., the time elapsed since the previous event) and either a MIDI event or a meta event or a system exclusive (sysex) event. If two events occur simultaneously, the delta time must be zero.

The encoding of numbers like the delta time deserves special consideration. Actually, it is a variable-length encoding that can lead to some data compression. Only the lowest 7 bits of a byte are used to encode a number. The 8th bit serves as the variable length encoding: If some number requires n bits in standard binary encoding, then we need $\lceil n/7 \rceil$ bytes to store the bit pattern. The 8th bit of each of these bytes is set to 1, except the least significant byte for which the 8th bit is set to 0.

A *MIDI event* is any type of MIDI channel event preceded by a delta time. A channel event consists of a status byte and one or two data bytes. The status byte specifies the function. Since standard MIDI has 16 channels, each particular function has 16 different status bytes: for example, the function *note on* exists with values 90 to 9F. Table 7.4 shows only functions for channel 0. Bit 8 is set for status bytes and cleared for data bytes. The note number 0 stands for note C at 8.176 Hz, number

Table 7.4: Some MIDI Channel Events (for Channel 0)

status byte	function	data byte 1	data byte 2
80	note off	note number	velocity
90	note on	note number	velocity
A0	polyphonic aftertouch	note number	aftertouch pressure
B0	control mode change	controller function	data
C0	program change	change type	—
D0	channel aftertouch	aftertouch pressure	—
E0	pitch wheel change	pitch wheel LSB	pitch wheel MSB

1 for C# at 8.662 Hz, number 2 for D at 9.177 Hz up to note number 127 for G at 12,543.854 Hz. The velocity typically means the volume of a note (higher velocity = louder), but in case of note-off events it can also describe how quickly or slowly a note should end.

A *meta event* is tagged by the initial byte with value FF. The next byte specifies the type of meta event before the number of bytes of the subsequent metadata is given in variable-length encoding. It is not required for every program to support each meta event. The data of a time signature event consists of 4 bytes. The first byte is the

Table 7.5: Code and Meaning of Some Typical MIDI Meta Events

type	name	description
01	text event	any type of text
02	copyright notice	(c) year copyright-owner
03	track name	name of track
04	instrument name	name of instrument
05	lyrics	each syllable of lyrics is own event
2F	end of track	indicates end of track (length: 0x00)
51	set tempo	μsec. per quarter note (length: 0x03)
58	time signature	$\rightarrow$ see text (length: 0x04)
59	key signature	$\rightarrow$ see text (length: 0x02)

numerator, the second byte is the denominator given by the exponent of a power of two, the third byte expresses the number of MIDI ticks in a metronome click, and the fourth byte contains the number of 32nd notes in 24 MIDI ticks.

The data of a key signature event consists of 2 bytes. The first byte specifies the number of flats or sharps from the base key C and the second byte indicates major

(0) or minor (1) key. Flats are represented by negative numbers, sharps by positive numbers. For example, C minor would be expressed as FD 01, E major as 04 00, A flat major as FC 00.

A *sysex event* is tagged by the initial byte with value F0 or F7 before the length of the subsequent data is given in variable length encoding. The end of the data should be tagged with an F7 byte. Since these events are tailored to specific MIDI hardware or software, a more detailed description is omitted here.

Example 7.2 (Smoke on the Water).
SMF files are stored in binary format. Here an annotated hexadecimal dump of the binary MIDI file is given for the same piece of music considered in Example 7.1. The abbreviation dt *in the annotation stands for "delta time".*

```
4D 54 68 64                "MThd"
00 00 00 06                length of header - 8
00 00                      track format 0
00 01                      #tracks = 1
01 E0                      480 ticks per quarter note

4D 54 72 6B                "MTrk"
00 00 00 9E                length of track body: 158 byte
00 FF 51 03 07 A1 20       dt=0, meta: set tempo (500 msec/quarter)
00 FF 59 02 FE 00          dt=0, meta: key signature (B major)
00 FF 58 04 04 02 30 08    dt=0, meta: time signature (4/4)
00 FF 03 12                dt=0, meta: track name (18 byte)
53 6D 6F 6B 65 20          "Smoke "
6F 6E 20 74 68 65 20       "on the "
77 61 74 65 72             "water"

01 90 43 69                dt=1,    note 43 = G3 on, volume 69
83 5F 80 43 00             dt=479,  note 43 off
01 90 46 50                dt=1,    note 46 = A#3 on, volume 50
83 5F 80 46 00             dt=479,  note 46 off
01 90 48 5F                dt=1,    note 48 = C4 on, volume 5F
85 4F 80 48 00             dt=719,  note 48 off
01 90 43 50                dt=1,    note 43 = G3 on, volume 50
81 6F 80 43 00             dt=239,  note 43 off
81 71 90 46 50             dt=241,  note 46 = A#3 on, volume 50
81 6F 80 46 00             dt=239,  note 46 off
81 71 90 49 50             dt=241,  note 49 = C#4 on, volume 50
81 6F 80 49 00             dt=239,  note 49 off
01 90 48 5F                dt=1,    note 48 = C4 on, volume 5F
87 3F 80 48 00             dt=959,  note 48 off
01 90 43 69                dt=1,    note 43 = G3 on, volume 69
83 5F 80 43 00             dt=479,  note 43 off
01 90 46 50                dt=1,    note 46 = A#3 on, volume 50
83 5F 80 46 00             dt=479,  note 46 off
01 90 48 5F                dt=1,    note 48 = C4 on, volume 5F
85 4F 80 48 00             dt=719,  note 48 off
01 90 46 50                dt=1,    note 46 = A#3 on, volume 50
81 6F 80 46 00             dt=239,  note 46 off
81 71 90 43 50             dt=241,  note 43 = G3 on, volume 50
8D 0F 80 43 00             dt=1679, note 43 off
1A FF 2F 00                dt=26, meta: end of track
```

Typically, SMF files are generated by playing a MIDI instrument or using specific software that records the tune in its own format. Finally, the specific format is converted to SMF by some software tool. In Example 7.2 the tool ABC Converter *was applied to convert the abc file to binary SMF. The binary file was then converted with a hex editor into hexadecimal representation.*

7.2.4   MusicXML 3.0

The abbreviation XML stands for *eXtensible Markup Language*, which allows for storing and transmitting data in a semantically structured manner [13]. For this purpose an XML-based language is defined by a DTD (document type definition), which is publicly available via WWW. Each XML document should contain the URL to its DTD at the beginning of the file.

The DTD specifies the structure of the XML document and which elements and attributes are "understood" by the application for which the DTD is designed. Additional elements and attributes can be used in the document, but they are ignored by applications using a specific DTD without these extensions.

An element appears within an XML document with a start tag <elem> and end tag </elem>, if elem is the name of the element. Everything between the start and end tag belongs to this element. Nesting of elements is of course possible and a key concept for structuring the data. Each element may have arbitrarily many attributes which are placed within the start tag. For example, element <elem attr1="12.34" attr2="music data"> contains two attributes named attr1 and attr2 whose values are assigned in textual form.

XML documents are intended for machine processing; it is a common misconception that XML files should be processed or generated by humans. A thoughtful choice of element and attribute names may also make the files digestible by humans, but this is not a necessary requirement. Nevertheless, most DTDs for specific application domains use names that give strong hints for a semantic interpretation of the data. This is also the case for the DTD of *MusicXML 3.0*, which was designed to exchange sheet music data between programs [4] (also see http://www.musicxml.com/).

Example 7.3 (Smoke on the Water).
As can be seen in the subsequent example, XML documents have a lavish use of space. The example has been generated from the tiny Example 7.1 with the tool abc2xml, *that converts* abc *files to* MusicXML. *Since the XML document is actually interpretable by humans, we refrain from a detailed description here.*

```
<?xml version='1.0' encoding='utf-8'?>
<!DOCTYPE score-partwise PUBLIC
  "-//Recordare//DTD MusicXML 3.0 Partwise//EN"
  "http://www.musicxml.org/dtds/partwise.dtd">
<score-partwise>
  <movement-title>Smoke on the water</movement-title>
  <identification>
    <creator type="composer">
      Ritchie Blackmore,Ian Gillian,Roger Glover,Jon Lord,Ian Paice
    </creator>
    <encoding>
```

```
      <encoder>abc2xml version 66</encoder>
      <encoding-date>2015-07-09</encoding-date>
    </encoding>
  </identification>
  <part-list>
    <score-part id="P1"><part-name/></score-part>
  </part-list>
  <part id="P1">
    <measure number="1">
      <direction placement="above">
        <direction-type>
          <metronome>
            <beat-unit>quarter</beat-unit>
            <per-minute>120.00</per-minute>
          </metronome>
        </direction-type>
        <sound tempo="120.00" />
      </direction>
      <attributes>
        <divisions>120</divisions>
        <key><fifths>-2</fifths><mode>major</mode></key>
        <time><beats>4</beats><beat-type>4</beat-type></time>
      </attributes>
      <note>
        <pitch><step>G</step><octave>4</octave></pitch>
        <duration>120</duration><voice>1</voice><type>quarter</type>
      </note>
      <note>
        <pitch><step>B</step><alter>-1</alter><octave>4</octave></pitch>
        <duration>120</duration><voice>1</voice><type>quarter</type>
      </note>
      <note>
        <pitch><step>C</step><octave>5</octave></pitch>
        <duration>180</duration><voice>1</voice><type>quarter</type><dot />
      </note>
      <note>
        <pitch><step>G</step><octave>4</octave></pitch>
        <duration>60</duration><voice>1</voice><type>eighth</type>
      </note>
    </measure>
    <measure number="2">
      <note>
        <rest /><duration>60</duration><voice>1</voice><type>eighth</type>
      </note>
      <note>
        <pitch><step>B</step><alter>-1</alter><octave>4</octave></pitch>
        <duration>60</duration><voice>1</voice><type>eighth</type>
      </note>
      <note>
        <rest /><duration>60</duration><voice>1</voice><type>eighth</type>
      </note>
      <note>
        <pitch><step>D</step><alter>-1</alter><octave>5</octave></pitch>
        <duration>60</duration><voice>1</voice><type>eighth</type>
        <accidental>flat</accidental>
      </note>
      <note>
        <pitch><step>C</step><octave>5</octave></pitch>
        <duration>240</duration><voice>1</voice><type>half</type>
      </note>
    </measure>
    <measure number="3">
      <note>
        <pitch><step>G</step><octave>4</octave></pitch>
        <duration>120</duration><voice>1</voice><type>quarter</type>
      </note>
      <note>
```

```
        <pitch><step>B</step><alter>-1</alter><octave>4</octave></pitch>
        <duration>120</duration><voice>1</voice><type>quarter</type>
      </note>
      <note>
        <pitch><step>C</step><octave>5</octave></pitch>
        <duration>180</duration><voice>1</voice><type>quarter</type><dot />
      </note>
      <note>
        <pitch><step>B</step><alter>-1</alter><octave>4</octave></pitch>
        <duration>60</duration><voice>1</voice><type>eighth</type>
      </note>
    </measure>
    <measure number="4">
      <note>
        <rest /><duration>60</duration><voice>1</voice><type>eighth</type>
      </note>
      <note>
        <pitch><step>G</step><octave>4</octave></pitch>
        <duration>60</duration><voice>1</voice><type>eighth</type>
        <notations><tied type="start" /></notations>
      </note>
      <note>
        <pitch><step>G</step><octave>4</octave></pitch>
        <duration>360</duration><voice>1</voice><type>half</type><dot />
        <notations><tied type="stop" /></notations>
      </note>
      <barline location="right">
        <bar-style>light-light</bar-style>
      </barline>
    </measure>
  </part>
</score-partwise>
```

7.3   From Signal to File

A common first step to convert analog audio signals into digital format is called *pulse code modulation (PCM)*, whose result is a sequence of digital words that may be stored word by word to a binary file as a so-called *raw audio file* (Section 7.3.1). This kind of format tacitly assumes that users know the number of channels, sampling rate, number of bits per sample, and so forth. Without this information, a correct interpretation of the sequence of bits within the file is quite unlikely. Therefore, many audio file formats divide the file content into a header and body part. The header provides all the information needed to correctly decipher the bit sequence in the file whereas the body contains the audio data, possibly in raw audio format, which leads to the popular WAVE file format (Section 7.3.2).

Since audio formats like WAVE need the header information for a correct sensing they are not suited for streaming audio data digitally via Internet or radio broadcast since audio decoders must be able to enter the stream at arbitrary positions. The simple solution to this requirement is the decomposition of the audio data into so-called frames, each of them preceded by a copy of that part of the header information necessary for the decoder to interpret the frame's audio data. Synchronization points within the stream are indicated by a specific bit pattern. Examples for such formats are *MPEG-1 Layer 3*, which is better known as *MP3 format* (Section 7.3.3), and *MPEG-2/AAC*.

7.3.1 Pulse Code Modulation and Raw Audio Format

The process of converting sound music in the form of time-continuous waveforms with continuous amplitudes into raw audio format via pulse code modulation (PCM) consists of three steps: sampling, quantization, and encoding.

Suppose the amplitude of the signal can be described by some continuous function $a(t) \in \mathbb{R}$ with time $t \geq 0$. The discretizing of continuous time is achieved by *sampling* the signal at equidistant time steps $t_i = i \cdot dt$ for $i = 0, 1, \ldots$ with some $dt > 0$ which leads to a sequence of amplitude values $a_i = a(t_i)$ for $i = 0, 1, \ldots$.

Next, the discretizing of the real-valued amplitudes is obtained by *quantization* [7, p. 116f.] where the set Q of possible amplitude values is partitioned into a set of intervals $Q_k = (q_k, q_{k+1}]$ with $k = 0, \ldots, L-1$. Since audio signals are bipolar, the quantization range $Q = [-v, v]$ is always chosen symmetrically to zero with $v = \max\{|a_i| : i = 0, \ldots, \} > 0$. If $a_i \in Q_k$, then amplitude a_i is assigned to quantization level $k \in \{0, \ldots, L-1\}$ with $L = 2^n$, where n is the number of bits available for encoding a sample. The quantization is termed uniform if the widths $|Q_k|$ of the intervals are equal for all $k = 0, \ldots, L-1$. In this case we obtain a *linear PCM (LPCM)*.

Finally, the L levels need an *encoding*: The assignment of binary code words to the quantization levels is arbitrary in principle: actually, there are $(2^n)!$ possibilities. For example, for $n = 3$ bits per sample we have $(2^3)! = 8! = 40,320$ possible code tables. Typically, either the binary two's complement or the offset binary encoding is used. In the first case, negative (positive) amplitude values are assigned negative (positive) digital numbers. In the latter case, only nonnegative digital numbers are used with the convention that the interval with the smallest analog values has digital (offset) value zero.

Example 7.4 (Raw audio). *Suppose we want to record the analog signal given by the superposition of sine waves*

$$a(t) = 3\sin(3\pi t - 0.3) + 2\sin(2\pi t + 0.2) + \sin(1.5\pi t - 0.1)$$

in the range $t \in [0, 1.6)$ (see Figure 7.3). If we sample the signal every $dt = 0.05$ time units we obtain $1.6/dt = 32$ values $a_i = a(t_i)$ with $t_i = 0.05 \cdot i$ for $i = 0, 1, \ldots, 31$.

Using $n = 4$ bits to encode each sample value leads to a division of the range of possible values into $L = 2^4 = 16$ intervals of equal width. The range is defined symmetrically to zero as $[-v, v]$ with $v = \max\{|a_i| : i = 0, \ldots, 31\} = 5.73297$. Table 7.6 contains the 16 interval centers of the quantization levels in its first column. The second and third columns show the assignment of binary code words to each quantization level using two's complement and offset binary encoding.

Using offset binary encoding and a hexadecimal representation of the 32 code words of our exemplary sequence leads to the raw audio data

7 a d f f f e b 9 6 4 3 3 3 5 6 8 9 9 9 8 7 6 5 5 6 7 9 a b c b

that can be stored to a file. Figure 7.3 illustrates the conversion from the analog waveform to raw digital audio data.

As evident from the preceding example, raw audio data can be interpreted cor-

Table 7.6: Quantization Levels and Possible Code Tables for a 4-Bit LPCM

interval center	signed binary code word	offset binary code word
−5.37466	1000	0000
−4.65804	1001	0001
−3.94142	1010	0010
−3.22480	1011	0011
−2.50818	1100	0100
−1.79155	1101	0101
−1.07493	1110	0110
−0.35831	1111	0111
0.35831	0000	1000
1.07493	0001	1001
1.79155	0010	1010
2.50818	0011	1011
3.22480	0100	1100
3.94142	0101	1101
4.65804	0110	1110
5.37466	0111	1111

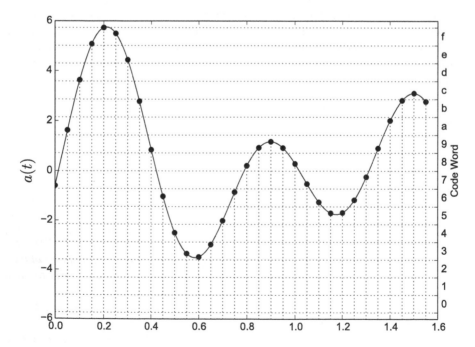

Figure 7.3: Exemplary conversion of analog waveform to raw digital audio format. Black dots represent the sampled value a_t at step t whereas the dashed horizontal lines indicate the borders of the quantization intervals.

rectly only if additional information like the number of bits per sample or the type of binary encoding is supplied. In case of a closed system, this information might be specified implicitly, but if the audio file is to be transferred to arbitrary destinations or stored in databases and later retrieved from arbitrary customers, the information must be provided in additional files. Alternatively, the audio file format can be modified for storing format specification and data in a single file (e.g. WAVE file format).

7.3.2 WAVE File Format

The WAVE file format (extension `wav`) is divided into a fixed-sized header and a body of variable length. The structure of the header (bytes 0 to 35) and the beginning of the body is given in Table 7.7. The type `uint`n indicates data to be interpreted as unsigned integers with n bytes, where the byte order is little-endian, i.e., the least significant byte (LSB) precedes the most significant byte (MSB). The values for the characters are just the ASCII character codes.

Table 7.7: Structure of the Header of a WAVE File

offset	length	type	content	description
0	4	char4	"RIFF"	tags file as member of RIFF format family
4	4	uint4		remaining length of file
8	4	char4	"WAVE"	indicates WAVE format
12	4	char4	"fmt "	marks start of format details
16	4	uint4	0x10	remaining length of header (= 16 bytes)
20	2	uint2		audio format
22	2	uint2		number of channels (mono = 1, stereo = 2)
24	4	uint4		samples per second per channel
28	4	uint4		average bytes per second
32	2	uint2		alignment size for audio data items
34	2	uint2		bits per sample
36	4	char4	"data"	marks start of audio data
40	4	uint4		length of audio data block
44	...			audio data

Some entries need elucidation: At offset 20, the audio format stored in the body can be specified by different identifiers (IDs). The ID 0x0001 stands for LPCM, the ID 0x0055 for MP3, and many other formats are possible. This observation reveals that the WAVE format may be considered as a *container format* that is able to encapsulate different audio formats and serves simply as a container for the transport of the audio data. Apparently, the LPCM raw audio format is most frequently used. In this book we follow this understanding by using WAVE and LPCM format interchangeably.

At offset 32, the number of bytes necessary to store a single sample is required as a multiple of 8 bits. In case of LPCM raw audio format, this value can be calculated from the data provided in the header: add 7 to the number of bits per sample (at offset 34), divide by 8, take the integer part and multiply the result by the number of channels (at offset 22).

At offset 34, the number of bits per sample is also an indicator of the type of binary encoding of LPCM audio data (starting at offset 44). Offset binary encoding is used up to 8 bits per sample, otherwise two's complement.

Example 7.5 (WAVE format). *Suppose we want to store the LPCM audio data of Example 7.4 in a file with WAVE format. Since the WAVE format insists on multiples of 8 bit per audio sample we must pad our 4-bit code words with 4 leading zeros. Therefore, the length of the audio data block (offset 40) is 32 bytes and the total size of the file sums up to 76 bytes, which in turn leads to the value 68 at offset 4. Since we store a mono recording (single channel), the value 1 is set at offset 22. Although we have 4 bits per sample (offset 34), we need 8 bits of memory for storing the data (offset 32). If we assume that a time unit represents 1 millisecond, we have drawn a sample every 50 microseconds or 20,000 samples per second (offset 24). Since each sample is stored with 8 bits, we have 20,000 bytes per second (offset 28). Now we are in the position to compile the WAVE file where each byte is represented in hexadecimal form:*

```
52 49 46 46 44 00 00 00 57 41 56 45 66 6d 74 20
10 00 00 00 01 00 01 00 00 00 20 4e 00 00 20 4e
01 00 04 00 64 61 74 61 20 00 00 00 07 0a 0d 0f
0f 0f 0e 0b 09 06 04 03 03 03 05 06 08 09 09 09
08 07 06 05 05 06 07 09 0a 0b 0c 0b
```

7.3.3 MP3 Compression

Raw audio files like those in the WAV file format are very large (about 10 MB per minute) and they cannot be compressed significantly by traditional techniques since the values of raw audio data are almost uniformly distributed over the file. As a consequence, a zipped WAV file is only marginally smaller than the uncompressed original. Therefore, MP3 (short for *MPEG-1, Layer III*) takes another approach to the compression problem: "Rather than just seeking out redundancies like zip does, MP3 provides a means of analyzing patterns in an audio stream and comparing them to human hearing and perception. Also unlike zip compression, MP3 actually discards huge amounts of information, preserving only the data absolutely necessary to reproduce an intelligible signal." [5, p. 2]. Thus, the reduction is achieved not only by eliminating redundance but also by omitting irrelevant data.

The typical compression ratio is about 1:10, meaning that a 3-minute song of about 30 MB of raw data shrinks to just 3 MB of compressed data, which can be stored to a file with suffix .mp3. The amount of data discarded during compression is configurable in the MP3 encoder so that the user may decide about his/her individual optimal balance between file size and quality of the music playback after MP3 decompression. If configured properly, music playback from MP3 data is practically indistinguishable from the uncompressed original.

How can this be achieved? The key to the high compression rate of MP3 is the joint application of two different compression techniques: First, it sorts out negligible signal data based on a *psychoacoustic model* of the human ear. Second, redundancies in the frequency domain of the reduced data set are exploited by means of Huffman encoding.

The psychoacoustic technique exploits the imperfection of the human acoustic system [5, p. 24f.]. Typically, humans cannot hear frequencies below 20 Hz and above 20 kHz. Therefore signal data below or above these thresholds need not be stored anyway. Moreover, humans do not perceive all frequencies in that range equally well. Most people are less sensitive to low and high frequencies whereas they are most sensitive to frequencies between 2 and 4 kHz. This threshold of audibility can be expressed by a function of sound pressure (aka volume) versus frequency. This function is not static; rather, its characteristic may be affected by so-called *auditory* and *temporal masking*. Auditory masking describes the effect that a certain signal cannot be distinguished from a stronger signal (say, plus 10 db) if the pitches are only slightly different (say, 100 Hz difference); here, the stronger signal masks the weaker signal so that the latter signal need not be stored. In case of temporal masking, a strong, abruptly ending tone provokes a short pause of a few milliseconds in which the human hear is unable to perceive very quiet signals which need not be stored accordingly. Temporal masking also appears in the opposite direction: the last few milliseconds of a quiet tone are wiped out by a sudden strong subsequent signal.

The identification of such masking effects and the adjustment of the audibility curve is a complex task and causes some computational effort which is fortunately only necessary in the encoding and not in the decoding phase. The psychoacoustic analysis is made in the frequency domain after a fast Fourier transform and it finally provides an adjusted audibility curve which is later used by the second compression technique for the decision which signals can be omitted.

In the beginning of MP3 compression, the data stream is separated into 32 spectral bands whose bandwidths are not equally wide but are adapted to the human acoustic system (in the range 1 to 4 kHz the bands are denser than below and above). Each spectral band is divided into frames containing either 384 or 1152 samples. Long frames are for sub-bands with low frequencies whereas the other sub-bands use short frames. Next, a modified discrete cosine transformation (MDCT) finally leads to a division into 576 frequency bands. Now the adjusted audibility curve is used to decide which signals must be stored and which can be neglected. The remaining frequency data are then quantized with fixed or variable bit depth. Next, the frequency samples are Huffman encoded: short code words are assigned to frequently occurring frequencies whereas rarely occurring frequencies get longer code words. Since this assignment is specific to each frame, the code tables of all frames must be included in the final MP3 data stream.

The end user does not need to know the structure of an MP3 file, but we are curious about the technical realization. An MP3 file is simply a sequence of many encoded frames. Each frame contains a header of 32 bit (see Table 7.8) and a body with the compressed signal data.

At the beginning of the frame header, there is a synchronization (sync) pattern of

Table 7.8: The Structure of the Frame Header of an MP3 File

bit	size	description
31	11	synchronization pattern (all set to 1)
20	2	MPEG audio version (10: MPEG2; 11: MPEG1)
18	2	MPEG layer (01: III; 10: II; 11: I)
16	1	protection (0: on; 1: off)
15	4	bit rate index (see lookup Table 7.9)
11	2	sampling rate index (see lookup Table 7.10)
9	1	padding (0: off; 1: on)
8	1	private (0: off; 1: on)
7	2	channels (00: stereo; 01: joint stereo; 10: dual channel; 11: mono)
5	2	joint stereo extension
3	1	copyright (0: off; 1: on)
2	1	original (0: copy; 1: original)
1	2	emphasis (now obsolete)

Table 7.9: Lookup Table of the Bitrate Index (kbit/sec) for MPEG Encodings

bit pattern	MPEG1-I	MPEG1-II	MPEG1-III	MPEG2-I	MPEG2-II/III
0000	free	free	free	free	free
0001	32	32	32	32	8
0010	64	48	40	48	16
0011	96	56	48	56	24
0100	128	64	56	64	32
0101	160	80	64	80	40
0110	192	96	80	96	48
0111	224	112	96	112	56
1000	256	128	112	128	64
1001	288	160	128	144	80
1010	320	192	160	160	96
1011	352	224	192	176	112
1100	384	256	224	192	128
1101	416	320	256	224	144
1110	448	384	320	256	160
1111	bad	bad	bad	bad	bad

11 bits which is important for broadcasted data streams. When such a data stream is entered at arbitrary time, the MP3 decoder first seeks the sync pattern within the stream. If such a bit pattern is found, it may be the sync pattern but also some data from the body part of the frame. Thus, the decoder must check its hypotheses of a detected sync pattern by verifying that this sync pattern appears also at the position in the stream where the next frame should start. The more often this hypothesis cannot be rejected, the more likely the event that a true sync pattern has actually been found.

Table 7.10: Lookup Table of the Sampling Rate Index (in Hz) for MPEG Encodings

bit pattern	MPEG1	MPEG2	MPEG2.5
00	44100	22050	11025
01	48000	24000	12000
10	32000	16000	8000
11	reserved	reserved	reserved

The MP3 encoding is only a special case of audio file formats from the MPEG family. In case of MP3, the MPEG version is no. 1 and the MPEG layer is no. 3. Thus, the next four bits are set to 1101. If protection is activated, then a 16-bit checksum is placed directly after the header. Bit and sampling rates are given by indices to fixed lookup tables. In MPEG encodings, it may happen that a frame requires one byte fewer than the "standard size". In this case the padding bit is set. The private bit may trigger some application-specific behavior. The next bits indicate if the tune is a mono or stereo recording and which kind of stereo mode is used. If copyright bit is switched on, then it is officially illegal to copy the track and the next bit indicates if this file is the original file or a copy of it. The last bits of the header are now obsolete.

In most cases, the MP3 file is preceded by an *ID3 tag* that contains meta information like the name of artist and the title of tune that may be displayed by an MP3 player. The ID3 tag has a header (10 bytes) and a body. The structure of the header is presented in Table 7.11.

The first three bytes indicate an ID3 tag by the ASCII codes of the string "ID3" and the next two bytes are the ID3v2 revision number in little-endian format. The following byte contains 8 flags whose meaning will not be discussed here. The next four bytes contain the size of the ID3 tag's body in bytes but in a 7-bit encoding, i.e., for each byte the most significant bit is always set to zero and it is ignored so that $4 \times 7 = 28$ bits are available for representing the body size. If the size is stored in bytes b_6 to b_9, then the size (in bytes) is given by $b_9 + 128 \times (b_8 + 128 \times (b_7 + 128 \times b_6))$. The MP3 data follow directly after the ID3 tag. Further information about ID3 tags can be found online (see `http://id3.org/`).

Table 7.11: Header of ID3 Tags

offset	length	type	description
0	3	`char`	"ID3" (0x49 0x44 0x33)
3	2	`uint2`	ID3v2 revision number
5	1	`byte`	ID3v2 flags
6	4	`byte`	tag size (7 bit format)

7.4 From File to Sheet

Most tools that can work with a digital audio format also have a driver that converts the digital audio file to vector or bitmap graphics (like PDF or PNG). Figure 7.2 is a typical example, where digital sheet music in abc format was converted to vector graphics by the tool abc2ps.

Nevertheless, there are software systems that are exclusively devoted to typeset digital sheet music. Here we briefly discuss MusicTeX, which is an extension of the widely known typesetting system LaTeX.

7.4.1 MusicTeX Typesetting

MusicTeX is a macro package for the TeX or LaTeX typesetting system and intended to typeset polyphonic music. After processing with a LaTeX compiler, the output can be stored in eps or pdf format for display on a screen or printing.

The command and macro names of this package are defined by a mixture of French and English words, whose semantics are explained in detail in the package documentation [14]. Here, only the basics of typesetting music with MusicTeX are described.

The preamble of the LaTeX document must contain \usepackage{musictex} and every piece of music within the body of the document must be bracketed by a \begin{music} ... \end{music} pair.

In Example 7.6 the number of instruments is set to one using the MusicTeX macro nbinstruments. The instrument only needs a single system of staff (*fr. portée*) specified by the macro nbporteesi. The suffix i stands for the roman number 1 and indicates that this macro is associated with instrument number 1. If there were a second instrument, the macro name would be nbporteesii with suffix ii and so forth. The default clef is the violin clef and needs not be stated explicitly. Likewise, the default signature is without any flats or sharps. Otherwise the command generalsignature is used to indicate the number of flats (< 0) or sharps (> 0). The meter is chosen as four-four time and indicated by the command generalmeter. Notes are specified by their length, pitch and direction of stem. Table 7.12 provides an overview. Pitches are determined by letters a to z if they are written under the G clef, lower pitches under the F clef are denoted by letters A to N. Bars are indicated by barre. A sequence of notes is bracketed by the commands notes and enotes.

Table 7.12: Specifying Notes in MusicTeX

command	note type
\wh p	whole note at pitch p
\hu p	half note at pitch p, stem up
\hl p	half note at pitch p, stem down
\qu p	quarter note at pitch p, stem up
\ql p	quarter note at pitch p, stem down
\cu p	eighth note at pitch p, stem up
\cl p	eighth note at pitch p, stem down

An alternative to MusicTeX is the package MusiXTeX that delivers a more aesthetic rendering. Based on these packages, the typesetting system lilyPond convinces with a simpler command language.

Example 7.6 (Smoke on the Water).
This example presents how the few notes in Figure 7.2 can be specified with MusicTEX.

```
\begin{music}
\def\nbinstruments{1}              % single instrument
\def\nbporteesi{1}                 % instrument has single staff
\generalmeter{\meterfrac{4}{4}}    % four-four time
\generalsignature{-2}              % key signature: 2 flats
\debutmorceau                      % start
\normal                            % normal spacing
\notes\Uptext{\metron{\qu}{120}}\enotes
\notes \qu g \ql i \qlp j \cu g \enotes%
\barre
\notes \ds \cl i \ds \qsk \cl {_k} \hl j \enotes%
% \qsk skips a virtual quarter note to the right
\barre
\notes \qu g \ql i \qlp j \cl i \enotes%
\barre
\notes \ds \itenl 0g \cu g  \qsk \tten 0 \hup g  \enotes%
% \itenl \tten initiate and terminate a tie
\hfil\finmorceau
\end{music}
```

7.4.2 Transcription Tools

If the source is digital audio format, the data has to be converted to digital sheet music before the tools for generating written sheet music can be deployed. Some software systems that claim to convert *polyphonic* digital audio to MIDI (and possibly other formats) and details how to accomplish this task can be found in Chapter 17.

7.5 From File to Signal

If digital sheet music should be made to sounding music, it must be converted to digital audio data first. Typically, there is a system-immanent sound device that assigns a specific sound from a sound library to each note. In case of MIDI, specific instruments may be assigned to each channel and there are even extensions to the MIDI specification for including a sound library to the MIDI file.

If digital audio files are given, the conversion path to analog audio is simply inverse to the path described in Section 7.3. In case of MP3, the data reduction by the psychoacoustic model cannot be reversed so that only the Huffman compression is undone.

7.6 Further Reading

A considerable collection of information about musical notations and tools is provided through the web site `http://www.music-notation.info/`. Less common musical file formats are presented in [12]. The deepest information about the file formats can be extracted from their formal specifications (see e.g. `https://www.midi.org/` for MIDI or the standardization documents ISO/IEC 11172-3 and ISO/IEC 13818-3 for MP3) which should be consulted before implementing own converters.

Bibliography

[1] D. Bainbridge and T. Bell. The challenge of optical music recognition. *Computers and the Humanities*, 35(2):95–121, 2001.

[2] A. H. Bullen. Bringing sheet music to life: My experiences with OMR. *The code{4}lib Journal*, Issue 3, 2008-06-23, 2008.

[3] R. B. Dannenberg. Music representation issues, techniques, and systems. *Computer Music Journal*, 17(3):20–30, 1993.

[4] M. D. Good. MusicXML: The first decade. In J. Steyn, ed., *Structuring Music through Markup Language: Designs and Architectures*, pp. 187–192. IGI Global, Hershey (PA), 2013.

[5] S. Hacker. *MP3: The Definite Guide*. O'Reilly, Sebastopol (CA), 2000.

[6] D. M. Huber. MIDI. In G. Ballou, ed., *Handbook of Sound Engineers*, pp. 1099–1130. Focal Press, Burlington (MA), 4th edition, 2008.

[7] N. S. Jayant and P. Noll. *Digital Coding of Waveforms*. Prentice Hall, Englewood Cliffs (NJ), 1984.

[8] Melodyne.   `http://www.celemony.com/en/melodyne/what-is-melodyne`, accessed 10-Mar-2016.

[9] J. Ratcliff. *Timecode: A User's Guide*. Focal Press, Burlington (MA), 3rd edition, 1999.

[10] A. Rebelo, I. Fujinaga, F. Paszkiewicz, A. R. S. Marcal, C. Guedes, and J. S. Cardoso. Optical music recognition: State-of-the-art and open issues. *International Journal on Multimedia Information Retrieval*, 1(2):173–190, 2012.

[11] T. Sauer. *Notations 21*. Mark Batty Publisher, New York, 2009.

[12] E. Selfridge-Field, ed. *Beyond MIDI: The Handbook of Musical Codes*. MIT Press, Cambridge (MA), 1997.

[13] A. Skonnard and M. Gudgin, eds. *Essential XML*. Pearson Education, Indianapolis (IN), 2002.

[14] D. Taupin. MusicTEX: Using TEX to write polyphonic or instrumental music (version 5.17). `https://www.ctan.org/pkg/musictex`, 2010, accessed 26-May-2015.

[15] Widisoft. `http://www.widisoft.com/`, accessed 10-Mar-2016.

Chapter 8

Music Data: Beyond the Signal Level

DIETMAR JANNACH, IGOR VATOLKIN
Department of Computer Science, TU Dortmund, Germany

GEOFFRAY BONNIN
LORIA, Université de Lorraine, Nancy, France

8.1 Introduction

In Chapter 5 we discussed a number of features that can be extracted from the audio signal, including rhythmic, timbre, or harmonic characteristics. These features can be used for a variety of applications of Music Information Retrieval (MIR), including automatic genre classification, instrument and harmony recognition, or music recommendation.

Beside these signal-level features, however, a number of other sources of information exist that explicitly or indirectly describe musical characteristics or metadata of a given track. In recent years, for example, more and more information can be obtained from Social Web sites, on which users can, for instance, tag musical tracks with genre or mood-related descriptions. At the same time, various music databases exist which can be accessed online and which contain metadata for millions of songs. Finally, some approaches exist to derive "high-level", interpretable musical features from the low-level signal to be able to build more intuitive and better usable MIR applications.

This chapter gives an overview of the various types of additional information sources that can be used for the development of MIR applications. Section 8.2 presents a general approach to predict meaningful semantic features from audio signal. Section 8.3 deals with features that can be obtained from digital symbolic representations of music and Section 8.4 provides a short introduction to the analysis of music scores. In Section 8.5, methods to extract music-related data from the Social Web are discussed. The properties of typical music databases are outlined in Section 8.6. Finally, Section 8.7 introduces lyrics as another possible information source to determine musical features for MIR applications.

8.2   From the Signal Level to Semantic Features

The automated classification of music and the organization of digital music collections are typically done for human listeners. It therefore seems to be helpful for users if they, for example, can understand why a set of tracks belongs to the same class. In general, to make the outcomes of an automated process better interpretable by end users, one possible goal is to derive "high-level" music data descriptors from signal-level features. Furthermore, the analysis of such interpretable features may in turn be helpful for the automated recommendation of new music, the identification of properties of certain music styles or artists, and even the automatic composition of pieces adapted to the style of a particular composer or a personal music taste, as will be discussed in Chapter 24.

8.2.1   Types of Semantic Features

The *semantic descriptors* we are interested in are typically related to music theory. Table 8.1 shows five groups of such descriptors together with examples of concrete semantic features and the related low-level signal characteristics which are often used for the estimation of the corresponding semantic descriptors. For example, features that describe inharmonic properties of semitones such as tristimulus and inharmonicity may characterize the noisiness of onsets and be helpful to recognize instruments. The chroma vector, as another example, is a basis for the estimation of key and mode (see Chapter 19 for details).

Table 8.1: Groups of Semantic Features with Examples

Group	Examples	References	Related low-level features
Instrument and vocal characteristics, playing styles, digital effects	Occurrence and share of strings in a given frame, vocal roughness	Chapter 18	Tristimulus, inharmonicity, Section 5.13
Harmony	Key and mode, chords	Chapters 3, 19	Chroma and extended variants, Section 5.3.1
Melody	Rising or falling melody, share of minor and major thirds in a melodic line, number of melodic transpositions	Chapter 3	Chroma and extended variants, Section 5.3.1
Tempo, rhythm, and dynamics	Number of beats per minute, number of bars in four-four meter, number of triplets, variance of loudness	Chapters 3, 20	Rhythmic features, Section 5.4
Emotional and contextual impact on a listener	Levels of arousal and valence; emotions fear, anger, joy, sadness; moods earnest, energetic, sentimental	Chapter 21	Root mean square energy, Equation (2.48); fluctuation patterns, Section 5.4.3

The boundaries between low-level and semantic features are often blurred. Consider, for example, the chroma as described in Equation (5.18). The idea to map all related frequencies to a semitone bin is very close to the signal level, but the progress of a chroma component with the largest value over time may describe the melody line, which we would consider as a semantic feature based on music theory.

Generally, there exists no agreement on which features should be described as "high-level". In [5], the features are categorized by their "closeness" to a listener. Signal-level features therefore include descriptors like timbre, energy, or pitch. In contrast, rhythm, dynamics, and harmony are musical characteristics considered to be more meaningful and closer to a user. The highest-level features according to [5] are referred to as "human knowledge" and relate to the personal music perception (emotions, opinions, personal identity, etc.). These features are particularly hard to assess. In yet another categorization scheme, [39] describes seven "aspects" of musical expression: temporal, melodic, orchestrational, tonality and texture, dynamic, acoustical, electromusical and mechanical. Rötter et al. [35] finally list 61 high-level binary descriptors suitable for the prediction of personal music categories.

8.2.2 Deriving Semantic Features

Many semantic features can be directly estimated from the digital score as will be discussed in the next section. However, the score might not always be available. For audio recordings, supervised classification methods – including those introduced in Chapter 12 – can be applied to derive semantic features. To train the classification models, ground truth labels are required for the classification instances (typically frames of time signal) which are represented by features. For example, the labels may indicate the occurrence of a particular instrument or a mood in the frame. This information can be provided by music experts or collected from web databases like The Echo Nest or AllMusicGuide; see Section 8.6.

Supervised classification can be applied in an incremental manner, where already calculated characteristics are used to predict the next ones. This approach is similar to classification chains proposed in [31], where the result of a classification model becomes itself a feature for the prediction of additional classes. An individual model in such an approach would predict, e.g., a mood or the occurrence of a particular instrument.

A general procedure called *Sliding Feature Selection* (SFS) [42] is sketched in Figure 8.1. Here, in each step, classification models are built (preferably with an ensemble of classifiers), and for each model only the most relevant features are kept after multi-objective feature selection, which minimizes the number of features and the classification error simultaneously.[1] The number of features on a level i is given by N_i. Note that on each level the new features do not replace the previous ones but extend the pool of available descriptors. For a better interpretability of the final models, it can then be reasonable to remove the low-level signal features in the last training step.

Not all possible sequences of extraction steps are, however, meaningful. For example, we may expect that temporal and rhythmic properties do not necessarily

---

[1]Multi-objective optimization is introduced in Section 10.4, multi-objective feature selection in Section 15.7, and classification methods in Chapter 12.

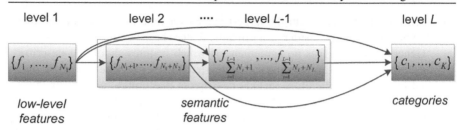

Figure 8.1: Sliding Feature Selection.

improve or simplify the recognition of instruments. Example 8.1 shows several possible sequences of SFS. The final step in each sequence obviously comprises the prediction of the aspect that we are actually interested in.

Example 8.1 (Levels of Sliding Feature Selection).

- *low-level features* ↦ *instruments* ↦ *moods* ↦ *genres*
- *low-level features* ↦ *instrument groups (keys, strings, wind)* ↦ *individual instruments* ↦ *styles* ↦ *genres* ↦ *personal preferences*
- *low-level features* ↦ *harmonic properties* ↦ *moods* ↦ *styles*
- *low-level features* ↦ *harmonic properties* ↦ *rhythmic patterns* ↦ *moods* ↦ *styles* ↦ *genres*

The individual levels of the SFS chain (cf. Figure 8.1) can be combined with *feature construction* techniques; see Section 14.5. In the described generic approach, new features can be constructed through the application of mathematical operators (like sum, product, or logarithm) on their input(s). As an example, consider a chroma vector which should help us identify harmonic properties. In the first step, new characteristics can be constructed by summing up the strengths of the chroma amplitudes for each pair of chroma semitones. The "joint strength of C and G" – as a sum of the amplitudes of C and G – would, for example, measure the strength of the consonant fifths C-G and fourths G-C. The overall number of these new features based on 12 semitone strengths is equal to $\frac{1}{2} \cdot 12 \cdot 11 = 66$. In the next step, the "strength of sad mood" could be predicted after applying the SFS chain using a supervised classification model which is trained on the subset of these 66 descriptors that leads to the best accuracy.

8.2.3 Discussion

The extraction of robust semantic characteristics from audio signals of several music sources is generally challenging, both in cases where various signal transforms (see Chapters 16-21) are applied or when supervised classification is used as introduced above. Nevertheless, the selection of the most relevant semantic features can be helpful to support a further analysis by music scientists or help music listeners to understand the relevant properties of their favorite music.

The usage of higher-level descriptors derived with the help of SFS may not necessarily improve the classification quality when compared to approaches that only use low-level features, as both methods start with the same signal data. Another challenge is the proper selection of training data in each classification step. For example, too many training tracks from the same album typically lead to a so-called *album effect*, where the characteristics of albums are learned instead of genres. In the context of the genre recognition problem, however, SFS was proven to be sufficiently robust in [44] and the experiments also showed that models that were trained on a subset of the features performed significantly better than models that relied on all available features.

8.3 Symbolic Features

In Chapter 7, the MIDI format was presented as one way of digitally and symbolically encoding music in a structured "how to play" form. In the MIR literature, several approaches exist that try to derive or reconstruct musical features from the MIDI encoding and use them in different applications. Note that *symbolic features* can be extracted from various digital formats. In this section, we, however, restrict our examples to MIDI files because of their popularity.

In [22], for example, the goal was to automatically determine the musical style for a given MIDI file, since style information is commonly used to classify and retrieve music. In their multi-step approach, the authors first propose a method to extract or approximate the main melody, which is not always trivial because there can be multiple channels, i.e., multiple notes sound at the same time. In the second step, chords are assigned to the melody based on music theoretical considerations. Finally, the resulting melody and chord patterns are matched with a set of classification rules that were learned using a larger set of training data.

Music classification based on melody lines was also the goal of the work proposed in [8] where hidden Markov models were trained on a set of folk songs from different countries. In contrast to [22], only monophonic melodies were considered.[2] In [12], the authors experimented with various machine learning approaches for musical style recognition based on MIDI files. Finally, in [24] the authors analyzed MIDI-encoded musical pieces with respect to several parameters including pitch, pitch distance, duration or melodic intervals or melodic bigrams and trained artificial neural networks for tasks such as author attribution or style identification.

Unlike the previous works, Cataltepe et al. in [4] first transform the MIDI files into audio and then combine the extracted audio features with the MIDI features for genre classification. To use the MIDI features for classification, they are first automatically extracted and then transformed into a string representation, based on which the similarity of two musical pieces can be determined [9].

Generally, a large number of musical features can be extracted from MIDI files. In [25], for example, 109 different features were determined and used for a genre

---

[2]The data files used in the experiments used two special symbolic formats. Using MIDI-encoded files would be possible in principle.

classification task based on a combination of neural networks and a k-nearest neighbors method (cf. Section 12.4.2). Their feature set covered aspects like instrumentation, musical texture, rhythm, melody, chords and others.

In [26], the jSymbolic software library was presented that can extract 160 different "high-level" features[3] from MIDI files. The more recent *music21* toolkit [11] is even capable of determining more than 200 features, supports various input formats, and is thereby able to process features that cannot be captured in the MIDI format, for example, enharmonic tones.

Given such a large set of features, the problem can arise, however, that some classification techniques do not work very well anymore ("curse of dimensionality") as the number of required labeled cases increases strongly. Possible ways of mitigating this problem suggested by the authors of [26] include the manual or automated selection of features (see Chapter 15) based on the application domain or the construction of intermediate representations such as histograms from which further features can be derived (see e.g., [41]).

Example 8.2 (Extraction of Symbolic Features for Classical, Pop, and Rock Pieces). *A set of 12 features from jSymbolic is provided in Table 8.2. The features were extracted for two classical pieces (Cla1: Bach, Toccata and Fuga in D minor, BWV 565; Cla2: Beethoven, Sonata in C sharp minor 'Moonlight', Op. 27 No. 2), two pop pieces (Pop1: Abba, Thank You for the Music; Pop2: Madonna, Hung Up), and two rock pieces (Roc1: Nightwish, Stargazers; Roc2: Scorpions, Wind of Change).*

We can observe that some of the features may help to identify a genre. Both classical pieces are characterized by a higher level of chromatic motion, rather rising melodic intervals, a higher fraction of tritones, and a higher variability of note duration. Pop tracks have a high fraction of octaves and rock tracks a positive fraction of electric guitar. Both pop and rock pieces have a larger amount of arpeggiation. Other features like rhythmic properties or the importance of the bass register seem to be less relevant. Note that in this example the number of tracks is very low. A reliable analysis of genre properties should be done with a significantly larger number of MIDIs.

By some authors, using features extracted from MIDI files is considered easy when compared to situations when only the audio signal is available [45]. For instance, some interpretable music characteristics like instruments or harmonic and melodic properties can be directly extracted from the score. This may be very hard for polyphonic audio recordings.

On the other side, symbolic formats also have their limitations. For new or less popular music pieces, the score may be not available, and it is harder to extract style properties of a concrete performer. Detecting higher-level musical structures or musical aesthetics as discussed in [23] and [24] can be challenging. The MIDI format is also not suited to express nuances of musical scores as mentioned in [11] such as the detection of enharmonic tones or the difference between an eighth note and a

---

[3]Those are features that considered to be "musical abstractions that are meaningful to musically trained individuals."

Table 8.2: Examples of Features from jSymbolic, Alphabetically Sorted

Feature	Cla1	Cla2	Pop1	Pop2	Roc1	Roc2
Amount of arpeggiation (fraction of related horizontal intervals)	0.467	0.545	0.484	0.655	0.728	0.577
Chromatic motion (fraction of melodic intervals corresponding to a semitone)	0.109	0.106	0.078	0.024	0.062	0.020
Combined strength of the two strongest rhythmic pulses	0.028	0.329	0.484	0.262	0.199	0.189
Direction of motion (fraction of melodic intervals that are rising rather than falling)	0.470	0.529	0.353	0.431	0.332	0.509
Electric guitar (fraction)	0	0	0	0	0.239	0.191
Importance of bass register (fraction of notes between MIDI pitches 0 and 54)	0.175	0.329	0.095	0.489	0.664	0.236
Melodic octaves (fraction)	0.088	0.056	0.074	0.150	0.048	0.068
Melodic tritones (fraction)	0.031	0.023	0.024	0.000	0.006	0.003
Pitch variety (number of pitches used at least once)	57	60	62	48	52	49
Repeated notes (fraction)	0.039	0.192	0.079	0.364	0.576	0.129
Rhythmic variability (standard deviation of bin values)	0.019	0.026	0.032	0.021	0.022	0.015
Variability of note duration (standard deviation, in s)	0.855	0.752	0.694	0.470	0.332	0.734

staccato quarter. Therefore, the authors of [11], for example, propose to combine MIDI features with other features, including lyrics, popularity information, or chord annotations that can be obtained from different sources.

8.4 Music Scores

In Section 8.3 we have seen that a form of a "digital score" like MIDI allows us to do various types of automated analysis like melody extraction, which in turn help us build more elaborate solutions, e.g., for music classification. There might, however, be situations where only the (printed) music score is available instead of a digital symbolic representation of the music. In order to exploit the information from the score in a music data analysis scenario, it is therefore necessary to visually analyze the music sheet, recognize the various symbols, and store them in a machine-processable form like MIDI.

The automated recognition of printed sheet music has been investigated by researchers for decades. Some early works in *optical music recognition* (OMR)[4] date back to the late 1960s as discussed, for example, in the survey of Carter et al. from 1988 [3]. At first glance, the problem appears to be a comparably simple document analysis problem, because the set of symbols are defined, there are staff lines, and there are some quite strict rules that can be used to validate and correct the hypotheses that are developed during the recognition process [34]. In practice, however, OMR is considered to be challenging because, for example, the individual symbols

---

[4]Other terms are *optical score reading* or *music image analysis*.

Figure 8.2: Fragment of a printed and scanned score.

can be highly interconnected (see Figure 8.2) and that they can vary in shape and size even within the same score [33].

An OMR process usually consists of several phases [32]. First, image processing is done, which involves techniques such as image enhancement, binarization, noise removal or blurring. The second step is symbol recognition, which typically includes tasks like staff line detection and removal, segmentation of primitive symbols and symbol recognition, where the last step is often done with the help of machine learning classifiers which are trained on labeled examples. In the following steps, the identified primitive symbols are combined to build the more complex musical symbols. At that stage, graphical and syntactical rules can be applied to validate the plausibility of the recognition process and correct possible errors. In the final phase, the musical meaning is analyzed and the symbolic output is produced, e.g. in terms of a MIDI file.

Over the years, a variety of techniques have been proposed to address the challenges in the individual phases, but a number of limitations remain in particular with respect to hand-written scores. At the same time, from a research and methodological perspective, better means are required to be able to compare and benchmark different OMR systems [32].

From a practical perspective, today a number of commercial OMR tools exist including both commercial ones like SmartScore[5] and open-source solutions like Audiveris.[6] According to [32], these tools produce good results for printed sheets, but have limitations when it comes to hand-written scores.

8.5   Social Web

During the last decade, *Social Web* platforms have become popular and nowadays link millions of users. Several of these social platforms support a number of social interactions about music which can be used for music data analysis tasks. These interactions, for example, include the collaborative annotation of music through *tags*, sharing of hand-crafted *playlists*, or the recording, publication and discussion of the users' music *listening activities*.

---

[5]http://www.musitek.com. Accessed 03 January 2016
[6]https://audiveris.kenai.com. Accessed 03 January 2016

8.5.1 Social Tags

One common feature provided by music websites like Last.fm is to let users assign tags to musical resources. Usually, such tags are freely chosen by the users and can be, for instance, the genre of an artist, the mood of a track, the year of release of an album, etc.

As these music websites are visited by millions of users, the number of tags available on these sites can be much higher than the amount of music annotations that could be done by music experts. Moreover, as these tags are assigned in a collaborative way, the subjectivity of each individual annotation can at least partially result in "inter-subjective" annotations.

However, as tags are freely chosen by non-expert users, they usually contain a lot of noise. For instance, tracks can be tagged with advertisements for other online services, or are misused by the users as a bookmark tool if the website allows to search music by tags. This noise can be ignored if a sufficient number of different users have tagged a given track. Unfortunately, tags tend to be concentrated on the most popular tracks [6]. This makes it difficult to use tags as an additional source of information for less popular or new music [7]. For more information about how social tags can be collected and used see [18] and [40].

8.5.2 Shared Playlists

Another particular source of knowledge about music are *playlists* that are created and shared by the users of music platforms. Websites and platforms allowing users to create and share playlists include Last.fm, 8tracks,[7] Art of the Mix,[8] and Spotify.[9]

One interesting piece of information contained in such playlists are relationships between tracks which were made by the playlist creators but cannot be captured solely from metadata or the audio signal. For instance, if two tracks are found one after the other in several playlists, then it can be deduced that both these tracks share something important, even if their content and metadata are completely different. These relationships can, of course, also correspond to the content and metadata, which can also be interesting, particularly when the content or metadata are not known. For instance, playlists often group tracks of the same genre, and this information can be used to infer the genres of the tracks.

The extraction of relationships between tracks from playlists is often based on the co-occurrences of the tracks (or artists). This, however, means that to be reasonably confident about the relationship between two tracks (or artists), they must appear together in a sufficient number of playlists. Therefore, this strategy allows us to capture only limited information for the less popular tracks, in particular when compared to what can be obtained for the same tracks based on their content or metadata.

In the following, we present an approach from [43] to derive genre information that works reasonably well even when the analyzed tracks occur seldom in a large

---

[7]http://www.8tracks.com. Accessed 03 January 2016
[8]http://www.artofthemix.org. Accessed 03 January 2016
[9]https://www.spotify.com. Accessed 03 January 2016

set of playlists, see Example 8.3. To measure the "degree of co-occurrence" of two artists, the concepts of *support* and *confidence* from the field of association rule mining can be used.

Vatolkin et al. in [43] define the normalized support $Supp(a_i, a_j)$ of two artists a_i and a_j in a collection of playlists P as the number of playlists in which both a_i and a_j appeared divided by the number of playlists $|P|$. The confidence value $Conf(a_i, a_j)$ relates the support to the frequency of an artist a_i, which helps to reduce the overemphasis on popular artists that comes with the support metric.

Let us now assume that our problem setting is a binary classification task with the goal to predict the genre of an unknown artist (or, analogously, the genre of a track where we know the artist). We assume that each artist is related to one predominant genre. Our training data can therefore be seen to contain T_p annotated "positive" examples of artists for each genre $(ap_1, ..., ap_{T_p})$ and T_n artists who do not belong to a given genre ("negative" artists $an_1, ..., an_{T_n}$). To learn the classification model we now look at our playlists and determine for each "positive" artist ap_i those artists that appeared most often together with ap_i. Similarly, we look for co-occurrences for the negative examples for a given genre.

When we are now given a track of some artist a_x to classify, we can determine with which other artists a_x co-occurred in the playlists. Obviously, the higher the co-occurrence of a_x with artists that co-occurred also with some ap_i, the higher the probability that a_x has the same predominant genre as ap_i. Otherwise, if a_x often co-occurs with artists that do not belong to the genre in question, we see this as an indication that a_x does not have this predominant genre either. Technically, the co-occurrence statistics are collected in the training phase and used as features to learn a supervised classification model (see Chapter 12). In [43], experiments with different classification techniques were conducted and the result showed that the approach based on playlist statistics outperformed an approach based on audio features for 10 out of 14 tested genres. The results also showed that using *confidence* is favorable in estimating the strength of a co-occurrence pattern in most cases.

Example 8.3 (Extraction of Artist Co-Occurrences in Playlists). Table 8.3 shows those five artists (provided in the table header) that most frequently co-occur with four "positive" artists for the genres Classical, Jazz, Heavy Metal, and Progressive Rock based on Last.fm playlist data.

Even if the top co-occurring artists are very popular, this method can be helpful to classify less popular artists. For example, after the comparison of support values for Soulfly, *the most probable assignment would be the genre Heavy Metal given the statistics of the data set used in [43]:* $Supp(Soulfly, Beethoven) = 5.841E\text{-}5$, $Supp(Soulfly, Miles Davis) = 2.767E\text{-}5$, $Supp(Soulfly, Metallica) = 516.539E\text{-}5$, $Supp(Soulfly, Pink Floyd) = 66.810E\text{-}5$.

One underlying assumption of the approach is that playlists are generally homogeneous in terms of their genre. Also, this method does not take into account that artists can be related to different genres over their career. Finally, a practical challenge when using public playlists is that artists are often spelled differently or even wrongly, consider, e.g., "Ludwig van Beethoven", "Beethoven", "Beethoven,

Table 8.3: Top 5 Co-Occurrences for Artists with the Predominant Genres Classical, Jazz, Heavy Metal, and Progressive Rock

Chopin, Frederic	Baker, Chet	AC/DC	The Alan Parsons Project
Beethoven, Ludwig van	Davis, Miles	Metallica	Pink Floyd
Bach, Johann Sebastian	Simone, Nina	Iron Maiden	Genesis
Mozart, Wolfgang Amadeus	Holiday, Billie	Guns'n'Roses	Queen
Radiohead	Coltrane, John	Led Zeppelin	Dire Straits
Tchaikovsky, Pyotr	Fitzgerald, Ella	The Beatles	Supertramp

Ludwig van", "L.v.Beethoven", etc. A string distance measure can help to identify identical artists, when the distance is below some threshold. In [43], for example, the Smith–Waterman algorithm [36] is applied to compare artist names.

Generally, shared playlists can be used for music-related tasks other than genre classification. They can, for instance, provide a basis for automated playlist generation and next-track music recommendation; see for example [1], [16] and Chapter 23.

8.5.3 Listening Activity

Some of the today's Web music platforms record the details of the tracks that are played by their users. For instance, the users of Last.fm can let the system record the artist name, title, and timestamp of each track they played, which is referred to as "scrobbling".

The resulting data is called the listening logs and can be exploited to derive various types of information. One possible type of useful information, e.g. for music recommendation, is the popularity of the tracks (or artists), which can be calculated simply by counting the overall number of occurrences of the tracks (or artists) in the logs. Similarly, the currency of a track can be computed by also exploiting the timestamps.

Another example of information contained in the logs is the listening duration for each track. This information can be used to determine whether a track was played entirely or "skipped" [2]. Again, in a playlist generation scenario, these skips can represent a negative signal regarding the compatibility of two tracks and an automated playlisting application can try to avoid such patterns. Note that the hand-crafted playlists discussed in the previous section do not contain such negative information.

However, as the only available information is some track identifier (e.g. the artist and track names) and a timestamp, it is impossible to be sure that a track was fully played because the user enjoyed it or if it was played because the user was busy and could not click on the "skip" button. It is also impossible to know that a track was skipped because the user thought it did not fit to the previous track or if it was skipped because the user simply wanted a change of atmosphere. For more information about how the listening activity of users can be collected and analyzed; see [2] and [29].

8.6 Music Databases

Over the last years, a number of free and commercial music databases have become available on the Web. These databases, which can usually be accessed programmatically via standardized Web interfaces, contain a variety of information that can be used in music-related applications like music recommendation, playlist generation or the structuring of music collections. The existing music databases can be categorized along different dimensions.

- Creation and maintenance: Some music databases like *MusicBrainz* are created and curated by music enthusiasts, others like *Gracenote* are maintained by commercial service providers and major music labels.[10]

- Genre scope: Some databases are devoted to very specific musical genres like Heavy Metal[11] or Latin music [15]. Others cover a broad spectrum and provide information about millions of musical tracks.

- Content: Most databases focus on artist and basic track metadata like duration, release date, chart positions, as well as community-provided information like tags. A few databases like *The Echo Nest*[12] contain information about musical features like the (average) tempo, energy or loudness of the individual tracks. Some databases like *AllMusic*[13] also provide mood annotations and community ratings; see also Table 21.6 in Chapter 21.

Today's music databases can be huge. *Gracenote*, for example, as of 2014 claims to have information about more than 180 million different tracks in their database. On *The Echo Nest*, details for over 35 million tracks can be accessed and for many of them, detailed musical features are available. Finally, even the community-curated *MusicBrainz* website and database hosts information of about 13 million tracks.

From the MIR perspective, the database by *The Echo Nest* is probably the most interesting one as it contains – beside the track and artist metadata mentioned above – detailed information about features to which one would otherwise only have access after a computationally intensive extraction phase. The features extracted from the audio signal include, for example, the duration, begin and end of fade-in and fade-out parts, the mode, loudness, segment information, MFCCs, or the tempo. Most of these feature values are accompanied by a confidence value. From the audio-based signals, a number of additional features are derived using an internal logic including "danceability", energy, or "acousticness".

Finally, many of the mentioned music databases and Web platforms provide a number of additional functionalities that can be used when developing music applications. Typical features include the automated generation of playlists from seed songs, the calculation of tracks, artists, or genres that are similar to a currently played one, automatic recognition of tracks based on sound samples, or a service for correcting artist misspellings.

---

[10]http://www.musicbrainz.org, http://www.gracenote.com. Accessed 03 January 2016
[11]http://www.metal-archives.com. Accessed 03 January 2016
[12]http://the.echonest.com. Accessed 03 January 2016, acquired by Spotify in 2014
[13]http://allmusic.com. Accessed 03 January 2016

Apart from these public music databases and services, the database used by *Pandora*,[14] the probably most popular Internet radio station in the United States at the moment, is worth mentioning. The Internet radio is based on the data created in the *Music Genome Project*. In contrast to databases which derive features from audio signals, each musical track in the *Pandora* database is annotated by hand by musical experts in up to 400 different dimensions ("genes").[15] The available genes depend on the musical style and can be very specific like "level of distortion on the electric guitar" or "gender of the lead vocalist".[16] The annotation of one track is said to last 20 to 30 minutes; correspondingly, the size of the database – approximately 400,000 tracks – is limited when compared to other platforms.

8.7 Lyrics

Many music tracks, particularly in the area of popular music, are "songs", i.e., they are compositions for voice and performed by one or more singers. Correspondingly, these tracks have accompanying *lyrics*, which in turn can be an interesting resource to be analyzed and used for music-related applications. For example, instead of trying to derive the general mood of a track based only on the key or tempo, one intuitive approach could be to additionally look at the lyrics and analyze the key terms appearing in the text with respect to their sentiment.

In the literature, a number of approaches exist that try to exploit lyric information for different MIR-related tasks. In [14], for example, the authors combine acoustic and lyric features for the problem of "hit song" prediction. Interestingly, at least in their initial approach, the lyrics-based prediction model that used Latent Semantic Analysis (LSA) [13] for topic detection was even slightly better than the acoustics-based one; the general feasibility of hit song prediction is, however, not undisputed [30].

Also the work of [21] is based on applying an LSA technique on a set of lyrics. In their work, however, the goal was to estimate artist similarity based on the lyrics. While the authors could show that their approach is better than random, the results were worse than those achieved with a similarity method that was based on acoustics, at least on the chosen dataset. Since both methods made a number of wrong classifications, a combination of both techniques is advocated by the authors.

Instead of finding similar artists, the problem of the Audio Music Similarity and Retrieval task in the annual Music Information Retrieval eXchange (MIREX) is to retrieve a set of suitable tracks, i.e., a short playlist, for a given seed song. In [20], the authors performed a user study in which the participants had to subjectively evaluate the quality of playlists generated by different algorithms. Several participants of the study stated that they themselves build playlists based on the lyrics of the tracks or liked certain playlists because of the similarity of the content of their lyrics. This indicates that lyrics can be another input that can be used for automated playlist generation. As lyrics alone are, however, not sufficient and other factors like track

---

[14]http://www.pandora.com. Accessed 03 January 2016
[15]http://www.pandora.com/about/mgp. Accessed 03 January 2016
[16]http://en.wikipedia.org/wiki/Music_Genome_Project. Accessed 03 January 2016

popularity have to be taken into account, lyrics-based features have to be combined with other inputs, e.g. in a faceted scoring approach as proposed in [16].

Experiencing music is strongly connected to emotions – as discussed in depth in Chapter 21 – and automated mood detection (classification) is a central task in Music Information Retrieval. Some works try to determine the mood of musical tracks with the help of their lyrics [10, 47]. Instead of an LSA technique, the authors of these works use Term-Frequency / Inverse Document Frequency (TF-IDF) representations of the lyrics as an input to their mood classification tasks. TF-IDF representations are commonly used for document retrieval tasks in the Information Retrieval Literature. The idea is to determine importance weights for the (subset of relevant) terms appearing in a document, resulting in TF-IDF vectors. The weights are determined by multiplying two factors. The Term-Frequency component TF assigns higher scores to terms that appear more often in a document, assuming that these words are more important. The IDF component assigns higher values to terms that appear infrequently in the whole document corpus, assuming that rarely used words are more discriminative than others.[17]

Table 8.4 shows an example of TF-IDF vectors for three Christmas-related pop songs. The term "christmas" obtains very high weights for the given song collection because the term is occurring several times in each track (TF weight) and at the same time is only rarely used in all other songs (IDF weight). Term vectors like these can then be used for different MIR-related purposes. For example, they can serve as feature vectors in a mood classification problem.

Alternatively, the angle between two vectors (cosine similarity) can be used to retrieve similar tracks for a given seed track. Other similarity measures are discussed in Section 11.2. The examples in Table 8.4 show that in the retrieval scenario a few overlapping terms like "snow" can be sufficient to retrieve tracks that have at least some similarity with a seed track. Tracks that have no word in common will be considered to be completely unrelated.[18]

Table 8.4: Example for TF-IDF Vectors

Terms/Track	christmas	feed	...	bell	everyday	snow
Do they know it's Christmas	0.863	0.379	...	0.057	0.000	0.054
I wish it could be Christmas everyday	0.736	0.000	...	0.197	0.400	0.140
Let it snow	0.000	0.000	...	0.000	0.000	0.862

TF-IDF Calculation Details [17]

The calculation of the TF-IDF vectors for a collection of text documents d typically begins with a pre-processing step. In our case, each document contains

---

[17]Mathematically, different ways to compute the weights are possible. For an example, see [10].

[18]Compared to Latent Semantic Analysis techniques mentioned above, TF-IDF-based approaches cannot uncover hidden (latent) relationships between terms.

the lyrics of one track. In this phase, irrelevant so-called "stop-words" like articles are removed. Furthermore, *stemming* can be applied, a process which replaces the terms in the document with their word stem.

We then compute a normalized term-frequency value $TF(i, j)$, which represents how often the term i appears in document j. Normalization should be applied to avoid that longer text documents lead to higher absolute term-frequency values. Different normalization schemes are possible. For instance, we can compute the normalized frequency value of a term by dividing it by the highest frequency of any other term appearing in the same document. Let $maxFrequencyOtherTerms(i, j)$ be the maximum frequency of terms other than i appearing in document j. If $freq(i, j)$ represents the unnormalized frequency count, then

$$TF(i, j) = \frac{freq(i, j)}{maxFrequencyOtherTerms(i, j)}. \tag{8.1}$$

The IDF component of the TF-IDF encoding reduces the weight of a term proportional to its appearance in documents across the entire collection. Let N be the number of documents in d and $n(i)$ be the number of documents in which term i appears. We can calculate the Inverse Document Frequency as

$$IDF(i) = log\frac{N}{n(i)} \tag{8.2}$$

and the final TF-IDF score as $TF\text{-}IDF(i,j) = TF(i, j) \cdot IDF(i)$.

The resulting term vectors can be very long and sparse as every word appearing in the documents corresponds to a dimension of the vector. Therefore, additional pruning techniques can be applied, e.g., by not considering words that appear too seldom or too often in the collection.

An approach that combines lyric and acoustic information is presented in [37]. In this work, the application scenario is to identify and retrieve musical tracks based on the user's singing voice. In contrast to previous approaches that only rely on melody identification (as done in "query by humming" approaches), the authors first try to recognize the lyrics and identify the track based on the lyrics. In a second step, melody information is extracted to verify the lyrics-based retrieval result and to thereby further increase the retrieval accuracy. A similar "query-by-singing" approach was later proposed in [28], which was, however, not combined with an acoustic retrieval method.

Finally, a few works exist that aim at the automatic transcription of lyrics from the audio signal, e.g., [28]. The problem is often considered to be challenging because of the polyphonic background music and the differences between spoken and sung voices as mentioned in [27]. One particular problem in that context is the detection of *phonemes* (a "unit of speech" in a language) as basic building blocks for the lyric transcription problem. A comparison of using different supervised classification techniques and different features sets for this task can be found in [38].

Overall, lyrics have been successfully used as an add-on information source in various MIR applications, including mood and emotion detection, see [46] or [19], song classification and identification or hit song prediction. Given the recent developments in the area of sentiment analysis and the increasing availability of lyric databases as well as "ground truth" information about moods, e.g., on the *AllMusic* platform and other music databases, further advances can be expected in the area.

8.8 Concluding Remarks

In this chapter, we reviewed a variety of different types of information and data sources that can be applied in music data analysis tasks. In particular the increasing availability of public music databases and the collective knowledge available on Social Web platforms will, in our view, open a variety of new opportunities in the future to end up, e.g., with better music recommendation and music classification techniques.

Bibliography

[1] G. Bonnin and D. Jannach. Automated generation of music playlists: Survey and experiments. *ACM Computing Surveys*, 47:1–35, 2014.

[2] K. Bosteels, E. Pampalk, and E. E. Kerre. Evaluating and analysing dynamic playlist generation heuristics using radio logs and fuzzy set theory. In *Proc. of the International Society for Music Information Retrieval Conference (ISMIR)*, pp. 351–356. International Society for Music Information Retrieval, 2009.

[3] N. Carter, R. Bacon, and T. Messenger. The acquisition, representation and reconstruction of printed music by computer: A review. *Computers and the Humanities*, 22(2):117–136, 1988.

[4] Z. Cataltepe, Y. Yaslan, and A. Sonmez. Music genre classification using midi and audio features. *EURASIP Journal of Applied Signal Processing*, 2007(1):150–150, 2007.

[5] Ò. Celma and X. Serra. FOAFing the music: Bridging the semantic gap in music recommendation. *Journal of Web Semantics: Science, Services and Agents on the World Wide Web*, 6(4):250–256, 2008.

[6] Ò. Celma. *Music Recommendation and Discovery: The Long Tail, Long Fail, and Long Play in the Digital Music Space*. Springer, 2010.

[7] Ò. Celma and P. Lamere. Music recommendation tutorial. International Society for Music Information Retrieval Conference (ISMIR), September 2007.

[8] W. Chai and B. Vercoe. Folk music classification using hidden Markov models. In *Proc. of the International Conference on Artificial Intelligence (ICAI)*, Las Vegas, 2001.

[9] R. Cilibrasi, P. Vitányi, and R. De Wolf. Algorithmic clustering of music based on string compression. *Computer Music Journal*, 28(4):49–67, 2004.

[10] H. Corona and M. P. O'Mahony. An exploration of mood classification in the million songs dataset. In *Proc. of the 12th Sound and Music Computing Conference (SMC)*. Music Technology Research Group, Department of Computer Science, Maynooth University, 2015.

[11] M. S. Cuthbert, C. Ariza, J. Cabal-Ugaz, B. Hadley, and N. Parikh. Hidden beyond MIDI's reach: Feature extraction and machine learning with rich symbolic formats in music21. In *Proc. of the NIPS 2011 Workshop on Music and Machine Learning*, 2011.

[12] R. B. Dannenberg, B. Thom, and D. Watson. A machine learning approach to musical style recognition. In *Proc. of the International Computer Music Conference (ICMC)*, pp. 344–347. Michigan Publishing, 1997.

[13] S. Deerwester, S. T. Dumais, G. W. Furnas, T. K. Landauer, and R. Harshman. Indexing by latent semantic analysis. *Journal of the American Society for Information Science*, 41(6):391–407, 1990.

[14] R. Dhanaraj and B. Logan. Automatic prediction of hit songs. In *Proc. of the International Conference on Music Information Retrieval (ISMIR)*, pp. 488–491, 2005.

[15] C. L. dos Santos and J. Silla, Carlos N. The Latin music mood database. *EURASIP Journal on Audio, Speech, and Music Processing*, 2015(1):1–11, 2015.

[16] D. Jannach, L. Lerche, and I. Kamehkhosh. Beyond "hitting the hits": Generating coherent music playlist continuations with the right tracks. In *Proc. of the 9th ACM Conference on Recommender Systems (RecSys)*, pp. 187–194, New York, 2015. ACM Press.

[17] D. Jannach, M. Zanker, A. Felfernig, and G. Friedrich. *Recommender Systems: An Introduction*. Cambridge University Press, 2011.

[18] P. Lamere. Social tagging and music information retrieval. *Journal of New Music Research*, 37(2):101–114, 2008.

[19] C. Laurier, J. Grivolla, and P. Herrera. Multimodal music mood classification using audio and lyrics. In *Proc. of the 7th International Conference on Machine Learning and Applications (ICMLA)*, pp. 688–693. IEEE Computer Society, 2008.

[20] J. H. Lee. How similar is too similar?: Exploring users' perceptions of similarity in playlist evaluation. In *Proc. of the 12th International Society for Music Information Retrieval Conference (ISMIR)*, pp. 109–114. University of Miami, 2011.

[21] B. Logan, A. Kositsky, and P. Moreno. Semantic analysis of song lyrics. In *Proc. of the IEEE International Conference on Multimedia and Expo (ICME)*, volume 2, pp. 827–830. IEEE, 2004.

[22] S. Man-Kwan and K. Fang-Fei. Music style mining and classification by melody. *IEICE Transactions on Information and Systems*, 86(3):655–659, 2003.

[23] B. Manaris, T. Purewal, and C. McCormick. Progress towards recognizing and classifying beautiful music with computers: MIDI-encoded music and the Zipf-Mandelbrot law. In *Proc. of IEEE SoutheastCon 2002*, pp. 52–57. IEEE, 2002.

[24] B. Manaris, J. Romero, P. Machado, D. Krehbiel, T. Hirzel, W. Pharr, and R. B. Davis. Zipf's law, music classification, and aesthetics. *Computer Music Journal*, 29(1):55–69, 2005.

[25] C. Mckay and I. Fujinaga. Automatic genre classification using large high-level musical feature sets. In *Proc. of the International Conference on Music Information Retrieval (ISMIR)*, pp. 525–530, 2004.

[26] C. Mckay and I. Fujinaga. jSymbolic: A feature extractor for MIDI files. In *Proc. of the International Computer Music Conference (ICMC)*, pp. 302–305. Michigan Publishing, 2006.

[27] M. McVicar, D. Ellis, and M. Goto. Leveraging repetition for improved automatic lyric transcription in popular music. In *Proc. IEEE International Conference on Acoustics, Speech and Signal Processing (ICASSP 2014)*, pp. 3117–3121. IEEE, 2014.

[28] A. Mesaros and T. Virtanen. Automatic recognition of lyrics in singing. *EURASIP J. Audio Speech Music Process: Special Issue on Atypical Speech*, 2010:4:1–4:7, January 2010. http://dx.doi.org/10.1155/2010/546047.

[29] J. L. Moore, S. Chen, D. Turnbull, and T. Joachims. Taste over time: The temporal dynamics of user preferences. In *Proc. of the International Society for Music Information Retrieval Conference (ISMIR)*, pp. 401–406, 2013.

[30] F. Pachet and P. Roy. Hit song science is not yet a science. In *Proc. of the International Conference on Music Information Retrieval (ISMIR)*, pp. 355–360, 2008.

[31] J. Read, B. Pfahringer, G. Holmes, and E. Frank. Classifier chains for multi-label classification. In *Proc. of the European Conference on Machine Learning and Knowledge Discovery in Databases (ECML PKDD), Part II*, pp. 254–269, 2009.

[32] A. Rebelo, I. Fujinaga, F. Paszkiewicz, A. R. S. Maral, C. Guedes, and J. S. Cardoso. Optical music recognition: State-of-the-art and open issues. *International Journal of Multimedia Information Retrieval*, 1(3):173–190, 2012.

[33] F. Rossant. A global method for music symbol recognition in typeset music sheets. *Pattern Recognition Letters*, 23(10):1129–1141, 2002.

[34] F. Rossant and I. Bloch. A fuzzy model for optical recognition of musical scores. *Fuzzy Sets and Systems*, 141(2):165–201, 2004.

[35] G. Rötter, I. Vatolkin, and C. Weihs. Computational prediction of high-level descriptors of music personal categories. In B. Lausen, D. van den Poel, and A. Ultsch, eds., *Algorithms from and for Nature and Life*, pp. 529–537. Springer, 2013.

[36] T. Smith and M. Waterman. Identification of common molecular subsequences.

Journal of Molecular Biology, 147:195–197, 1981.

[37] M. Suzuki, T. Hosoya, A. Ito, and S. Makino. Music information retrieval from a singing voice using lyrics and melody information. *EURASIP Journal of Applied Signal Processing*, 2007(1), 2007.

[38] G. Szepannek, M. Gruhne, B. Bischl, S. Krey, T. Harczos, F. Klefenz, C. Dittmar, and C. Weihs. *Classification as a Tool for Research*, chapter Perceptually Based Phoneme Recognition in Popular Music, pp. 751–758. Springer, 2010.

[39] P. Tagg. Analyzing popular music: Theory, method and practice. *Popular Music*, 2:37–65, 1982.

[40] D. Turnbull, L. Barrington, and G. R. Lanckriet. Five approaches to collecting tags for music. In *Proc. of the International Society for Music Information Retrieval Conference (ISMIR)*, volume 8, pp. 225–230, 2008.

[41] G. Tzanetakis and P. Cook. Musical genre classification of audio signals. *IEEE Transactions on Speech and Audio Processing*, 10(5):293–302, 2002.

[42] I. Vatolkin. *Improving Supervised Music Classification by Means of Multi-Objective Evolutionary Feature Selection*. PhD thesis, Department of Computer Science, TU Dortmund, 2013.

[43] I. Vatolkin, G. Bonnin, and D. Jannach. Comparing audio features and playlist statistics for music classification. In A. F. X. Wilhelm and H. A. Kestler, eds., *Analysis of Large and Complex Data*, pp. 437–447, 2016.

[44] I. Vatolkin, G. Rudolph, and C. Weihs. Evaluation of album effect for feature selection in music genre recognition. In *Proc. of the 16th International Society for Music Information Retrieval Conference (ISMIR)*, pp. 169–175, 2015.

[45] C. Xu, M. Maddage, X. Shao, F. Cao, and Q. Tian. Musical genre classification using support vector machines. In *Proc. of the IEEE International Conference on Acoustics, Speech, and Signal Processing (ICASSP)*, volume 5, pp. 429–432. IEEE, 2003.

[46] D. Yang and W.-S. Lee. Music emotion identification from lyrics. In *Proc. of the 11th IEEE International Symposium on Multimedia (ISM)*, pp. 624–629. IEEE Computer Society, 2009.

[47] M. Zaanen and P. Kanters. Automatic mood classification using TF*IDF based on lyrics. In *Proc. of the 11th International Society for Music Information Retrieval Conference (ISMIR)*, pp. 75–80. International Society for Music Information Retrieval, 2010.

Part II

Methods

Chapter 9

Statistical Methods

CLAUS WEIHS
Department of Statistics, TU Dortmund, Germany

9.1 Introduction

Statistical models and methods are the basis for many parts of music data analysis. In this chapter we will lay the statistical foundations for the methods described in the next chapters. An overview is given over the most important notions and theorems in statistics, needed in this book. The notion of probability is introduced as well as random variables. We will define, characterize and represent stochastic distributions in general and give examples relevant for music data analysis. We will show how to estimate unknown parameters and how to test hypotheses on the distribution of variables. Typical statistical models for the relationship between different random variables will be introduced, and the estimation of their unknown parameters and the properties of predictions from such models will be discussed. We will introduce the most important statistical models for signal analysis, namely time series models and, finally, a first impression of dimension reduction methods will be given.

9.2 Probability

9.2.1 Theory

The notion of probability is basic for statistics, but often used intuitively. In what follows, we will give an exact definition. Notice that probabilities are defined for fairly general sets of observations of variables. There is no need that these results are quantitative so that they could be used in calculations. Instead, we will define probability on subsets of a set of possible observations (sample space). These subsets have to have some formal properties (formalized as so-called σ-algebras) in order to guarantee general and consistent usability of probabilities.

Let us start with a motivating example. In this example the notion of probability is intuitively used by now.

Example 9.1 (Semitones). *Consider the set of the 12 semitones ignoring octaves:*

$\Omega = \{C, C\#, \ldots, B\}$. *In 12-tone music, the idea is that all 12 semitones are equiprobable, e.g., $P(D) = \frac{1}{12}$. In different historical periods and different keys, these probabilities might be different.*

Here, it is implicitly assumed that all relevant probabilities exist and can be easily calculated from basic probabilities. For example, one might want to calculate the probability of groups of tones by adding the individual probabilities of the tones in the group. Can this be done in general? In order to do this, we have to be able to calculate the probability of unions of sets for which the probabilities are already known. This is formalized in the following definition.

Definition 9.1 (Probability).   A *sample space* Ω is the set of all possible observations ω. The action of randomly drawing elements from a sample space is called *sampling*. The outcome of sampling is a *sample*.
    A random *event* A is a subset of Ω. $\Omega - A$, consisting of all elements of Ω which are not in A, is called the *complementary event* of A in Ω. In order to be able derive probabilities for all sets from the probabilities of elementary sets, we restrict ourselves to sets $\mathscr{A}$ of subsets of Ω (called σ-*algebra*) with $\Omega \in \mathscr{A}$, $(\Omega - A) \in \mathscr{A}$ for all $A \in \mathscr{A}$, and $\bigcup_{i=1}^{K} A_i \in \mathscr{A}$ for all K and all sets $A_1, A_2, \ldots \in \mathscr{A}$.
    Then, a *probability function P* is defined as any real-valued function on $\mathscr{A}$ with values in the interval $[0, 1]$, i.e. $P : \mathscr{A} \rightarrow \mathbb{R}$ with $A \mapsto P(A) \in [0, 1]$ iff $P(A) \geq 0 \, \forall A \in \mathscr{A}$, $P(\Omega) = 1$, and for all sets of pairwise disjoint events $A_1, A_2, \ldots (A_i \cap A_j = \emptyset, i \neq j)$ it is true that $P\left(\bigcup_{i=1}^{K} A_i\right) = \sum_{i=1}^{K} P(A_i)$ for all K. The values of a probability function are called *probabilities*.

Example 9.2 (Semitones).   Let us reconsider the situation in Example 9.1. Here, the sample space is given by the set $\Omega = \{C, C\#, \ldots, B\}$. A random sample from Ω can be any subset of Ω if the set $\mathscr{A}$ of subsets of Ω consists of all possible subsets of Ω. For 12-tone music, e.g., the probability function P is only specified by $P(C) = P(C\#) = \ldots = P(B) = \frac{1}{12}$. The probability of groups of basic events is derived from the properties of the probability function. For example, $P(\{C, B\}) = P(C) + P(B) = \frac{1}{6}$.

Let us now look at a somewhat more involved example.

Example 9.3 (Chords).   The probability of individual chords appearing in a piece of music could be used for comparison of different music styles. The pure standard 12-bar blues has the form (I, I, I, I, IV, IV, I, I, V, V, I, I), where bars are separated by commas (for chord notation cp. Section 3.5.4). Then, only 3 chords have a probability $p > 0$, namely tonic ($p_I = 8/12$), subdominant ($p_{IV} = 2/12$), and dominant ($p_V = 2/12$). Note that there are many variants of this scheme. One "standard" jazz version, e.g., which is much more sophisticated, is of the form (I7, IV7 IVdim, I7, Vm7 I7, IV7, IVdim, I7, III7 VI7, IIm7, V7, III7 VI7, II7 V7). Obviously, here, the 9 different chords I7, IV7, IVdim, Vm7, III7, VI7, IIm7, V7, II7 are involved. In this example, the sample spaces consists of all possible chords for both blues versions, but the probabilities of these chords are different for standard and the jazz blues. Even, it may be so that the same songs are played in the different schemes. Note that in the

two previous examples the progression of elements (tones or chords) is ignored for
the moment.

Let us now extend our ability to work with probabilities. Often, we are interested
in the probability of an event if another event has already happened. In the example
below we would like to know the probability of a certain tone height when it is
already known that it is in the literature of a fixed voice type. This leads to what is
called conditional probability. The following definition also contains related terms.

Definition 9.2 (Conditional Probability and Independence). Let A, B be two events
in $\mathscr{A}$. Then, the *conditional probability* of A conditional on the event B is defined
by: $P_B(A) = P(A \mid B) := P(A \cap B)/P(B)$ if $P(B) > 0$.

A and B are called *stochastically independent events* iff $P(A \cap B) = P(A)P(B)$,
i.e. the probability of the intersection of two events is the product of the probabilities
of these events. Equivalently, the conditional probability $P(A \mid B)$ does not depend
on B, i.e. $P(A \mid B) := P(A \cap B)/P(B) = P(A)$.

The K events $A_i, i = 1, 2, \ldots, K$, with $P(A_i) > 0$ build a *partition* of Ω, iff $A_i \cap A_j = \emptyset, i \neq j$, and $\bigcup_{i=1}^{K} A_i = \Omega$, i.e. iff the events are separated and cover the whole Ω.

Note that the division by $P(B)$ in the definition of conditional probability guarantees that $P(B \mid B) = 1$.

Example 9.4 (Semitones (cont.)). *Consider again Example 9.1, i.e. the set of semi-*
tones $\Omega = \{C, C\#, \ldots, B\}$. *In this case, obviously there is a partition with* $K = 12$.
Let us now distinguish different octaves. As notation we use C (65.4 Hz), c (130.8
Hz), c' (261.6 Hz), c" (523.2 Hz) etc. (Helmholtz notation, cp. Table 7.1). For
singing, obviously, the probability of a semitone in the singing voice depends on the
type of voice. For example, with very few exceptions soprano voices stay between c'
and c''' (261.6–1046 Hz). The "Queen of Night" in "The Magic Flute" of Mozart
is one exception going up to f''', 1395 Hz. Also, bass voices stay between F and
f' (87.2–348 Hz), again with some exceptions like the "Don-Cossack" choir going
down to $F_1 = 43.6$ *Hz. This means that the probability of a tone outside such ranges*
is (nearly) zero conditional on the voice type.

Formally, this can be modeled by so-called combined events $A = \{(tone_1, type_1),$
$(tone_2, type_2), \ldots, (tone_p, type_p)\}$ *over the classical singers' literature, say, where,*
e.g., $tone_i \in \{F_1, \ldots, C, D, \ldots, f'''\}$ *and* $type_i \in \{soprano, mezzo, alto, tenor, bari-*
tone, bass\}. Then, we may be interested in conditional probabilities, e.g.,
$P((tone, type) \mid type = soprano)$ *and* $P((tone, type) \mid type = bass)$. *For example, for*
$A = \{(c'', type)\}$ *and* $B = \{(tone, soprano)\}$, $P(A \mid B) = P((c'', type) \mid type = soprano)$
$= P(c'', soprano)/P(soprano)$. *Obviously, there are tones sung by different voices*
and tones sung only by one voice. For example, only tones $\in \{c', c\#', d', d\#', e',*
*f'\}$ *are sung by both, soprano and bass. Therefore, the events* $A = \{(tone, soprano)\}$
and $B = \{(tone, bass)\}$ *are independent for any fixed tone* $\notin \{c', c\#', d', d\#', e', f'\}$
since $P(A \cap B) = P(A)P(B) = 0$ *because* $A \cap B = \emptyset$ *and either* $P(A)$ *or* $P(B)$ *is zero.*

Also, the subsets $A_1 = \{(tone, soprano)$ *for all tones*$\}$, $A_2 = \{(tone, mezzo)$ *for*
all tones$\}$, *...,* $A_6 = \{(tone, bass)$ *for all tones*$\}$ *build another example for a partition.*

Note that conditional probabilities are not symmetric, i.e. $P(A|B) \neq P(B|A)$. Often, it is reasonably easy to get one of these types of probabilities, say $P(B|A)$, but one is interested in the other type $P(A|B)$. For example, consider the case that we want to calculate the conditional probability $P(A_i \mid B)$ of the type of voice A_i given a certain semitone B. However, the only probabilities we have from literature are $P(B \mid A_i)$. Such problems can be solved by means of the following fundamental properties of probabilities which can be derived from the notions of conditional probability, independence, and partition.

Theorem 9.1 (*Total Probability*). Let $A_i, i = 1, 2, \ldots, K$, be a partition of Ω with $P(A_i) > 0$. Then, for every $B \in \mathscr{A}$: $P(B) = \sum_{i=1}^{K} P(B \mid A_i)P(A_i)$.

The probability of an event B can be, thus, calculated by means of the conditional probabilities of B conditional to a partition of the sample space Ω.

Theorem 9.2 (*Bayes Theorem*). Let $A_i, i = 1, 2, \ldots, K$, be a partition of Ω with $P(A_i) > 0$. Then, for every event $B \in \mathscr{A}$ with $P(B) > 0$ it is valid that:

$$P(A_i \mid B) = \frac{P(B \mid A_i)P(A_i)}{\sum_{j=1}^{K} P(B \mid A_j)P(A_j)}.$$

The conditional probabilities of the events A_i of a partition conditional to an event B, thus, can be calculated by the conditional probabilities of B conditional to A_i and the so-called *a-priori probabilities* $P(A_i)$.

Example 9.5 (Semitones (cont.)). *Consider a special case of Example 9.4, e.g., the set of semitones $\Omega = \{C, C\#, \ldots, b'''\}$. Consider the case that we want to calculate the conditional probability $P(A_i \mid B)$ of the type of voice $A_i \in \{soprano, mezzo, alto, tenor, baritone, bass\}$, i.e. $K = 6$, given that a certain semitone, e.g., e', is sung in a song. This can be realized by means of the Bayes Theorem using conditional probabilities $P((e',type) \mid A_i)$ of the semitone e' for the different voice types taken from literature and the so-called a-priori probabilities $P(A_i)$ also taken from literature.*

You may object that the probabilities $P(e' \mid A_i)$ and $P(A_i)$ are not available in the literature. In such a case, you might want to estimate such probabilities from a library of songs and then apply the Bayes Theorem. Obviously, this would lead us away from probabilities to their empirical analogues called frequencies.

9.2.2   Empirical Analogues

Probabilities cannot be observed, but so-called frequencies, which are then taken as 'estimators' for probabilities. In order to observe frequencies, we have to conduct so-called 'experiments' in which events are counted. These counts are called *absolute frequencies*. Such experiments could be the counting of events like, e.g., the appearance of certain notes or certain chords in certain pieces of music. *Relative frequencies* are the corresponding proportions summing up to 1.

Definition 9.3 (Frequencies). The number of appearances of event A in $N > 0$ repetitions of an experiment is called the *absolute frequency* $H_N(A)$ of the event A. The *relative frequency* is defined by $h_N(A) := H_N(A)/N$.

Example 9.6 (Semitones (cont.)). *Consider again Example 9.1, i.e. the set of semitones* $\Omega = \{C, C\#, \ldots, B\}$. *In a composition, let the planned probabilities be* $P = \{1/3, 0, 0, 0, 1/3, 0, 0, 1/3, 0, 0, 0, 0\}$.

In one piece of music, however, the relative frequencies are realized as $h = \{1/3, 0, 0, 0, 1/6, 1/6, 0, 1/3, 0, 0, 0, 0\}$, *i.e. the distribution of the notes in the piece is not exactly following the specification, e.g., because of musical reasons. Over a large number of pieces, however, the relative frequencies should be near to the probabilities if the planned probabilities are realistic.*

9.3 Random Variables

9.3.1 Theory

Most often, events are first mapped on real numbers, before they are used in statistical analyses in order to be able to apply standard calculus. This is realized by so-called random variables. Probabilities work on events, random variables on real numbers. For example, the two musical modes, minor and major, might be mapped to $\{0, 1\}$ or $\{-1, 1\}$ or the 12 semitones $\{C, C\#, \ldots, B\}$ to $\{1, 2, \ldots, 12\}$. Random variables, though, should represent probabilities. Because of this, there is an explicit connection between these two concepts, the so-called measurability property. Random variables might take only integers like in the above examples or general real values. Therefore, we will introduce two kinds of random variables, namely discrete and continuous random variables. The values of random variables are ordered on the real axis so that the so-called (cumulative) *distribution* of their values represents the progression of probabilities from the lowest to the highest values of the random variable. So-called *densities* represent probabilities of points or intervals as illustrated below.

Definition 9.4 (Random Variable, Distribution, and Density). A *random variable* is a function from the sample space Ω into $\mathbb{R}$, for which a distribution function can be calculated. This is called *measurability property*.

The *distribution function* F_X of a random variable X is defined as
$F_X(x) := P(X \le x) := P(\{\omega | X(\omega) \le x\})$ for every $x \in \mathbb{R}$.
That $P\{\omega | X(\omega) \le x\})$ can be calculated defines *measurability*.

A random variable X is called *discrete* iff it can maximally take countably many values $\{x_1, x_2, x_3, \ldots\}$. If a random variable is discrete, then also the corresponding distribution function is called discrete.

A *discrete distribution function* has the form $F_X(x) = \sum_{x_i \le x} P(X = x_i)$.
The function $f_X(x) := P(X = x_i)$ if $x = x_i$, and $f_X(x) := 0$ otherwise, is called *discrete density function* or *probability function* of X. A discrete distribution function can be written as: $F_X(x) = \sum_{x_i \le x} f_X(x_i)$.

A random variable X with a range of genuine real values is called *continuous* iff the distribution function can be represented as follows by means of a so-called *density function* $f_X(x)$: $F_X(x) = \int_{-\infty}^{x} f_X(z)dz$.

Obviously, for continuous random variables X, the distribution function $F_X(x)$ is not only continuous but even differentiable since $f_X(x) = F_X'(x)$. To illustrate that densities represent the probabilities of points or intervals first realize that $f_X(x) := P(X = x_i)$ for discrete densities if x_i is a values taken by the random variable X. For continuous distributions, the probability of individual points is always zero. Thus, a density value in a certain point x in the image of X is not equal to the probability of x. However, densities represent the probability of intervals in that $P(a \leq x \leq b) = F_X(b) - F_X(a) = \int_a^b f_X(x)dx$.

Example 9.7 (Discrete Distributions). *Let us come back to the motivating examples for random variables. A discrete random variable X_{mode} would have the distribution function $F_{mode}(x) = 0$ for $x < -1$, $F_{mode}(x) = 0.5$ for $-1 \leq x < 1$, and $F_{mode}(x) = 1$ for $1 \leq x$. Obviously, this distribution function is not continuous. Also, the corresponding density takes the values $f_{mode}(-1) = f_{mode}(1) = 0.5$ and $f_{mode}(x) = 0$ otherwise. Analogous arguments are true for the distribution of the semitones.*

Examples for continuous distributions will be given later.

You might have noticed that the definition of distribution functions needs probabilities. The most important property of random variables is, though, that distribution functions and their densities can even be characterized without any recourse to probabilities. This way, random variables and their distributions in a way replace the much more abstract concept of probabilities in practice. The following properties of random variables characterize their distributions in general.

Theorem 9.3 (*Representation of Distribution and Density Functions*). Let F_X be the distribution function of a random variable X. Then,

1. $F_X(-\infty) := \lim_{x \to -\infty} F_X(x) = 0$,
2. $F_X(\infty) := \lim_{x \to \infty} F_X(x) = 1$,
3. F_X is monotonically increasing: $F_X(a) \leq F_X(b)$ for $a < b$,
4. F_X is continuous from the right: $\lim_{0 < h \to 0} F_X(x+h) = F_X(x)$.

Every function F from $\mathbb{R}$ into the interval $[0,1]$ with the above properties 1–4 defines a *distribution function*.

Every function f from $\mathbb{R}$ into the interval $[0,1]$ defines a *discrete density function*, iff for a maximally countable set $\{x_1, x_2, x_3, \ldots\}$:

1. $f(x_i) > 0$ for $i = 1, 2, 3, \ldots$,
2. $f(x) = 0$ for $x \neq x_i$, $i = 1, 2, 3, \ldots$,
3. $\sum_i f(x_i) = 1$.

Every function $f : \mathbb{R} \to [0, \infty)$ defines a *density function of a continuous distribution* iff $f(x) \geq 0 \ \forall x$ and $\int_{-\infty}^{\infty} f(x)dx = 1$.

That such properties are sufficient to characterize distributions can be most easily seen for discrete density functions because of their direct relationship to probabilities.

So, any function of the types in the above definition represents a distribution function or density and therefore implicitly a system of probabilities. Statistics concentrates, therefore, most often on functions of this type. Distributions of random variables are in the core of statistics.

In order to easily study more examples, let us first introduce the empirical analogues of random variable, distribution, and density.

9.3.2 Empirical Analogues

Random variables and their distributions are theoretical constructs needed to derive theoretical properties of methods, their analogues called *features* and *empirical distributions and densities* can be observed in reality. On the one hand, the observed features do not have ideal properties; on the other hand, they are assumed to behave similarly and this has to be checked as exemplified below.

Definition 9.5 (Features and Empirical Distributions). The empirical analogue of a random variable is called a *feature*. A random variable has a theoretical distribution, the observations of a feature have an *empirical distribution*, which is built of frequencies. An *empirical discrete density* is the set $\{h_1, \ldots, h_K\}$ of relative frequencies of the K different possible values of the feature.

In order to compare empirical with theoretical distributions, the frequencies are graphically represented by so-called bar charts or histograms and compared with the graph of the density function corresponding to the theoretical distribution.

Definition 9.6 (Bar Chart and Histogram). A *bar chart* is a diagram representing the distribution of a discrete random variable of the empirical distribution of a feature by means of vertical bars (narrow rectangles with meaningless width) not adjoining each other whose height (!) represents the frequencies of the feature values indicated on the x-axis.

A *histogram* is a graphical representation of the relative frequencies of the values of a continuous random variable X. For this, the possible values of X are divided into classes $A_1, A_2, \ldots, A_K$. The borders of the classes are plotted on the x-axis. For each class a box is drawn limited by the class borders on the x-axis. The area (!) of each box represents the relative frequency of the class. The height r_k of the box of the k-th class is $r_k = h_k/b_k$, where h_k is the relative frequency of the class and b_k the class width.

Note that only for a class width $b_k = 1$ the box height is equal to the relative frequency ($r_k = h_k$). The total area of the boxes is equal to 1, since the sum of the relative frequencies is equal to 1.

Example 9.8 (Example Distributions: Semitones). *In Example 9.6 we introduced a discrete uniform distribution on the semitones $\{C, E, G\}$, where every semitone in this set has the same probability $1/3$. The realized frequencies might not coincide with the theoretical distribution, as, e.g., in Figure 9.1, left.*

Example 9.9 (Example Distributions: MFCCs). *Let us now introduce an example*

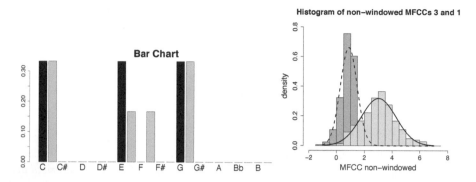

Figure 9.1: Comparison of two distributions; left: discrete case, theoretical distribution in black, realized distribution in grey; right: continuous case, non-windowed MFCC 3 grey and MFCC 1 lightgrey.

data set often used in this section. The data is composed of 13 MFCC variables (non-windowed and windowed) (see Section 5.2.3) and 14 chroma variables. All variables are available for 5654 guitar and piano tones.

Let us now briefly indicate how these variables are calculated. This paragraph is not necessary to understand the density example here, but illustrates the relationship to signal analysis (see Chapters 4, 5). Each single analyzed tone has a length of 1.2 seconds and is given as a WAVE signal (see Section 7.3.2) with sampling rate 44,100 Hz and samples x_i, $i \in \{1, \ldots, 52,920\}$. The non-windowed MFCCs are calculated over the whole tone, i.e. we have one value of each MFCC variable per tone. For the other features, the signal is framed by half overlapping windows containing 4096 samples each. This results in 25 different windows, the last window not being complete. We aggregate the windows to so-called blocks of 5 overlapping windows each. This way, one block is composed of 12,288 observations and corresponds to around 0.25 seconds. In the names of the variables the individual blocks are noted, e.g., 'MFCC 1 block 1' means the 1st MFCC calculated for block 1. As chroma features we rely on the so-called Pitchless Periodogram describing the distribution of the fundamental frequency and of 13 overtones of a tone. The periodogram is called pitchless because the value of the pitch of the tone, i.e. of its fundamental frequency, is ignored in the representation, only the periodogram heights p_i (see Section 9.8.2, Definition 9.47) of the fundamental frequency and its overtones are presented on an equidistant scale, $i \in \{0, \ldots, 13\}$. This way, the overtone structure is represented on the same scale for all fundamental frequencies (cp. [3]). The windowed MFCCs and the chroma variables are calculated for each of the 5 blocks of each tone.

Two continuous distributions can be compared in one density plot. At the right, Figure 9.1 shows two densities found for the non-windowed MFCCs 1 (light grey) and 3 (grey). The histograms are approximated by best fitting normal densities (see Section 9.4.3). Obviously, the normal densities fit the histograms quite well.

9.4 Characterization of Random Variables

9.4.1 Theory

Random variables are characterized by their distribution. Often, however, it is not appropriate to use a function as a characterization, since for most of us it is not easy to fully take into account functions as a whole. Therefore, most of the time algebraic summaries are used as characteristics of a distribution. There are at least two kinds of such characteristics, the first corresponding to so-called moments, as expected value, standard deviation, skewness, and kurtosis, and the second corresponding to certain so-called quantiles splitting the distribution into, in a sense, equally sized parts. We will introduce both kinds of characteristics starting with the moments.

Definition 9.7 (Expected Value, Standard Deviation, Skewness, and Kurtosis). Let X be a random variable with density f_X. The *expected value* μ_X or $E[X]$ of X is defined by:

- $E[X] := \mu_X := \sum_i x_i f_X(x_i)$ for discrete X and
- $E[X] := \mu_X := \int_{-\infty}^{\infty} x f_X(x) dx$ for continuous X,

iff the sum and the integral are absolutely convergent.

X is called *symmetrically distributed* around its expected value, if $f_X(\mu_X - x) = f_X(\mu_X + x) \ \forall \, x \in \mathbb{R}$.

The *variance* σ_X^2 or $\text{var}[X]$ of X is defined by:

- $\text{var}[X] = \sigma_X^2 := E[(X - \mu_X)^2] = \sum_i (x_i - \mu_X)^2 f_X(x_i)$ for discrete X and
- $\text{var}[X] = \sigma_X^2 := E[(X - \mu_X)^2] = \int_{-\infty}^{\infty} (x - \mu_X)^2 f_X(x) dx$ for continuous X.

The *standard deviation* σ_X of X is defined as $\sigma_X = \sqrt{\text{var}[X]}$.

One often used characteristic of a distribution characterizes the "location" and the "variation" of the distribution by the

$$2\text{-summaries characteristic} = (\text{expected value, standard deviation}).$$

The expected value is a 1st-order moment, the variance is 2nd order. Other important moments characterizing the asymmetry and the curvature of a distribution are 3rd- and 4th- order moments $E[(X - \mu)^3]$ and $E[(X - \mu)^4]$ leading to the *skewness* γ_{1X} of X defined by $\gamma_{1X} = \frac{E[(X-\mu)^3]}{\sigma_X^3}$ and the *(excess) kurtosis* γ_{2X} of X defined by $\gamma_{2X} = \frac{E[(X-\mu)^4]}{\sigma_X^4} - 3$.

Negative values of the skewness indicate distributions with more weight on high values (steepness at the right), positive values stand for steepness at the left. Symmetric distributions like the normal or Student's t-distribution have skewness 0 (see Definition 9.15).

The "minus 3" at the end of excess kurtosis formula is often explained as a correction to make the kurtosis of the normal distribution equal to zero (cp. Section 9.4.3). The "classical" interpretation of the kurtosis, which applies only to symmetric and unimodal distributions (those whose skewness is 0), is that kurtosis measures both the "peakedness" of the distribution and the heaviness of its tail. A distribution with positive kurtosis has a more acute peak around the mean and fatter tails. An example of such distributions is the Student's t-distribution. A distribution with negative kurtosis has a lower, wider peak around the mean and thinner tails. Examples of such distributions are the continuous or discrete uniform distributions (see Definition 9.15).

Example 9.10 (Problem with Expected Value). *Notice that for all the characteristics of a distribution it is assumed that their values make sense in the context of the application. This might well not be the case, though. For example, reconsider Example 9.6, where* $P = \{1/3, 0, 0, 0, 1/3, 0, 0, 1/3, 0, 0, 0, 0\}$. *Then, the expected value of the corresponding random variable X with values* $\{1, 2, \ldots, 12\}$ *is* $E[X] := \sum_i x_i f_X(x_i) = 1/3(1+5+8) = 14/3 = 4\,2/3$. *Obviously, this value is not interpretable in the context of the application since it corresponds to a note between D# and E. This is typically a problem of the expected value and the standard deviation. The (empirical) quantiles can be defined so that they always take realized values (see Section 9.4.2).*

The other often-used characteristic of a distribution is dividing the distribution into four, in a sense, equally sized parts.

Definition 9.8 (Quantiles, Quartiles, and Median). Let X be a random variable with distribution function F_X. The *q-quantile* ξ_q of X is defined as the smallest number $\xi \in \mathbb{R}$ with $F_X(\xi) \geq q$. The *median* med_X, $\text{med}(X)$ or $\xi_{0.5}$ of X is the 0.5-quantile. The *lower* and the *upper quartile* of X are defined as $q_4(X) := \xi_{0.25}$ and $q^4(X) := \xi_{0.75}$, correspondingly. Every value for which the density f_X takes a (local) maximum, is called a *modal value* or *mode* of X denoted by modus_X or $\text{mod}(X)$.

An often-used characteristic is the so-called

$$5\text{-summaries characteristic} = (\text{minimum}, q_4(X), \text{med}_X, q^4(X), \text{maximum}),$$

dividing the distribution into four in that sense equally sized parts that 25% of the distribution lies between each two neighboring characteristics. For example, the lowest 25% of the distribution lies between minimum and $q_4(X)$.

Example 9.11 (Uniform Distribution (see Section 9.4.3)). *Reconsider the 12-tone music case with* $\{C, C\#, \ldots, B\}$ *having all the same probability* $1/12$. *These notes are mapped to* $\{1, 2, \ldots, 12\}$ *leading to expected value* $E_X = (1 + \ldots + 12)/12 = 6.5$, *standard deviation* $\sigma_X = \sqrt{((1-6.5)^2 + \ldots + (12-6.5)^2)/12} = \sqrt{(5.5^2 + \ldots + 0.5^2)/6}$ ≈ 3.5, *median* $\text{med}_X = 6$, *and quartiles* $q_4(X) = 3, q^4(X) = 9$.

The 5-summaries characteristic might be much more illustrative than the above 2-summaries characteristic. The latter, however, is much better suited for the derivation of theoretical properties, e.g., of transformations of random variables. In particular, *standardization* is very important since standardized variables always have 'standard form' with expected value 0 and variance 1.

Theorem 9.4 (Linear Transformation and Standardization). Let X be a random variable. Then:

- $E[a + bX] = a + b\mu_X$ and $\text{var}[a + bX] = b^2 \text{var}[X]$. Therefore:

- *Centering* by μ_X leads to $E[X - \mu_X] = 0$, $\text{var}[X - \mu_X] = \text{var}[X]$,
- *Normalization* by σ_X leads to $E[X/\sigma_X] = \mu_X/\sigma_X$, $\text{var}[X/\sigma_X] = 1$, and
- *Standardization* by centering and normalization leads to
  $E[(X - \mu_X)/\sigma_X] = 0$, $\text{var}[(X - \mu_X)/\sigma_X] = 1$, i.e. the random variable
  $(X - \mu_X)/\sigma_X$ always has expected value 0 and variance 1.

Note that these rules are valid independent of the underlying distribution.

9.4.2 Empirical Analogues

In practice, not only discrete and continuous features are distinguished, but also nominal, ordinal and cardinal ones. The main differences between these types of features correspond to whether their values can be ordered by size and whether their values can be used in calculations. Obviously, the values of random variable corresponding to music mode should not be used in calculations (what is the sum of "minor" and "major"?) and cannot be ordered (is "minor" smaller than "major" or vice versa?). In contrast, the values of a random variable corresponding to the tones $\{C, C\#, \ldots, B\}$ are ordered, but should not be summed up, etc. However, the MFCC features in Example 9.9 are ordered and quantitative as defined below.

Definition 9.9 (Feature Types). A feature is called *qualitative* if it represents a property which is assigned to a subject or object by means of non-quantitative methods. A feature is called *quantitative* if its values are genuine measurements (which can be added, multiplied, etc.)

Quantitative features are also called *metric* or *cardinal*. Qualitative features are subdivided into two types: *ordinal* features whose possible values can be ordered by size, although they are not allowed to be added or multiplied, and *nominal* features which do not even allow ordering.

One says that features are observed on a nominal, ordinal, or cardinal *scale*, respectively.

Note that qualitative features are typically discrete because of the non-quantitative measurement method. Also, digital measurements are most of the time discrete because of their finite measurement accuracy. Finally, note that discrete features with many possible outcomes are often nevertheless modeled as continuous variables because the methods derived for continuous variables are often more powerful.

Since qualitative features cannot be used in calculations, different location and dispersion measures have to be used for the characterization of such features. In order to make that clear, we will first look at the empirical analogues of the characteristics given above.

Definition 9.10 (Empirical Location Measures). A location measure characterizes the "center" of the observations $\{x_1, \ldots, x_N\}$ of the feature X. The most important examples are the following (compare their theoretical analogues above).

The *(arithmetical) mean* is defined as $\bar{x} := (x_1 + \ldots + x_N)/N$.

The *(empirical) median* med_x = "central value" = 50%-value is defined as any value for which 50% of the observed values are greater or equal and 50% smaller or equal. The median can be any central value of the ordered list of the observed values. Let $x_{(i)} := i$-th value of the ordered list, then, e.g.,

1. $med_x := (x_{(N/2)} + x_{(N/2+1)})/2$ if N is even,

2. $med_x := x_{((N+1)/2)}$ if N is odd, or

3. $med_x := x_j, j = \lceil N/2 \rceil$, in general, where $\lceil N/2 \rceil$ is the smallest integer $\geq N/2$.

Note that definition 3 does not coincide with the 1st but with the 2nd one. In what follows, we will always use definition 3, since then the median is always an observed value. Note, however, that in practice very different median definitions are used.

The *(empirical) modal value / mode* mod_x is defined as the most frequent value in $\{x_1, \ldots, x_N\}$. The mode does not necessarily lie in the center of the observations. Nevertheless, it appears to be a good representative of the data. Notice that the mode might not be unique!

Definition 9.11 (Dispersion Measures). Dispersion measures represent the variability in the observations $\{x_1, \ldots, x_N\}$ of the feature X. The most important examples are the following (notice that only the first one has a theoretical analogue introduced above).

The *empirical standard deviation* s_x is defined as the square root of the *empirical variance* $var_x :=$ "average of the squared deviations from the arithmetical mean" $= ((x_1 - \bar{x})^2 + \ldots + (x_N - \bar{x})^2)/(N-1)$. Note the division by $N-1$, not by N, according to the estimation properties below.

The *quartile difference* is defined by

$qd := q^4 - q_4 := q_{0.75} - q_{0.25} :=$ 3rd quartile - 1st quartile, where

$q_p := p$-quantile

    := any observed value of the feature X until $p \cdot 100\%$ of the empirical

      distribution is reached as exactly as possible

    = (e.g.) $x_{(j)}, j := \lceil N \cdot p \rceil$, where

$x_{(j)} = j$-th element of the ordered list of observations.

The *range* is defined as $R := max - min = x_{(N)} - x_{(1)}$.

For nominal features, another dispersion measure Φ is in use. Φ is constructed to be maximal (= 1) if the observed values of the feature vary maximally, i.e. are uniformly distributed, and Φ is minimal (= 0) for no dispersion, i.e. if only one value is observed. So the dispersion measure Φ ranks the variation of observed values between minimum and maximum dispersion. This leads to the following definition.

Definition 9.12 (Dispersion Measures Φ). For the (relative) frequency distribution $\{h_1, \ldots, h_K\}$ of the different observed values $\{a_1, \ldots, a_K\}$ let a_{mod} be a value with maximal frequency (mode) and $h_{(a_{mod})}$ the corresponding relative frequency. Let $\Phi_{min} := 2(1 - h_{(a_{mod})})$ and $\Phi_{max} := \sum_{k=1}^{K} | h_k - \frac{1}{K} |$. Then, $\Phi := \frac{\Phi_{min}}{\Phi_{min} + \Phi_{max}}$ is called Φ-*dispersion measure* for nominal features.

Note that the factor 2 in the definition of Φ_{min} is somewhat arbitrary since it could be varied without changing the maximal/minimal property of the measure. The factor

Table 9.1: Applicable Location and Dispersion Measures

scale	$\bar{x}$	med_x	mod_x	s_x	qd	R	Φ
cardinal	yes	yes	(yes)	yes	yes	yes	(yes)
ordinal	no	yes	yes	no	yes	yes	yes
nominal	no	no	yes	no	no	no	yes

2 is chosen so that the extreme value of Φ_{min} is equal to the extreme value of Φ_{max}, since $\Phi_{min} = 2(1 - \frac{1}{K})$ for $h_k = \frac{1}{K}, i = 1, \ldots, K$, and $\Phi_{max} = 1 - \frac{1}{K} + (K-1)\frac{1}{K} = 2(1 - \frac{1}{K})$ for $h_j = 1$ and $h_k = 0$ for all $k \neq j$.

Table 9.1 gives an overview of the applicability of the introduced measures for nominal, ordinal, and cardinal features. Note that the mode most of time does not make much sense for cardinal features since the values will often not be repeated. Also, the Φ-dispersion measure only makes sense for cardinal features if their values repeat, e.g., after aggregation to predefined classes. Moreover, in order to be meaningful for ordinal features, quantiles should only take observed values, like in the above definition 3 of the median and in our definition of the p-quantile q_p. Note that sometimes computer programs use other definitions of quantiles which might not make sense for ordinal features.

Example 9.12 (Uniform Distribution). Reconsider the 12-tone music Example 9.11. Assume that we have observed the following notes in a short piece of music $\{12, 8, 5, 8, 5, 5, 2, 1\}$. *The (not very sensible) mean is then* $\bar{x} = (12 + 8 + 8 + 5 + 5 + 5 + 2 + 1)/8 = 5.75$ *and the empirical standard deviation* $s_x = \sqrt{((12 - 5.75)^2 + 2(8 - 5.75)^2 + 3(5 - 5.75)^2 + (2 - 5.75)^2 + (1 - 5.75)^2)/7}$ $= \sqrt{(6.25^2 + 2 \cdot 2.25^2 + 3 \cdot 0.75^2 + 3.75^2 + 4.75^2)/7} \approx 3.5$, *which is reasonably close to* $E_X = 6.5$ *and nearly equal to* $\sigma_X \approx 3.5$, *correspondingly. The median is* $med_x = 5$ *and the quartiles* $q_4 = 2, q^4 = 8$, *all close to the theoretical values. Notice however, that empirical and theoretical values are expected to be similar only for a large number of observations N. In our example, though, we spread the observations well over their range so that also for our small number of observations empirical and theoretical values are similar.*

Based on the above empirical measures, the 5-summaries characteristic is often used for illustration.

Definition 9.13 (Boxplot). A *box- (and whisker-) plot* is defined to be a box with (vertical) borderlines in the lower and upper quartile q_4 and q^4, the median *med* as an inner (vertical) line, (horizontal) lines (whiskers) from the quartiles to the most extreme value inside so-called fences, i.e. $\geq q_4 - 1.5 \cdot qd$ and $\leq q^4 + 1.5 \cdot qd$, qd being the above defined quartile difference. All points outside the whiskers are called outliers, marked by o (see Figure 9.2).

Obviously, the center 50% of the observations are located inside the box. For so-

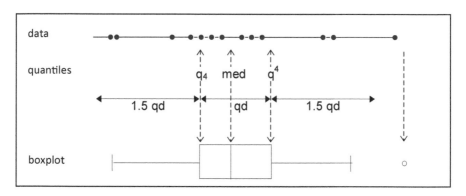

Figure 9.2: Scheme of boxplot.

called "skewed" distributions, the parts left and right of the median are of different size. The choice of $3 \cdot qd$ as the maximal length of the two whiskers together leads to only 0.7% outliers in the case of a normal distribution (cp. Section 9.4.3). Note that boxplots may well be drawn vertically and side by side for different features making them easily comparable.

Example 9.13 (Characteristics of Discrete and Continuous Distributions in Music: Chords). *Consider again Example 9.3. Let us compare the variation of chords in the standard 12-bar blues (I, I, I, I, IV, IV, I, I, V, V, I, I) and in the "standard" jazz version (I7, IV7 IVdim, I7, Vm7 I7, IV7, IVdim, I7, III7 VI7, IIm7, V7, III7 VI7, II7 V7). We assume that only full schemes are observed in corresponding pieces of music. Obviously, in the first blues scheme, 3 different chords I, IV, V are involved in contrast to 9 different chords I7, IV7, IVdim, Vm7, III7, VI7, IIm7, V7, II7 in the jazz version. Let us now calculate the Φ-dispersion measure for both cases. For the 1st scheme, $\{h_1 = 2/3, h_2 = 1/6, h_3 = 1/6\}$ are the relative frequencies of the different observed values $\{a_1 = I, a_2 = IV, a_3 = V\}$ and $a_{mod} = I$ with relative frequency $h_{(a_{mod})} = 2/3$. Therefore, $\Phi_{min} := 2(1 - h_{(a_{mod})}) = 2/3$ and $\Phi_{max} = |2/3 - 1/3| + 2 \cdot |1/6 - 1/3| = 2/3$. Then, $\Phi = \frac{\Phi_{min}}{\Phi_{min} + \Phi_{max}} = 0.5$. For the 2nd scheme $h = \{7/24, 3/24, 3/24, 1/24, 2/24, 2/24, 2/24, 3/24, 1/24\}, h_{(a_{mod})} = 7/24, \Phi_{min} := 2(1 - 7/24) = 1.42, \Phi_{max} = (7/24 - 1/9) + 3(3/24 - 1/9) + 3(1/9 - 2/24) + 2(1/9 - 1/24) = 0.44, and \Phi = 0.76. This shows that the 2nd scheme varies more, as expected!*

Example 9.14 (Characteristics of Discrete and Continuous Distributions in Music: Boxplots of distributions). *Looking again at the data in Example 9.9, we compare the non-windowed MFCCs 1–4 by means of boxplots (see Figure 9.3). Obviously, MFCC 1 has the highest values, MFCC 3 is heavy tailed, and at least MFCCs 2 and 4 are not symmetric.*

MFCC non-windowed

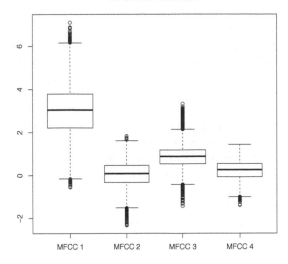

Figure 9.3: Comparison of different distributions by means of boxplots.

9.4.3 Important Univariate Distributions

Let us now discuss important univariate distributions, i.e. distributions of a single random variable. We start with *discrete distributions*.

Definition 9.14 (Typical Discrete Distributions). We only consider two important discrete distributions here.

Every discrete density function of the type $f(x) = \frac{1}{K}$ with $x = x_1, x_2, \ldots, x_K$ and $f(x) = 0$ else, where $K \in \mathbb{N}$, defines a density of a *discrete uniform distribution*. A random variable with such a density is called *discrete uniformly distributed*. Thus, in a uniform distribution all possible outcomes have the same probability. The expected value of a discrete uniform distribution is $E[X] = \frac{\sum_{i=1}^{K} x_i}{K}$ and $E[X] = \frac{K+1}{2}$ in the special case of $x_i = i$. In this case $\text{var}[X] = \frac{K^2-1}{12}$.

Every discrete density function of the type $f(x) = \binom{K}{x} p^x q^{K-x}$ for $x = 0, 1, \ldots, K$ and $f(x) = 0$ else, where $K \in \mathbb{N}$, $0 \le p \le 1$ and $q := 1 - p$, defines a density of a *binomial distribution* (with parameters K, p). A random variable with such a density is called *binomially distributed*. A binomial distribution is the distribution of the sum of K independent decisions between two possibilities (denoted by $\{0, 1\}$). The expected value of a binomial distribution is $E[X] = Kp$ and the variance $\text{var}[X] = Kpq$.

Example 9.15 (Discrete Uniform Distribution). *Reconsider the 12-tone music Example 9.11, an example for a discrete uniform distribution with $x_i = i$ and $K = 12$. For such a distribution, the expected value is $E_X = \frac{K+1}{2} = 6.5$ and $\text{var}[X] = \frac{K^2-1}{12} = 143/12$. This coincides with the values we calculated above, since $\sigma_X = \sqrt{143/12} \approx 3.45$.*

Example 9.16 (States of Musical Tones (cp. [5]). *Let us look at audio input observed as a sequence of about 30 windows per second. We would like to characterize the* *development of the signal over time, i.e. from one window to the next. For this, we model this development as a sequence of certain musical states $x_1, x_2, \ldots, x_T$, one for each window. In order to keep the model simple, we just distinguish the states* attack *(atck) and* sustain *(sust) (for alternative models cp. the end of this example). The corresponding state graph is shown in Figure 9.4. It models the development in time as a so-called* hidden Markov chain *of states, where each state only depends on the preceding state. In our graph, music is modeled as a sequence of sub-graphs, one for each solo note, which are arranged so that the process enters the start of the $(n+1)$-st note as it leaves the n-th note. From the figure, one can see that each note begins with a short sequence of states meant to capture the attack portion of the note (atck). This is followed by another sequence of states with "self-loops" meant to capture the main body of the note (sust, sustain), and to account for the variation in note duration we may observe.*

If we chain together m states changed with probability p, i.e. remain unchanged with probability $q = 1 - p$, then the total number of states visited, T, i.e. the number of audio frames, spent in the sequence of m states has a so-called negative binomial *distribution $P(T = t) = \binom{t-1}{m-1} p^m q^{t-m}$ for $t = m, m+1, \ldots$, indicating the probability of m "successes" in T runs, where a "success" means a state change. The expected value of T is given by $E[T] = \frac{m}{p}$ and the variance by $\mathrm{var}[T] = \frac{mq}{p^2}$. Unfortunately, the parameters m and p are unknown, in general. In order to "estimate" these parameters having seen several performances of the music piece in question, we could choose them individually for each note so that the empirical mean and variance of T agree with the true mean and variance as given in the above formulas: $\bar{x}_T = \frac{m}{p}$ and $s_T^2 = \frac{m(1-p)}{p^2}$. This is the so-called* method of moments.

In reality, one has to use a wider variety of note models than depicted in the figure, with variants for short notes, notes ending with optional rests, notes that are rests, etc., though all are following the same essential idea.

Let us now continue with important continuous distributions.

Definition 9.15 (Typical Continuous Distributions). A continuous density function of the type $f(x) = \frac{1}{b-a}, x \in [a,b]$, and $f(x) = 0$ else, where $a, b \in \mathbb{R}$, defines a density of a *continuous uniform distribution* or *rectangular distribution* on the interval $[a,b]$. A random variable with such a density is called *continuous uniformly distributed*. The expected value of a continuous uniform distribution is $E[X] = \frac{a+b}{2}$, and the variance $\mathrm{var}[X] = \frac{1}{12}(b-a)^2$.

A continuous density function of the type $f(x) = \frac{1}{\sqrt{2\pi}\sigma} e^{-\frac{1}{2}\left(\frac{x-\mu}{\sigma}\right)^2}$, where $\sigma > 0$ and $\mu \in \mathbb{R}$, defines a density of a *normal distribution* with the parameters μ, σ^2. A random variable X with such a density is called *normally distributed*. μ is the expected value and σ^2 the variance of the normal distribution.

Let $X_i, i = 1, \ldots, N$, be independent identically $\mathcal{N}(\mu, \sigma^2)$ distributed (for independence of random variables see Definition 9.18). Then, the random variable $t_{N-1} := \frac{\bar{X} - \mu}{s_X/\sqrt{N}}$ is called *t-distributed* with $(N-1)$ degrees of freedom, when $s_X :=$ empirical standard deviation of observations $x_1, \ldots, x_N$ of the $X_i, i = 1, \ldots, N$, esti-

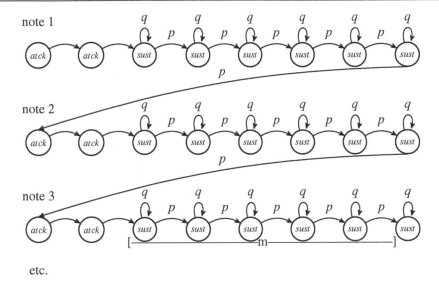

Figure 9.4: Scheme of state changes.

mating the standard deviation $\sigma_X, X \sim \mathcal{N}(\mu, \sigma^2)$ (see Section 9.4.2). One can show that $E[t_{N-1}] = 0$ if $N > 2$, and $\mathrm{var}[t_{N-1}] = (N-1)/(N-3)$ if $N > 3$.

The sum of squares of n independent standard normal distributions is called χ^2-*distribution* (chi-squared distribution) with n degrees freedom χ_n^2.

The ratio of two independent scaled χ^2-distributions with n and m degrees of freedom

$$F_{n,m} = \frac{\chi_n^2/n}{\chi_m^2/m}$$

is called *F-distribution* with n, m degrees of freedom.

Please notice that t would be $\mathcal{N}(0,1)$−distributed if the true standard deviation σ_x would be used instead of s_x. The variance of the t-distribution is somewhat greater than the variance of the $\mathcal{N}(0,1)$−distribution. For $N \to \infty$ the t-distribution converges towards the $\mathcal{N}(0,1)$−distribution. Examples for χ^2- and F-distributions can be found in Section 13.5.

Example 9.17 (Continuous Distributions: Automatic Composition). *In automatic composition, Xenakis [6, pp. 246-249] experimented with amplitude and/or duration values of notes obtained directly from a probability distribution (e.g., uniform or normal). Also many other distributions, not introduced here, were tried.*

Example 9.18 (Continuous Distributions: MFCCs). *Let us come back to the data introduced in Example 9.9. We look at the quantitative variable non-windowed MFCC 1. From Figure 9.5 it should be clear that this variable can be very well approxi-*

Histogram of non–windowed MFCC 1

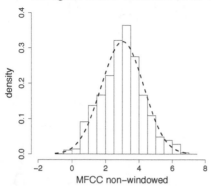

Figure 9.5: Comparison of empirical distribution (histogram) with theoretical distribution (normal density, dashed line).

mated by a normal distribution with expected value = empirical mean and variance = empirical variance.

9.5   Random Vectors

9.5.1   Theory

Most of the times a user is interested in more than one random variable, i.e. in a random vector, in particular in the relationship between different random variables. A multivariate distribution is the distribution of a vector of random variables. The most important multivariate distribution is the multivariate normal distribution.

Definition 9.16 (Multivariate Normal Distribution).   A random vector $\boldsymbol{X} = (X_1 \ldots X_m)^T$ is said to follow a *multivariate normal distribution* iff its density has the form:

$$f_X(x_1,\ldots,x_m) = \frac{1}{\sqrt{(2\pi)^m \mid \boldsymbol{\Sigma_X} \mid}} e^{-\frac{1}{2}(x-\boldsymbol{\mu_X})^T \boldsymbol{\Sigma_X^{-1}}(x-\boldsymbol{\mu_X})},$$

where $\boldsymbol{\Sigma_X}$ is the positive definite (and thus invertible) covariance matrix of the random vector $\boldsymbol{X}$, $\mid \boldsymbol{\Sigma_X} \mid$ is the determinant of $\boldsymbol{\Sigma_X}$, and $\boldsymbol{\mu_X}$ is the vector of expected values of the elements of X. The covariance matrix will be defined below.

Note that normal distributions are also defined in the case of singular $\boldsymbol{\Sigma_x}$. This case will, however, not be discussed here.

Example 9.19 (Multivariate Normality in Classes).   *Imagine you want to distinguish classes like genres or instruments by means of the values of a vector of influential variables* $\boldsymbol{X}$. *This is called a supervised classification problem (cp. Chapter 12). A typical assumption in such problems is that the influential variables* $\boldsymbol{X}$ *follow individual (multivariate) normal distributions for each class. In the case of m influential variables* $\boldsymbol{X} = (X_1 \ldots X_m)^T$ *this leads to a density for class c of the following kind:*

$$f_X(x_1,\ldots,x_m) = \frac{1}{\sqrt{(2\pi)^m \mid \boldsymbol{\Sigma_X}(c) \mid}} e^{-\frac{1}{2}(x-\boldsymbol{\mu_X}(c))^T \boldsymbol{\Sigma_X}(c)^{-1}(x-\boldsymbol{\mu_X}(c))}.$$

Densities of 2D normal distributions

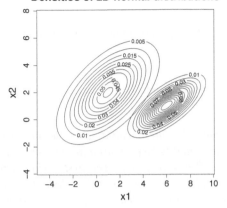

Figure 9.6: Two well distinguishable bivariate normal distributions in the random variables x1 and x2.

In classification, it is important that the distributions in the classes are as different as possible, i.e. that expected values $\mu_X(c)$ and covariance matrices $\Sigma_X(c)$ are as different as possible for different c. An example for such a situation is illustrated in Figure 9.6.

In what follows we will introduce covariances and their scaled versions, the correlations.

Definition 9.17 (Covariance and Correlation). Let X, Y be random variables on the same sample space. The *covariance* of X and Y is defined as
$\text{cov}(X,Y) := \sigma_{XY} := E[(X - \mu_X)(Y - \mu_Y)]$, the *correlation coefficient* of X and Y as
$\rho_{XY} := \frac{\text{cov}(X,Y)}{\sigma_X \sigma_Y}$, if $\sigma_X, \sigma_Y > 0$.
The *covariance matrix* $\text{COV}_{X,Y}$ of the random vector $(X \ Y)^T$ is defined as

$$\text{COV}_{X,Y} := \begin{pmatrix} \sigma_X^2 & \sigma_{XY} \\ \sigma_{XY} & \sigma_Y^2 \end{pmatrix}, \text{ the } \textit{correlation matrix } \text{COR}_{X,Y} \text{ as}$$

$$\text{COR}_{X,Y} := \begin{pmatrix} 1 & \rho_{XY} \\ \rho_{XY} & 1 \end{pmatrix}, \text{ and the random variables } X \text{ and } Y \text{ are called } \textit{uncorre-}$$
lated iff $\text{cov}(X,Y) = 0$.
Covariance and correlation matrices are analogously defined for $K > 2$ random

$$\text{variables } X_1, \ldots, X_K, \text{ e.g., } \text{COV}_{X_1,\ldots,X_K} := \begin{pmatrix} \sigma_{X_1}^2 & \cdots & \sigma_{X_1 X_K} \\ \vdots & \ddots & \vdots \\ \sigma_{X_K X_1} & \cdots & \sigma_{X_K}^2 \end{pmatrix}.$$

Let us discuss a special case of random vectors, where the entries are so-called *independent* (note the analogue to independent subsets in Definition 9.2).

Definition 9.18 (Independence of Random Variables). Random variables $X_1, \ldots, X_N$ with densities $f(X_1), \ldots, f(X_N)$ are called independent iff
$f(X_1, \ldots, X_N) = f(X_1) \cdot \ldots \cdot f(X_N)$.

In this case, covariances are zero and expected values and variances of (functions of) random vectors can be easily calculated.

Theorem 9.5 (Expected Values and Independence). For independent random variables $X_1, \ldots, X_N$ it is true that:

- $E[X_1 \cdot \ldots \cdot X_N] = E[X_1] \cdot \ldots \cdot E[X_N]$,
- $\operatorname{cov}(X_i, X_j) = 0, i \neq j$, i.e. X_i and X_j are uncorrelated,
- $\operatorname{var}[\sum_{i=1}^{N} X_i] = \sum_{i=1}^{N} \operatorname{var}[X_i]$.

Example 9.20 (Independence). *In Example 9.16, the music data model might be composed of three variables b_t, e_t, and s_t assumed to be (conditionally) independent given the state x_t: $P(b_t, e_t, s_t | x_t) = P(b_t | x_t) P(e_t | x_t) P(s_t | x_t)$. The first variable, b_t, measures the local "burstiness" of the signal, particularly useful in distinguishing between note attacks and steady-state behavior (sustain) distinguished in Figure 9.4. The 2nd variable, e_t, measures the local energy, useful in distinguishing between rests and notes. And the vector-valued variable s_t represents the magnitude of different frequency components given the state x_t. For each of the three components a distribution may be fixed independently.*

The above Bayes Theorem 9.2 can also be formulated for densities. For this, we first have to define the generalization of conditional probabilities for densities:

Definition 9.19 (Conditional Density). Let X, Y be random variables on the same sample space with a bivariate density $f(x,y)$. Then, $f(x \mid y) := \frac{f(x,y)}{f(y)}$ and $f(y \mid x) := \frac{f(x,y)}{f(x)}$ are called *conditional densities* of X given Y and vice versa, where $f(y) := \int_{-\infty}^{\infty} f(x,y)dx$, $f(x) := \int_{-\infty}^{\infty} f(x,y)dy$ are the so-called *marginal densities* of Y, X corresponding to the joint density $f(x,y)$.

With this definition, the Bayes theorem for densities can be formulated:

Theorem 9.6 (Bayes Theorem for Densities). Let X, Y be random variables on the same sample space. Then, $f(x \mid y) = \frac{f(y|x)f(x)}{f(y)} = \frac{f(y|x)f(x)}{\int_{-\infty}^{\infty} f(y|x)f(x)dx}$.

The Bayes theorem can be well generalized to the multivariate case, as demonstrated in the following example.

Example 9.21 (Application of Bayes Theorem for Densities). *This version of the Bayes theorem will be very important in Chapter 12, i.e. in supervised classification of classes like genres or instruments. A typical assumption in classification is that an individual (multivariate) normal distribution $f(\mathbf{x} \mid c)$ of the influential variables $\mathbf{X}$ are valid for each class c (see Example 9.19). With the Bayes theorem, it is then possible to calculate the discrete density of class c, i.e. its probability, given an observation $\mathbf{x}$ by means of the density of the observation given the class:*
$$f(c \mid \mathbf{x}) = \frac{f(\mathbf{x}|c)f(c)}{f(\mathbf{x})} = \frac{f(\mathbf{x}|c)f(c)}{\sum_{c=1}^{G} f(\mathbf{x}|c)f(c)} \qquad \text{if } G \text{ classes are distinguished.}$$

9.5.2 Empirical Analogues

Naturally, there are also empirical analogues for the covariance and the correlation coefficient.

Definition 9.20 (Empirical Covariance and Correlation). The *(Pearson) empirical correlation coefficient* r_{XY} of cardinal features X and Y is defined as

$$r_{XY} := \frac{\sum_{i=1}^{N}(x_i - \bar{x})(y_i - \bar{y})}{\sqrt{\sum_{i=1}^{N}(x_i - \bar{x})^2 \sum_{i=1}^{N}(y_i - \bar{y})^2}} = \frac{s_{XY}}{s_X s_Y},$$

where $\bar{x}, \bar{y}$ are the (arithmetical) means of the observations of the features X, Y, $s_{XY} := \hat{\text{cov}}(X,Y) := \frac{1}{N-1}\sum_{i=1}^{N}(x_i - \bar{x})(y_i - \bar{y})$ is the *empirical covariance* of the features X and Y and s_X, s_Y are the above empirical standard deviations of X and Y. The features X and Y are called *empirically uncorrelated* iff $r_{XY} = 0$.

Notice that the correlation coefficient characterizes the strength of the linear relationship only. If r_{XY} is close to 1, then we expect a positive linear relationship, i.e. Y increases proportionally with X. If r_{XY} is near -1, then we expect a negative linear relationship, i.e. Y decreases proportionally when X increases. If r_{XY} is near 0, then we do not expect any linear relationship. If X or Y does not vary, the correlation coefficient is not meaningful, and thus not defined. Also, a correlation coefficient cannot capture a nonlinear relationship between X and Y. On the one hand, it may well be that r_{XY} is close to 0, e.g., for an exact quadratic relationship. On the other hand, nonlinear relationships can also produce quite high correlation coefficients so that scatterplots should be preferred for the characterization of relationships between two features.

Definition 9.21 (Scatterplot). A *scatterplot* is a graphical representation of a pair of cardinal features $(X \ Y)^T$, where one feature is represented on the x-axis and the other on the y-axis. Each observation of this pair of features is represented by a point $(x \ y)^T$.

Moreover, notice that a high correlation between features X and Y does not imply a causal relationship. Neither Y is necessarily influenced by X nor vice versa. It may well be that such a high correlation is caused by a third so-called latent background feature that strongly influences both, X and Y. In such cases the correlation is called *spurious*. In any case, found correlations have to be meaningful, otherwise we call them *nonsense correlations*.

In contrast to the correlation coefficient, the covariance is not dimensionless and the interpretation of its size is problem depending. Therefore, the covariance is not that well suited for the comparison of the validity of relationships.

Example 9.22 (Linear Relationships and Scatterplots). *Looking again at the data in Example 9.9, the scatterplot of non-windowed MFCC 1 vs. MFCC 1 in block 1 illustrates a high empirical correlation of 0.93 (see left part of Figure 9.7). This means the MFCC 1 over the whole tone is highly linearly related to MFCC 1 in block 1. In contrast, between the two non-windowed MFCCs 1 and 2 there is only a*

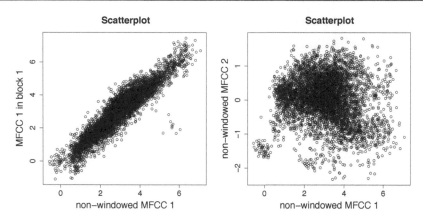

Figure 9.7: Scatterplot of two MFCC variables each, left: high correlation, right: low correlation.

slight relationship if any (see right part of Figure 9.7). Note that here the empirical correlation is −0.11.

Obviously, the above covariances and correlations are only defined for cardinal features. For ordinal features, ranks are used instead of the original observations for the calculation of covariances and correlations.

Definition 9.22 (Ranking). *Ranking* refers to the data transformation in which numerical or ordinal values are replaced by their ranks when the data are sorted. Typically, the ranks are assigned to values in ascending order. Identical values (so-called *rank ties*) are assigned a rank equal to the average of their positions in the ascending order of the values.

For example, if the numerical data 3.4, 5.1, 2.6, 7.3 are observed, the ranks of these data items would be 2, 3, 1 and 4, respectively. As another example, the ordinal data high, low, and middle pitch would be replaced by 3, 1, 2.

Using ranks instead of the original observations for the calculation of correlations leads to the following definition.

Definition 9.23 (Spearman's Rank Correlation). The *Spearman (rank) correlation coefficient* of two features is defined as the Pearson correlation coefficient between the corresponding ranked features. For a sample of size N, the N raw observations x_i, y_i are converted to ranks p_i, q_i, and the correlation coefficient is computed from these:

$$r_{XY} := \frac{\sum_{i=1}^{N}(p_i - \bar{p})(q_i - \bar{q})}{\sqrt{\sum_{i=1}^{N}(p_i - \bar{p})^2 \sum_{i=1}^{N}(q_i - \bar{q})^2}}.$$

Example 9.23 (Spearman Rank Correlation). *Looking again at the data in Example 9.22, the Spearman rank correlation of the non-windowed MFCC and MFCC 1*

Table 9.2: Contingency Table for Binary Features (left) and for General Nominal Features (right); (a • Index Indicates Summing up over this Index)

	$y = 1$	$y = 0$	total
$x = 1$	H_{11}	H_{10}	$H_{1\bullet}$
$x = 0$	H_{01}	H_{00}	$H_{0\bullet}$
total	$H_{\bullet 1}$	$H_{\bullet 0}$	N

	y_1	$\cdots$	y_m	total
x_1	H_{11}	$\cdots$	H_{1m}	$H_{1\bullet}$
$\cdots$		$\cdots$		$\cdots$
x_n	H_{n1}	$\cdots$	H_{nm}	$H_{n\bullet}$
total	$H_{\bullet 1}$	$\cdots$	$H_{\bullet m}$	N

in block 1 shows with 0.91 a similarly high value as the Pearson correlation with 0.93. Also between the two non-windowed MFCCs 1 and 2 there is nearly the same low rank correlation (−0.12) as the Pearson correlation (−0.11). This shows the close connection of the two concepts of correlation.

For nominal features so-called contingency coefficients are in use instead of correlation coefficients. A typical example is the so-called ϕ coefficient.

Definition 9.24 (ϕ coefficient). The ϕ *coefficient* (also referred to as the "mean square contingency coefficient") is a measure of association for, e.g., two binary features. For the definition of the ϕ coefficient, consider the so-called *contingency table*. The contingency table for binary features x and y is defined as in Table 9.2 (left), where $H_{11}, H_{10}, H_{01}, H_{00}$, are the absolute frequencies "cell counts" that sum to N, the total number of observations.

Then, the ϕ *coefficient* is defined by

$$\phi := \frac{H_{11}H_{00} - H_{10}H_{01}}{\sqrt{H_{1\bullet}H_{0\bullet}H_{\bullet 0}H_{\bullet 1}}}.$$

In case of two general nominal features with n and m levels the contingency table looks as in Table 9.2 (right). Then, the ϕ *coefficient* is defined by

$$\phi := \sqrt{\frac{1}{N}\sum_{i=1}^{n}\sum_{j=1}^{m}\frac{(H_{ij} - H_{eij})^2}{H_{eij}}} = \sqrt{\sum_{i=1}^{n}\sum_{j=1}^{m}\frac{(H_{ij} - H_{i\bullet}H_{\bullet j}/N)^2}{H_{i\bullet}H_{\bullet j}}},$$

where H_{ij} is the observed frequency in the (ij)-th cell of the contingency table and $H_{eij} = H_{i\bullet}H_{\bullet j}/N$ is the so-called *expected absolute frequency* in the cell for stochastically independent variables. The above formula for binary features is a special case of this more general formula.

This measure is similar to the Pearson correlation coefficient in its interpretation. In fact, the Pearson correlation coefficient calculated for two binary variables will result in the above ϕ coefficient.

Example 9.24 (Melody Generation[1]). *Automatic melody generation is often carried out by means of a Markov chain on note or pitch values. In a Markov chain, the*

---

[1]cp. http://en.wikipedia.org/wiki/Pop_music_automation. Accessed 13 March 2016.

Table 9.3: Transition Matrix (left) and Contingency Table (right)

note	A	C#	Eb
A	0.1	0.6	0.3
C#	0.25	0.05	0.7
Eb	0.7	0.3	0

note	A	C#	Eb	total
A	44	207	98	349
C#	79	22	226	327
Eb	225	98	0	323
total	348	327	324	999

value at time point t only depends on the preceding value at time point t − 1. The transitions from one value to the next are controlled by so-called transition probabilities gathered in a transition probability matrix. This matrix is constructed row-wise, note by note, by vectors containing the probabilities to switch from one specific note to any other note (row sums = 1, see Table 9.3(left)). Note values are generated by an algorithm based on the transition matrix probabilities. From the resulting Markov chain we can generate a contingency table with the numbers of the realized transitions (x = starting tone, y = next tone). Based on the contingency Table 9.3(right), the φ coefficient is calculated as 0.75. That there is dependence is expected because of the transition probabilities. This dependency should decrease, though, when taking y = "overnext realized tone", and indeed then the φ coefficient is calculated as 0.41.

9.6 Estimators of Unknown Parameters and Their Properties

As we have already seen, many of the used types of densities have parameters, which are most of the time unknown. In order to determine them, we have to observe the random variable and have to calculate so-called estimates of the unknown parameters from the observations. In the following, we will first give a general definition and then demonstrate how to estimate expected values and variances.

Definition 9.25 (Estimators). Let the independent random variables $X_1, \ldots, X_N$ all have the same density $f_X(x, \boldsymbol{\theta})$. Let $\tau(\boldsymbol{\theta})$ be a function of a vector of unknown parameters $\boldsymbol{\theta} = (\theta_1 \ \ldots \ \theta_K)^T$. A *(point-)estimator* is a function $T(X_1, \ldots, X_N)$, whose value is used to represent the unknown $\tau(\boldsymbol{\theta})$ as well as possible. An *interval estimator* is a pair of functions $T_1(X_1, \ldots, X_N)$ and $T_2(X_1, \ldots, X_N)$ with $T_1(X_1, \ldots, X_N) < T_2(X_1, \ldots, X_N)$ so that the probability that $\tau(\boldsymbol{\theta})$ lies between T_1 and T_2 is equal to a pre-fixed probability $\gamma \in (0, 1)$ called the *confidence level*, i.e. $P_{\boldsymbol{\theta}}(T_1(X_1, \ldots, X_N) < \tau(\boldsymbol{\theta}) < T_2(X_1, \ldots, X_N)) = \gamma$. T_1 and T_2 are called *lower* and *upper confidence limits* for $\tau(\boldsymbol{\theta})$. Typical values for γ are $\gamma = 0.9, 0.95, 0.99$. An interval $(T_1(x_1, \ldots, x_N), T_2(x_1, \ldots, x_N))$ of values of an interval estimator is called *two-sided $100\gamma\%$-confidence interval* for $\tau(\boldsymbol{\theta})$.

A point estimator $T(X_1, \ldots, X_N)$ is called *unbiased* estimator for $\tau(\boldsymbol{\theta})$ iff $E_{\boldsymbol{\theta}}[T] = E_{\boldsymbol{\theta}}[T(X_1, \ldots, X_N)] = \tau(\boldsymbol{\theta})$. An unbiased estimator $T(X_1, \ldots, X_N)$ for $\tau(\boldsymbol{\theta})$ is called *best unbiased estimator* iff $\text{var}_{\boldsymbol{\theta}}(T) = E_{\boldsymbol{\theta}}[(T - \tau(\boldsymbol{\theta}))^2]$ is minimal for all $\boldsymbol{\theta}$ over all unbiased estimators.

95%–Confidence Interval for the Mean

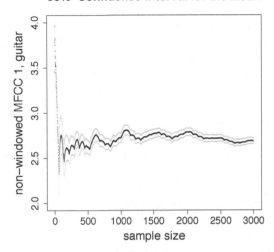

Figure 9.8: Confidence interval (grey lines) for the mean (black line) when sample size is increasing: non-windowed MFCC 1, guitar.

Note that $\tau(\boldsymbol{\theta})$ may be equal to $\boldsymbol{\theta}$ itself. The above formulation thus allows for a transformation of $\boldsymbol{\theta}$ to be estimated. As an example, let us now apply this general definition to the estimation of the expected value and the variance.

Theorem 9.7 (Estimation of Expected Value and Variance). Let $X_1,\ldots,X_N$ be independent random variables with identical expected values μ and variances σ^2. Let the *mean* be the random variable $\bar{X} := \frac{1}{N}\sum_{i=1}^{N} X_i$. Then,

- $E[\bar{X}] = \mu$ and $\mathrm{var}[\bar{X}] = \sigma^2/N$.
- Let x_i be a realization of X_i. Then, $\hat{\mu} = \bar{x} = \frac{1}{N}\sum_{i=1}^{N} x_i$ is an estimator for the expected value of the X_i with shrinking variance for $N \to \infty$.
- An analogue estimator for the variance is $\hat{\sigma}^2 = \frac{1}{N}\sum_{i=1}^{N}(x_i - \bar{x})^2$.
- An unbiased estimator for the variance is $s^2 = \frac{1}{N-1}\sum_{i=1}^{N}(x_i - \bar{x})^2$.
- An $(1-\alpha)100\%$ confidence interval for μ with unknown σ and independent identically $\mathscr{N}(\mu,\sigma^2)$-distributed random variables X_i is given by:

$$\left[\bar{x} - t_{N-1;1-\alpha/2}\frac{s}{\sqrt{N}} \quad , \quad \bar{x} + t_{N-1;1-\alpha/2}\frac{s}{\sqrt{N}}\right],$$

where s is the above estimator for the standard deviation of the X_i, $t_{N-1;1-\alpha/2}$ the $(1-\alpha/2)$ quantile of a t-distribution with $N-1$ degrees of freedom, and α is typically 0.05 or 0.01.

Example 9.25 (Effect of Increasing Sample Size). *Let us look again at the data in Example 9.9 and study the effect of increasing the sample size. We study the estimates of the expected value and standard deviation as well as the corresponding confidence intervals of the non-windowed MFCC 1, guitar only, in dependence of the sample size (see Figure 9.8). We see that the mean is nearly stable from sample size 1200 on, whereas the confidence interval is continuously shrinking.*

A very general estimation principle applicable to any kind of densities is Maximum Likelihood Estimation.

Definition 9.26 (Maximum Likelihood Estimation). Suppose there is a sample $x_1, x_2, \ldots, x_N$ of N independent observations of a random variable X following a distribution with a density with an unknown parameter θ. Both the observations x_i and the parameter θ can be vectors. We are looking for an estimator $\hat{\theta}$ of θ which makes the observed sample as likely as possible. For this, we consider the joint density function for all observations

$$f(x_1, x_2, \ldots, x_N \mid \theta) = f(x_1 \mid \theta) \cdot f(x_2 \mid \theta) \cdot \ldots \cdot f(x_N \mid \theta)$$

for the independent sample $x_1, x_2, \ldots, x_N$ of X. Now we consider the observed values $x_1, x_2, \ldots, x_N$ to be fixed, θ being now the function's variable allowed to vary freely. This function is called the *likelihood*:

$$L(\theta \mid x_1, \ldots, x_N) = f(x_1, x_2, \ldots, x_N \mid \theta) = \prod_{i=1}^{N} f(x_i \mid \theta).$$

In practice it is often more convenient to work with the logarithm of the likelihood function, called the *log-likelihood*:

$$\log L(\theta \mid x_1, \ldots, x_N) = \sum_{i=1}^{N} \log f(x_i \mid \theta).$$

The *maximum-likelihood estimator (MLE)* $\hat{\theta}_{mle}$ of θ maximizes $\log L(\theta \mid x)$:

$$\{\hat{\theta}_{mle}\} \subseteq \{\underset{\theta \in \Theta}{\arg \max}\ \log L(\theta \mid x_1, \ldots, x_N)\} \text{ if such a maximum exists.}$$

The MLE estimate is the same for maximizing L or $\log L$, since log is a monotonically increasing function.

Example 9.26 (ML Estimators of Expected Value and Variance). *Sometimes, the MLE is equal to other well-known estimators. For normal distributions, the MLE of the expected value is $\bar{x}$ if the variance σ is known, and the MLE of the variance is $\hat{\sigma}^2$ as defined in Theorem 9.7. Note that both the unbiased estimator s^2 and the MLE $\hat{\sigma}^2$ have interesting properties so that they both are used in the practice of music data analysis to estimate an unknown variance. Also note that the difference between the two estimators will be smaller, the bigger the number of observations N.*

9.7   Testing Hypotheses on Unknown Parameters

On the one hand, the unknown parameters of distributions might be estimated as demonstrated in the previous section. On the other hand, one might want to formulate hypotheses on such parameters which should be tested on the basis of observed samples.

Example 9.27 (Test on Location Differences). *As an example, let us, again, come back to the MFCC data in Example 9.9. From Example 9.13 it can be suspected that a) the non-windowed MFCCs 1 and 2 differ in location in contrast to b) the MFCCs 2 and 4. We want to apply statistical tests to study these hypotheses.*

In the following, we will first give a general definition of statistical hypotheses and tests and then come back to the example.

Definition 9.27 (Hypotheses and Error Types). A *statistical hypothesis* or *null-hypothesis H_0* for an unknown parameter θ of a distribution is a conjecture on this parameter. The *alternative hypothesis* is called H_1. A statistical hypothesis can be one-sided, like hypotheses with left-sided alternative $H_0 : \theta \geq \theta_0, H_1 : \theta < \theta_0$ or hypotheses with right-sided alternative $H_0 : \theta \leq \theta_0, H_1 : \theta > \theta_0$, or it can be two-sided like $H_0 : \theta = \theta_0, H_1 : \theta \neq \theta_0$.

A *statistical test* of a statistical hypothesis H_0 is a decision rule resulting in the *rejection* or non-rejection of the statistical hypothesis. A statistical test is based on a so-called *test statistic* which can be calculated by means of the observations of the studied random variable.

The *critical region* (rejection area) of a statistical test is that subset of the possible sample values, where the statistical hypothesis is rejected. A *critical value* of a test is a threshold of the test statistic corresponding to a border of the critical region. *Critical values* are set to values of the distribution of the test statistic corresponding to the statistical hypothesis H_0 so that a pre-fixed portion α, the so-called *significance level*, of the possible values of this distribution fall outside of the critical values.

The statistical hypothesis is *rejected* if a critical value is exceeded by the realized value of the test statistic. A *type I error* occurs if the hypothesis H_0 is rejected based on the value of the test statistic though it is true.

A *p-value* of a test is the probability that a value of the test statistic is 'more extreme' than the realized value in the distribution derived from the statistical hypothesis H_0. Assuming the statistical hypothesis H_0 is true, for a realized sample value x of X, the p-value is given by $P(X \geq x|H_0)$ for right-sided alternatives, $P(X \leq x|H_0)$ for left-sided alternatives, and $2\min\{P(X \leq x|H_0), P(X \geq x|H_0)\}$ for two-sided hypotheses. If the p-value is $\leq \alpha$, the *significance level*, then the hypothesis is rejected and the result is called *significant*. Therefore, with statistical tests the probability of rejecting a true hypothesis is the significance level α.

A *type II error* occurs if the hypothesis H_0 is not rejected though it is wrong.

The above definition does not specify the test statistic. There are many such statistics for different purposes. We restrict ourselves to so-called location tests, where a hypothesis on the location of expected values is tested. Other tests will be discussed, e.g., in Section 13.5. The most prominent statistical test is the t-test. We consider different variants.

Definition 9.28 (t-Tests). *One sample t-test:* If all X_i are independently $\mathcal{N}(\mu, \sigma^2)$-distributed with unknown variance, $i = 1, \ldots, N$, then:

$$t = \frac{\bar{X} - \mu_0}{\sqrt{s^2/N}} \sim t_{N-1},$$

where s is the unbiased estimator of the standard deviation σ, and the *test statistic t* is *t*-distributed with $N - 1$ degrees of freedom.

A typical corresponding pair of statistical hypotheses is:
$H_0 : \mu = \mu_0$ vs. $H_1 : \mu \neq \mu_0$.

Two sample t-test: If all X_i are independently $\mathcal{N}(\mu_X, \sigma_X^2)$-distributed with unknown variance, $i = 1, \ldots, N$, and all Y_i are independently $\mathcal{N}(\mu_Y, \sigma_Y^2)$-distributed with unknown variance, $i = 1, \ldots, M$, then analogous to the one sample case the test statistic

$$t = \frac{(\bar{X} - \bar{Y}) - \delta_0}{\sqrt{s_X^2/N + s_Y^2/M}}$$

can be used for the comparison of two expected values with unknown variances, where s_X and s_Y are the unbiased estimators of the standard deviations and N and M are the corresponding sample sizes.

Here, we obviously do not test on equal expected values, but on a difference δ_0, and for $H_0 : \mu_X - \mu_Y = \delta_0$ the test statistic t is t-distributed with k degrees of freedom, where

$$k = \left\lfloor \frac{\left(\frac{s_X^2}{N} + \frac{s_Y^2}{M}\right)^2}{\frac{1}{N-1}\left(\frac{s_X^2}{N}\right)^2 + \frac{1}{M-1}\left(\frac{s_Y^2}{M}\right)^2} \right\rfloor.$$

Some typical hypotheses and alternatives as well as critical regions of two-sample t-tests are:

(a) $H_0 : \mu_X - \mu_Y = \delta_0$ vs. $H_1 : \mu_X - \mu_Y \neq \delta_0$ (two-sided)
     Reject if: $|t| > t_{1-\alpha/2}(k)$, i.e. $\pm t_{1-\alpha/2}(k)$ are the critical values of the test

(b) $H_0 : \mu_X - \mu_Y \geq \delta_0$ vs. $H_1 : \mu_X - \mu_Y < \delta_0$ (one-sided)
     Reject if: $t < -t_{1-\alpha}(k)$, i.e. $-t_{1-\alpha}(k)$ is the critical value of the test

(c) $H_0 : \mu_X - \mu_Y \leq \delta_0$ vs. $H_1 : \mu_X - \mu_Y > \delta_0$ (one-sided)
     Reject if: $t > t_{1-\alpha}(k)$, i.e. $t_{1-\alpha}(k)$ is the critical value of the test

The t-test is the most often applied test for location differences. Unfortunately, the test has the problem that the tested data should stem from normal distributions which is often not plausible. In case of non-normality, so-called *nonparametric tests* should be applied, in particular the following test.

Definition 9.29 (Wilcoxon Test). Let the continuous distribution functions of the random variables X and Y only differ by a shift a: $F_Y(x) = F_X(x-a)$. Especially let the two variances be equal: $\sigma_X = \sigma_Y$ (variance homogeneity). Moreover, let the two samples $X_1, \dots, X_N$ of X and $Y_1, \dots, Y_M$ of Y be independent.

The *Wilcoxon–Mann–Whitney test* of the hypotheses
$H_0 : a = 0$ vs. $H_1 : a \neq 0$
uses the *Wilcoxon rank-sum statistic* $W_{N,M} = \sum_{i=1}^{N} R(X_i)$
with $R(X_i) := $ rank of X_i in the ordered pooled sample, i.e. the ordered version of the union of both samples.

The exact critical values w for a significance level α can be derived by means of a recursion formula. The calculation effort, however, quickly increases for large values N, M. Therefore, for $N > 10$ or $M > 10$, say, often the following normal approximation is used for the determination of (approximate) critical values: $W_{N,M} \approx$
$\mathcal{N}\left(\frac{N(N+M+1)}{2}, \frac{NM(N+M+1)}{12}\right)$.

Obviously, if $F_Y(x) = F_X(x-a)$, with this test the following hypothesis / alternative pair is tested: $H_0 : \mu_X = \mu_Y$ vs. $H_1 : \mu_X \neq \mu_Y$.

The Wilcoxon–Mann–Whitney test is also interpreted as a test for the equality of medians. Moreover, the test can be easily reformulated for the one-sided hypotheses: $H_0 : a \leq 0$ vs. $H_1 : a > 0$ and $H_0 : a \geq 0$ vs. $H_1 : a < 0$.

Example 9.28 (Test on Location Differences). *Let us come back to our motivating Example 9.27. We would like to test that a) the non-windowed MFCCs 1 and 2 differ in location in contrast to b) the MFCCs 2 and 4. Then, the null-hypotheses take the form:*
a) $H_0 : E[\text{non-windowed MFCC 1}] = E[\text{non-windowed MFCC 2}]$ and
b) $H_0 : E[\text{non-windowed MFCC 2}] = E[\text{non-windowed MFCC 4}]$.
Note that we want to reject the first null-hypothesis, whereas for the second we are not sure about rejection. Let us look at the results of the two sample t-test and the Wilcoxon test. In both cases, a two-sided hypothesis in Definition 9.28 with $\delta_0 = 0$ is tested. In case a), the t-statistic takes the very high absolute value 159 and the p-values of the t-test and the Wilcoxon test are $< 2.2 \cdot 10^{-16}$. In case b), the absolute value of the t-statistic is 17, i.e. much lower. Nevertheless, this is also high enough for p-values $< 2.2 \cdot 10^{-16}$, also in the Wilcoxon test.
This shows one of the drawbacks of statistical tests in that with very high numbers of observations (we have $N = M = 5654$ here), significance has to be expected even for small differences.
A non-significant result would be reached, e.g., if one uses a one-sided hypothesis of type (c) with $\delta_0 = -0.17$. Then, the value of the t-statistic would be -0.69 and the p-value 0.24. Anyway, in sensible applications we have to choose δ_0 a priori in a way that we test for a relevant difference!

Naturally, there are also tests on other parameters of distributions, e.g., on variances. These will not be discussed here.

Let us finish this section with some comments on so-called *multiple testing*. If k tests with hypotheses $H_{01}, \ldots, H_{0k}$ are carried out on the same data set, one might want to test the "global" null-hypothesis

$$H_0 : H_{01}, \ldots, H_{0k} \text{ are all valid} \quad \text{vs.} \quad H_1 : H_{0i} \text{ is not valid for at least one } i$$

on the "global" significance level α. In such cases, the significance levels of the k individual tests have to be adapted. A conservative possibility to do this adequately is the usage of the significance level $\alpha_k = \alpha/k$ for each individual test (*Bonferroni correction*). Such corrections are essential because of the following argument.

If k independent tests each are carried out on the significance level α, then the probability to incorrectly reject any of the hypotheses is α, i.e. for each test the probability to reject the hypothesis correctly is $1 - \alpha$. Since the tests are independent, the probability to reject all k hypotheses correctly is the product of the individual probabilities, namely $(1 - \alpha)^k$. Therefore, the probability to reject at least one of the hypotheses incorrectly is $1 - (1 - \alpha)^k$. With an increasing number of tests, this error probability is increasing. For example, for $\alpha = 0.05$ and $k = 100$ independent tests it takes the value $1 - (1 - 0.05)^{100} = 0.994$. In other words, testing 100 independent correct hypotheses leads almost surely to at least one wrong significant result. This makes significance level corrections like the one above necessary. Note that $1 - (1 - 0.05/100)^{100} \approx 0.04878 < 0.05$ for the Bonferroni correction.

9.8 Modeling of the Relationship between Variables

In Section 9.5 we introduced measures for the intensity of linear relationships between two random variables or features. Let us now look at models for the relationship between two or more variables. We look at so-called directed models, where one response variable is influenced by influential variables.

We will mainly deal with the two most important statistical modeling cases, i.e. the classification and the regression case:

- In *classification problems* an integer-valued response $Y \in \mathbb{Z}$ with finitely many possible values $y^1, \ldots, y^G$ has to be predicted by a so-called *classification rule* based on N observations of $z_i = \begin{bmatrix} x_i^T & y_i \end{bmatrix}^T, i = 1, \ldots, N$, where the vector x summarizes the influential factors.

- In *regression problems* a typically real-valued response $Y \in \mathbb{R}$ has to be predicted by a so-called *regression model* based on N observations of $z_i = \begin{bmatrix} x_i^T & y_i \end{bmatrix}^T, i = 1, \ldots, N$, where the vector x summarizes the influential factors.

In both cases, the influential factors are assumed to be real-valued.

Classification methods will be thoroughly discussed in Chapter 12, but regression problems will be discussed in what follows.

In regression we always assume that all variables including the response are quantitative so that calculations with their observations are allowed. In the general (nonlinear) case we consider the following model:

Definition 9.30 (Nonlinear Multiple Statistical Model). A *nonlinear multiple statistical model* is defined by

$$Y = f(X_1, \ldots, X_K; \beta_1, \ldots, \beta_L) + \varepsilon$$

for a *response* Y dependent on K *influential factors* $X_1, \ldots, X_K$ and the *unknown coefficients* $\beta_1, \ldots, \beta_L$ as well as an *error term* ε. The function f is assumed to be at least twice continuously differentiable in all arguments.

Note that the number of influential variables and the number of unknown coefficients is generally not the same. This is even true for simple linear models as will be seen in what follows. Also note that the differentiability assumption is needed for the estimation of the unknown parameters.

9.8.1 Regression

As a motivation of what will follow, consider the following example.

Example 9.29 (Fit Plot and Residual Plot). *Let us come back to Example 9.22. There, we observed a high correlation between the non-windowed MFCC 1 and MFCC 1 in block 1. This indicates a linear relationship of the non-windowed MFCC*

1 and MFCC 1 in block 1. We are interested in the linear model between the two variables in order to be able to predict the non-windowed MFCC 1 by MFCC 1 in block 1, i.e. by MFCC 1 in the beginning of the tone.

Let us, therefore, start with the simplest regression model for a linear relationship between two variables.

Definition 9.31 (Linear Regression Model for One Influential Variable). The simple so-called *2-variables regression model* is of the form:

$$y_i = \beta_0 + \beta_1 x_i + \varepsilon_i, \quad i = 1, \ldots, N,$$

where y_i = observation of the response variable, β_0 = intercept, β_1 = slope, x_i = observation of the influencing variable, and ε_i = error term.

In such a model it is typically assumed that $E[\varepsilon_i] = 0$, $\mathrm{var}[\varepsilon_i] = \sigma^2$, and ε_i are i.i. (independently identically) $\mathcal{N}(0, \sigma^2)$-distributed (see below). For random variables being independently identically distributed the term *i.i.d.* is used.

The least-squares estimator is then estimating the unknown coefficients β_0, β_1 by solving an optimization problem:

Definition 9.32 (Linear Least Squares (LS) Estimator). *Simple regression of y on x* is realized by minimization of the sum of squared errors

$$\sum_{i=1}^{N} \varepsilon_i^2 = \sum_{i=1}^{N} (y_i - \beta_0 - \beta_1 x_i)^2.$$

The *least-squares (LS) estimators* then have the form:
$\hat{\beta}_1 = r_{XY} \frac{s_Y}{s_X}$, where $r_{XY} = \frac{s_{XY}}{s_X s_Y}$ is the usual empirical correlation coefficient between Y and X, and $\hat{\beta}_0 = \bar{y} - \hat{\beta}_1 \bar{x}$.

Let us now generalize this result for more than one influencing variable:

Definition 9.33 (Multiple Linear Regression Model). The *multiple linear regression model* has the form $y = X\beta + \varepsilon$, where y = vector of the response variable with N observations, X = matrix with entries x_{ik} for observation i of influential variable k, β = vector of $K + 1$ unknown regression coefficients, and ε = error vector of length N. Notice that typically

$$X = \begin{bmatrix} 1 & x_{11} & \cdots & x_{1K} \\ 1 & x_{21} & \cdots & x_{2K} \\ \vdots & \vdots & \ddots & \vdots \\ 1 & x_{N1} & \cdots & x_{NK} \end{bmatrix}$$

so that the first influential "variable" is assumed constant, i.e. a constant term is included in the model. The following *assumptions* are assumed to be valid:

(A.1) X is non-stochastic with $\mathrm{rank}(X) = K + 1$, i.e. all columns are linearly independent,

(**A.2**) $E[\varepsilon_i] = 0$, i.e. $\mu_i = E[y_i] = x_i^T \beta$,

(**A.3**) the errors are i.i.d. with variance σ^2,

(**A.4**) the errors are normally distributed, i.e. the y_i are independently $\mathcal{N}(\mu_i, \sigma^2)$-distributed.

Note that in (A.1) it is assumed that the influential variables can be observed without measurement error. Also, if this is not true, the model is nevertheless very often successfully utilized. Moreover, (A.1) implies that the matrix $X^T X$ is invertible. If the columns of the matrix X are 'nearly linearly dependent' so that the inversion of $X^T X$ leads to numerical problems, then the features are often called *collinear*.

Definition 9.34 (Multiple Regression). *Multiple regression* is realized by minimization of the sum of squared errors

$$\sum_{i=1}^{N} \varepsilon_i^2 = \varepsilon^T \varepsilon = (y - X\beta)^T (y - X\beta).$$

This leads to the results:

Theorem 9.8 (LS-Estimator). The *LS-estimator* has the form $\hat{\beta} = (X^T X)^{-1} X^T y$ and the corresponding variance estimator $\hat{\sigma}^2 = \frac{SSR}{N-K}$, where *SSR* is the sum of squared residuals, i.e. $SSR := (y - X\hat{\beta})^T (y - X\hat{\beta})$ for the LS-estimator $\hat{\beta}$.

Note that the above formula for the LS-estimator is numerically bad, and should not be directly used for its calculation. Up-to-date computer programs avoid the inverse $(X^T X)^{-1}$.

Let us continue with important properties of LS-estimates. We will see that under reasonable assumptions the LS-estimator is best unbiased, i.e. it is unbiased (see Section 9.6) and has minimum variance. Then, we will derive confidence intervals for the true model coefficients. Note that minimum variance of the LS-estimator guarantees minimum length of confidence intervals. Last but not least, we will show, how the LS-estimator simplifies in the case of uncorrelated influential variables which can be guaranteed in some time series models below.

Theorem 9.9 (Properties of LS-Estimates). Under the assumptions (A.1) – (A.3) the LS-estimator is *unbiased with minimum variance* among the linear estimators of the unknown coefficient vector β.

Under assumption (A.4) the $(1 - \alpha) \cdot 100$-*confidence interval* for the unknown coefficient β_i has the form: $\left[\hat{\beta}_i - t_{crit} \sqrt{\text{vâr}(\hat{\beta}_i)}, \ \hat{\beta}_i + t_{crit} \sqrt{\text{vâr}(\hat{\beta}_i)} \right]$,

where $t_{crit} := t_{N-K;(1-\alpha/2)}$ is the $(1 - \alpha/2)$-quantile of the *t*-distribution with $N - K$ degrees of freedom and α is typically 0.05 or 0.01.

If the columns of X are uncorrelated, then $\hat{\beta} = (X^T X)^{-1} X^T y = D X^T y$, where D is a diagonal matrix. In this case, the estimate of the coefficient for the influential variable x_k is independent of the observations of the other variables.

Significance of an estimate is now defined by means of its confidence interval.

Definition 9.35 (Significance of LS-Estimates). An LS-estimate $\hat{\beta}_i$ is called *significant* at level α if the $(1 - \alpha) \cdot 100$-*confidence interval* is not including 0.

Significance thus indicates that one can be "$(1 - \alpha) \cdot 100\%$ sure" that the true regression coefficient is not zero, i.e. that the corresponding influential variable has really an influence on response Y.

After having identified a linear model, one is often interested in its goodness of fit as a measure of its relevance.

Definition 9.36 (Goodness of Fit). The *goodness of fit* of a model is defined as the part of the variance in the data explained by the model:

$$R^2 = 1 - \frac{\mathrm{v\hat{a}r}(\boldsymbol{\varepsilon})}{\mathrm{v\hat{a}r}(\boldsymbol{y})}.$$

A model is adequate if the variance of the residuals is small in comparison to the variance of Y. Therefore, if $R^2 = 0$, then $\mathrm{var}[\varepsilon] = \mathrm{var}[Y]$, and the regression model does not explain any part of the data, i.e. there is no linear relationship between Y and the X_i. In contrast, if $R^2 = 1$, then $\mathrm{var}[\varepsilon] = 0$ and the model is error-free, i.e. all y lie on a line (plane, hyperplane) determined by the x_i.

The goodness of fit assesses the fit of the model, i.e. how adequate the model is for the data, from which it is "learned". In practice, however, it is more important to judge the "prediction quality" of a model, i.e. how near so-called model predictions are to the true response. To judge this, we have to define predictions and, since we only have the one observed sample, hold out some observations during estimation to have observations with true responses to judge the corresponding predictions. Hold out methods will be further discussed in Chapter 13.

Definition 9.37 (Point Prediction and Prediction Intervals). The *point prediction* of a response Y for values $x_{01}, \ldots, x_{0K}$ of the influential factors $X_1, \ldots, X_K$ is defined as

$$
\begin{aligned}
\hat{y}_0 &:= f(x_{01}, \ldots, x_{0K}; \hat{\beta}_1, \ldots, \hat{\beta}_L) \\
&= \boldsymbol{x}_0^T \hat{\boldsymbol{\beta}} \text{ for linear models.}
\end{aligned}
$$

In case of linear models, point predictions are called *linear predictions*.

$(1 - \alpha) \cdot 100$-*prediction intervals* are intervals around point predictions, which cover the "true" value of the response to be predicted with probability $(1 - \alpha)$.

In the case of multiple linear models and assumption (A.4), $(1 - \alpha) \cdot 100$-prediction intervals have the form:

$$\left[\hat{y}_0 - t_{crit} \hat{\sigma} \sqrt{1 + \boldsymbol{x}_0^T (\boldsymbol{X}^T \boldsymbol{X})^{-1} \boldsymbol{x}_0}, \hat{y}_0 + t_{crit} \hat{\sigma} \sqrt{1 + \boldsymbol{x}_0^T (\boldsymbol{X}^T \boldsymbol{X})^{-1} \boldsymbol{x}_0} \right],$$

where $t_{crit} := t_{N-K; (1-\alpha/2)}$ is the $(1 - \alpha/2)$-quantile of the t-distribution with $N - K$ degrees of freedom.

Prediction quality can be, e.g., defined by the *coverage* of the prediction interval, which should be as close to $(1 - \alpha) \cdot 100\%$ as possible, i.e. the prediction interval should, if possible, cover exactly $(1 - \alpha) \cdot 100\%$ of the distribution of Y_0. Additionally, the *length of the prediction interval* should be as small as possible, i.e. the uncertainty about the location of the true value of y_0 should be as low as possible.

In order to assess the coverage one has to have sufficient new $\boldsymbol{x}_0$ with known y_0 at one's disposal for checking the condition whether the realized y_0 lies in the prediction interval. Methods for generating such $\boldsymbol{x}_0$ and other prediction quality measures will be discussed in Chapter 13.

Regression results are often illustrated by means of various plots. We introduce the most well-known ones, namely the fit plot and the residual plot.

Definition 9.38 (Fit Plot and Residual Plot). For one influencing variable X a *fit plot* is a scatterplot of $(X \ Y)^T$ adding the line $\hat{y} = \boldsymbol{x}^T \hat{\boldsymbol{\beta}}$ of values fitted by the model.
   A *residual plot* is a scatterplot of the values of the model line $\hat{y}$ on the x-axis versus the estimated model error (residual) $\hat{\varepsilon}$ on the y-axis.

Example 9.30 (Fit Plot and Residual Plot). Let us come back to our motivating Example 9.29. We are interested in the linear regression model of the non-windowed MFCC 1 (mfcc_unwin_1) on MFCC 1 in block 1 (mfcc.block1_1). Because of the high correlation we expect a good fit. And indeed, the goodness-of-fit is $R^2 = 0.87$, which is reasonably near 1 (see Figure 9.9, left). The corresponding model has the form: (non-windowed MFCC 1) $= 0.458 + 0.840 \cdot (MFCC\ 1\ in\ block\ 1)$. The 95%-confidence interval of, e.g., the coefficient of MFCC 1 in block 1 is $[0.831, 0.849]$, i.e. this coefficient is highly significantly different from 0. If one adds another 10 MFCCs to the model, i.e. MFCC 2 in block 1 . . . MFCC 11 in block 1, then the goodness-of-fit increases by only 0.01 to 0.88. Note that the coefficients of all these 11 influential factors are significant for $\alpha = 0.0001$. This can be interpreted as MFCC 1 in block 1 explaining nearly all variation in non-windowed MFCC 1. The other MFCC i in block 1 improve explanation by a very small but significant part.
   Finally, let us have a look at the residual plot of the simple regression of non-windowed MFCC 1 on MFCC 1 in block 1 (see Figure 9.9, right). Obviously, there is no structure in the residuals except some high residuals for fitted values between 2 and 3. This does not change when taking 11 influential factors instead.

9.8.2 Time Series Models

Time series are observations of time-dependent variables. In this section we assume that all variables are quantitative. For time series special kinds of statistical models are in use representing the time dependence of a random variable.

Definition 9.39 (Time Series). A time-dependent series $y[t]$, $t = 1, \ldots, T$, of observations of a quantitative variable Y is called a *time series*. Note that we assume here that the observations are equidistant, i.e. that the time intervals between each two observations are equal.

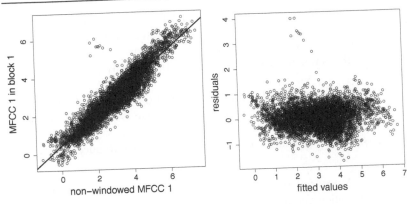

Figure 9.9: Fit plot (left) and residual plot (right) of simple regression of "mfcc_unwin_1" on "mfcc.block1_1".

Obviously, time is giving the data a "natural" structure and the time dependence is decisive for the interpretation.

Example 9.31 (Music Observations as Time Series). Music is nothing but vibrations generated by music instruments played over time. Therefore, musical signals can be represented as time series in a natural way. Indeed, in Example 9.16 we already saw a model for musical tone progression over time. However, this model considered a sequence of discrete states instead of observations of quantitative signals. In this section we will introduce time series models for quantitative vibrating signals like the waveform of an audio signal or the MFCCs and the chroma features introduced in Example 9.9.

There are lots of different models for time series data. Here, however, we will concentrate on autoregressive and periodical models which are most often used for modeling musical time series.

Definition 9.40 (Time Series Models). The model $y[t] = \beta_1 + \beta_2 y[t-1] + \varepsilon[t]$, $|\beta_2| < 1$, where $\varepsilon \sim$ i.i.$\mathcal{N}(0,\sigma^2)$, is called a *(stationary) 1st-order autoregressive model (AR(1)-model)*. In such a model, the value of Y in time period t linearly depends on its value in time period $t-1$, i.e. the value of Y with time *lag* 1.

The model $y[t] = \beta_1 + \beta_2 y[t-1] + \ldots + \beta_{p+1} y[t-p] + \varepsilon[t]$, where $\varepsilon \sim$ i.i.$\mathcal{N}(0,\sigma^2)$, is called *p-th-order autoregressive model (AR(p)-model)*. If all roots of the so-called characteristic polynomial have absolute value greater than 1, meaning that $|z| > 1$ for all z with $1 - \beta_2 z - \beta_3 z^2 - \ldots - \beta_{p+1} z^p = 0$, this model is *stationary*. Obviously, p is the maximal involved time lag.

A model is called *stationary* if its predictions have expected values, variances, and covariances that are invariant against shifts on the time axis, i.e. that do not depend on time. This means that $E[\hat{Y}[t]]$ and var$[\hat{Y}[t]]$ are constant for all t and the cov$(\hat{Y}[t],\hat{Y}[s])$ only depend on the difference $t - s$.

Please notice that the conditions $|\beta_2| < 1$ and "all roots of the characteristic polynomial $1 - \beta_2 z - \beta_3 z^2 - \ldots - \beta_{p+1} z^p$ have absolute value greater than 1" both guarantee the stationarity of the autoregressive model. Moreover, the two conditions are equivalent for AR(1)-models, since for $1 - \beta_2 z = 0$ obviously $|z| > 1$ is equivalent with $|\beta_2| < 1$.

Indeed, an AR(1)-model only represents a damped waveform if $|\beta_2| < 1$. The notion *autoregression* is based on the fact that a variable is regressed on itself in a previous time period.

Autoregressive models relate to *autocorrelation* already introduced in Sections 4.8 and 2.2.7.

Definition 9.41 (Autocorrelation). The *autocorrelation coefficient of order p* is defined as the correlation coefficient of $y[t]$ with its lag $y[t-p]$. One can show that in a stationary AR(1)-model the coefficient of $y[t-1]$ is equal to the 1st-order autocorrelation coefficient. The *empirical 1st-order autocorrelation coefficient* looks as follows:

$$r_{y[t],y[t-1]} := \frac{\sum_{k=2}^{T}(y[k] - \bar{y})(y[k-1] - \bar{y})}{s_y^2}$$

in case of stationarity.

Note that in the autocorrelation function in Section 4.8 it is implicitly assumed that $y[t]$ is centered at 0 and the variance normalization is ignored (see also the discussion in Section 2.2.7).

Example 9.32 (Stationary AR(1)-Models). *Figure 9.10 illustrates that an AR(1)-model with a positive coefficient β_2 (positive autocorrelation) causes a slow oscillation after a short "attack" phase and a negative coefficient β_2 causes a "nervous" oscillation (negative autocorrelation), both around $\frac{\beta_1}{1-\beta_2}$. Independent of the starting value of the oscillation, in the long run the model will "converge" to this value in that the expected value of the model prediction will be constant $\frac{\beta_1}{1-\beta_2}$ for t big enough.*

Note that positive autocorrelation relates to low-pass filtering and negative autocorrelation to high-pass filtering (cp. Example 4.2).

Unbiasedness can be proven for the estimates in "static" linear models which are independent of time (cp. Section 9.6). In contrast, time dependency as in time series models, often also called dynamics, typically leads to situations where unbiasedness cannot be expected. This is then replaced by so-called asymptotic properties, i.e. properties which are only valid for $T \to \infty$. Typical such properties are consistency and asymptotic normality, which are valid for least-squares estimates of the coefficients of stationary AR(p)-models.

Definition 9.42 (Consistency and Asymptotic Normality). Let $\boldsymbol{\theta}$ be an unknown parameter vector of a statistical distribution. An estimator $\boldsymbol{t}_T$ of $g(\boldsymbol{\theta}) \in \mathbb{R}^q$ based on T repetitions of the corresponding random variable is called *consistent* iff for all $\eta > 0$: $P(\|\boldsymbol{t}_T - g(\boldsymbol{\theta})\| > \eta) \to 0$ for $T \to \infty$, which is often written as $\mathrm{plim}(\boldsymbol{t}_T) = g(\boldsymbol{\theta})$, where plim stands for *probability limit*.

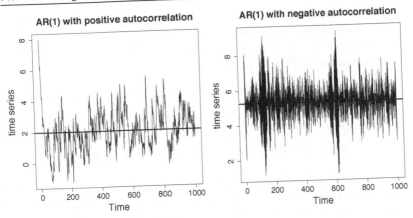

Figure 9.10: Stationary AR(1)-processes; left: positive autoregression ($\beta_1 = 0.2, \beta_2 = 0.9$), right: negative autoregression ($\beta_1 = 10, \beta_2 = -0.9$).

An estimator t_T of $g(\boldsymbol{\theta}) \in \mathbb{R}^q$ based on T repetitions is called *asymptotically normal* iff there is a sequence of nonsingular matrices A_T and vectors a_T so that $(A_T t_T - a_T)$ converges in distribution to a multivariate normal distribution $\mathcal{N}(\boldsymbol{0}, \boldsymbol{\Sigma})$ for $T \to \infty$, where $\boldsymbol{\Sigma}$ is nonsingular. In the simplest case $a_T = A_T \cdot E(t_T)$, where A_T is a scalar $\in \mathbb{R}$.

Having now defined all terms, let us finally state that least-squares estimates of the coefficients of stationary AR(p)-models are *consistent*. Also, these estimates are *asymptotically normal* if $E[y[t]^4]$ is finite.

Example 9.33 (Waveform of Superposition of Autoregressive Sources (see [1])). *For audio scene analysis, we may wish to decompose the waveform $y[t], t = 1, \ldots, T$, of a mixed audio signal into the waveforms of its constituent sources. We can model the variability of these waveforms by autoregressive models. Specifically, suppose that the waveform $s_i[t], t = 1, \ldots, T$, of the i-th source satisfies a p-th-order autoregressive model $s_i[t] = \sum_{j=1}^{p} a_{ij} s_i[t-j] + \varepsilon_i[t]$ for samples at times $t > p$. The particular realization of the i-th source's waveform is determined by p initial values of the autoregressive model denoted by $s_i[t], t = 1, \ldots, p$. Therefore, there are different possible realizations of the process depending on the p initial values, which can not only parameterize variations in amplitude (by scaling $s_i[t]$), but also variations in phase (by shifting $s_i[t]$) and timbre (by re-weighting $s_i[t]$). Finally, signals of variable duration T can be described by simply evolving the model for different numbers of time steps.*

Let us now switch to the second kind of time series models discussed here, the periodical models. First, we have to define some fundamental terms.

Definition 9.43 (Period and Frequency). A function $g(t)$ is called *periodical* with a *period* $P \neq 0$ iff $g(t + P) = g(t)$ for all $t \in \mathbb{R}$. The *base period* of a periodical function g is the smallest period P. Notice that each integer multiple of the base period is again a period of the function g. The *frequency* f of $g(t)$ is the inverse of the base period P, i.e. $f = \frac{1}{P}$. Typically, frequencies are represented in the unit Hertz (Hz), i.e. in number of oscillations per second.

Examples for periodic functions are the so-called harmonic oscillations.

Definition 9.44 (Harmonic Models). A simple *(harmonic)* oscillation is defined by the model

$$y[t] = \beta_1 + \beta_2 \cos(2\pi \frac{f}{f_s} t) + \beta_3 \sin(2\pi \frac{f}{f_s} t) + \varepsilon[t],$$

where $\varepsilon[t] \sim$ i.i.$\mathcal{N}(0, \sigma^2)$, f is the *frequency* of the oscillation, $f_s :=$ *sampling rate* := number of observations in a desired time unit, and β_2, β_3 are the *amplitudes* of cosine and sine, respectively. If the time unit is a second, n is also measured in Hz. Frequently, oscillations with different frequencies are superimposed. This leads to a model of the form:

$$y[t] = \beta_1 + \sum_{k=1}^{K} (\beta_{2k} \cos(2\pi \frac{f_k}{f_s} t) + \beta_{2k+1} \sin(2\pi \frac{f_k}{f_s} t)) + \varepsilon[t].$$

Note that such harmonic oscillations are not damped, i.e. go on in the same way forever. Decisive for the adequacy of the model is the correct choice of the *frequencies* f_k. Harmonic oscillations have the favorable property that the influence of so-called *Fourier frequencies* $f_\mu = \mu \frac{f_s}{T}$ can be determined independently of each other, when $\mu = 1, \ldots, \frac{T}{2}$. Then $0 \le f_\mu \le \frac{f_s}{2}$. $\frac{f_s}{2}$ is also called *Nyquist frequency*. Note that $\frac{f_\mu}{f_s} = \frac{\mu}{T}$. These oscillations do not influence each other, they are uncorrelated. Therefore, important frequencies of this type can be determined independently of each other, e.g., one can individually check those frequencies which make sense by substantive arguments. In the following, we will always denote the Fourier frequencies by $f_\mu = \mu \frac{f_s}{T}$. Note, however, that important frequencies will usually not be Fourier frequencies. Therefore, in general we cannot make use of the regression simplification for uncorrelated influential variables mentioned in Theorem 9.9, since the estimate of the amplitude of a non-Fourier frequency depends on the other frequencies.

Example 9.34 (Polyphonic Sound (see [2])). *In general, the oscillations of the air generated by a music instrument are harmonic, i.e. they are composed of a fundamental frequency and so-called overtone frequencies, which are multiples of the fundamental frequency. The tones related to the fundamental frequency and to the overtone frequencies are called partial tones. In a polyphonic sound the partial tones of all involved tones are superimposed. For the identification of the individually played tones, a general model is proposed for J simultaneously played tones with varying numbers M_j of partial tones. This model will be introduced here in the special case of constant volume over the whole sound length:*

$$y[t] = \sum_{j=1}^{J} \sum_{m=1}^{M_j} \left\{ a_{j,m} \cos\left(2\pi(m + \delta_{j,m}) \frac{f_{0j}}{f_s} t \right) + b_{j,m} \sin\left(2\pi(m + \delta_{j,m}) \frac{f_{0j}}{f_s} t \right) \right\} + \varepsilon[t],$$

K = number of simultaneously played tones,
M_j = number of partial tones of the j-th tone,
Θ = amplitude vector with elements $a_{j,m}$ and $b_{j,m}$

$$= \quad (a_{1,1} \;\; b_{1,1} \;\; a_{1,2} \;\; b_{1,2} \;\; \cdots \;\; a_{1,M_1} \;\; b_{1,M_1}$$
$$a_{2,1} \;\; b_{2,1} \;\; a_{2,2} \;\; b_{2,2} \;\; \cdots \;\; a_{2,M_2} \;\; b_{2,M_2}$$

$$\cdots$$

$$a_{J,1} \;\; b_{J,1} \;\; a_{J,2} \;\; b_{J,2} \;\; \cdots \;\; a_{J,M_J} \;\; b_{J,M_J})^T$$

$\delta_{j,m}$ = shift parameter (detuning) of the m-th partial tone of the j-th tone,
f_{0j} = fundamental frequency of the j-th tone,
f_s = sampling rate, and
$\varepsilon[t]$ = error term, $t = 1, \ldots, T$.

One might want to change this model so that we do not allow a detuning in the fundamental frequency, i.e. we set $\delta_{j,1} = 0$. Therefore, the fundamental frequency of the j-th tone is only determined by f_{0j}. The other partial tones are, however, allowed to be out of tune. This is especially relevant for singers in the range of the so-called singer's formant in which the singer has to be particularly loud in order to be heard when accompanied by an orchestra.

We now define another representation of a time series by means of the "strengths" of the frequencies in the time series. These strengths can be determined by the so-called Fourier transform of a time series. The Discrete Fourier transform was introduced for deterministic complex signals in Section 4.4.1. In the following, it is discussed for (stochastic) real signals. For the comparison with Section 4.4.1, note that $e^{i\phi} = \cos(\phi) + i \cdot \sin(\phi)$ and $e^{-i\phi} = \cos(\phi) - i \cdot \sin(\phi)$.

Definition 9.45 (Discrete Fourier Transform of Time Series). The *discrete Fourier transform (DFT)* of a real time series $(y[t])_{t=1,\ldots,T}$ is defined for Fourier frequencies $0 \le f_\mu \le \frac{f_s}{2}$, f_s = sampling rate, as follows:

$$F_y\left[f_\mu\right] := C\left[f_\mu\right] - iS\left[f_\mu\right] := \sum_{t=1}^{T} y[t] \cos(2\pi \frac{f_\mu}{f_s} t) - i \sum_{t=1}^{T} y[t] \sin(2\pi \frac{f_\mu}{f_s} t).$$

The discrete Fourier transform defines the so-called frequency representation of a time series equivalent to the time representation. Note that the Fourier frequencies are called *center frequencies of the DFT bins* in Chapter 4 (see below, Definition 4.10). The back-transformation can be realized as follows.

Definition 9.46 (Inverse of the Discrete Fourier Transform). The *inverse of the discrete Fourier transform* has the form for $t = 1, \ldots, T \le \frac{f_s}{2}$:

$$y[t] = \bar{y} + \frac{2}{f_s} \sum_{\mu=1}^{\tilde{M}} C\left[f_\mu\right] \cos(2\pi \frac{f_\mu}{f_s} t) + \frac{2}{f_s} \sum_{\mu=1}^{\tilde{M}} S\left[f_\mu\right] \sin(2\pi \frac{f_\mu}{f_s} t) \text{ if } f_s = 2\tilde{M} + 1,$$

$$y[t] = \bar{y} + \frac{2}{f_s} \sum_{\mu=1}^{\tilde{M}-1} C\left[f_\mu\right] \cos(2\pi \frac{f_\mu}{f_s} t) + \frac{2}{f_s} \sum_{\mu=1}^{\tilde{M}-1} S\left[f_\mu\right] \sin(2\pi \frac{f_\mu}{f_s} t) + \frac{1}{f_s} C\left[\frac{f_s}{2}\right] \cos(\pi t)$$

if $f_s = 2\tilde{M}$.

$1,\ldots,N$, $k = 1,\ldots,K$, of the PCs are called *scores*. The vector $\mathbf{z}_k$ of the scores of the k-th PC Z_k has the form $\mathbf{z}_k = \mathbf{X}\mathbf{g}_k$ so that $z_{ik} = (x_{i1} - \bar{x}_1)g_{1k} + \ldots + (x_{iK} - \bar{x}_K)g_{Kk}$.

Notice that orthogonality of score vectors is equivalent to empirical uncorrelation of score vectors. Moreover, notice that the length restriction of the loading vectors is necessary since the empirical variance of the score vectors increases quadratically with the length of the loading vector. PCs are often interpreted as *implicit (latent) features*, since they are not observed but derived from the original features.

The loadings of the PCs are constructed by means of a property of *covariance matrices*.

Theorem 9.10 (Construction of Principal Components). The empirical covariance matrix $S := \frac{\mathbf{X}^T\mathbf{X}}{N-1}$ of the mean centered features in $\mathbf{X}$ can be transformed by means of the so-called *spectral decomposition* into a diagonal matrix[2] where a matrix $\mathbf{G}$ is constructed so that $\mathbf{G}^T\mathbf{S}\mathbf{G} = \mathbf{\Lambda}$, where $\mathbf{G}^T\mathbf{G} = \mathbf{I}$ and $\mathbf{\Lambda}$ is a diagonal matrix whose elements are 0 except on the main diagonal: $\lambda_{11} \geq \ldots \geq \lambda_{KK} \geq 0$.

This matrix $\mathbf{G} := [\mathbf{g}_1 \ \ldots \ \mathbf{g}_K]$ satisfies the properties of a loading matrix since $\mathbf{\Lambda} = \mathbf{G}^T\mathbf{S}\mathbf{G} = \frac{\mathbf{G}^T\mathbf{X}^T\mathbf{X}\mathbf{G}}{N-1} = \frac{\mathbf{Z}^T\mathbf{Z}}{N-1}$. Thus, the columns of $\mathbf{Z} := [\mathbf{Z}_1 \ \ldots \ Z_k]$, i.e. the score vectors of the PCs, are uncorrelated and $\hat{\text{var}}_{\mathbf{z}_1} = \lambda_{11} \geq \ldots \geq \lambda_{KK} = \hat{\text{var}}_{\mathbf{z}_K} \geq 0$.

All K PCs together span the same K-dimensional space as the original K features. A PCA can, however, be used for *dimension reduction*. For this, the number of dimensions relevant to "explain" most of the variance in the data are determined by a *dimension reduction criterion*. A typical simple criterion is the *share r_p of the first p PCs* on the "total variation in the data", i.e. the ratio of the variance of the first p PCs and the total variance. Since the PCs are empirically uncorrelated, the empirical variances just add up to the total variation so that

$$r_p = \frac{\hat{\text{var}}_{\mathbf{z}_1} + \hat{\text{var}}_{\mathbf{z}_2} + \ldots + \hat{\text{var}}_{\mathbf{z}_p}}{\hat{\text{var}}_{\mathbf{z}_1} + \hat{\text{var}}_{\mathbf{z}_2} + \ldots + \hat{\text{var}}_{\mathbf{z}_K}}.$$

Since the first PCs represent the biggest share of the total variance, a criterion like $r_p \geq 0.95$ often leads to a drastic dimension reduction so that an adequate graphical representation is possible. The "latent observations" z_{ik}, i.e. the scores, of the first few PCs are then often plotted in order to identify structures or groups in the observations. Notice that the absolute distances between the score values are not interpretable.

Also notice that PCA is *not scale invariant* so that the result may change also in interpretation if the units of the features change, e.g., from MHz, db, and seconds to Hz, Joule, and milliseconds. Typical kinds of PCA are *PCA on the basis of covariances*, where "natural units" are assumed, and *PCA on the basis of correlations*, where all features are standardized to variance 1.

A further disadvantage of the PCA is the fact that weighted sums are most of the time not interpretable. In such cases, results might even be useless for a user.

---

[2]The spectral decomposition can be seen a *singular value decomposition (SVD)* for quadratic and symmetric matrices. SVDs are used in recommendation systems.

Another possible way of dimension reduction is the selection of relevant PCs by means of their significance in a linear model of a response on the PCs of the influential variables. Such a regression is called *principal component regression* using the model $y = \sum_{k=1}^{K} \beta_k z_k + \varepsilon$. Only PCs with significant coefficients are then taken to be relevant to explain the response. Notice that in this model the significance of the coefficients can really be assigned to the corresponding PCs since the PCs are uncorrelated. Also notice that these PCs do not have to be the first PCs!

The scores and loadings of principal components are often illustrated by means of score plots or biplots. A score plot, most of the time, shows the scores of only the first two principal components for convenience. A so-called *biplot* additionally shows the directions of the original variables in the score plot. To achieve this, a second coordinate system is added to the score plot with the same origin showing the vectors of the loadings of the original variables. In order to utilize the same plotting space as the score plot, both loadings are scaled by the same factor. For each original variable this vector is pointing to a multiple of

$$(\text{loading of component 1} \quad \text{loading of component 2})^T$$

for this variable. Biplots are often used to interpret structure in score plots by identifying original variables related to the structure (see next example).

Example 9.37 (Scores Plots and Bi-Plots (Example 9.36 cont.). *Let us once more consider the data in Example 9.9, this time the* 14 *chroma variables of block 1. A principal component analysis of these data leads to a share of 93% of the variance of the first two principal components on the basis of covariances. A scatterplot of the first two principal components can be seen in Figure 9.12. Obviously, there is structure in the plot, namely, there are (nearly) strict bounds for the realizations.*

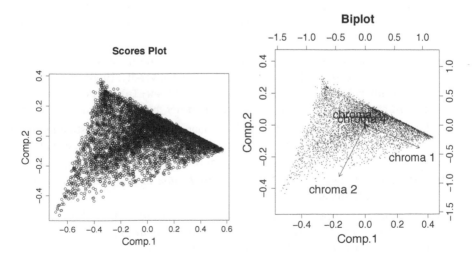

Figure 9.12: First 2 principal components of 14 chroma vectors - left: scores plot; right: biplot.

Original Features

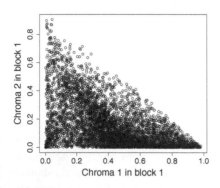

Figure 9.13: First 2 original chroma vectors.

The biplot shows that the triangle legs nearly correspond to the directions of the first two chroma elements (for the fundamental and the first overtone). The other chroma elements all have very small contributions. This leads to the idea of plotting the first two original chroma elements against each other (see Figure 9.13) leading to a graphics very similar to the score plot of the first two principal components but somewhat rotated. This graphics is easily understood by the fact that the sum of the two components never exceeds 1.

9.9   Further Reading

A very good introduction to the theory of statistics, but without music examples, can be found in [4].

Bibliography

[1] Y. Cho and L. Saul. Learning dictionaries of stable autoregressive models for audio scene analysis. In L. Bottou and M. Littman, eds., *Proceedings of the 26th International Conference on Machine Learning*, pp. 169–176, Montreal, 2009. Omnipress.

[2] M. Davy and S. Godsill. Bayesian harmonic models for musical pitch estimation and analysis. Technical Report 431, Cambridge University, Engineering Department, Cambridge, 2002. Published in http://www-labs.iro.umontreal. ca/~pift6080/H08/documents/papers/davy_bayes_extraction.pdf.

[3] M. Eichhoff, I. Vatolkin, and C. Weihs. Piano and guitar tone distinction based on extended feature analysis. In A. Giusti, G. Ritter, and M. Vichi, eds., *Classification and Data Mining*, pp. 215–224. Springer, 2013.

[4] A. Mood, F. Graybill, and D. Boes. *Introduction to the Theory of Statistics.* McGraw-Hill, Singapore, 1974.

[5] C. Raphael. Music plus one and machine learning. In J. Fürnkranz and T. Joachims, eds., *Proceedings of the 27th International Conference on Machine Learning (ICML-10)*, pp. 21–28, Haifal, 2010. Omnipress.

[6] I. Xenakis. *Formalized Music.* Pendragon Press, New York, 1992.

Chapter 10

Optimization

GÜNTER RUDOLPH
Department of Computer Science, TU Dortmund, Germany

10.1 Introduction

In music data analysis there are numerous occasions to apply optimization techniques to achieve a better performance, e.g. in the classification of tunes into genres, in recognizing instruments in tunes, or in the generation of playlists. Many musical software products and applications are based on optimized methods in the field of music data analysis which will be detailed in later chapters. This chapter concentrates on the introduction of terminology in the field of optimization and the description of commonly applied optimization techniques. Small examples illustrate how these methods can be deployed in music data analysis tasks.

Before introducing the basic concepts in a formal manner, we provide a brief abstract overview of the topics covered in this chapter. Let the map $f : X \to Y$ describe the input/output behavior of some system from elements $x \in X$ to elements $y \in Y$. Here, it is assumed that the map is deterministic and time-invariant (i.e., every specific input always yields the same specific output). The task of optimization is to find one or several inputs $x^* \in X$ such that the output $f(x^*) \in Y$ exhibits some extremal property in the set Y. In most cases Y is a subset of the set of real numbers $\mathbb{R}$ and the extremal property requires that $f(x^*)$ is either the maximum or the minimum value in $Y \subset \mathbb{R}$, where the map $f(x)$ is termed a real-valued *objective function*. The task to find such an element x^* is known under the term single-objective optimization. The situation changes if the map is vector-valued with elements from $\mathbb{R}^d$ where $d \geq 2$. In this case a different extremal property has to be postulated, which in turn requires different optimization methods. This problem is known under the term multi-objective optimization.

The main difference between single- and multi-objective optimization rests on the fact that two distinct elements from Y are not guaranteed to be comparable in the latter case since Y is only partially ordered. To understand the problem to full extent it is important to keep in mind that the values $f_1(x), f_2(x), \ldots, f_d(x)$ of the $d \geq 2$ real-valued objective functions represent incommensurable quantities that cannot be

263

minimized simultaneously in general: While f_1 may measure the total computation time for a tune classification in seconds, f_2 may measure the accuracy of the genre prediction in percent, f_3 the number of features used by the classifier, and so forth. Moreover, the objectives are conflicting: Improvement in one specific objective may result in worsening another objective and vice versa. As a consequence, the notion of the "optimality" of some solution needs a more general formulation as in the single-criterion case. It appears to be reasonable to regard those elements as being optimal which cannot be improved with respect to one criterion without getting a worse value in another criterion. Elements with this property are said to be Pareto-optimal in this context. These concepts are rendered more precisely in Section 10.2, before we describe prevalent solution methods for single-objective problems in Section 10.3 and multi-objective problems in Section 10.4. Since the field of optimization is a well-developed discipline with many facets that cannot be covered in a single chapter, additional pointers to the literature are provided in Section 10.5 for further reading.

10.2   Basic Concepts

Definition 10.1. Let X be some set and $Y \subseteq \mathbb{R}^d$ with $d \in \mathbb{N}$. The map $f : X \to Y$ is termed the *objective function* and X the *feasible region*. Every element $x \in X$ is called a *feasible solution*. If the feasible region is described by $p \in \mathbb{N}$ equalities and/or $q \in \mathbb{N}$ inequalities of the form $h(x) = 0 \in \mathbb{R}^p$ resp. $g(x) \geq 0 \in \mathbb{R}^q$, then these equalities and inequalities are called *constraints*.

Notice that the objective function and/or the constraints may be defined by computer procedures about whose input/output behavior nothing (black box scenario) or only little (gray box scenario) is known. In general, it cannot be assumed that the extent of constraint violation can be specified numerically by the deviation from the target value or the exceedance of the limit value. It may happen that there is simply an oracle returning either 'yes' or 'no' upon on the inquiry about feasibility of some solution fed to the oracle.

Definition 10.2. Let $f : X \to \mathbb{R}$ be the objective function and $N(x) \subset X$ some neighborhood of $x \in X$.

a) A feasible solution $x^* \in X$ is termed a *globally optimal solution* or *global solution* of $f(\cdot)$ if for all $x \in X$ holds $f(x^*) \leq f(x)$. Then the value $f(x^*)$ is called the *global minimum*.

b) A feasible solution $x^* \in X$ is termed a *locally optimal solution* or *local solution* of $f(\cdot)$ if for all $x \in N(x^*)$ holds $f(x^*) \leq f(x)$. Then the value $f(x^*)$ is called a *local minimum*.

The task of finding an optimal solution of a scalar-valued objective function is termed a *single-objective optimization problem*.

Typically, the elements of X are given by n-tuples $x = (x_1, x_2, \ldots, x_n)$ and each component x_i with $i = 1, \ldots, n$ may take different numerical values from $\mathbb{B}$, $\mathbb{Z}$ or $\mathbb{R}$. Therefore $x_1, \ldots, x_n$ are considered as variables and termed *decision variables*.

Notice that it is sufficient to consider only minimization problems since every max-imization problem can be equivalently reformulated and solved as a minimization problem and vice versa:

$$\max\{f(x) : x \in X\} = -\min\{-f(x) : x \in X\} \text{ with}$$

$$\operatorname{argmax}\{f(x) : x \in X\} = \operatorname{argmin}\{-f(x) : x \in X\}.$$

Before defining the optimality concept of multi-objective optimization some termi-nology is required:

Definition 10.3. Let $f : X \to \mathbb{R}^d$ be the vector-valued objective function with $d \geq 2$. The feasible space X is also termed the *decision space*, whereas its image $F := f(X) = \{f(x) : x \in X\}$ is called the *objective space*. Elements $x \in X$ are also termed *decision vectors* and their images $f(x) \in F$ *objective vectors*. An objective vector $f(x_1)$ is said to *dominate* objective vector $f(x_2)$, denoted $f(x_1) \prec f(x_2)$, if

$$\forall i = 1, \ldots, d : f_i(x_1) \leq f_i(x_2) \quad \text{and} \quad \exists k = 1, \ldots, d : f_k(x_1) < f_k(x_2).$$

A decision vector dominates another decision vector if their images do. Two distinct decision vectors $f(x_1)$ and $f(x_2)$ are called *comparable* if either $f(x_1) \prec f(x_2)$ or $f(x_2) \prec f(x_1)$. Otherwise they are termed *incomparable*, denoted $f(x_1) \parallel f(x_2)$.

For example, in case of two objectives (to be minimized simultaneously) we have

$$\begin{pmatrix} 3 \\ 5 \end{pmatrix} \prec \begin{pmatrix} 4 \\ 6 \end{pmatrix}, \quad \begin{pmatrix} 3 \\ 5 \end{pmatrix} \prec \begin{pmatrix} 3 \\ 6 \end{pmatrix} \quad \text{but} \quad \begin{pmatrix} 3 \\ 5 \end{pmatrix} \parallel \begin{pmatrix} 2 \\ 6 \end{pmatrix}.$$

The possible incomparableness of solutions requires the definition of an own concept of optimality.

Definition 10.4. Let $f : X \to \mathbb{R}^d$ be the vector-valued objective function gathering $d \geq 2$ scalar-valued objective functions. The task to minimize these d objective func-tions simultaneously is termed the *multi-objective optimization problem*. A feasible solution or decision vector $x^* \in X$ is said to be *Pareto-optimal* if there is no $x \in X$ whose image $f(x)$ dominates $f(x^*)$, i.e., if $\nexists x \in X : f(x) \prec f(x^*)$. All Pareto-optimal decision vectors are collected in the *Pareto set* X^*. The images $f(x^*)$ of Pareto-optimal decision vectors $x^* \in X^*$ are termed *efficient solutions* or *Pareto-optimal objective vectors*. All Pareto-optimal objective vectors are gathered in the *efficient set* or *Pareto frontier* or *Pareto front*, denoted F^*.

Although the optimality of solutions is now well defined, it is by no means clear which element of the solution sets is actually sought for by the decision maker. Therefore, frequently only an approximation or incomplete representation of the Pareto front is presented to the decision maker first before spending more effort in identifying further solutions in the region of the decision maker's interest. The ne-cessity of a finite approximation of the Pareto frontier becomes evident by the fact that the cardinality of the solution sets may be innumerable. For example, if $X \subset \mathbb{R}^n$ and $Y \subset \mathbb{R}^d$, the dimensions of the Pareto set and the Pareto frontier may be as large as $\min(n, d-1)$. In most practical cases we have $n \gg d$ so that the solution sets are typically $(d-1)$-dimensional manifolds. Thus, one typically obtains (possibly disconnected) curves for bi-objective problems $(d=2)$, surfaces for tri-objective problems $(d=3)$, 3D objects for quad-objective problems $(d=4)$, and so forth. The visualization of the Pareto front does not pose any problem up to three objectives. In case of higher-dimensional objective spaces, some pointers to the literature are given in Section 10.5.

10.3   Single-Objective Problems

The difficulty of single-objective optimization depends on the type and number of decision variables, the shape of the feasible region, the degree of nonlinearity in each variable and the kind of coupling (e.g. correlation between variables) between the variables. Thus, whenever a reduction in the number of variables or a decoupling of variables etc. is possible, this opportunity should be unhesitatingly accepted as the optimization task can be expected to become easier.

Example 10.1 (Hyperparameter tuning of a classifier). *Suppose we aim at minimizing the error rate of a classifier for instrument recognition (cp. Chapter 18), whose performance depends on several real-valued hyperparameters $x_1, \ldots, x_n$ (cp. Chapter 13). If we know that the error rate $f(x)$ is actually an additively decomposable objective function*

$$f(x_1, \ldots, x_n) = \sum_{i=1}^{n} f_i(x_i) \tag{10.1}$$

with $x_i \in X_i$ and $f_i : X_i \to \mathbb{R}$ for $i = 1, \ldots n$, the optimization problem could be decomposed into the optimization of n simpler problems: Since the minimum of additively decomposable functions is obtained by minimizing each partial function $f_i(x_i)$ independently over X_i, the n-dimensional problem reduces to n independent minimization problems over a single variable. Unfortunately, this special case rarely occurs in practice.

10.3.1   Binary Feasible Sets

The field of *pseudo-Boolean optimization* is specialized to Boolean input vectors $x \in \mathbb{B}^n = \{0, 1\}^n$ and a real-valued output $f(x) \in \mathbb{R}$. The distinction between local and global minima is typically realized by a specific neighborhood structure.

Definition 10.5.   The subset $N_k(\tilde{x}) = \{x \in \mathbb{B}^n : \rho(x, \tilde{x}) \leq k\} \subseteq \mathbb{B}^n$ is called the *k-bit neighborhood* of $\tilde{x} \in \mathbb{B}^n$ for $k \in \{1, \ldots, n\}$ where

$$\rho(x, \tilde{x}) = \|x - \tilde{x}\|_1 = \sum_{i=1}^{n} |x_i - \tilde{x}_i|$$

is the distance between x and $\tilde{x}$ based on the *L1*-norm.

Notice that the value of this distance is just the number of different values at each bit position of x and $\tilde{x}$. Therefore $\rho(\cdot,\cdot)$ coincides with the *Hamming distance* (cp. Section 11.2).

The insertion of this kind of neighborhood in Definition 10.2 for $X = \mathbb{B}^n$ and $k = 1$ specifies the notion of local and global optimality in pseudo-Boolean optimization.

Example 10.2 (Feature Selection). *The classification error of some classifier during the training phase depends on the training data and the features calculated for each element of the training data. Suppose there are n different features available and let $x_i = 1$ indicate that feature $i \in \{1,\ldots,n\}$ is used during classification whereas $x_i = 0$ indicates its non-consideration. Thus, vector $x \in \mathbb{B}^n$ represents which features are applied for classification and if $f : \mathbb{B}^n \to \mathbb{R}_+$ measures the error depending on the used features, the task of* feature selection *consists of finding that set of features for which the classifier exhibits least error. Formally, we like to find $x^* = \operatorname{argmin}\{f(x) : x \in \mathbb{B}^n\}$.*

Since $\mathbb{B}^n$ has finite cardinality 2^n the exact optimum can be found by a complete enumeration of all possible input/output pairs, but this approach is not efficient for large dimensionality n and therefore prohibitive for almost all practical cases. Nevertheless, there are efficient exact optimization methods if the objective function exhibits special properties.

Example 10.3 (Feature Selection (cont'd)). *Suppose we know that the error rate $f(x)$ in Example 10.2 is an additively decomposable objective function so that Equation (10.1) reduces to a linear pseudo-Boolean function*

$$f(x_1,\ldots,x_n) = \sum_{i=1}^{n} c_i \cdot x_i$$

with constant vector $c \in \mathbb{R}^n$. The global minimum is attained at $x^* \in \mathbb{B}^n$ with $x_i^* = 0$ if $c_i \geq 0$ and $x_i^* = 1$ if $c_i < 0$. Even if vector c is not explicitly known and only the objective function values are available, $n + 1$ objective function evaluations are sufficient for determining the global solution: start with an arbitrary solution $x \in \mathbb{B}^n$ and invert all bits positions sequentially from left to right (or vice versa). After each inversion, determine the objective function value. If the value is better, then proceed with the new solution, otherwise stay with the old solution without this particular inversion. After n inversions with subsequent function evaluations the global minimum has been found.

Notice that the method of Example 10.3 is specialized to a special class of problems and does not lead to (local) optimal solutions in general. Therefore, unless we know special properties of the objective function justifying the deployment of a highly specialized and efficient optimization method, it is advisable to use a method that guarantees at least the detection of a local minimum. Such methods are known as *local search methods*.

Two popular versions are local search with *first improvement heuristic* and with *best improvement heuristic*. These methods search for a better solution in a prescribed local neighborhood of the current solution. For that purpose they enumerate all solutions within the finite neighborhood in an arbitrary order. In case of the first improvement heuristic (Algorithm 10.1), the new solution is accepted as soon as a better solution than the current one is found. In contrast, the best improvement heuristic (Algorithm 10.2) first evaluates all solutions in the neighborhood and accepts that one with the best improvement.

Algorithm 10.1: Local Search: First Improvement

1: initialize $x_0 \in \mathbb{B}^n$; set $t = 0$; choose $k \in \{1, 2, \dots, n\}$
2: **repeat**
3:    let $N_k(x_t) = \{x(1), \dots, x(N)\}$, set $i = 1$
4:    **repeat**
5:       $x^* = \operatorname{argmin}\{f(x(i)), f(x_t)\}$
6:       increment i
7:    **until** $f(x^*) < f(x_t)$ **or** i = N
8:    $x_{t+1} = x^*$
9:    increment t
10: **until** $x_t = x_{t-1}$

Notice that the cardinality of a k-bit neighborhood of some $x \in \mathbb{B}^n$ is

$$|N_k(x)| = \sum_{i=0}^{k} \binom{n}{i}$$

growing exponentially fast for increasing k. Therefore, local search methods typically work with small neighborhoods, i.e., with small k.

Algorithm 10.2: Local Search: Best Improvement

1: initialize $x_0 \in \mathbb{B}^n$; set $t = 0$; choose $k \in \{1, 2, \dots, n\}$
2: **repeat**
3:    find $x^* = \operatorname{argmin}\{f(x) : x \in N_k(x_t)\}$
4:    $x_{t+1} = x^*$
5:    increment t
6: **until** $x_t = x_{t-1}$

Example 10.4 (Feature Selection (cont'd)). *Suppose the error rate of Example 10.2 depends on $n = 3$ potential features and is given by the (unknown) objective function*

$$f(x) = x_1 + x_2 + x_3 - 4x_1 x_2 x_3 + 1, \tag{10.2}$$

which is not additively decomposable. Let us employ a 1-bit neighborhood for minimizing the objective function from Equation (10.2) with the first improvement heuristic where $N_1(x)$ is enumerated by inverting the bits of x from left to right. Unless *we do not start at $x^* =$ 111 with global minimum $f(111) = 0$, only the starting point $x =$ 011 leads to the global minimum, whereas all other starting points lead to the local optimum $f(000) = 1$.*

The situation changes when using the best improvement heuristic that determines the objective function values for all elements in the neighborhood before choosing that element with maximum improvement (using some tie-breaking mechanism if necessary).

When we apply this method to the objective function from Equation (10.2) with a 1-bit neighborhood, then all starting points in $N_1(x^*)$ with $|N_1(x^*)| = 4$ lead to the global solution $x^* =$ 111, whereas the remaining starting points $\mathbb{B}^3 \setminus N_1(x^*)$ with cardinality $2^3 - 4 = 4$ end up in the local solution. A 50:50 chance for a uniformly distributed starting point seems acceptable. But if the error rate would be given by the generalized version

$$f(x) = \sum_{i=1}^{n} x_i - (n-1) \prod_{i=1}^{n} x_i + 1 \tag{10.3}$$

of the objective function from Equation (10.2), we would have $n + 1$ starting points leading to the global solution and $2^n - (n + 1)$ starting points leading to the local solution. Larger neighborhoods improve the relation between good and bad starting points, but they also require substantially more objective function evaluations per iteration.

These basic local search methods may be extended in various ways: Suppose that the neighborhood radius k is not fixed but depends on the iteration counter $t \geq 0$, i.e., $k_t \in \{1, \ldots, n\}$ for $t \in \mathbb{N}_0$. In this case the two local search methods above are variants of *variable neighborhood search*.

If the local search method is restarted with a new starting point randomly chosen after a certain number of iterations or depending on some other restart strategy, these methods are called *local multistart* or *local restart* strategies.

If the new starting point of a local restart strategy is not chosen uniformly at random from $\mathbb{B}^n$ but in a certain distance from the current local solution, the resulting restart approach is termed *iterated local search*.

Apart from these local methods there are also sophisticated global methods provided by solvers like CPLEX or COIN based on branch-and-bound or related approaches. If such solvers are at your disposal and the problem dimension is not too high it would be unjustifiable to ignore these methods.

If such solvers are not at your disposal or these solvers need too much time and global solutions are not required then we might resort to metaheuristics like evolutionary algorithms (EAs), whose algorithmic design is inspired by principles of biological evolution.

A feasible solution $x \in \mathbb{B}^n$ of an EA is considered as a chromosome of an individual that may be perturbed at random by a process called *mutation*. The objective function $f(x)$ is seen as a *fitness function* to be optimized. Typically, an individual $x \in \mathbb{B}^n$ is mutated by inverting each of the n bits independently with mutation probability $p \in (0, 1) \subset \mathbb{R}$. This mutation operation can be expressed in terms of an *exclusive or* operation $\oplus$ between individual x and a binary random vector z whose entries indicate if a bit should be inverted ($z_i = 1$) or not ($z_i = 0$). Thus, let z be a random vector consisting of n independent Bernoulli random variables with distribution $P(z_i = 1) = p$ and $P(z_i = 0) = 1 - p$ where $0 < p < 1$. Then $y = x \oplus z$ is considered the offspring of parent x newly formed by mutation, where $\oplus$ denotes the bitwise *exclusive-or (XOR)* operation. The simplest version of an EA is given by the so-called *(1+1)-EA*.

Algorithm 10.3: (1+1)-EA with Binary Encoding

1: initialize $x_0 \in \mathbb{B}^n$; set $t = 0$; choose $p \in (0,1)$
2: **repeat**
3:    draw random vector z
4:    $y = x_t \oplus z$
5:    $x_{t+1} = \operatorname{argmin}\{f(y), f(x_t)\}$
6:    increment t
7: **until** stopping criterion fulfilled

Typical stopping criteria used in EAs are the exceedance of a specific maximum number of iterations or the observation that there was no improvement in the objective function value within a prescribed number of iterations.

In most cases the mutation probability is set to $p = 1/n$, resulting in a single bit mutation *on average*. Nevertheless, provided that $0 < p < 1$ any number of mutations from 0 to n is possible with nonzero probability. This observation reveals that the $(1+1)$-EA may be seen as a randomized version of the variable neighborhood search instantiated with the first improvement heuristic.

But the crucial ingredient that distinguishes evolutionary algorithms from other optimization methods is the deployment of a *population* of individuals in each iteration of the algorithm.

Algorithm 10.4: Algorithmic Skeleton of Evolutionary Algorithm

1: initialize population of $\mu > 1$ individuals
2: evaluate all individuals by fitness function
3: **repeat**
4:    select individuals (parents) for reproduction
5:    vary selected individuals to obtain new individuals (offspring)
6:    evaluate offspring by fitness function
7:    select individuals for survival from offspring and possibly parents based on fitness
8: **until** stopping criterion fulfilled

There are many degrees of freedom for an instantiation of the algorithmic skeleton above. Typically, the variation of the parents is done by *recombination* of two parents (also called *crossover*) with subsequent *mutation*. Whereas mutation can be realized as in the $(1+1)$-EA, the crossover operation requires some inspiration from biology. A simple version (called 1-point crossover) chooses two distinct parents at random, draws an integer k uniformly at random between 1 and $n-1$, and compiles a preliminary offspring by taking the first k entries from the first and the last $n-k$ entries from the second parent. This may be generalized in an obvious manner to multiple crossover points. An extreme case is termed *uniform crossover* where each entry is chosen independently from the first or second parent with the same probability. Of course, these recombination operations are not limited to a binary encoding; rather, they may be applied accordingly for any kind of encoding based on Cartesian products.

The selection operations are independent from the encoding as they are typically solely based on the fitness values. Suppose the population consists of μ individuals. An individual is selected by *binary tournament selection* if we draw two parents at random from the population and keep the one with the better fitness value. This process may be iterated as often as necessary. But notice that a finite number of iterations does not guarantee that the current best individual gets selected. If it got lost, then the worst of the selected individuals is replaced by the current best individual. This kind of 'repair mechanism' is termed 1-*elitism*. In general, if some selection method guarantees that the current best individual will survive the selection process, then it is called an *elitist selection* procedure. Elitism is guaranteed if we generate λ offspring from μ parents and select the μ best (based on fitness) from parents and offspring; this method is called $(\mu+\lambda)$-*selection*. If the μ new parents are selected only from the λ offspring (where $\lambda > \mu$), elitism is not guaranteed and this method is termed (μ,λ)-*selection* or *truncation selection*.

10.3.2 Continuous Feasible Sets

The field of *continuous optimization* is specialized to real input vectors $x \in \mathbb{R}^n$ and a real-valued output $f(x) \in \mathbb{R}$. The distinction between local and global minima is typically realized by the usual ε-neighborhood $N_\varepsilon(\tilde{x}) := \{x \in X \subseteq \mathbb{R}^n : \|x - \tilde{x}\|_2 < \varepsilon\}$ where $\varepsilon > 0$ and $\|\cdot\|_2$ denotes the Euclidean norm.

10.3.2.1 Analytical Solution

If the objective function and the constraints are explicitly given and continuously differentiable, then it might be possible to find the optima analytically by solving a system of (nonlinear) equations deduced from the so-called KKT conditions [1]. Evidently, the analytical approach is the best choice provided it leads to a solution. In general, this approach may be tedious and mathematically challenging. Even worse, this approach may fail since highly nonlinear functions / constraints prevent an explicit solution of the nonlinear set of equations resulting from the KKT conditions. In this case, and also in case of non-differentiability as in a black- or gray-box sce-

nario, this approach is not applicable and one has to resort to iterative optimization
methods.

10.3.2.2 Descent Methods with Derivatives

Descent methods exist in a wide variety and they may be divided into derivative-
based and derivative-free or direct descent methods. Needless to say, if derivatives
are available they should not be ignored as they provide information about possible
directions of descent.

Definition 10.6. A unit vector $d \in \mathbb{R}^n$ is termed a *direction of descent* in $x \in X \subset \mathbb{R}^n$
for $f : X \to \mathbb{R}$ if there exists a step size $s_0 > 0$ such that for all step sizes s smaller
than s_0 a step in direction d leads to a smaller function value; formally, if

$$\exists s_0 > 0 : \forall s < s_0 : f(x + s \cdot d) < f(x).$$

The algorithmic pattern of descent methods is $x^{(k+1)} = x^{(k)} + s^{(k)} d^{(k)}$ for $k \geq 0$
and some starting point $x^{(0)} \in X$. First, a descent method determines a descent direction for the current position and moves in that direction until no further improvement
can be made. Then it determines a new direction of descent and the process repeats.

The big variety of descent methods differ on individual choices of the step sizes
$s^{(k)} \in \mathbb{R}^+$ and the descent directions $d^{(k)} \in \mathbb{R}^n$. If the objective function is differentiable it is easily verified whether a chosen direction is a direction of descent.

Theorem 10.1. If $f : X \to \mathbb{R}$ is differentiable, then $d \in \mathbb{R}^n$ is a direction of descent
if and only if $d^T \nabla f(x) < 0$ for all $x \in S$.

Popular representatives of the derivative-based approach are gradient methods
which use the negative gradient as direction of descent and differ by their step size
rules.

Gradient Method According to Theorem 10.1 the negative gradient is a direction
of descent since $d^T \nabla f(x) = -\nabla f(x)^T \nabla f(x) = -\|\nabla f(x)\|^2 < 0$ for all $x \in S$ if $d = -\nabla f(x) \neq 0$. Therefore, the gradient method instantiates the algorithmic pattern of
descent methods to $x_{t+1} = x_t - s_t \nabla f(x_t)$ where the step size $s_t = \alpha^k$ with $\alpha, \gamma \in (0,1)$
and

$$k = \min\{i \in \mathbb{N}_0 : f(x_t + \alpha^i \cdot d) \leq f(x_t) + \gamma \cdot \alpha^i \cdot d' \nabla f(x_t)\}$$

is chosen according to the so-called *Armijo rule*. Alternative step size rules are, for
example, the *Goldstein* or *Wolfe–Powell* rules. Regardless of the step size rule the
gradient method can only locate local minima.

If the objective function is representable as a sum of N sub-functions, i.e.,

$$f(x) = \sum_{i=1}^{N} f_i(x) \quad \text{for } f_i : \mathbb{R}^n \to \mathbb{R} \text{ and } i = 1, \dots, N,$$

a specific and typically randomized variant of the gradient method may come into
operation. In principle, the *stochastic gradient method* (Algorithm 10.5) may be
considered as an *inexact gradient method* that uses approximations $\nabla f(x) + e(x)$ of

the gradient with some unknown additive error function $e(x)$. Inexact gradients may accelerate the convergence velocity if the objective function is ill-conditioned. But if the objective function is sufficiently well conditioned, inexact gradients may slow down the approach to the optimum considerably.

Algorithm 10.5: Stochastic Gradient Method

1: initialize $x_0 \in \mathbb{R}^n$; set $t = 0$
2: **repeat**
3:     draw random permutation π of size N and set $x = x_t$
4:     **for** $i = 1$ **to** N **do**
5:         choose $s > 0$
6:         $x = x - s \cdot \nabla f_{\pi(i)}(x)$
7:     **end for**
8:     $x_{t+1} = x$
9:     increment t
10: **until** stopping criterion fulfilled

This undesired behavior can be compensated by the stochastic gradient method if the number N of sub-functions is sufficiently large: especially in the beginning of the optimization, when the current solution usually is far away from the optimum, many of the sub-functions' gradients will point into almost the same direction and much time can be saved by evaluating the entire gradient only partially. This situation arises, for example, in case of classifier training.

Example 10.5 (Classifier training with stochastic gradient method). *Suppose we are given a training set $\{(x_i, y_i) : i = 1, \ldots N\}$ for some classification task. For example, the input vector $x_i \in \mathbb{R}^n$ contains feature values of the ith tune in our music database and the output vector $y \in \mathbb{R}^d$ determines the membership of x_i (resp. the ith tune) to a particular class depending on the membership of the output vector in a particular set from a collection of disjoint sets.*

The classifier can be represented by the model $\varphi(x; w)$ that maps input x to the target value y for a given parametrization w of the classifier. Ideally, if w is chosen appropriately, then every input x_i of our training set is mapped to its associated output y_i. The performance of the classifier may be assessed by the magnitude of errors made in the mapping. The squared error of sample (x_i, y_i) is $f_i(w) = \|\varphi(x_i; w) - y_i\|^2$ for given w. Summing the error over the entire training set leads to the error function named total sum of squared errors (TSSE) given by

$$f(w) = \sum_{i=1}^{N} f_i(w) = \sum_{i=1}^{N} \|\varphi(x_i; w) - y_i\|^2$$

with gradient

$$\nabla f(w) = \sum_{i=1}^{N} \nabla f_i(w) = 2 \sum_{i=1}^{N} (\varphi(x_i; w) - y_i) \nabla \varphi(x_i; w).$$

Thus, the standard gradient method would be

$$w_{t+1} \;=\; w_t - s_t \cdot \nabla f(w_t) \;=\; w_t - s_t \cdot \sum_{i=1}^{N} \nabla f_i(w_t),$$

whereas the stochastic gradient method may use any partial sum of the sub-functions' gradients for updating the parametrization w. The update can be made after the classification of each tune or after a certain number of tunes has been classified. The order of the tunes should be randomly shuffled to provide the chance to escape from local optima.

Newton's Method If the second partial derivatives, gathered in the so-called Hessian matrix $\nabla^2 f(x)$ of the objective function, are also available, then Newton's method may be applied. Its advantage is its rapid convergence to the optimum under certain conditions. In any case, Newton's method should only be used if the Hessian matrix is positive definite, otherwise its sequence of iterates is not guaranteed to converge.

Algorithm 10.6: Newton's method.

1: initialize $x_0 \in \mathbb{R}^n$; set $k = 0$
2: **repeat**
3:     solve $\nabla^2 f(x_k)(x_{k+1} - x_k) = -\nabla f(x_k)$ yielding x_{k+1}
4:     increment k
5: **until** stopping criterion fulfilled

Rearrangement of line 3 in Algorithm 10.6 reveals that

$$x_{k+1} = x_k - \left[\nabla^2 f(x_k)\right]^{-1} \nabla f(x_k).$$

Thus, the so-called Newton direction $d = -\left[\nabla^2 f(x_k)\right]^{-1} \nabla f(x_k)$ is the direction of descent and the step size is $s_k = 1$ for all $k \in \mathbb{N}$.

If the Hessian matrix is not positive definite, then the *Levenberg–Marquardt Modification of Newton's Method* takes some measures to turn the Hessian matrix to a p.d. matrix again. Therefore this (more complicated) method is available in most software libraries and its deployment should be preferred over the standard method.

10.3.2.3 *Descent Methods without Derivatives*

Iterative methods moving along gradient or Newton direction are endowed with some greediness to reach the closest local optimum as quick as possible. This property is a hindrance in finding the global minimum in case of different local minima. Therefore it might be useful to take leave of the concept of gradient and higher derivatives. If the objective function is not differentiable, other concepts are necessary anyway.

Direct Search Methods that base all decisions on where to place the next step only on information gained from objective function evaluations without attempting to approximate partial derivatives are termed *direct search methods*. Many variants have been proposed since at least the 1950s. The simplest version unfolding the main idea is called *compass search*.

The compass search defines the set D of potential descent directions by the coordinate axes, which results in $2n$ directions. It chooses a direction from D and tests if a move in this direction with the current step size leads to an improvement. If so, the move is made. Otherwise, another direction of D is probed. If none of the directions in D leads to an improvement, the step size is made smaller and the process repeats with the same set D of potential descent directions. The algorithms stops if the step size gets smaller than a chosen limit $\varepsilon > 0$.

Algorithm 10.7: Compass Search Method

 1: initialize $x_0 \in \mathbb{R}^n$ and $s_0 > 0$; choose $\gamma \in (0,1)$, $\varepsilon > 0$; set $k = 0$
 2: let $D = \{\pm e_i : i = 1, \ldots, n\}$ with unit vectors $e_i \in \mathbb{R}^n$
 3: **repeat**
 4:   **if** exists $d \in D : f(x_k + s_k \cdot d) < f(x_k)$ **then**
 5:     $x_{k+1} = x_k + s_k \cdot d$
 6:     $s_{k+1} = s_k$
 7:   **else**
 8:     $x_{k+1} = x_k$
 9:     $s_{k+1} = \gamma s_k$
10:   **end if**
11:   increment k
12: **until** $s_k < \varepsilon$

This method is guaranteed to converge to a local optimum under mild conditions. More advanced methods extend the set D of search directions, introduce additional rules for increasing the step size, re-align the main search direction by an estimated direction based on a regression over a number of previous successful search steps, and others.

Evolutionary Algorithms A different concept of direct search is realized by evolutionary algorithms (EA). They maintain a population of individuals (candidate solutions) in each iteration. In search space $\mathbb{R}^n$ the mutation operation is typically modeled by adding a multinormal random vector with zero mean vector and some positive definite covariance matrix. Other distributions like the Cauchy distributions have been proposed but are rarely used. Recombination can be k-point or uniform crossover, but now it is also possible to create other operations like averaging the positions of the parents (intermediate recombination). There are many variants of evolutionary algorithms for problems in $\mathbb{R}^n$. Even the *CMA-EA (covariance matrix adaptation EA)*, considered as state-of-the-art EA, has many variants.

10.3.3   Compound Feasible Sets

If the optimization problem is given explicitly in terms of mathematical formulas and consists of some discrete and some continuous variables, then mathematical solvers like CPLEX should be considered first. If they are not applicable or need too much time to find the solution, a possible remedy is the deployment of an evolutionary algorithm.

Example 10.6 (Musical optimization problems with mixed types of variables). *Suppose we like to optimize a method for classifying music in genres. The classification method may have some real parameters which should be adjusted for optimal classification results. Since the classification quality also depends on the set of features we might aim at optimizing the selection of features, modeled by 0/1 variables, and the parameters of the classification method simultaneously.*

Another optimization problem with mixed types of variables is given by the task to find the optimal feature set (binary variables) and the optimal window size (continuous variable) for extracting the feature value simultaneously.

The EA for this kind of problems requires appropriate variation operators only. Mutation is realized independently for the binary and the continuous part of the individual. Similarly, recombination between parents is done independently for the different type of variables. This approach tacitly assumes that both types of variables are uncorrelated. This assumption is rarely true, but a properly derived method for a correlated variation, which would accelerate convergence, is not available currently.

10.4   Multi-Objective Problems

In multi-objective optimization the methods are typically classified as *a priori* or *a posteriori* methods. *A priori* methods transform the original problem into some surrogate problem based on a number of assumptions so that only a single solution to the surrogate problem is possible; as a consequence, the solution is selected implicitly based on the assumptions *before* other possible nondominated solutions are known. A well-known example is the scalarization of the vector-valued (multi-) objective function to a scalar-valued (single-) objective function by a *weighted sum*.

Let $f(x) = (f_1(x), \dots, f_d(x))^T$ denote the original objective function. The surrogate problem

$$f^s(x) = \sum_{i=1}^{d} w_i f_i(x)$$

with nonnegative weights w_i summing up to 1 provides the opportunity to apply single-objective optimization methods. Different solutions may be obtained by running single-objective optimizations with different weight settings. But although each optimal solution of f^s is Pareto-optimal, not all Pareto-optimal solutions are optimal solutions of f^s. Therefore it is possible that only a tiny subset of all possible Pareto-optimal solutions is found by this approach.

In *a posteriori* methods the solution is selected by the decision maker *after* a finite representative subset of the Pareto front has been found. For the approximation of the entire Pareto front population-based evolutionary algorithms like NSGA-2 (Algorithm 10.8) and SMS-EMOA (Algorithm 10.9) are commonly used and widely accepted. Both EAs use their population as an approximation of the Pareto front.

Algorithm 10.8: NSGA-2

1: draw multiset P with μ elements $\in \mathbb{R}^n$ at random
2: **repeat**
3:    generate μ offspring $\in \mathbb{R}^n$ from P by variation yielding Q
4:    $P = P \cup Q$
5:    build ranking $R_1, \ldots, R_h$ from P by nondominated sorting
6:    $P = \emptyset, i = 1$
7:    **while** $|P \cup R_i| \leq \mu$ **do**
8:       $P = P \cup R_i$
9:       increment i
10:   **end while**
11:   **if** $|P| < \mu$ **then**
12:      $k = |P \cup R_i| - \mu$
13:      discard k individuals from R_i with least crowding distance
14:      $P = P \cup R_i$
15:   **end if**
16: **until** stopping criterion fulfilled

The variation operators need not be changed for the multi-objective case, but the selection methods must cope appropriately with unavoidable incomparableness of solutions. The ranking of the individuals is achieved in two stages. In the first stage the population P is partitioned in h disjunct nondominated sets $R_1, \ldots, R_h$ via

$$R_1 = \text{ND}_f(P, \preceq) \text{ and } R_k = \text{ND}_f\left(P \setminus \bigcup_{i=1}^{k-1} R_i, \preceq\right) \text{ for } k = 2, \ldots, h \text{ if } h \geq 2$$

where $\text{ND}_f(A, \preceq) = \{x \in A : \nexists v \in f(A) : v \prec f(x)\}$ is the set of individuals in a set $A \subseteq P$ whose images are not dominated in the set $f(A)$.

This procedure is known by the term *nondominated sorting*. Evidently, every element from R_j is dominated by some individual in R_i if $i < j$. Figure 10.1 illustrates how a population is ranked by nondominated sorting.

In the second stage it is necessary to establish an order within each of the sets $R_1, \ldots, R_h$. This can be done in different ways but it must be kept in mind that the first stage may result in a trivial partitioning consisting of a single set $(h = 1)$ which typically happens if the population is very close to the Pareto front and/or if the number d of objectives is very large. In this case only the second stage is responsible for a proper selection of solutions and hence for the steering of the population's evolution.

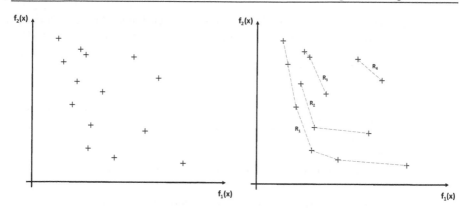

Figure 10.1: Left: Population of $\mu = 14$ individuals in 2-dimensional objective space. Right: Population after nondominated sorting. The dashed lines indicate which individuals belong to which set $R_1, \ldots, R_h$ where $h = 4$ in this particular example.

In case of NSGA-2 the *crowding distance* is used in the second stage. The crowding distance of an individual measures the proximity of other solutions in objective space. The individual with smallest value has close neighbors which are considered sufficient to approximate this part of the Pareto front so that the individual with least crowding distance can be deleted. This process is iterated (with or without recalculation of distances after deletion) as often as necessary.

Another qualifier for deleting individuals from crowded regions in the objective space is based on a commonly accepted measure [14] for assessing the quality of an approximation of the Pareto front since it simultaneously measures the closeness to the Pareto front and the spread along the Pareto front with a single scalar value:

Definition 10.7. Let $v^{(1)}, v^{(2)}, \ldots v^{(\mu)} \in \mathbb{R}^d$ be a nondominated set and $r \in \mathbb{R}^d$ such that $v^{(i)} \prec r$ for all $i = 1, \ldots, \mu$. The value

$$H(v^{(1)}, \ldots, v^{(\mu)}; r) = \Lambda_d \left(\bigcup_{i=1}^{\mu} [v^{(i)}, r] \right)$$

is termed the *dominated hypervolume* with respect to *reference point r*, where $\Lambda_d(\cdot)$ measures the volume of a set in $\mathbb{R}^d$. The *hypervolume contribution* of some element $x \in R_k$ is the difference $H(R_k; r) - H(R_k \setminus \{x\}; r)$ between the dominated hypervolume of set R_k and the dominated hypervolume of set R_k without element x.

Thus, the hypervolume contribution of an individual is that amount of dominated hypervolume that would get lost if this individual is deleted. Therefore the SMS-EMOA deletes individuals with least hypervolume contribution from crowded regions. Figure 10.2 illustrates how to determine the dominated hypervolume of a given population and the hypervolume contribution of each nondominated individual of the population.

Now we are in the position to describe the SMS-EMOA in its entirety.

Algorithm 10.9: SMS-EMOA

  1: draw multiset P with μ elements $\in \mathbb{R}^n$ at random
  2: **repeat**
  3:      generate offspring $x \in \mathbb{R}^n$ from P by variation
  4:      $P = P \cup \{x\}$
  5:      build ranking $R_1, \ldots, R_h$ from P
  6:      $\forall i = 1, \ldots, d : r_i = \max\{f_i(x) : x \in R_h\} + 1$
  7:      $\forall x \in R_h : h(x) = H(R_h; r) - H(R_h \setminus \{x\}; r)$
  8:      $x^* = \operatorname{argmin}\{h(x) : x \in R_h\}$
  9:      $P = P \setminus \{x^*\}$
10: **until** stopping criterion fulfilled

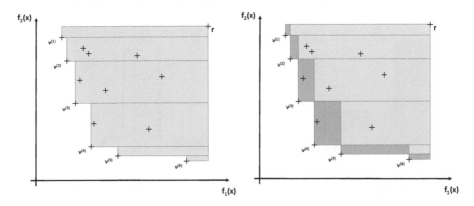

Figure 10.2: Left: The nondominated individuals (set R_1) of the population given in Figure 10.1 are labeled with $v^{(1)}, \ldots, v^{(6)}$ and a reference point r is chosen that is dominated by each of the nondominated individuals. The union of the rectangles $[v^{(i)}, r]$ for $i = 1, \ldots, 6$ is the (shaded) area which is dominated by the population up to the limiting reference point r. The volume of the area is the dominated hypervolume. Right: The darker shaded rectangles characterize the amount of dominated hypervolume that is exclusively contributed by each particular nondominated individual.

Example 10.7 (Multi-objective genre classification). *Suppose we aim at constructing a classifier for genre classification that exhibits maximum percentages of accuracy and precision but uses a minimum number of features for the classification task. These three objectives are conflicting since a too small number of features decreases accuracy and precision whereas accuracy and precision cannot be maximized simultaneously. Moreover, these objectives have incommensurable quantities. Therefore, we have a true multi-objective optimization problem and the evolutionary algorithms introduced in this section may be used to find an approximation of the Pareto front.*

Once this has been achieved, a closer inspection of the approximation set yields insight about the tradeoff between the conflicting objectives.

As already mentioned, the approximation of the Pareto front by a multi-objective evolutionary algorithm is only a preparatory step in the decision process. The final choice of the solution from the approximation that is to be realized is made by the user (i.e., the decision maker).

But this kind of preparatory activity can already be applied prior to running the optimization itself, namely in the phase of building the optimization model. As demonstrated in Example 10.8, one might use the a posteriori approach to assess the reasonableness of the objectives specified in the optimization model.

Example 10.8. *An (unfortunately costly) procedure is to estimate the tradeoff between measures after several multi-objective optimization experiments as applied to music classification in [13]. Figure 10.3(a) shows an example of the non-dominated front of solutions, where two objectives have to be minimized. The ideal solution is marked with an asterisk, and the reference point with a diamond. To measure*

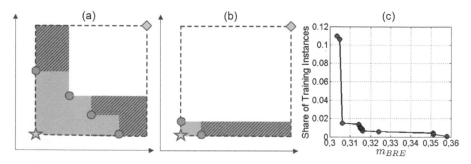

Figure 10.3: Two theoretical examples for non-dominated fronts (a,b) after [13] and a practical example minimizing the balanced classification error and the size of the training set (c). For further details see the text.

the tradeoff between both evaluation measures, we may estimate the shaded area between the ideal solution v_{ID} and the front built with solutions $v_1, \ldots, v_K$. This share of the area exquisitely dominated by v_{ID} can be, in general, estimated as (cf. Definition 10.7):

$$\varepsilon_{ID} = \frac{H(v_{ID};r) - H(v_1, \ldots, v_K;r)}{H(v_{ID};r)} \cdot 100\%, \qquad (10.4)$$

where r is the reference point. Larger ε_{ID} corresponds to a broader distributed non-dominated front and means that the optimization of both criteria is reasonable in contrast to smaller ε_{ID}; an example of the latter case is illustrated in Figure 10.3(b). Here, the optimization of one of both measures may be sufficient.

Another possibility to check if two measures should be optimized simultaneously is to first calculate the maximum hypervolume exclusively dominated by an individual solution from the non-dominated front. Then, the share of the hypervolume exquisitely dominated by other solutions (marked as area with diagonal lines in Figures 10.3(a) and (b) in relation to the hypervolume of the front can be measured:

$$\varepsilon_{MAX} = \frac{H(\mathbf{v}_1, \ldots, \mathbf{v}_K) - \max\{H(\mathbf{v}_k) : k = 1, \ldots, K\}}{H(\mathbf{v}_1, \ldots, \mathbf{v}_K)} \cdot 100\%. \tag{10.5}$$

Small ε_{MAX} means that there exists one solution in the non-dominated front whose exclusive contribution to the overall hypervolume of the front is significantly larger than for other solutions. Because this solution can be found using proper weights for a single-objective weighted sum approach (see Section 10.4), the multi-objective optimization is not necessary.

Figure 10.3, (c) shows an example after 10 statistical repetitions for the simultaneous minimization of the training set size and m_{BRE} using the model (d) from Figure 13.1 and SMS-EMOA as optimization algorithm (see Algorithm 10.9). As we can observe, approximately 11% of all training classification windows are enough to produce the smallest error $m_{BRE} = 0.30$. A further reduction of the training set size leads to higher errors up to $m_{BRE} = 0.36$.

10.5 Further Reading

Additional useful methods for pseudo-Boolean optimization (binary variables) can be found in [3, 4]. An overview of methods with integer variables has been compiled in [7], whereas the theoretical foundation and many deterministic methods for continuous variables are detailed in [1]. The theoretical framework for direct search is presented in [9]. Evolutionary algorithms are thoroughly introduced in [6], whereas most recent variants of EAs in continuous search space are described in [12]. The theoretical background on multiobjective optimization can be extended from material given in [8]. The basics of multiobjective EAs may be found in [5], whereas the SMS-EMOA has been proposed in [2]. The visualization of Pareto fronts especially in higher dimensions is discussed in [10, 11].

Bibliography

[1] M. S. Bazaraa, H. D. Sherali, and C. M. Shetty. *Nonlinear Programming: Theory and Algorithms.* Wiley, Hoboken (NJ), 3rd edition, 2006.

[2] N. Beume, B. Naujoks, and M. Emmerich. SMS-EMOA: Multiobjective selection based on dominated hypervolume. *European Journal of Operational Research*, 181(3):1653–1669, 2007.

[3] E. Boros and P. L. Hammer. Pseudo-Boolean optimization. *Discrete Applied Mathematics*, 123(1-3):155–225, 2002.

[4] C. Buchheim and G. Rinaldi. Efficient reduction of polynomial zero-one optimization to the quadratic case. *SIAM Journal on Optimization*, 18(4):1398–1413, 2007.

[5] K. Deb. *Multi-Objective Optimization Using Evolutionary Algorithms.* Wiley, 2001.

[6] A. Eiben and J. Smith. *Introduction to Evolutionary Computing*. Springer, 2nd edition, 2007.

[7] R. Hemmecke, M. Köppe, J. Lee, and R. Weismantel. Nonlinear integer programming. In M. Jünger et al., eds., *50 Years of Integer Programming 1958–2008*, pp. 561–618. Springer, Berlin Heidelberg, 2010.

[8] J. Jahn. *Vector Optimization: Theory, Applications, and Extensions*. Springer, 2004.

[9] T. G. Kolda, R. M. Lewis, and V. Torczon. Optimization by direct search: New perspectives on some classical and modern methods. *SIAM Review*, 45(3):385–482, 2003.

[10] P. Korhonen and J. Wallenius. Visualization in the multiple objective decision-making framework. In J. Branke et al., eds., *Multiobjective Optimization: Interactive and Evolutionary Approaches*, pp. 195–212. Springer, Berlin Heidelberg, 2008.

[11] A. V. Lotov and K. Miettinen. Visualizing the Pareto frontier. In J. Branke et al., eds., *Multiobjective Optimization: Interactive and Evolutionary Approaches*, pp. 213–243. Springer, Berlin Heidelberg, 2008.

[12] G. Rudolph. Evolutionary strategies. In G. Rozenberg, T. Bäck, and J. Kok, eds., *Handbook of Natural Computing*, pp. 673–698. Springer, Berlin Heidelberg, 2013.

[13] I. Vatolkin. Exploration of two-objective scenarios on supervised evolutionary feature selection: A survey and a case study (application to music categorisation). In A. Gaspar-Cunha et al., eds., *Proc. of 8th International Conference on Evolutionary Multi-Criterion Optimization (EMO 2015), Part II*, pp. 529–543. Springer, 2015.

[14] E. Zitzler and L. Thiele. Multiobjective optimization using evolutionary algorithms: A comparative case study. In A. Eiben et al., eds., *Conference on Parallel Problem Solving from Nature (PPSN V)*, pp. 292–301, Berlin Heidelberg, 1998. Springer.

Chapter 11

Unsupervised Learning

CLAUS WEIHS
Department of Statistics, TU Dortmund, Germany

11.1 Introduction

In this chapter we will introduce two kinds of methods for *unsupervised learning*, namely *unsupervised classification* and *independent component analysis*.

Unsupervised classification (also called cluster analysis or clustering) is the task of grouping a set of objects in such a way that objects in the same group (called cluster) are more similar (in some sense or another) to each other than to those in other clusters. Cluster analysis typically includes the definition of a distance measure and a threshold for cluster distinction. In this section we will discuss agglomerative hierarchical clustering methods like single linkage, complete linkage, and average linkage as well as the Ward method. Also, we introduce partitioning methods like the k-means method and self-organizing maps (SOMs). Moreover, we discuss different distance measures for different data scales and for features as well as the relation of clustering and outlier detection.

For independent component analysis (ICA) the aim is to "separate" the underlying independent components having produced the observations. One typical musical application is transcription where the relevant part of music to be transcribed (e.g. human voice) has to be separated from other sounds (e.g. piano accompaniment). In this case, ICA ideally generates two "independent" parts, namely the human voice and the accompaniment.

Clusters can be useful for various applications. Sometimes, it is particularly important that the found classes are well separated (when further used for supervised classification). Often, we would like to replace a big number of observations or variables by representatives (data reduction), which could be, e.g., cluster centers. Also, groups of missing values (cp. Section 14.2.3) should often be replaced by good representatives.

Definition 11.1 (Clustering). Given a set of observations or variables, a *clustering* is a partition of such objects into groups, so-called *clusters*, so that the *distance* of the objects inside a group is distinctly smaller than the distance of objects of different groups. We speak of *homogeneity* inside clusters and *heterogeneity* between clusters. The *separation quality* of a clustering is defined as the average heterogeneity of pairs of clusters.

The basis of any cluster analysis is the definition of the distance between objects. Typically, by means of a cluster analysis we look for a partition of objects into classes in order to reach one of the following two targets: data reduction and a better data overview or the finding of unknown object groups for the clarification of the studied issue.

The *methodical approach* can be summarized as follows:

- Observations are appointed to all objects.
- The distances between the objects are calculated based on the matrix of observations.
- The clustering criteria are applied to the distances finally leading to a clustering.

Obviously decisive for the "success" of a clustering is the definition of the distance between observations. Please notice that not only observations may be clustered but also features.

11.2   Distance Measures and Cluster Distinction

In this section we discuss typical distance measures and thresholds for cluster distinction. Distances are introduced for different data scales and for features. Moreover, specialized distances adequate for music applications are discussed.

The notion of distance is the most important basis for unsupervised classification since there is no validation mechanism as for objects with known groups. Obviously, the choice of the distance measure determines whether two objects naturally go together. Therefore, the right choice of the distance measure is one of the most decisive steps for the determination of cluster properties. The distance measure should not only adequately represent the relevant scaling of the data, but also the study target to obtain interpretable results.

For every individual problem the adequate distance is to be decided upon. In statistical musicology the main problem is often to find an adequate transformation of the input time series as an adequate basis for distance definition (see below). Also, local modeling is proposed in order to account for different sub-populations, e.g. instruments.

Some classical distance measures in unsupervised classification are discussed in the following. Any of such distances between two data points can then be used for defining the distance between groups of data. These will be discussed in the next section. In practice, most of the time there are different plausible distance measures for an application. Then, quality criteria are needed for distance measure selection. In unsupervised learning, one might want to use background information about reasonable groupings to judge the partitions, or one might want to use indices like the ratio between the between- and within-cluster variances which is also optimized in *discriminant analysis* in the supervised case ([1], p. 226).

In what follows a somewhat systematic sequence of examples is given illustrating problem specific distances.

- The Euclidean distance is by far the most chosen distance for metric features. For vectors of k *cardinal features*, often the *Euclidian distance* of two observations in k dimensions is used defined by

$$d_E(x_1, x_2) := \sqrt{\sum_{j=1}^{k}(x_{1j} - x_{2j})^2}.$$

- On the one hand, one should notice, however, that the Euclidean distance is well-known for being outlier sensitive. This might motivate switching to another distance measure like, e.g., the *Manhattan distance* or *City-Block distance* ([6])

$$d_C(x_1, x_2) := \sum_{j=1}^{k}|x_{1j} - x_{2j}|.$$

- On the other hand, one might want to discard correlations between the features and to restrict the influence of single features. This might lead to transformations by means of covariance or correlation matrices S, i.e. to *Mahalanobis distances* ([6])

$$d_M(x_1, x_2) := \sqrt{(x_1 - x_2)^T S^{-1}(x_1 - x_2)}.$$

- For *ordinal features*, often the Euclidian distance of the ranks of the k features is used. This means that the observations are first ordered, feature by feature, and then the ranks of the observations are used as if they were the observations themselves.

- The values of *nominal features* are often first coded by real values, either in a "natural" way or "optimally" according to the application. For example, sometimes the coding -1 and 1 is often in use for the "extreme" values. Optimal coding is often achieved by special methods like *multidimensional scaling* ([1], p. 249). The coded features are then used as if they were cardinal.

- For *dichotomous/binary features*, often the so-called *Hamming distance* is used:

$$d_H(x_1, x_2) := \text{no. of entries with non-matching values in observations } x_1, x_2.$$

Note that d_H could also be directly used for vectors of general *nominal features*.

- Other distance measures in use for binary features are the *Jaccard index* defined by

$$d_J(x_1, x_2) := \frac{\text{no.(non-matching entries in } x_1 \text{ and } x_2)}{\text{no.(double-positives + non-matching)}}$$

and the *simple matching index* defined by

$$d_S(x_1, x_2) := \frac{\text{no. of non-matching entries}}{\text{total no. of entries}}.$$

Note that the *Jaccard index* assumes an asymmetric situation in that only the double positive results (both entries $= 1$) determine similarity! In contrast, in the *simple matching index*, both the double positives and the double negatives count for similarity.

- In each of the above cases the matrix D of the distances of pairs of observations $1, \ldots, n$ is called the *distance matrix*:

$$
D := \begin{bmatrix}
0 & d_{12} & \ldots & d_{1n} \\
d_{21} & 0 & \ldots & d_{2n} \\
\vdots & \vdots & \ddots & \vdots \\
d_{n1} & d_{n2} & \ldots & 0
\end{bmatrix}.
$$

- Similarity of features is often measured by means of the correlation coefficient. One possibility to define the *distance between two features* is the so-called (*linear indetermination*, i.e.

$$
1 - \text{coefficient of determination} = 1 - \text{correlation}^2,
$$

which is interpreted as that part of one feature which is not determined by the other feature.

- The basis for this distance measure is the *correlation matrix (contingency matrix)*:

$$
Cor := \begin{bmatrix}
1 & r_{12} & \ldots & r_{1h} \\
r_{21} & 1 & \ldots & r_{2h} \\
\vdots & \vdots & \ddots & \vdots \\
r_{h1} & r_{h2} & \ldots & 1
\end{bmatrix},
$$

where r_{ij} is the correlation (contingency) between features i and j.

- This way, the distance matrix (*matrix of indetermination*) of the features $1, \ldots, k$ is defined by

$$
D := ((1 - r_{ij}^2)) = \begin{bmatrix}
0 & 1 - r_{12}^2 & \ldots & 1 - r_{1k}^2 \\
1 - r_{21}^2 & 0 & \ldots & 1 - r_{2k}^2 \\
\vdots & \vdots & \ddots & \vdots \\
1 - r_{k1}^2 & 1 - r_{k2}^2 & \ldots & 0
\end{bmatrix}.
$$

Example 11.1 (Distances). *An example for metric observations will be given in the next section. We will concentrate here on examples for distances between qualitative observations and between features:*

- *Scale ("major" vs. "minor") as well as rhythm ("three-four time" vs. "four-four time") are both binary nominal features. One might be interested in the distance between classical composers according to scale or rhythm. For this, we might rank movements of their compositions by the number of their performances, e.g. over the radio, and compare the pieces on corresponding ranks 1–10, say. In such cases, the simple matching index might be adequate to group together composers well known for pieces with similar emotions.*

- *Distances between two features correspond to that part of one feature which is not determined by the other. This way, features which can explain each other well, i.e. are highly correlated, are clustered together. See Section 11.5 for examples.*

Let us now discuss typical *transformations*, e.g. before using the Euclidean distance. In *musical applications*, e.g., the Euclidean distance is used for clustering the "log-odds ratio" of the probabilities of notes for various compositions ([1], p. 238) leading to a clear separation of "early music" from the rest. Note the transformation of the frequencies $p_j, j = 0, 1, \ldots, 11$, of the notes (modulo 12) by means of the log-odds ratio, i.e. to $\xi_j = \log(p_j/(1 - p_j))$.

Another transformation used in musical applications is the entropy of melodic shapes and spectral entropies ([1], pp. 93-96). This lead to a clear separation of Bach's Cello suites from "Das Wohltemperierte Klavier" ([1], p. 242).

[1] also proposes a specific smoothing for tempo curves (HISMOOTH) ([1], pp. 141–144). This leads to a similar clustering for different group distances ([1], p. 243).

In all these applications, so-called "complete linkage" and "single linkage" are used for defining distances between groups of observations. Let us now look at such distances systematically.

11.3 Agglomerative Hierarchical Clustering

For the understanding of cluster analysis it is important to formalize the term classification. In general, grouping methods generate a *partition*, a special covering of the objects. The methods introduced in the following consist of a sequence of iterated partitions, a so-called *hierarchy* of partitions.

Definition 11.2 (Partition and Hierarchy). A *partition* K^t with classes $K_1^t, \ldots, K_k^t$ is a covering of all observations, in which each object belongs to exactly one class:

$$K_i^t \cap K_j^t = \emptyset, i \neq j.$$

A *hierarchy* K is a sequence of partitions $K^t, t = 1, \ldots, m$. On each stage of a hierarchy the classes build a partition.

Definition 11.3 (Types of Classification Methods). Classification methods, which start the hierarchy with a partition of one-element subsets making the partitions rougher and rougher, are called *agglomerative methods*. Classification methods, which start the hierarchy with the roughest partition, i.e. the full set, making the partitions finer and finer, are called *divisive methods*.

11.3.1 Agglomerative Hierarchical Methods

Here, we will only discuss agglomerative methods. In order to get rougher and rougher partitions in agglomerative methods, successively classes in the actual partition are combined until the full set of observations is reached. Then, that partition of the hierarchy is identified for further use which has the best quality measure (see below).

Agglomerative hierarchical methods have the following structure:

Stage 0: (Initialization)

$$\text{Let } K^0 := \{K_1^0 \cup K_2^0 \cup \ldots \cup K_n^0\} := \{\{x_1\} \cup \{x_2\} \cup \ldots \cup \{x_n\}\},$$

where n = no. of observations of, e.g., feature x.

Let $v(K_1, K_2)$ be a *heterogeneity measure* for classes or groups of observations.

Stage 1: Choose $K^1 := \{K_1^1 \cup K_2^1 \cup \ldots \cup K_{n-1}^1\}$, where those two classes are combined which have minimal distance at stage 0. Such minimally heterogeneous classes have the property:

$$v(K_{i1*}^0, K_{i2*}^0) = \min_{\substack{i1,i2 \in \{1,2,\ldots,n\} \\ i1 \neq i2}} v(K_{i1}^0, K_{i2}^0).$$

Stage t ($\leq n-1$): Choose $K^t := \{K_1^t \cup K_2^t \cup \ldots \cup K_{n-t}^t\}$, where those two classes are combined which have minimal distance on stage $(t-1)$, i.e.

$$v(K_{i1*}^{t-1}, K_{i2*}^{t-1}) = \min_{\substack{i1,i2 \in \{1,2\ldots,n-t+1\} \\ i1 \neq 12}} v(K_{i1}^{t-1}, K_{i2}^{t-1}).$$

The grouping criterion is obviously dependent on the *heterogeneity measure* v. Such measures should characterize the dissimilarity between the classes of one partition. They, thus, represent something like the *distance between the classes*.

Definition 11.4 (Heterogeneity Measures). The three most important heterogeneity measures are directly based on the distance between individual elements. Here are the most often used *distance measures between two classes*:

1. *Distance of the two most dissimilar elements of the classes*

$$v_1(K_{i1}, K_{i2}) := \max_{j \in K_{i1}, k \in K_{i2}} d_{jk},$$

where d_{jk} is one of the distances above of the two elements $x_j \in K_{i1}$ and $x_k \in K_{i2}$. Cluster methods with this heterogeneity measure are called *complete linkage methods* or *farthest neighbor methods*.

Problem: the heterogeneity tends to be *overestimated*.

2. *Distance of the two most similar elements in the classes*

$$v_2(K_{i1}, K_{i2}) := \min_{j \in K_{i1}, k \in K_{i2}} d_{jk}.$$

Cluster methods with this heterogeneity measure are called *single linkage methods* or *nearest neighbor methods*.

Problem: the heterogeneity tends to be *underestimated*.

3. *Mean distance of the elements of the classes*

$$v_3(K_{i1}, K_{i2}) := \frac{1}{|K_{i1}||K_{i2}|} \sum_{j \in K_{i1}} \sum_{k \in K_{i2}} d_{jk}.$$

Cluster methods with this heterogeneity measure are called *average linkage methods*.

Based on these heterogeneity measures, the quality of a partition can be evaluated.

Definition 11.5 (Quality measures of partitions). As a *quality measure of a partition*, often the inverse mean heterogeneity of the classes in the partition is taken. Let K be a partition, $|K| :=$ no. of clusters in partition K, and $n =$ no. of observations. Then,

$$g_v := \frac{\sum_{K_{i1} \in K} \sum_{K_{i2} \in K, i2 < i1} v(K_{i1}, K_{i2})}{\frac{|K|(|K|-1)}{2}}$$

is called *mean class heterogeneity*, where v is a heterogeneity measure.

Another *quality measure* of a partition is the so-called *Calinski measure* which relates the between-cluster variation SSB to the within-cluster-variation SSW:

$$Ca := \frac{SSB/(|K|-1)}{SSW/(n-|K|)} = \frac{\sum_{i=1}^{|K|} n_i(\bar{x}_i - \bar{\bar{x}})^2/(|K|-1)}{\sum_{i=1}^{|K|} \sum_{j=1}^{n_i} (x_{ij} - \bar{x}_i)^2/(n-|K|)},$$

where n_i is the no. of elements in cluster $i = 1, \ldots, |K|$, $\bar{x}_i$ is the empirical mean in cluster i, and $\bar{\bar{x}}$ is the overall mean of all data.

Obviously, g_v evaluates only the distances between clusters, whereas Ca is judging between-cluster variation by within-cluster variance. Both quality measures g_v and Ca should be maximized.

11.3.2 Ward Method

Let us finish our introduction to hierarchical agglomerative cluster methods with *Ward's method*. Ward's minimum variance criterion minimizes the total within-cluster variance. At each step the pair of clusters with minimum between-cluster distance is merged. This leads to the minimum increase in total within-cluster variance after merging. The increase is a weighted squared distance between cluster centers:

$$\frac{m_j m_k}{m_j + m_k} d_E(\bar{x}_j, \bar{x}_k)^2,$$

where $\bar{x}_j =$ mean vector of cluster j and $m_j =$ no. of observations in cluster j. Obviously, this expression has to be minimized in each iteration step.

At the initial step, all clusters are singletons (clusters containing a single point). To apply a recursive algorithm to this objective function, the initial distance between individual objects must be proportional to the squared Euclidean distance. The initial cluster distances in Ward's minimum variance method are therefore defined to be the squared Euclidean distance between points.

11.3.3   Visualization

As a visual representation of a hierarchy of partitions, a so-called *dendrogram* is used.

Definition 11.6 (Dendrogram).   In a *dendrogram* the individual data points are arranged along the bottom of the dendrogram and referred to as leaf nodes. Data clusters are formed by any hierarchical cluster method leading to the combination of individual observations or existing data clusters with the combination point referred to as a *node*. A dendrogram shows the leaf nodes and the combination nodes. The combinations are indicated by lines. At each dendrogram node we have a right and left sub-branch of clustered data. The vertical axis of a dendrogram is labeled *distance* and refers to a distance measure between observations or data clusters. The height difference between a node and its sub-branch nodes can be thought of as the distance value between a node and its right or left sub-branch nodes, respectively.

Data clusters can refer to a single observation or a group of data. As we move up the dendrogram, the data clusters get bigger, but the distance between data clusters may vary. One way to identify a "natural" clustering (partition) is to cut the dendrogram in its longest branch, this means at a place where sub-branch clusters have the biggest distance to the nodes above.

Example 11.2.   *Let us now introduce an example data set often used in this section. The data is composed of MFCC variables (non-windowed and windowed), chroma variables, and envelopes in time and spectral space. All variables are available for 4309 guitar and 1345 piano tones. Blocks are composed of 12,288 observations each for a signal sampled with 44,100 Hz. This means that one block corresponds to around 0.25 seconds. We have studied the tones 4530–4640, including 55 guitar and 56 piano tones, and the features MFCC 1 in the first block ("mfcc_block1_1") and in the last (fifth) block ("mfcc_block5_1") of the tones. The idea is that these blocks are able to distinguish between guitar and piano tones since the beginning and the end of the tones are important for distinction. Figure 11.1 shows the partitions of different hierarchical clustering methods based on Euclidean distances. Note that the different symbols represent the different clusters and the filling state (empty or filled) distinguish guitar and piano. Obviously, complete linkage and Ward reproduce the instruments much better than average and single linkage. Note that the number of clusters was fixed by means of cutting the dendrogram at a certain height so that 5 clusters were produced each. No automatic rule was followed here. The dendrograms are somewhat confusing because of the high number of observations and the labels of the tones. We will thus restrict ourselves to a dendrogram representing only tones 4575–4595 analyzed by complete linkage clustering (Figure 11.2). Notice that the classes are perfectly split into different clusters, since the guitar tones (4575–4584) are in different clusters than the piano tones (4585–4595). Cutting at distance = 3 would lead to perfect clusters.*
     Note that the illustration of clustering by scatterplots, as in Figure 11.1, is only

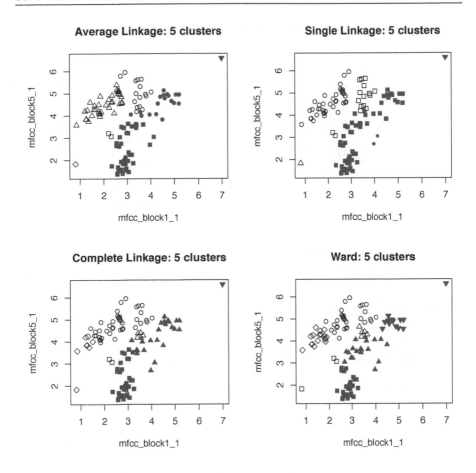

Figure 11.1: Partitions with different hierarchical clustering methods.

possible since the original data space was 2D. Dendrograms, however, can also be used in higher dimensions.

11.4   Partition Methods

Let us now switch to non-hierarchical partitioning methods. Such methods do not use a hierarchy of partitions but directly generate the one partition used for clustering.

11.4.1   k-Means Methods

The collective term *k-means-methods* stands for iterative methods for the determination of partitions satisfying certain requirements on inter-heterogeneity and intra-homogeneity, for which the number of classes k is preset.

A *k-means-algorithm*, e.g., consists of the following steps:

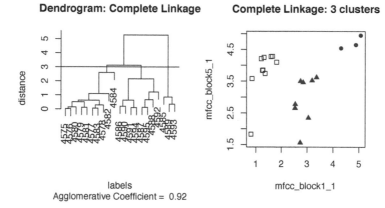

Figure 11.2: Left: Dendrogram for complete linkage, tones 4875–4895, right: corresponding scatterplot.

Algorithm 11.1: k-means clustering

1: Choice of k cluster centers $z_1^0, \dots, z_k^0$.
2: Assignment of all objects to the nearest center.
3: Replacement of the centers by the mean of the observations assigned to the cluster.
4: Stop if iteration comes to an end or branch to step 2.

The initial *cluster centers* are vectors of the same dimension as the observations, which are central in the observations. In practical applications it might be sensible to utilize *prior knowledge* for the setting of initial centers by hand. Another possibility is the drawing of k elements from a *uniform distribution* on the indices of the observations. For the choice of the initial cluster centers there exist also different pre-optimization methods aiming at the improvement of the convergence speed of the iteration.

The *second step* of the algorithm leads to a partition of the objects into k classes. In iteration t the sets $C_1^{t-1}, \dots, C_k^{t-1}$ contain for each class the indices of objects assigned to it. Formally, these sets are determined as follows:

$$C_h^{t-1} = \bigcup_{i=1}^{n} i \cdot T_h(x_i), \, h = 1, \dots, k, \text{ where}$$

$$T_h(x) = \begin{cases} 1 & , ||x - z_h^{t-1}|| = \min_{1 \le j \le k} ||x - z_j^{t-1}|| \\ \emptyset & , \text{else} \end{cases}.$$

In cases with more than one $||x - z_h^{t-1}||$ being minimum, the assigned cluster is randomly chosen from the competing ones. This way, the set $\{C_1^{t-1}, \dots, C_k^{t-1}\}$ becomes a so-called *minimal-distance partition* of the set $\{1,\dots,n\}$ since

$$C_i^{t-1} \cap C_j^{t-1} = \emptyset, i \neq j, \quad \text{and} \quad \bigcup_{j=1}^{K} C_j^{t-1} = \{1,\dots,n\}.$$

In the *third step*, the centers are replaced by location measures of the temporary clusters, i.e. of the assigned observations. Which location measures are chosen, depend on the preset L_p-*criterion* to be minimized. Each center z_h^t is defined as that $x_{(h)}^L$ for which the following expression is minimal:

$$\sum_{i \in C_h^{t-1}} \left(\sum_{j=1}^{m} |x_{ij} - x_{(h)j}^L|^p \right)^{\frac{1}{p}}.$$

For some parameters p we get the location measure x^L from Table 11.1.

Table 11.1: Location Measures for the Replacement of the Centers

p	x^L
1	median
2	mean
∞	(max + min) / 2

The algorithm might stop if the maximal relative change in the cluster centers is small enough defined by a preset threshold, if the clusters do not change by the latest iteration, or if a preset maximum number of iterations is reached (early stopping).

11.4.2 Self-Organizing Maps

Let us discuss a second non-hierarchical algorithm for clustering, namely *Self-Organizing Maps (SOMs)*. Here, the aim is to visualize big high-dimensional data sets in order to enable interpretation. Clustering is an essential part of the visualization. The basics of this method were developed by Kohonen (1989).

The SOM-map consists of a regular 2D-grid of processing units, called "neurons". Some multidimensional observation, eventually a vector consisting of features, is associated with each unit. The map attempts to represent all the available observations with optimal accuracy using a restricted set of neurons. These neurons are ordered on the grid so that similar neurons are close to each other and dissimilar ones far from each other.

The weights of the neurons are initialized either to small random values or sampled uniformly from the subspace spanned by the two largest principal component eigenvectors of the observations. With the latter alternative, learning might be much faster because the initial weights may already give a good approximation of SOM weights.

The network must be fed a large number of example vectors that represent, as close as possible, the expected kinds of vectors. The examples are usually presented several times in iterations.

When a training example x_t is fed to the network, its Euclidean distance to all weight vectors is computed. The neuron whose weight vector is most similar to the input is called the best matching unit (BMU). The weights of the BMU and neurons close to it in the SOM lattice are adjusted towards the input vector. The magnitude of the change decreases with time and with distance (within the lattice) from the BMU. The update formula for a neuron with weight vector m_t is

$$m_{v,t+1} = m_{v,t} + \alpha(t)\theta(u,v,t)(x_t - m_{v,t}),$$

where t is the step index, u is the index of the BMU for x_t, $\alpha(t)$ is a monotonically decreasing learning coefficient, and v is assumed to pass through all neurons.

The neighborhood function $\theta(u,v,t)$ depends on the lattice distance between the BMU (neuron u) and neuron v. In the simplest form it is 1 for all neurons close enough to BMU and 0 for others, but a Gaussian function is a common choice, too. Regardless of the functional form, the neighborhood function shrinks with time. At the beginning, when the neighborhood is broad, the self-organizing takes place on a global scale. When the neighborhood has shrunk to just a couple of neurons, the weights converge to local estimates. In some implementations the learning coefficient α and the neighborhood function θ decrease steadily with increasing t, in others they decrease in a step-wise fashion, once every n steps. One possible neighborhood function is $\theta(u,v,t) = e^{-\lambda(t)\|m_{u,t}-m_{v,t}\|}$, where $\lambda(t)$ is again a learning coefficient decreasing with time t.

This process is repeated for each input vector for a (usually large) number of cycles. The network develops associating output nodes with groups or patterns in the input data set. If these patterns can be named, the names can be attached to the associated nodes in the trained net.

While representing input data as vectors has been emphasized here, it should be noted that any kind of object which can be represented digitally, which has an appropriate distance measure associated with it, and in which the necessary operations for training are possible can be used to construct a self-organizing map. This includes matrices, continuous functions or even other self-organizing maps. Summarizing, Algorithm 11.2 is a possible SOM algorithm.

The SOM-map is a representation of the input data with own distances between the nodes which can be used for clustering by any of the above methods. To do so, distances are calculated of the weights of the nodes of the map. These distances are then used for clustering with any of the above clustering methods. In a way, a SOM-map is just a pre-step to clustering making a high-dimensional space 2D.

Another way to indicate cluster borders, the so-called *U-Matrix*, is constructed on top of the SOM-map.

Definition 11.7 (U-Matrix). Let v be a neuron on the SOM-map, $NN(v)$ be the set of immediate neighbors of v on the map, m_v the weight vector associated with neuron

---

Algorithm 11.2: SOM Construction

1: Set $t := 1$, $T :=$ maximum no. of iterations.

2: Choose a lattice (map) and initialize the weights of its neurons.

3: Repeat for each vector x_t in the input data set:

     a. Evaluate each node in the map by the Euclidean distance of its weight vector to the input vector.

     b. Track the node that produces the smallest distance (this node is the best matching unit BMU u).

     c. Update the nodes v in the neighborhood of the BMU (including the BMU itself) by pulling them closer to the input vector

$$m_{v,t+1} = m_{v,t} + \alpha(t)\theta(u,v,t)(x_t - m_{v,t}).$$

4: Increase t and repeat from step 3 while no convergence and $t < T$.

---

v, then

$$U_v := \sum_{u \in NN(v)} ||m_v - m_u||$$

is the height of the *U-matrix* in node v. The U-matrix is a display of the U-heights on top of the grid positions of the neurons on the map.

The U-matrix delivers a "landscape" of the distance relationships of the input data in the data space. Properties of the U-matrix are:

- the position of the projections of the input data points reflects the topology of the input space, according to the underlying SOM algorithm;
- weight vectors of neurons with large U-heights are very distant from other nodes;
- weight vectors of neurons with small U-heights are surrounded by other nodes;
- outliers in the input space are found in "funnels";
- "mountain ranges" on a U-Matrix point to cluster boundaries; and
- "valleys" on a U-Matrix point to cluster centers.

The U-matrix realizes the so-called emergence of structure of features corresponding the distances within the data space. Outliers, as well as possible cluster structures, can be recognized for high dimensional data spaces. The proper setting and functioning of the SOM algorithm on the input data can also be visually checked.

Example 11.3. Let us now look at the results of k-means and SOM clustering for the above example. Figure 11.3 shows on the left the partition of the k-means method with 5 clusters, the same number of clusters as for the above hierarchical methods. The plot should be interpreted in the same way as Figure 11.1. If we associate a cluster with the more frequent class, k-means delivers just slightly better results (7 elements with wrong class in clusters) than Ward (8 errors).

Let us now evaluate the result of k-means clustering by means of the Calinski

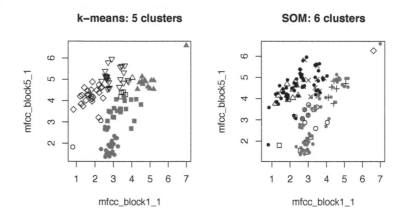

Figure 11.3: Left: k-means with 5-means clusters, right: SOM with 6 clusters indicated at weights of nodes.

Table 11.2: Quality Measures for Different Number of Clusters k

k	2	3	4	5	6	7	8
Ca	87	124	121	139	139	145	148

quality measure Ca without knowledge of the "true" classes. To this end, we have repeated the k-means algorithm 50 times for $k = 2, \ldots, 8$ with random starting vectors, and took the partition with maximum Ca-index for each k. This led to the results in Table 11.2. Obviously, the 5-means method is not optimal, since its Ca is not maximal.

For SOM clustering, let us first consider a similar scatterplot as for the other clustering methods (Figure 11.3, right). Notice, however, that the symbols of the 6 clusters are not marking the original observations but the (nearby) weight vectors for the SOM-nodes. The original observations are included by black and grey dots indicating the true classes. In order to understand the structure of the SOM-map, we look at the U-matrix and at a map with the assigned original observations separately. The U-matrix is given on the left of Figure 11.4 together with the proposed boundaries of 6 clusters. Note the "funnel" in the lower left corner representing the singleton corresponding to the upper right individual in Figure 11.3. Also note that higher weights correspond to less-intensive greys. On the right of Figure 11.4 the location of the original observations is indicated by their class numbers in the nodes of the SOM. Obviously, the reproduction of the original classes was similarly successful as k-means. Moreover, note that some of the nodes do not contain any original observation. Notice that the 6 clusters are generated by means of complete linkage clustering applied to the weights of the SOM-map. The resulting dendrogram can be seen in Figure 11.5. Note the nodes are numbered consecutively row by row

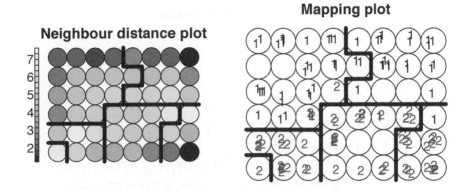

Figure 11.4: Left: U-matrix with boundaries of 6 clusters, right: original classes in SOM.

starting in the lower left corner (node 1) and ending in the upper right corner (node 48).

11.5 Clustering Features

Let us continue with an example of feature clustering.

Example 11.4. *We would like to cluster the 16 MFCC features and the 14 chroma features of block 1. We apply single linkage clustering to the distance matrix of indermination D, leading to the dendrogram in Figure 11.6. Obviously, when splitting at distance = 1.31 (indicated by the horizontal line), only cluster 1 contains both MFCC and chroma features. Indeed, the first MFCC feature is located in a cluster otherwise containing only chroma features. This shows that the MFCC features are much more similar to themselves than to the chroma features, except the first MFCC feature.*

11.6 Independent Component Analysis

Let us finish with an unsupervised learning problem, somewhat related to unsupervised classification, which might be called *unsupervised separation*. Here, the aim is to "separate" the underlying independent components having produced the observations.

Let us start with an example: As a first step of the transcription algorithm in Chapter 17, the relevant part of music to be transcribed (e.g. the human voice) has

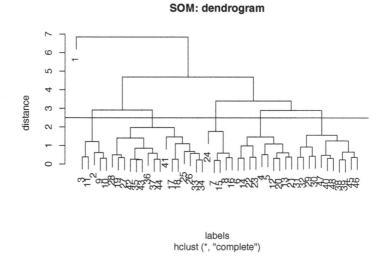

Figure 11.5: Dendrogram of nodes of SOM-map generated by complete linkage clustering.

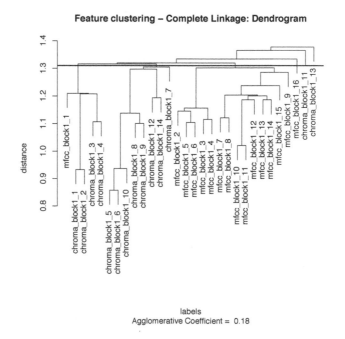

Figure 11.6: Dendrogram of features generated by single linkage clustering.

to be separated from other sounds (e.g. piano accompaniment). The outcomes of a separation are time series of the parts of the music corresponding to the different generating components. To achieve this sound source separation task, the commonly used standard method is Independent Component Analysis (ICA) as proposed by Hyvärinen [3].[1]

Imagine that you are in a room where two instruments are playing simultaneously. You have two microphones, which stand in different locations. The microphones generate two recorded time signals, which we could denote by x_{1t} and x_{2t} with x_1 and x_2 the amplitudes and t the time index. Each of these recorded signals is a weighted sum of the signals emitted by the two instruments, which we denote by s_{1t} and s_{2t}. We could express this as a linear equation system:

$$x_{1t} = a_{11}s_{1t} + a_{21}s_{2t}$$
$$x_{2t} = a_{12}s_{1t} + a_{22}s_{2t}$$

where a_{ij} are some parameters that depend on the distances of the microphones from the instruments. The aim is now to estimate the two original instrument signals s_{1t} and s_{2t} using only the recorded signals x_{1t} and x_{2t}. This is a music equivalent of the so-called "cocktail-party" problem for speech.

More generally, let us assume that we observe N linear mixtures $x_1, \ldots, x_N$ of N independent components

$$x_j = a_{1j}s_1 + \ldots a_{Nj}s_N, j = 1, \ldots, N.$$

We have now dropped the time index t in the ICA model, and instead assume that each mixture x_j as well as each independent component s_i is a random variable, instead of a proper time signal. The observed values x_{jt}, e.g. the microphone signals in the instruments identification problem above, are then a sample of the random variable x_j.

In matrix formulation, the data matrix X is considered to be a linear combination of non-Gaussian independent components, i.e.

$$X = SA,$$

where the columns of S contain the independent components and A is a linear mixing matrix. In short, ICA attempts to "un-mix" the data by estimating an un-mixing matrix W where $XW = S$.

In this formulation the two most important assumptions in ICA are already stated, i.e. *independence and non-Gaussianity*. One approach to solving $X = SA$ is to use some information on the statistical properties of the signals s_i to estimate the a_{ij}. Actually, and perhaps surprisingly, it turns out that it is enough to assume that the s_i are statistically independent. This is not an unrealistic assumption in many cases, and it need not be exactly true in practice.

The other key to estimating the ICA model is non-Gaussianity. Actually, without non-Gaussianity the estimation is not possible at all. Indeed, in the case of Gaussian

---

[1]This section is composed from [4] and [5].

variables, we can only estimate the ICA model up to an orthogonal transformation. In other words, the matrix A is not "identifiable" for Gaussian independent components.

In order to extract the independent components/sources, we search for an un-mixing matrix W that maximizes the non-Gaussianity of the sources. In so-called *Fast-ICA*, non-Gaussianity is measured using approximations to *neg-entropy* J which are fast to compute. Neg-entropy is defined as the difference of the entropies of a standard normal variable and a general random variable. Note that this is nonnegative since normals have maximum entropy.

Entropy is the basic concept of information theory. The entropy of a random variable can be interpreted as the degree of information that an observation of the variable gives. The more random, i.e. unpredictable and unstructured, the variable is, the larger its entropy. For a discrete random variable Y, *entropy H* is defined as

$$H(Y) = -\sum_i P(Y = a_i) \log_2 P(Y = a_i),$$

where the a_i are the possible values of Y. This very well-known definition can be generalized for continuous-valued random variables and vectors, in which case it is often called *differential entropy*. The differential entropy H of a random vector y with density $f(y)$ is defined as

$$H(y) = -\int f(y) \log_2 f(y) dy.$$

Neg-entropy is then defined as:

$$J(y) = H(y_{gauss}) - H(y),$$

where y_{gauss} is a Gaussian random variable with the same covariance matrix as y. Due to the above-mentioned properties, neg-entropy is always non-negative, and it is zero iff y has a Gaussian distribution.

Because of the complex calculation of neg-entropy, in *FastICA* simple approximations to neg-entropy are used which will not be discussed here. The maximization of J obviously produces a kind of maximum non-Gaussianity. Before maximization, first the data are pretransformed in the following way:

1. The data are centered by subtracting the mean of each column of the data matrix X and

2. the data matrix is then "whitened" by projecting the data onto its principal component directions, i.e. $X \rightarrow XG$, where G is a loading matrix (see Section 9.8.3). The number of components can be specified by the user. This way, we already have uncorrelated components.

The ICA algorithm then estimates another matrix W so that

$$XGW = S.$$

W is chosen to maximize the neg-entropy approximation under the constraint that W is an orthonormal matrix. This constraint ensures that uncorrelatedness of the estimated components is preserved. We will not discuss the optimization strategy here.

Finally, note that there are some ambiguities for ICA components.

1. We cannot determine the order of the independent components, meaning that the independent components that produced the observations may be found in any order.

2. The signs of ICA components may be different from the signs of the independent components to be found.

3. The means of ICA components are zero so that ICA components may be shifted in relation to the independent components having produced the data.

Example 11.5 (ICA on the first 4 MFCC components of block 1). We checked whether ICA is able to reconstruct the first two MFCC components when we have observed only 4 linear combinations of the first 4 MFCC components x_1, x_2, x_3, x_4 of block 1:

$$y_1 = x_1 + x_2, y_2 = x_1 - x_2, y_3 = x_1 + x_2 + x_4, \text{ and } y_4 = x_1 + x_2 + x_3.$$

From Figure 11.7, it becomes clear that we found a correspondence of the 1st data component "mfcc_block1_1" with the 2nd ICA component, and with a correspondence of the 2nd data component "mfcc_block1_2" with the 4th ICA component. Note the sign change in the 2nd ICA component. Also note that the ICA components are centered in contrast to the original components.

For a possibly more relevant example see Chapter 17.

11.7 Further Reading

We have seen that data type is an important indicator for distance selection. However, distance measures can also be related to other aspects like, e.g., application. For example, *time series* representing music pieces need special distances ([7]).

Moreover, *variable selection* is a good candidate to identify the adequate space for distance determination for both supervised and unsupervised classification. For an overview of variable selection methods in classification see, e.g., Dash and Liu ([2]).

Last but not least, the observed variables are often not ideal as a basis for classification. Instead, *transformations* may be much more sensible which directly relate to a re-definition of the distance measure (see also the music examples at the end of Section 11.2).

Outliers can be seen as special types of clusters. As illustrated in Example 11.2, outliers are often identified by some clustering methods, like the extreme upper right observation in Figure 11.1 by all hierarchical methods except Ward clustering, and the extreme lower left observation identified by Average and Single Linkage clustering. At the same time, it should be clear that the concept of outliers is not identical to the concept of clusters which should also be clear by the results of the different clustering methods discussed above.

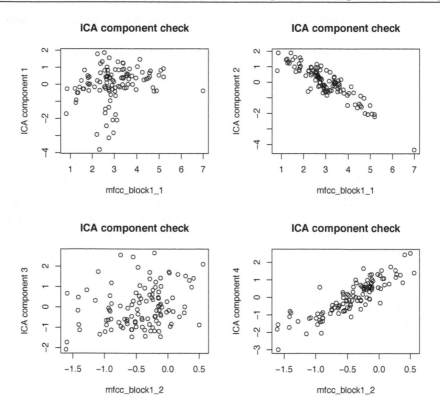

Figure 11.7: ICA component check.

Bibliography

[1] J. Beran. *Statistics in Musicology.* Chapman&Hall/CRC, 2004.

[2] M. Dash and H. Liu. Feature selection for classification. *Intelligent Data Analysis*, 1:131–156, 1997.

[3] A. Hyvärinen, J. Karhunen, and E. Oja. *Independent Component Analysis.* Wiley, 2001.

[4] A. Hyvärinen and E. Oja. Independent component analysis. *Neural Networks*, 13:411–430, 2000.

[5] J. L. Marchini and C. Heaton. Package fastICA. http://cran.stat.sfu.ca/web/packages/fastICA/fastICA.pdf, 2014.

[6] P.-N. Tan, M. Steinbach, and V. Kumar. *Introduction to Data Mining.* Addison-Wesley, 2005.

[7] C. Weihs, U. Ligges, F. Mörchen, and D. Müllensiefen. Classification in Music Research. *Advances in Data Analysis and Classification (ADAC)*, 1:255–291, 2007.

Chapter 12

Supervised Classification

CLAUS WEIHS
Department of Statistics, TU Dortmund, Germany

TOBIAS GLASMACHERS
Institute for Neuroinformatics, Ruhr-Universität Bochum, Germany

12.1 Introduction

Classification of entities into categories is omnipresent in everyday life as well as in engineering and science, in particular music data analysis. As a problem per se it is analyzed with mathematical rigor within the disciplines of statistics and machine learning. Classification simply means to assign one (or more) of finitely many possible class labels to each entity.

Formally, a classifier is a map $f : X \to Y$, where X is the input space containing characteristics of the entities to classify, also called *instances*, and Y is the set of categories or classes. For example, X may consist of characteristics of all possible pieces of music, while Y may consist of genres. Thus, a genre classifier is a function that assigns each piece of music to a genre. For other classification examples, see Section 12.3.

The problem of constructing f can be approached by fixing a statistical model class f_θ. The process of estimating the a priori unknown parameter vector θ from data is called training or learning. Parameters in θ are, e.g., mean and variance in the case of a normal distribution. Learning a classifier model from data has advantages and disadvantages over manual (possibly algorithmic) construction of the function f: constructing a model programmatically allows for a rather direct incorporation of expert knowledge, however, it is cumbersome and time consuming. In many classification problems data of problem characteristics are easier to obtain and to encode in machine readable form than expert knowledge, and one can hope to save considerable effort by letting a learning machine figure out a good model by itself based on a corpus of data, e.g., characterizing a large collection of music. This proceeding has the added benefit of increased flexibility: the model can be improved as more data becomes available without the need for further expensive, time-consuming engineering.

The questions of which model class to use for which problem and how to estimate or learn its parameters based on data are core topics of statistics and machine learning.

12.2 Supervised Learning and Classification

For sure, supervised learning is the most prominent machine learning paradigm for technical applications. It is generally assumed that the data consists of a finite collection of (example) inputs, called the training data set. In our example the characteristics of each single piece of music are available in a vector of input data x_i, with the whole music collection forming the data set $\{x_1, \ldots, x_n\}$, containing n enumerated inputs.

In a supervised learning setting these inputs are augmented with labels or target values for the classifier. In other words, for each characteristic $x_i \in X$ of a piece of music we also need to provide the information to which genre $y_i \in Y$ it belongs, or equivalently, which answer $f(x_i)$ we desire. This added label information y_i is thought of as being provided by an expert or "supervisor" – therefore the term *supervised learning*. The augmented training data consists of input-label pairs $\{(x_1, y_1), \ldots, (x_n, y_n)\}$. Now the task of supervised learning is to generalize the finite set of input-label correspondences of the training data set to a rule $y = f(x)$ that is applicable to all $x \in X$, also the ones that were not seen during training. This allows us to classify pieces of music that are missing in our collection, and even ones that will be written in the future.

A standard assumption of supervised learning is that the data pairs (x_i, y_i) are independent and identically distributed (i.i.d.) samples from a "data generating" distribution P on $X \times Y$. An example of such a distribution is the process by which a user adds pieces to his private music collection. It may be impacted by listening to the radio, talking to friends, and many other factors. It is obvious that such a process is hard to model, while looking at the existing collection is much easier. Please note that the i.i.d. assumption (cp. Chapter 9) is only needed for valid statistical inference about the results like for testing hypotheses like 'Classification model 1 is significantly better than classification model 2'. Such questions will be discussed in Chapter 13. In this chapter here, we will be somewhat more descriptive, and allow some deviations from this assumption (see Example 12.1).

Importantly, the classifier is designed so as to perform well on any instance x from the unknown distribution. At this point we have to define what we mean by performing well: we have to quantify the severity of an event that is in general unavoidable, namely making a mistake. This means that given a previously unknown piece of music $x \in X$, our classifier predicts a value $\hat{y} = f(x)$ that differs from the true genre y – however, since x is not an element or our music collection the true label was not known to the classifier. Different types of mistakes may be differently severe. This is conveniently captured by the concept of a loss function $L(y, \hat{y})$ that outputs zero for correct predictions $\hat{y} = y$ and a non-negative value otherwise, understood as the cost of misclassification. For example, we may define the cost of mistaking a piece of music from the classic period for one from the romantic period as less than mistaking it for modern electronic music or heavy metal. A standard loss function for classification is the *0/1-loss* that assigns a cost of one uniformly to all types of mistakes. Other types of losses can be found in Section 12.4.4.

Now we can formally state the goal of learning, which is to minimize the expected loss, also called the *risk*

$$\mathscr{R}(f) = E\left[L(y, f(\mathbf{x}))\right] = \int_{X \times Y} L(f(\mathbf{x}), y) \, \mathrm{d}P(\mathbf{x}, y).$$

For our examples this is the average (severity of) error of the classifier over all possible pieces of music, weighted by their probability of being encountered. Of course, we would like to pick a classifier f with an as small as possible risk $\mathscr{R}(f)$.

The minimizer f^* of the risk functional over all possible (measurable) classifiers f is known as the *Bayes classifier*, and the corresponding risk $\mathscr{R}^* = \mathscr{R}(f^*)$ is the *Bayes risk*. Note that in general even the best possible model has a non-zero risk. This is plausible in the context of genre classification since the assignment of a piece of music to a genre may be subjective, and some pieces combine aspects of different genres. Thus, even the assignment rule that works best on average cannot make 100% correct predictions.

The goal of finding this minimizer is in general not achievable. This is because the risk cannot even be computed since the data generating distribution $P(\mathbf{x}, y)$ is not known. The available information about this distribution is limited to a finite sample, the *training set*. The learning task now is to find a classifier f for which we know with reasonable certainty that it comes as close as possible to f^*, given restricted knowledge about P. How to estimate the *0/1-loss* will be discussed in greater detail in Chapter 13. In this chapter we will estimate the loss by the error rate on a so-called *test set*, drawn independently but disjoint from the population of examples.

12.3 Targets of Classification

We will now try some systematics of the types of classification found in music data analysis. We will distinguish at least *two dimensions*: content and class type. Concerning content, one can at least distinguish genre classification, artist classification, singer identification, mood detection, and instrument recognition. Concerning class type, we will distinguish binary, multi-class, and multi-label classification. In *binary classification* we distinguish only 2 classes, in *multi-class classification* there are more 2 classes to be distinguished. These two types are *single-label* cases, since each observation (music piece) is assigned only one label. In *multi-label classification* more than one label can be assigned to each observation. Note that genre classification is discussed in this chapter, instrument recognition in Chapter 18, and mood / emotion detection in Chapter 21.

In principle, each content class can be combined with each class type. For example, we can try to distinguish two genres, like in this chapter "Classic" from "Non-Classic," or we can be somewhat more ambitious and try to distinguish "Classic,"

"Pop," "Hard-Rock" and other genres. This way, genre classification is either of type "binary" or of type "multi-class." Even the type "multi-label" might be adequate for genre classification, since in some cases the genre is by no means clear. For example, some "Hard-Rock" pieces might be also "Pop." The class type might influence the choice of the classification method, the content possibly not that much. An exception might be caused by the ease of interpretation of the results of some methods, like decision trees (see Section 12.4.3).

Note that the input features are all assumed to be metric for convenience, in order to be able to apply all classification methods introduced below. However, the type of input features may define a *third dimension* in order to distinguish (at least) signal-level from high-level data. Signal-level features are introduced in Chapter 5, high-level features in Chapter 8. Typical signal-level or low-level features are chroma, timbre, and rhythmic features. Typical high-level features are MIDI-features, music scores, social web features, and even lyrics. Note that high-level features might have to be transformed into metric features before becoming usable in the following classification methods. Also this third dimension can be deliberately combined with content. For example, one might want to identify genre from low-level audio features or from music scores.

In this chapter we will only consider genre classification as an example application based on the data set described below.

Example 12.1 (Music Genre Classification). *We will consider a two-class example. We will try to distinguish Classic from Non-Classic music comprising examples from Pop, Rock, Jazz and other genres. Our training set consists of 26 MFCC features corresponding to 10 Classic and 10 Non-Classic songs. The first 13 MFCC features are aggregated for 4s classification windows with 2s overlap (mean and standard deviation, named MFCC..m and MFCC..s, respectively, where the dots stand for the number of the MFCC). Note that this obviously leads to non i.i.d. data at least because of overlapping data in consecutive windows are dependent. This leads to 2361 training observations over all 20 music pieces. Our test set consists of the corresponding features for 15 Classic and 105 Non-Classic songs. There is no overlap between the artists of the training and tests sets. Below, we will compare the performance of 8 classification methods, as delivered by the software R ([10]). Also we will discuss examples of class separation by the different methods by means of plots.*

12.4   Selected Classification Methods

In this chapter we cannot introduce all available classification methods because of space restrictions. Instead, we will restrict ourselves to representatives of four important classes of methods, namely Bayes and approximate Bayes methods (represented by Linear Discriminant Analysis (LDA) and the Naive Bayes method), distance methods (represented by the Nearest Neighbors method), decision trees (represented by CART and C4.5), and linear / nonlinear large margin methods (represented by the Support Vector Machine, SVM). Moreover, we will also briefly discuss how to generate so-called ensembles of such methods and give Random Forests as an example. Finally, we will give a very brief introduction into neural networks.

12.4.1 *Bayes and Approximate Bayes Methods*

It would be optimal if the above-mentioned *Bayes risk* could be achieved by a classification rule. Unfortunately, this is typically the case only if some theoretical distributions are assumed in the X space. As an example, let us consider the following assumptions:

L1: The distributions of influential factors inside the classes are normal distributions with different expected values $\boldsymbol{\mu}_i$ but identical covariance matrix $\boldsymbol{\Sigma}$ for all classes $y^1, \ldots, y^G$. This leads to different densities f_i for the classes.

L2: The misclassification costs are equal for all classes.

L3: The a priori probabilities of the classes may be different.

For two classes this leads to the following *Bayes decision rule*:

Choose class 1 iff $\frac{f_1(x)}{f_2(x)} > \frac{\pi_2}{\pi_1}$ for a priori class probabilities $\pi_i, i = 1, 2$. Otherwise choose class 2.

Inserting the densities of the normal distribution, this is equivalent to

$$\frac{\exp(-0.5(x - \boldsymbol{\mu}_1)^T \boldsymbol{\Sigma}^{-1}(x - \boldsymbol{\mu}_1))}{\exp(-0.5(x - \boldsymbol{\mu}_2)^T \boldsymbol{\Sigma}^{-1}(x - \boldsymbol{\mu}_2))} > \frac{\pi_2}{\pi_1} \tag{12.1a}$$

$$-0.5(x - \boldsymbol{\mu}_1)^T \boldsymbol{\Sigma}^{-1}(x - \boldsymbol{\mu}_1) + 0.5(x - \boldsymbol{\mu}_2)^T \boldsymbol{\Sigma}^{-1}(x - \boldsymbol{\mu}_2) > \log\left(\frac{\pi_2}{\pi_1}\right) \tag{12.1b}$$

$$x^T \boldsymbol{\Sigma}^{-1}(\boldsymbol{\mu}_2 - \boldsymbol{\mu}_1) < \log\left(\frac{\pi_1}{\pi_2}\right) + 0.5\boldsymbol{\mu}_2^T \boldsymbol{\Sigma}^{-1}\boldsymbol{\mu}_2 - 0.5\boldsymbol{\mu}_1^T \boldsymbol{\Sigma}^{-1}\boldsymbol{\mu}_1. \tag{12.1c}$$

This procedure is called *Linear Discriminant Analysis (LDA)* of two classes.

A further simplification is achieved if the a priori probabilities of the classes are equal: Choose class 1 iff $f_1(x) > f_2(x)$. Then the above Inequality (12.1c) can also be written in the following way: Let $a = \boldsymbol{\Sigma}^{-1}(\boldsymbol{\mu}_2 - \boldsymbol{\mu}_1)$, then $a^T x < \frac{1}{2}a^T(\boldsymbol{\mu}_1 + \boldsymbol{\mu}_2)$. Unknown parameters are empirically estimated by means of the empirical covariance matrix $\hat{\boldsymbol{\Sigma}}$ and the empirical means $\bar{x}_1, \bar{x}_2$ leading to the 1st linear discriminant component $\hat{a} = \hat{\boldsymbol{\Sigma}}^{-1}(\bar{x}_2 - \bar{x}_1)$. Then, we use the rule:

$$\hat{a}^T x < \frac{1}{2}\hat{a}^T(\bar{x}_1 + \bar{x}_2),$$

meaning that the separation is linear, and a projection of the mean of the empirical group means is the estimated border between the classes.

This method can be generalized to more than two classes. Generally, we can show that the k-th class is chosen by the Bayes rule under the assumptions (L1), (L2), and (L3) iff the function

$$h_i(x) := (\boldsymbol{\Sigma}^{-1}\boldsymbol{\mu}_i)^T x - 0.5\boldsymbol{\mu}_i^T \boldsymbol{\Sigma}^{-1}\boldsymbol{\mu}_i + \log(\pi_i) \tag{12.2}$$

is maximal for class $i = k$. By this rule, the p-dimensional space is partitioned in that each p-vector is assigned to exactly one class except for the vectors on borders between classes. Such borders can be characterized by equating the functions $h_i(x)$ of these classes:

$$(\boldsymbol{\Sigma}^{-1}\boldsymbol{\mu}_i)^T x - 0.5\boldsymbol{\mu}_i^T \boldsymbol{\Sigma}^{-1}\boldsymbol{\mu}_i + \ln(\pi_i) = (\boldsymbol{\Sigma}^{-1}\boldsymbol{\mu}_j)^T x - 0.5\boldsymbol{\mu}_j^T \boldsymbol{\Sigma}^{-1}\boldsymbol{\mu}_j + \ln(\pi_j),$$

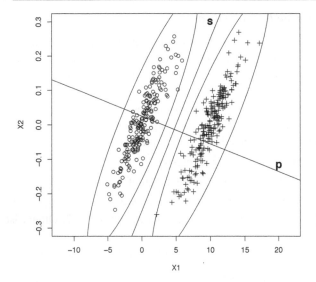

Figure 12.1: Observations of two classes (indicated by symbols "o" and "+") ideal class separation in two features X1 and X2, enclosing ellipses, separating hyperplane (line s), and orthogonal projection space (line p).

i.e. the borders $(\boldsymbol{\mu}_j - \boldsymbol{\mu}_i)^T \boldsymbol{\Sigma}^{-1} \boldsymbol{x} = \text{const}$ are *hyperplanes* in $\mathbb{R}^p$. Note that hyperplanes are lines in two dimensions, planes in three dimensions, etc. In the case of two classes in two dimensions, we are looking for that line separating the two classes "best." An idealized example can be found in Figure 12.1 where the two classes can be completely separated by a line. Please also notice the line indicating the projection direction. In reality the two classes will often overlap, though, so that an ideal separation with zero errors is not possible.

When we weaken assumption L1 so that different covariance matrices are allowed for the different classes, then we talk about *Quadratic Discriminant Analysis (QDA)*.

Let us now discuss approximate Bayes rules. LDA can be simplified by the so-called *Naive Bayes* assumption:

NB: All influential features are uncorrelated, i.e., $\boldsymbol{\Sigma} = diag(c_1, \ldots, c_p), c_j \in \mathbb{R}, j = 1, \ldots, p$.

This idea leads to a new method called the *Independence Rule (IR)* using the same decision rule as LDA but with $\boldsymbol{\Sigma} = diag(c_1, \ldots, c_p)$.

Dropping now the assumption of normality in the classes in L1, we arrive at the general *Naive Bayes (NB)* method. The probability model for many classifiers can be written as the conditional probability $P(y|x_1, \ldots, x_p)$ over a dependent class feature y with a small number of outcomes or classes, conditional on several influential features $x_1, \ldots, x_p$. Using Bayes' theorem (see Theorem 9.2), this can be written as $P(y|x_1, \ldots, x_p) = P(y) P(x_1, \ldots, x_p|y) / P(x_1, \ldots, x_p)$.

In practice, there is interest only in the numerator of this fraction, because the denominator does not depend on y and the values of the features x_i are given so that the denominator is effectively constant. Unfortunately, since $\boldsymbol{x} = (x_1, \ldots, x_p)$ is usually an unseen instance which does not appear in the training data, it may not be possible to directly estimate $P(x_1,\ldots,x_p|y)$. So, a simplification is made by assuming that the features $X_1,X_2,\ldots,X_p$ are conditionally independent of each other given the class, which is the above independence assumption NB. Then, the conditional distribution over the class feature y is $P(y|x_1,\ldots,x_p) = \frac{1}{Z}P(y)\prod_{i=1}^p P(x_i|y)$, where the evidence $Z = P(x_1,\ldots,x_p)$ is constant if the values of the features are known. The corresponding classifier, the so-called *Naive Bayes* classifier, is the function

$$NB(x_1,\ldots,x_p) = \operatorname*{argmax}_y P(y)\prod_{i=1}^p P(x_i|y).$$

Obviously, this rule is not restricted to two classes. A class prior may be estimated by assuming equiprobable classes, i.e. $\hat{P}(y) = 1/(\text{no. of classes})$, or by calculating an estimate for the class probability from a training set by

$$\hat{P}(y) = (\text{no. of samples in class } y)/(\text{total no. of samples}).$$

The conditional probabilities of the individual features x_i given the class y have to be estimated also from the data, e.g. by discretizing the data into groups for all involved quantitative attributes. Since the true density of a quantitative feature is usually unknown for real-world data, unsafe assumptions and, thus, unsafe probability estimations unfortunately often occur.

Discretization can circumvent this problem. With discretization, a qualitative attribute X_i^* is formed for each X_i. Each value x_i^* of X_i^* corresponds to an interval $(a_i,b_i]$ of X_i. Any original quantitative value $x_i \in (a_i,b_i]$ is replaced by x_i^*. All relevant probabilities are estimated with respect to x_i^*. Since probabilities of X_i^* can be properly estimated from corresponding frequencies as long as there are enough training instances, there is no need to assume the probability density function anymore. However, discretization might suffer from information loss. See also Section 14.2.1 for discretization methods.

The different implementations of the Naive Bayes classifier typically differ in different discretizations. Obviously, discretization can be effective only to the degree that $P(y|\mathbf{x}^*)$ is an accurate estimate of $P(y|\mathbf{x})$.

Example 12.2 (Music Genre Classification (cont.)). *Let us come back to Example 12.1 on music genre classification. We have applied* LDA, IR, *and* NB *to the above data leading to the error rates 16.2% for* LDA, *16.3% for* IR, *and 12.3% for* NB. *So, the idea of a simplification of the covariance matrix to a diagonal does not lead to an improvement by assuming a normal distribution (IR), but by using a discretization (NB). Let us try to visualize the class separation of the different methods by means of projections using not all 26 features for classification, but only the two most important, namely the means of the first two MFCCs,* MFCC01m *and* MFCC02m. *With these features we get the error rates 18.3% for* LDA, *18.4% for* IR, *and 18.6% for* NB. *Thus,* NB *suffers the most from feature selection. Looking at Figure 12.2 we see that the separation is similar for* LDA *and* NB, *except that* NB *leads to a somewhat nonlinear separation. Note that in the plots, the background colors indicate the posterior class probabilities through color alpha blending.* IR *leads to nearly exactly the same separation as* LDA.

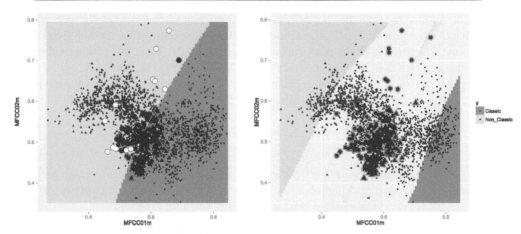

Figure 12.2: Class separation by *LDA* (left) and *NB* (right); true classes: Classic = circles, Non-Classic = triangles; estimated classes: Classic = darker region, Non-Classic = lighter region; errors indicated by bigger symbols.

12.4.2 Nearest Neighbor Prediction

Another kind of classification involves distance-based rules. The most well-known examples are the nearest-neighbor rules. Such rules are among the simplest and most intuitive predictive rules. They nevertheless turn out to be powerful predictors with interesting statistical properties. The simplest variant is the 1-Nearest-Neighbor (1-NN) prediction rule, also often referred to simply as the nearest-neighbor rule. Given an input x, it predicts an output $\hat{y}$ f x as follows: it searches for the nearest neighbor of x in the training set, i.e., the training point x_i with minimal distance to x. Let j $\arg\min_i\{d$ x,x_i $\}$ denote the index of the nearest neighbor,[1] where d is a given metric (e.g., Euclidean distance). Then the label of the neighbor x_j is used as a prediction for the label of x, i.e., f x y_j.

For this prediction rule to work well, we need the implicit assumption that points of equal class cluster together. It is assumed in particular that the label y_j is indeed the best prediction for a point x equal to or in the vicinity of x_j. However, this assumption is often wrong in the presence of noise or outliers. This shortcoming is addressed by the k-Nearest-Neighbor (k-NN) rule. Instead of relying on a single neighbor, this rule aggregates the labels of the k nearest neighbors of x. In the simplest case the output of the rule is a majority vote over the k neighbor labels. This allows the predictor to outvote isolated outliers. The result is a smoother and more reliable prediction. Indeed, increasing the parameter k has a regularizing effect. On the other hand, k should not be chosen too large. This would mean that also far-away points participate in the prediction, although there is no good reason to assume that such points share the same class label. Consequently this has a deteriorating effect on prediction performance.

---

[1]We ignore the issue of distance ties, i.e. of non-unique nearest neighbors. For the purpose of this introductory presentation ties can be broken with any rule, e.g., at random.

There are many elaborate variants of the nearest neighbor prediction scheme. An alternative to the choice of a fixed number of neighbors is to define the neighborhood directly based on the metric d, e.g., by thresholding distance to the query point by a fixed radius. Please notice the relation to the Naive Bayes idea in the previous section where continuous features where discretized using balls of nearby values.

It is also possible to modify the majority voting scheme. For example nearby points may be given more impact, either based on distance or on their distance rank relative to other neighbors. Elaborate tie-breaking mechanisms work with shared and averaged (non-integer) ranks that can be taken into account in a subsequent voting scheme. Finally and maybe most importantly, the ad hoc choice of the Euclidean metric can be replaced, e.g., with the Mahalanobis distance $d_M(x, x_i) := \sqrt{(x - x_i)^T S^{-1}(x - x_i)}$, where S is the sample covariance or correlation matrix (see Section 11.2).

Nearest neighbor predictors have the advantage that they essentially do not have a training step. On the downside, they require the storage of all training points for prediction making, and worse, at least in a naive implementation all distances between test and training points need to be computed. This makes predictions computationally slow.[2]

An important statistical property of the k-NN rule is that in the limit of infinite data it approaches the Bayes-optimal classifier *for all problems*. This property is known as universal consistency. It holds under quite mild assumptions, namely that the number of neighbors k_n as a function of data set size grows arbitrarily large ($\lim_{n \to \infty} k_n = \infty$) and the relative size of the neighborhood shrinks to zero ($\lim_{n \to \infty} k_n/n = 0$). This means that on the one hand the prediction rule is flexible enough to model the optimal decision boundary of any classification task. On the other hand, a relatively simple technical condition on the sequence k_n ensures that overfitting is successfully avoided in the limit of infinite data.

To summarize, the k-NN rule is an extremely simple yet powerful prediction mechanism for supervised classification. It does not have a training step, but it requires storing all training examples for prediction making. It is based on a metric measuring distances of inputs. Nearest neighbor models are most frequently applied to continuous features.

Example 12.3 (Music Genre Classification (cont.)). *We continue with the music genre classification example above, and look at different numbers of neighbors used for classification. With $k = 1, 11, 31$ we get the error rates $20.9\%, 18.4\%, 17.1\%$ using all 26 features. When only using the most important 2 features MFCC01m, MFCC02m, we get $26.4\%, 22.9\%, 20.5\%$, and the separation is shown in Figure 12.3. Obviously, the boundary between the two classes gets smoother when k is increasing. Notice that the boundaries are much more flexible than for the methods LDA and NB. Having in mind that the training set includes 2361 observations, $k = 31$ is not very large.*

---

[2]In low-dimensional input spaces it is possible to reduce prediction time considerably by means of binary space-partitioning tree data structures (e.g. KD-trees) from "brute force" search complexity of $\mathcal{O}(n)$ to only $\mathcal{O}(\log(n))$ operations (see [6]).

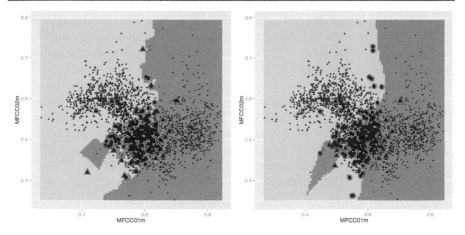

Figure 12.3: Class separation by 1-*NN* (left) and 31-*NN* (right); true classes: Classic = circles, Non-Classic = triangles; estimated classes: Classic = darker region, Non-Classic = lighter region; errors indicated by bigger symbols.

12.4.3 Decision Trees

Decision trees are one of the most intuitive models used in classification (and they can also be employed for regression and other tasks). Their basic idea is simple and quickly explained: The model is represented as a set of hierarchical "decision rules," organized usually in a binary tree structure (hence the name). Every rule is of the form $x_i < 3$ or $x_i =$ "guitar," for continuous or categorical features, respectively. When a new observation x needs to be classified, it is dropped down the tree and one either takes the left or right branch in each internal decision node of the tree, depending on the decision rule of the current node and the corresponding feature value x. Once a terminal node has been reached, a class label is assigned.

Class Labels at Terminal Nodes Let us now describe the training process which enables us to construct such a tree for a given data set. We will start with how class labels are assigned to the terminal nodes, assuming that the tree structure is already fixed. For each terminal node, we know exactly which portion of the training data is assigned to it. As the node is terminal, no further decisions are made w.r.t. the features, hence the only information we should use for the prediction is the distribution of class labels in that respective data portion and usually the majority class is assigned as a class for that terminal node. Note that by essentially the same mechanism we can also estimate posterior class probabilities: We simply have to count the proportion of observations belonging to each class for that terminal node and store the resulting table.

Splitting Rules The fitting process for the tree structure proceeds in a top-down, greedy manner. We start at the root, and consider all available data. Our task is now is to discover that (binary) decision rule which splits our set best for classification. What that means precisely, is usually expressed in a reduction in loss when employing the decision rule under consideration. Let us consider the two cases of quantitative and qualitative split features.

In the quantitative case, the *CART* (Classification And Regression Tree) method uses the following split technique. The observations are experimentally split into two child nodes by means of each realized value of the first feature and splits of the kind (value ≤ constant). The "yes-"cases are put into the left node, the "no-"cases into the right one. Let s be such a split in node t. *CART* then evaluates the reduction of the so-called *Gini-impurity* $i(t) := 1 - S$ with $S := \sum_{j=1}^{G} P^2(j|t)$, where S is the so-called *purity function* (which is maximal if all probability mass is concentrated on a single class) and $P(j|t) =$ probability of class j in node t. This reduction is calculated by means of the formula $\Delta i(s,t) = i(t) - p_L \cdot i(t_L) - p_R \cdot i(t_R)$, where p_L is the share of the cases in node t which are put into the left child node t_L, p_R analogously. *CART* chooses that split as the best for the chosen feature which produces the biggest impurity reduction. These steps are repeated for each feature. *CART* then orders the features corresponding to their ability to reduce the impurity and realizes the split with that feature and the corresponding split point with the biggest impurity reduction. This procedure is recursively repeated for each current terminal node. This way, *CART* constructs a very big tree with very many terminal nodes which are either pure or contain only a very small number of different classes.

In the case of qualitative split features, we could possibly try each possible feature value for splits of the type (value = constant). However, let us also consider the classification of the so-called C4.5 method. There, the node information content of a subtree below a node for a sample of size n is measured by its *entropy* $H(\boldsymbol{X})$:

$$H(\boldsymbol{X}) = - \sum_{c=1}^{G} \frac{|y^c|\boldsymbol{x}|}{n} \cdot \log_2 \left(\frac{|y^c|\boldsymbol{x}|}{n} \right), \tag{12.3}$$

where G is the number of classes and $|y^c|\boldsymbol{x}|$ is the number of observations from $\boldsymbol{X}$ which belong to class $c \in \{1, ..., G\}$. The efficiency of candidate nodes can be measured by the so-called *information gain*, aiming to reduce the information content carried by a node using a split s:

$$gain\,(\boldsymbol{X}, s) = H(\boldsymbol{X}) - \sum_{j=1}^{k} \frac{|\boldsymbol{X}_{js}|}{n} \cdot H(\boldsymbol{X}_{js}), \tag{12.4}$$

where $\boldsymbol{X}_{js}$ are the observations of $\boldsymbol{X}$ with the j-th value of the k outcomes of the split feature in split s. Note that an arbitrary number $k > 1$ of split branches is allowed in *C4.5*, not only 2 as in *CART*.

Several further enhancements led to the development of the decision tree algorithms *CART* and *C4.5* (for details see [3] and [9]), in particular handling of missing feature values (cp. Section 14.2.3) and tree pruning. Especially the latter technique is very important, since too large trees increase the danger of *overfitting*: if a model describes the data perfectly, from which it has been trained, but is not suitable anymore for reasonable classification of other instances (cp. Chapter 13). Moreover, understanding and interpretation of trees with many terminal nodes might be complicated.

Pruning Big decision trees are complex trees, measuring the *tree complexity* by means of the number of terminal nodes. The trade-off between goodness of fit on the training set and not too high complexity is measured by the so-called *cost-complexity* := (error rate on the training set) + β · (no. terminal nodes), where β = "penalty" per additional terminal node, often also called the *complexity parameter*.

The search for the tree of the "right size" starts with the *pruning* of the branches of the maximal tree (T_{max}) from the terminal nodes ("bottom up") as long as the training error rate stays constant (T_1). Then, we look for the so-called *weakest link*, i.e. for that node for which the pruning of the corresponding subtree below this node leads to the smallest increase of the training error. This is equivalent to looking for that node for which the increase of the "penalty parameter" is smallest for maintaining the cost-complexity at the same level. The subtree with the weakest link is pruned. This procedure is repeated until only the tree root is left. From the corresponding sequence of trees the tree with the lowest cross-validated error rate (see Section 13.2.3) is chosen to be the final tree. This leads to the tree with the smallest prediction error.

Example 12.4 (Music Genre Classification (cont.)). *Let us now look at the above music genre classification example by means of* CART *decision trees based on Gini-impurity (function* rpart *in the software* R*). We have tried unpruned and pruned trees. Unpruned trees with all 26 MFCC features lead to 21.5% error rate, pruned trees to 17.4%, both with default parameter values. Figure 12.4 starts with the correspond-ing pruned tree. Then, it shows the separation based on* MFCC01m *and* MFCC02m *for the unpruned case. Note that the separation is along the coordinate axes. Also notice the light area between the areas definitely assigned to one of the classes. In this area assignment is most uncertain. The unpruned and the pruned tree based on* MFCC01m *and* MFCC02m *lead to error rates* 18.5% *and* 19.2%.

12.4.4   Support Vector Machines

Support Vector Machines (SVMs) [14, 11, 12] are among the state-of-the-art machine learning methods for linear and non-linear classification. They are often among the strongest available predictors, and they come with extensive theoretical guarantees. SVMs are *kernel methods*: they are essentially linear models that can be turned into powerful, non-parametric predictors with the so-called kernel trick.

Linear SVMs We start with the most basic case, the linear SVM for binary classifi-cation. This machine separates two classes indicated by labels $y \in \{-1, +1\}$ with an affine function $f(x) = w^T x + b$, given by a weight vector $w \in \mathbb{R}^p$ and a bias or offset term $b \in \mathbb{R}$. An input x is classified according to $\text{sign}(f(x))$.[3] The SVM classifier is defined as the (affine) linear function f that maximizes the safety margin between the classes. Given f (or w and b) the margin of a correctly classified training exam-ple (x_i, y_i) is the distance of x_i from the decision boundary $\{f = 0\}$, which forms a hyperplane. The other way round, the margin of a function f is defined as the mini-mum of the margins over the training set. The SVM training step consists of finding

---

[3]We ignore the case $f(x) = 0$. Any prediction may be made in this case, e.g., at random.

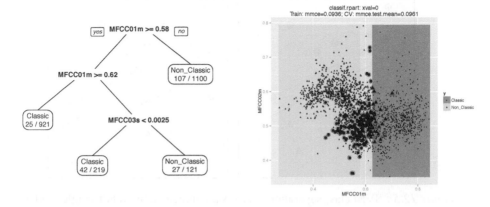

Figure 12.4: The tree of the pruned *rpart* model based on 26 MFCCs with node error rates (left) and class separation of an unpruned *rpart* model based on *MFCC01m* and *MFCC02m* (right); true classes: Classic = circles, Non-Classic = triangles; estimated classes: Classic = darker region, Non-Classic = lighter region; errors indicated by bigger symbols.

f so that all points are correctly classified and the safety margin between the classes is maximized; see Figure 12.5 for an illustration. The margins of some points coincide with the margin of the hyperplane, these points are called *support vectors*. The training problem is equivalent to requiring function values of at least $+1$ for positive class points and at most -1 for negative class points while moving the corresponding level sets $\{f = \pm 1\}$ as far away from the decision boundary $\{f = 0\}$ as possible. This amounts to minimization of the (squared) norm of $\boldsymbol{w} \in \mathbb{R}^p$:

$$\min_{\boldsymbol{w},b} \quad \frac{1}{2}\|\boldsymbol{w}\|^2 \quad \text{s.t.} \quad y_i \cdot (\boldsymbol{w}^T \boldsymbol{x}_i + b) \geq 1 \quad \forall i.$$

This is a quadratic problem (on a polyhedron) that can be solved efficiently.

The above optimization problem is well posed only for linearly separable data. This situation is not to be expected in practice: even if the decision boundary of the Bayes rule happens to be linear we cannot exclude the existence of outliers. Therefore the SVM has to allow for constraint violations. For this purpose we introduce slack variables ξ_i, one per training point, measuring the amount of constraint (or margin) violation. Since margin violations are undesirable, they are penalized in the objective function:

$$\min_{\boldsymbol{w},b} \quad \frac{1}{2}\|\boldsymbol{w}\|^2 + C \cdot \sum_{i=1}^{n} \xi_i \quad \text{s.t.} \quad y_i \cdot (\boldsymbol{w}^T \boldsymbol{x}_i + b) \geq 1 - \xi_i \text{ and } \xi_i \geq 0.$$

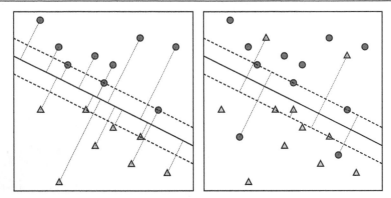

Figure 12.5: SVM class separation. Classes are indicated by light triangles and dark circles. Left: The hard-margin support vector machine separates classes with the maximum-margin hyperplane (solid line). Point-wise safety margins are indicated by dotted lines. The dashed lines parallel to the separating hyperplane indicate the safety margin. The five points located exactly on these margin hyperplanes are the support vectors. Note that the space enclosed by the margin hyperplanes does not contain any data points. Right: For a linearly non-separable problem, the support vector machine separates classes with a large margin, while it allows for margin violations, indicated by dotted lines.

The solution (w^*, b^*) of this problem is defined as the (standard) linear SVM. It has a single parameter, $C > 0$, trading maximization of the margin against minimization of margin violations. Although exact maximization of the margin is meaningless in the non-separable case, the SVM still achieves the largest possible margin given the constraints. It is therefore often referred to as a *large margin* classifier.

The SVM problem can be rewritten in unconstrained form as

$$\min_{w \in \mathbb{R}^p} \frac{1}{2} \|w\|^2 + C \cdot \sum_{i=1}^{n} L(f(x_i), y_i), \qquad (12.5)$$

where $L : \mathbb{R} \times Y \to \mathbb{R}_0^+$ defined by $L(f(x), y) := \max\{0, 1 - y \cdot f(x)\}$ is called *hinge loss*. The loss $L(f(x_i), y_i)$ coincides with the value of slack variable ξ_i, hence the hinge loss measures the amount of margin violation. Moreover, it is also an upper bound on the $0/1$-loss that is of actual interest for classification. There are other SVM variants which vary in the exact type of loss function applied, but for the sake of training efficiency, they all rely on losses that are convex in $f(x)$.

Kernels   Until now we have only considered linear decision functions f. Now we move to a general technique for turning linear predictors into non-linear ones. Let $X = \mathbb{R}^p$ be the input space. A function $k : X \times X \to \mathbb{R}$ is called a *(Mercer) kernel function* if it is symmetric and positive semi-definite. This means that for any finite collection $\{x_1, \ldots, x_n\}$ of points in X the so-called *kernel Gram matrix* $K \in \mathbb{R}^{n \times n}$, $K_{ij} = k(x_i, x_j)$, is positive semi-definite. This rather technical property implies the existence of a feature map $\phi : X \to \mathcal{H}$ into a feature space $\mathcal{H}$ with a scalar product

$\langle \cdot, \cdot \rangle$ given by the kernel function: $k(\boldsymbol{x}, \boldsymbol{x}') = \langle \phi(\boldsymbol{x}), \phi(\boldsymbol{x}') \rangle$. This proceeding remains implicit in the sense that the feature map ϕ does not need to be constructed since we only need the scalar products as specified by the kernel Gram matrix. Indeed, a linear method such as the SVM (just like linear regression and many others) can be formulated in terms of vector space operations (addition of vectors, multiplication of vectors with scalars) and inner products. Now the *kernel trick* amounts to replacing all inner products with a kernel function. This way of handling nonlinear transformations is in contrast to other transformation-based methods with an explicit (often manual) construction of a feature map and the application of the algorithm to the resulting feature vectors. The kernel approach is particularly appealing since (a) it can implicitly handle extremely high-dimensional and even infinite-dimensional feature spaces $\mathcal{H}$, and (b) for many feature maps of interest, the computation of the kernel is significantly faster than the calculation of the corresponding feature vectors.

The most important examples of kernel functions on $X = \mathbb{R}^p$ are as follows:

$$k(\boldsymbol{x}, \boldsymbol{x}') = \boldsymbol{x}^T \boldsymbol{x}' \qquad \qquad \text{linear kernel}$$

$$k(\boldsymbol{x}, \boldsymbol{x}') = (\boldsymbol{x}^T \boldsymbol{x}' + q)^d \qquad \qquad \text{polynomial kernel}$$

$$k(\boldsymbol{x}, \boldsymbol{x}') = \exp(-\gamma \|\boldsymbol{x} - \boldsymbol{x}'\|^2) \qquad \text{Gaussian kernel}$$

The Gaussian kernel (also often called *radial basis function (RBF) kernel*) corresponds to an infinite dimensional feature space $\mathcal{H}$.

Non-linear SVM Given a kernel k and a regularization parameter $C > 0$ the non-linear SVM classifier is defined as the solution $\boldsymbol{w} \in \mathcal{H}, b \in \mathbb{R}$ of the problem

$$\min_{\boldsymbol{w} \in \mathcal{H}, b \in \mathbb{R}} \frac{1}{2} \|\boldsymbol{w}\|^2 + C \cdot \sum_{i=1}^{n} L(f(\boldsymbol{x}_i), y_i),$$

where the squared norm $\|\boldsymbol{w}\|^2 = \langle \boldsymbol{w}, \boldsymbol{w} \rangle$ and the scalar product in the decision function are operations of the feature space $\mathcal{H}$, and the input $\boldsymbol{x}$ is replaced with its feature vector $\phi(\boldsymbol{x})$. i.e. $f(\boldsymbol{x}) = \langle \boldsymbol{w}, \phi(\boldsymbol{x}) \rangle + b$, where the scalar product $\langle \cdot, \cdot \rangle$ is given by the kernel function k. Fortunately, the so-called *representer theorem* guarantees that the optimum $\boldsymbol{w}^*$ is located in the span of the training data, i.e. $\boldsymbol{w}^* = \sum_{i=1}^{n} \alpha_i \phi(\boldsymbol{x}_i)$, leading to the following *decision function*:

$$f(\boldsymbol{x}) = \sum_{i=1}^{n} \alpha_i k(\boldsymbol{x}, \boldsymbol{x}_i). \tag{12.6}$$

Based on this decision function, the unknown class of an input $\boldsymbol{x}$ is predicted by the following *decision rule*: Predict class -1 iff $f(\boldsymbol{x}) < 0$ and class $+1$ iff $f(\boldsymbol{x}) > 0$. If $f(\boldsymbol{x}) = 0$, the class is undetermined, i.e., can be fixed deliberately.

For the Gaussian RBF kernel, Equation (12.6) is a weighted sum of Gaussians centered on the training inputs. This function class is so flexible that any desired output may be realized in the training points, which means that a solution without margin violations exists even for noisy data. The resulting classifier displays heavy overfitting (cp. Chapter 13), though. For such flexible kernels, the regularization terms play a vital role: by enforcing a smooth solution, the SVM is able to solve learning problems that require even highly non-linear decision functions while avoiding overfitting at the same time. This requires problem-specific tuning of the regularization trade-off parameter C.

Note that the sum in Equation (12.6) is usually sparse, i.e., many of the coefficients α_i are zero. This means that often only a small subset of the data needs to be stored in the model, namely the x_i corresponding to non-zero coefficients α_i. These points are the *support vectors* defined above.

Multiple Classes   Returning to our initial example of genre classification, it is apparent that the SVM's ability to separate two classes is insufficient. Many practical problems involve three or more classes. Therefore, the large margin principle has been extended to multiple classes.

The simplest and maybe most widespread scheme is the one-versus-all (OVA) approach. It is not at all specific to SVMs; instead it may be applied to turn any binary classifier based on thresholding of a real-valued decision function into a multi-class classifier. For this purpose, the G-class problem is converted into $G = |Y|$ binary problems. In the c-th binary problem, class c acts as the positive class (label $+1$), while the union of all other classes becomes the negative class (label -1). An SVM decision function f_c is trained on each of these G binary problems, thus the c-th decision function tends to be positive only for data points of class c. Now for a point x the OVA scheme produces G different predictions. If only one of the resulting values $f_c(x)$ is positive, then the machines produce a consistent result: all machines agree that the example is of class c, and hence the prediction $\hat{y} = c$ is made. However, either none or more than one function value may be positive. Then the G binary predictions are inconsistent. The above prediction rule is extended to this case by picking the function with largest value as the prediction:

$$\hat{y} = f(x) = \arg\max_{c \in Y} \left\{ f_c(x) \right\}. \tag{12.7}$$

There are other extensions of the large margin framework to multiple classes, e.g. replacing the hinge loss with a corresponding loss function for $G > 2$ classes. See, e.g., [16, 5, 8], for examples.

Example 12.5 (Music Genre Classification (cont.)). *Let us now look at the above music genre classification example by means of support vector machines. We have tried linear SVMs (lSVM) and SVMs with a Gaussian kernel with a width γ optimized on a grid (kSVM). In both cases, the cost parameter C was also optimized on a grid, taking that parameter value of a prefixed grid which minimizes the error estimate. For lSVM and kSVM we get $13.6\%, 19.6\%$ error rates based on all features. Using only the best 2 features we get $17.9\%, 25.5\%$ error rates. Figure 12.6 shows the separations for the two SVMs based on the best 2 features. Note the similarity of the plot for lSVM to the result of LDA and the flexibility of the separation by kSVM similar as for k-NN.*

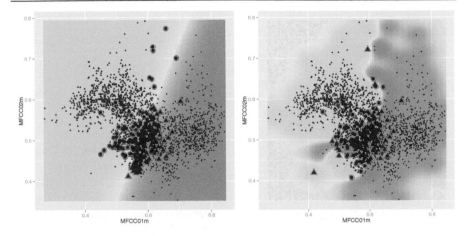

Figure 12.6: Class separation by *lSVM* (left) and *kSVM* (right); true classes: Classic = circles, Non-Classic = triangles; estimated classes: Classic = darker region, Non-Classic = lighter region; errors indicated by bigger symbols.

12.4.5 Ensemble Methods: Bagging

Sometimes, a single classification rule is not powerful enough to sufficiently predict classes of new data. Then, one idea is to combine several rules to improve prediction. This leads to so-called *ensemble methods*. *Bagging* is one of these methods combining several classification rules, e.g. based on different samples or on different classification ideas. Bagging might even be applied simultaneously to different classification methods.

One example of bagging is *Random Forests*, a combination of many decision trees (see, e.g., [4]). The construction of the different classification trees has stochastic components, leading to the term Random Forests.

First we have to fix how many trees should build the forest having in mind that the classification quality will be better with more trees involved, i.e. a big number of trees does not lead to overfitting (cp. Chapter 13). The adequate number of trees depends, however, on different parameters like the number of features and the number of classes. The trees are not determined from all available data, but for each tree a new sample (*bag*) is drawn with replacement (see also the bootstrap in Section 13.2.4). Since we use many different samples of this kind, the term *bagging* is used. The term might also be derived from Bootstrap Aggregation, i.e. the aggregation of a new classification rule from many simple rules. Moreover, in bagging in each tree, not all features are used but only a prefixed number of randomly drawn features. In *Random Forests* only the number of features used in each node is prefixed, the actually used features are randomly drawn for each node individually. Each tree is fully grown (no pruning) to have the smallest training error rate. A new object is classified by all trees in the forest and the class with the most votes is taken as the prediction. An advantage of bagging and *Random Forests* is that each feature has a high chance to contribute to separation since sometimes the more important features

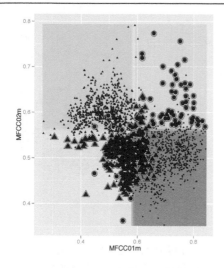

Figure 12.7: Separation of bagged trees based on *MFCC01m* and *MFCC02m*; true classes: Classic = circles, Non-Classic = triangles; estimated classes: Classic = darker region, Non-Classic = lighter region; errors indicated by bigger symbols.

Table 12.1: Test Error Rates (in %) for the Different Classifiers

data	LDA	IR	NB	31NN	unpru.	pruned	lSVM	kSVM	BDT
26 feat.	16	16	12	17	22	17	14	20	17
2 best	18	18	19	21	19	19	18	26	21

are not drawn. A disadvantage is that interpretation is much more complicated than for classification trees.

Example 12.6 (Music Genre Classification (cont.)). *For Bagged Decision Trees* (BDT) *(1000 replications with 50% randomly chosen features each) the estimated error rate is 16.9% based on all MFCCs and 21.4% based on the 2 best features. Figure 12.7 shows the separation based on the 2 best features only. Notice that the rectangles on the main diagonal are purely assigned to one class, and realizations of the other class are always marked as an error. In contrast, in the rectangles on the secondary diagonal the assignment is changing, meaning that the voting of the 1000 trees is ambiguous.*

Let us now compare the error rates of all introduced classifiers based on all 26 MFCCs and the two best. From Table 12.1 it is clear that the Naive Bayes method reproduces the true classes best (error rate 12%) based on all MFCCs. Based on only the two best MFCCs, 6 methods approximately produce the best error rates (near 18%). Obviously, feature selection, by taking only the two most important features into account, appears to be improvable. See Chapter 15 for a systematic introduction into feature selection methods.

12.4.6 Neural Networks

There are lots of classification methods not discussed in this chapter up to now. A very prominent one is the somewhat involved Artificial Neural Network (*ANN*), for which we only briefly introduce the model without discussing the typical estimation

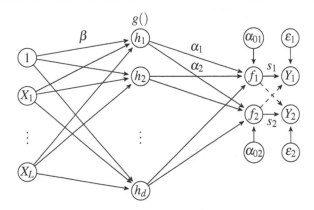

Figure 12.8: Model of a multi-layer neural network for classification. There are L input neurons and d hidden neurons in one hidden layer.

problems because of space restrictions. *ANNs* can be used to model regression problems with measured responses variable (cp. Section 9.8.1) and classification problems with class responses. We will concentrate here on the classification problem in the 2-class case. Let us start, however, with a very general definition of *ANNs* and specialize afterwards.

Definition 12.1 (Artificial Neural Network (*ANN*)). An *Artificial Neural Network (ANN)* consists of a set of processing units, the so-called *nodes* simulating neurons, which are linked analogous to the synaptic connections in the nervous system. The nodes represent very simple calculation components based on the observation that a neuron behaves like a switch: if sufficient neurotransmitters have been accumulated in the cell body, an action potential is generated. This potential is mathematically modeled as a weighted sum of all signals reaching the node, and is compared to a given threshold. Only if this limit is exceeded, the node "fires." Structurally, an *ANN* is obviously comparable to a *natural (biological) neural network* like, e.g., the human brain.

Let us now consider the special *ANN* exclusively studied in this chapter.

Definition 12.2 (Multi-Layer Networks). The most well-known neural network is the so-called *Multi-Layer Neural Network* or *Multi-Layer Perceptron*. This network is organized into *layers* of neurons, namely the input layer, any number of hidden layers, and the output layer. In a *feed-forward network*, signals are only propagated in one direction, namely from the input nodes towards the output nodes. As in any neural network, in a multi-layer network, a weight is assigned to every connection between two nodes. These weights represent the influence of the input node on the successor node.

For simplicity we consider a network with a single hidden layer (see Figure 12.8). In such an artificial neural network (*ANN*), linear combinations of the input signals

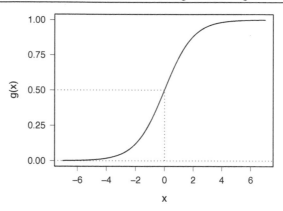

Figure 12.9: Logistic activation function.

$1, X_1, \ldots, X_L$ with individual weights β_l are used as input for each node of the hidden layer. Note that the input 1 corresponds to a constant. Each node then transforms this input signal using an *activation function* g to derive the output signal. In a 2-class classification problem, for each class 1 and 2, these output signals are then again linearly combined with weights $\alpha_{ig}, g \in \{1, 2\}$, to determine the value f_g of a node representing class g. In addition to the transformation g of the input signals $\boldsymbol{X} = (1\ X_1\ \ldots\ X_L)^T$, a constant term α_{0g}, the so-called *bias*, is added to the output, analogous to the intercept term of the linear model. Finally, in order to be able to interpret the outputs f_g as (pseudo-)probabilities, the real values of f_g are transformed by the so-called *softmax transformation* s_g which should be as near as possible to the value of the dummy variable Y_g being 1 if the correct class is g and 0 otherwise. Since both Y_g are modeled jointly, one of the probabilities should be near 1 iff the other should be near to 0. The model errors are denoted by $\varepsilon_g, g \in \{1, 2\}$. In what follows, we will introduce and discuss all these terms.

Definition 12.3 (Activation Function). The *activation function* is generally not chosen as a jump function "firing" only beyond a fixed activation potential, as originally proposed, but as a symmetrical sigmoid function with the properties:

$$\lim_{x \to -\infty} g(x) = 0, \quad \lim_{x \to \infty} g(x) = 1, \quad g(x) + g(-x) = 1.$$

A popular choice for the activation function is the *logistic function* (see Figure 12.9):

$$g(x) = \frac{1}{1 + e^{-x}}.$$

Another obvious choice for the activation function is the cumulative distribution function of any symmetrical distribution.

Definition 12.4 (Softmax Transformation). In order to be able to interpret the outputs f_g as (pseudo-) probabilities, the class-wise outputs $f_g(\boldsymbol{X}, \boldsymbol{\theta})$ are transformed by the so-called *softmax transformation* to the interval $(0,1)$ as follows:

$$s_g(\boldsymbol{X}, \boldsymbol{\theta}) = \frac{e^{f_g(\boldsymbol{X}, \boldsymbol{\theta})}}{e^{f_1(\boldsymbol{X}, \boldsymbol{\theta})} + e^{f_2(\boldsymbol{X}, \boldsymbol{\theta})}}, \quad g \in \{1,2\}.$$

Note that normalization leads to cross-dependencies of s_1 on f_2 and of s_2 on f_1 so that the (pseudo-) probability of one class is dependent on the corresponding (pseudo-)probability of the other class, which obviously makes sense.

Overall this leads to the following model for neural networks:

Definition 12.5 (Model for Neural Networks). The model corresponding to the multi-layer network with one hidden layer and two classes has the form:

$$Y_1 = s_1\left(\alpha_{01} + \sum_{i=1}^{d} \alpha_{i1} g(\boldsymbol{\beta}_i^T \boldsymbol{X} + \beta_{i0})\right) + \varepsilon_1 =: s_1(f_1(\boldsymbol{X}, \boldsymbol{\Theta})),$$

$$Y_2 = s_2\left(\alpha_{02} + \sum_{i=1}^{d} \alpha_{i2} g(\boldsymbol{\beta}_i^T \boldsymbol{X} + \beta_{i0})\right) + \varepsilon_2 =: s_2(f_2(\boldsymbol{X}, \boldsymbol{\Theta})),$$

where $\boldsymbol{X} = (X_1 \ \dots \ X_L)^T$ is the vector of input signals, $\boldsymbol{\beta}_i = (\beta_{i1} \ \dots \ \beta_{iL})^T$ is the vector of the weights of the input signals for the i-th node of the hidden layer, β_{i0} is the input weight of the constant, $\boldsymbol{\alpha} = (\alpha_{11} \ \dots \ \alpha_{d1} \ \alpha_{12} \ \dots \ \alpha_{d2})^T$ is the vector of the weights of the output signals of the nodes of the hidden layer, $(\alpha_{01} \ \alpha_{02})$ is the bias, ε is a random variable with expected value 0, and the whole vector of unknown *model coefficients* of this model is $\boldsymbol{\Theta} = (\alpha_{01} \ \dots \ \alpha_{d1} \ \alpha_{02} \ \dots \ \alpha_{d2} \ \beta_{10} \ \dots \ \beta_{d0} \ \boldsymbol{\beta}_1^T \ \dots \ \boldsymbol{\beta}_d^T)^T$.

The model coefficients have to be estimated (as statisticians would say) or learned (in the language of neural networks and machine learning) from data. The process of learning the coefficients of a neural network is also referred to as the *training* of the network whereas predictions are then used to *test* the network. The optimal number d of nodes in the hidden layer might be found by choosing the model with the least test error (cp. Chapter 13).

Training of the neural network corresponds to (nonlinear least-squares) *estimation* of the unknown parameters $\boldsymbol{\Theta}$. The squared model errors for the different classes are added in the (nonlinear least-squares) objective function. Replacing the squared error with a different loss function is straightforward. The optimization (training) is typically realized through stochastic gradient descent (cp. Theorem 10.1). Gradients w.r.t. $\boldsymbol{\Theta}$ are computed efficiently by the so-called *back-propagation* of error method. The objective function of the training problem is usually highly multi-modal, hence the training process yields only a local optimum. For more information on parameter estimation in neural networks see, e.g., [2].

Note that the above neural networks are not identifiable in that different parameter sets lead to the same fit. Since these parameter sets even do not have the same sign, the interpretation of parameter values is not possible. Only the prediction of an unknown class is well defined (see, e.g., [15, pp. 202–206, 389–395]). Such prediction is realized by the following *classification rule*:

For an input x with unknown class y, predict the class by $\hat{y} = \text{argmax}_g \, s_g(x, \hat{\Theta})$, where $\hat{\Theta}$ is the least-squares estimate of the unknown parameter vector Θ.

Using many hidden layers leads to *deep learning* (see, e.g., [1]). Convolutional neural networks (CNNs) are a particularly successful class of deep networks. Instead of using fully connected layers (each node in layer k feeds into each node in layer $k+1$), a CNN has a spatially constrained connectivity: each hidden node "sees" only a small patch of the input or previous hidden layer, its so-called receptive field. For music data, this means that each node in the first hidden layer processes a short time window. Each time window is processed by a number of neurons in parallel, so that different nodes can specialize on the extraction of different features. A second special property of CNNs is that neurons processing different time windows share the same weights in their input connections. In effect, the same features are extracted from each time window. The overall operation of such a processing layer is described compactly as a set of *convolutions*, see Equation (4.4), where the convolution kernels are encoded in the weights of the network and hence learned from data. Information from neighboring time windows is merged in subsequent layers. In such a deep learning architecture, low layers extract simple, low-level features, which are aggregated into complex, high-level information in higher layers. The last layer(s) of a CNN are fully connected. They compute a class prediction from the high-level features.

Obviously, this kind of modeling can be easily extended to $G > 2$ classes. For more information on neural networks see, e.g., [17].

12.5   Interpretation of Classification Results

The interpretation of the results of classification rules is crucial for applications. Surely, this might have individual needs in different applications. However, two aspects are often relevant, namely "feature selection" and "identification of problematic observations." One aim of feature selection is to find simple-to-understand classification rules, problematic observations are interesting for understanding the performance of a classification rule. Feature selection has already been mentioned in the application example in this chapter, but without going into details. The description of feature selection methods is postponed to Chapter 15. Identification of problematic observations is discussed in this section.

If an observation falls near a decision border, i.e. a border between different classes, the classifier might be somewhat unsure into which class this observation should be classified. If an observation distinctly lies on the "wrong side" of such a border, then it is severely misclassified by the classifier. Both kinds of observations together are defined here as problematic observations.

Distances from decision borders are measured in different ways for the different classifiers. Here, we concentrate on classifiers which deliver estimated class probabilities, like *lda*. Typically, if the probability of the correct class is smaller than 0.5, then this observation is misclassified. If this probability is extremely small, then this observation is "extremely misclassified." The most extreme examples might give an impression of very untypical regions of the corresponding class with respect to the classification rule. Other interesting observations are characterized by probabilities near 0.5 for two classes. For such observations, the classification rule is somewhat unsure about class assignment. Let us now demonstrate interpretation of classification rules by means of the outcomes of the classification methods in our example.

Example 12.7 (Music Genre Classification (cont.)). Obviously, a typical error rate for the classification based on all 26 MFCCs is around 17%, which, e.g., is achieved by LDA *and* BDT. *This means that around 17% of the 15387, i.e., around 2600 small time windows of length 4 seconds in the test set are misclassified. This does not mean that many major parts of the pieces of music or even many of the 120 pieces as a whole are misclassified. One task of interpretation would be, therefore, to identify longer misclassified successive parts of the involved music pieces. Then, one can identify the locations of such parts in the corresponding music pieces and try to find reasons for misclassification.*

In particular, we looked for successive misclassified parts of length > 32 seconds in the results of LDA *and* BDT. *This way, 33 parts were identified for* LDA *and 32 for* BDT. *For* LDA, *class probabilities are given for each small time window. If the probability for the correct class is smaller than 0.5, then this window is misclassified. For longer successive misclassified parts we took the means of the probabilities of the correct class in the involved small windows. If this mean is extremely small, then this part of music is "extremely misclassified." The most extreme examples will be discussed below. For* BDT, *1000 trees predicted the classes. In this case, if the number of false predictions is higher than the number of correct ones, then the small time window is misclassified and the ratio "(number of correct predictions) / 1000" gives an estimate of the probability of the correct class for each time window. For longer successive misclassified parts we, again, took the means of such probabilities.*

For both, LDA *and* BDT, *the identified successive misclassified music parts longer than 32 seconds appeared in 18 different music pieces, 13 of which are the same for the two classifiers. It is striking that many misclassified parts stem from 8 pieces of European Jazz which was not represented in the training set. Looking at the 5 most extremely misclassified music pieces from* LDA *and* BDT *each (probability of correct class < 3.5%), only 2 of them did not stem from this group, namely "Fake Empire" by the Indie-Rock Band The National and "Trilogy" by Emerson, Lake, and Palmer. Whereas the latter piece of music might have been suspected to be near to classical music, the former is misclassified in the beginning of the song, where only piano and voice are active.*

12.6 Further Reading

There are extensions to this chapter in this book. For music data examples with more than 2 classes see, e.g., Chapter 13, "Evaluation" and Chapter 18, "Instrument Recognition." Classification has to be applied regularly to a large number of features. In such cases, feature selection is often advisable in order to improve understanding. See Chapter 15 for a systematic introduction into feature selection methods. In this chapter we only used the error rate as an evaluation measure for the different classification methods. See Chapter 13 for a systematic introduction into various evaluation measures.

An ensemble method not discussed in this book because of space restrictions is *boosting*. Similar to bagging, with boosting an ensemble of classifiers is applied and aggregated. However, the different elements of the ensemble are also weighted by their quality in the overall decision (see, e.g., [7]).

The actual chapter only deals with classification problems with unambiguous labels. Problems with multi-labels, e.g. mentioned in the beginning of this chapter, can be treated, e.g., as described in [13]. Also, this chapter only deals with individual classification models. In music data analysis, though, sometimes so-called *hierarchical models* are used, meaning that the input of one classification model is determined by another classification model (see, e.g., Chapter 18).

Bibliography

[1] Y. Bengio. Learning deep architectures for AI. *Foundations and Trends in Machine Learning*, 2(1):1–71, 2009.

[2] C. M. Bishop. Neural networks and their applications. *Review of Scientific Instruments*, 65(6):1803–1832, 1994.

[3] L. Breiman. Bagging predictors. *Machine learning*, 24(2):123–140, 1996.

[4] L. Breiman. Random forests. *Machine Learning Journal*, 45(1):5–32, 2001.

[5] K. Crammer and Y. Singer. On the learnability and design of output codes for multiclass problems. *Machine Learning*, 47(2):201–233, 2002.

[6] J. H. Friedman, J. L. Bentley, and R. A. Finkel. An algorithm for finding best matches in logarithmic expected time. *ACM Transactions on Mathematical Software*, 3(3):209–226, 1977.

[7] T. Hastie, R. Tibshirani, and J. H. Friedman. *The elements of statistical learning: data mining, inference, and prediction*. New York: Springer-Verlag, 2001.

[8] Y. Lee, Y. Lin, and G. Wahba. Multicategory support vector machines: Theory and application to the classification of microarray data and satellite radiance data. *Journal of the American Statistical Association*, 99(465):67–82, 2004.

[9] J. R. Quinlan. *C4. 5: Programs for Machine Learning*, volume 1. Morgan Kaufmann, 1993.

[10] R Core Team. *R: A Language and Environment for Statistical Computing*. R Foundation for Statistical Computing, Vienna, Austria, 2014.

[11] B. Schölkopf and A. J. Smola. *Learning with Kernels: Support Vector Machines, Regularization, Optimization, and Beyond*. MIT Press, 2002.

[12] I. Steinwart and A. Christmann. *Support Vector Machines*. Springer, 2008.

[13] G. Tsoumakas and I. Katakis. Multi-label classification: An overview. *International Journal of Data Warehousing & Mining*, 3(3):1–13, 2007.

[14] V. Vapnik. *Statistical Learning Theory*. John Wiley and Sons, 1998.

[15] C. Weihs, O. Mersmann, and U. Ligges. *Foundations of Statistical Algorithms.* CRC Press, 2014.

[16] J. Weston and C. Watkins. Support Vector Machines for Multi-class Pattern Recognition. In M. Verleysen, ed., *Proceedings of the Seventh European Symposium on Artificial Neural Networks (ESANN)*, pp. 219–224. d-side publications, 1999.

[17] G. Zhang. Neural networks for classification: A survey. *IEEE Transactions on Systems, Man, and Cybernetics – Part C: Applications and Reviews*, 30(4):451–462, 2000.

Chapter 13

Evaluation

IGOR VATOLKIN
Department of Computer Science, TU Dortmund, Germany

CLAUS WEIHS
Department of Statistics, TU Dortmund, Germany

13.1 Introduction

Most models are not an exact image of reality. Often, models only give rough ideas of real relationships. Therefore, models have to be evaluated whether their image of reality is acceptable. This is true for both regression and classification models (cp. Section 9.8.1 and Chapter 12). What are, however, the properties a model should have to be acceptable? Most of the time, it is more important to identify models with good predictive power than to identify factors which significantly influence the response on the sample used for modeling (goodness of fit). This means that the predictions of a model should be acceptable, i.e. a model should be able to well approximate (unknown) responses for values of the influential factors which were not used for model building. For example, it is not acceptable that the goodness of fit of a regression model for valence prediction dropped from 0.58 to 0.30 and to 0.06 when the model was trained for film music and was validated on classical, respectively, popular pieces (cp. Section 21.5.6). Also, a classification model for instrument recognition determined on some instances of music pieces, has to be able to predict the playing instrument of a new piece of music from the audio signal (cp. Section 18.4). Such arguments lead to corresponding ideas for model selection, which will be discussed in this section.

If you wish to obtain an impression of the predictive power of a model without relying on too many assumptions, i.e. in the non-parametric case, you should apply the model on samples from the underlying distribution which were not used for model estimation. If the same sample would be used for estimations and checking predictive power, we might observe so-called *overfitting*, since the model is optimally fitted to this sample and might be less adequate for other samples. Unfortunately in practice, most of the time only one sample is available. So we have to look for other solutions.

New relevant data can only be generated by means of new experiments, which are often impossible to conduct in due time. So what should we do? As a solution to this dilemma, *resampling methods* have been developed since the late 1960s. The idea is to sample repeatedly from the only original sample we have available. These repetitions are then used to generate predictions for cases not used for model estimation. This way, we can at least be sure that the values in the sample can be realized by practical sampling.

In this chapter, we will briefly introduce such methods and refer to three kinds of evaluation tasks: model selection, feature selection, and hyperparameter tuning. Model selection is the main task needing evaluation and feature selection and hyperparameter tuning can be thought of as subtasks of finding optimal models.

Model Selection   In many cases, several models or model classes are candidates for fitting the data. Resampling methods and the related predictive power assessment efficiently support the selection process of the most appropriate and reliable model.

Feature Selection   Often an important decision for the selection of the best model of a given model type is the decision about the features to be included in the model (e.g., in a linear model). Feature selection is discussed in Chapter 14.

Hyperparameter Tuning   Most modeling strategies require the setting of so-called *hyperparameters* (e.g., the penalty parameter C in the optimality criterion of soft-margin linear support vector machines, see Section 12.4.4). Thus, tuning these hyperparameters is desirable to determine a model of high quality. This can be realized by some kind of (so-called) nested resampling introduced below in Section 13.4.

In the following, model quality is solely reflected by predictive power, which in our view is the most relevant aspect, although other aspects of model quality are also discussed in some settings. For example, interpretable models are preferable, which is obviously related to feature selection. It should also be noted that it is usually advisable to choose a less complex model achieving good results for small sample sizes, since more-complex models usually require larger data sets to have sufficient predictive power.

We will deal here with the two most important statistical modeling cases, i.e. supervised classification (see Chapter 12) and regression (see Chapter 9): In *classification problems* an integer-valued response $y \in \mathbb{Z}$ with finitely many possible values $y^1, \ldots, y^G$ has to be predicted by a so-called *classification rule* based on n observations of $z_i = \begin{bmatrix} x_i^T & y_i \end{bmatrix}^T$, $i = 1, \ldots, n$, where the vector x summarizes the influential factors. In *regression problems* a typically real-valued response $y \in \mathbb{R}$ has to be predicted. In both cases, the influential factors are assumed to be real-valued.

In this chapter, we mainly discuss model comparison and corresponding model selection. We assume that we are interested in problems related to so-called *learning samples* $L = \{z_1, \ldots, z_n\}$ of n observations, and that there is a set of competing candidate models available. Each of these model candidates is fitted to the learning sample L leading to a function $a(\cdot|L)$. These models are then compared with respect to certain interesting properties. In particular, models could be compared with respect to their ability to predict unknown response values y.

In order to identify the best model, the model candidates have to be compared by means of problem-specific quality measures. The value of such a measure should depend on the model as well as on the learning data. Thus, there has to be a function $p(a,L)$ that assesses the quality of the model prediction function $a(\cdot|L)$. Since L is a random sample, $p(a,L)$ is a random variable, whose variability is induced by the variability of the possible learning samples L generated from an underlying data distribution F.

Therefore, in order to identify the best model, it is natural to compare the distributions of the quality measures. For this, it would be best to draw random samples from the distribution of the quality measure for a model a by evaluating the performance measure $p(a,L)$ for different learning samples L. This will be realized by *resampling*.

Then, e.g., the statistical hypothesis of equal (expected) quality of model candidates can be tested by means of an adequate standard test if independent samples can be drawn from the distributions of the interesting quality measures. This way, we can also control the error probability of declaring a model wrongly to be best *(statistical guarantee)*.

In what follows we will introduce some evaluation quality measures $p(a,L)$ and some tests on their performance. In particular, we will consider the following hypothesis on the K competitive models meaning that in the mean all models have the same performance. Obviously, such a hypothesis should be rejected and one should come up with a sort of ranking of the different models.

Definition 13.1 (Quality Criterion Hypothesis). Consider a random sample $\{p_{k1},\ldots,$ $p_{kB}\}$ of B independent, identically distributed observations $p_{kb}=p(a_k,L^b), b=1,\ldots,$ B, of the performance measure p for model a_k, i.e. from B independent samples L^b drawn from the underlying data distribution F. Then, the null hypothesis for the quality criterion looks as follows ($E(p_k)$ denotes the expected value of the performance measure for model a_k):

$$\mathbf{H_0}: E(p_1) = \ldots = E(p_K) \quad \text{vs.} \quad \mathbf{H_1}: \exists k,k': E(p_k) \neq E(p_{k'}).$$

The remainder of this chapter is organized as follows. In the next section, several established resampling methods as general frameworks for model evaluation are introduced. Later on (Section 13.3), we will focus on evaluation measures, mainly different performance aspects (Sections 13.3.1-13.3.5), but also provide a discussion of several groups of measures beyond classification performance (Section 13.3.6). Section 13.4 provides a practical example how resampling can be applied for hyperparameter tuning. The application of statistical tests for the comparison of classifiers is discussed in Section 13.5. We conclude with some remarks on multi-objective evaluation (Section 13.6) and refer to further related works (Section 13.7).

Before we go into detail concerning resampling methods and performance measures let us introduce the two data sets and four classification models evaluated later as examples for the evaluation measures. On these training samples the below classification models are trained and the corresponding models are evaluated by the following performance measures on a larger test sample.

Example 13.1 (Training Samples).   *Table 13.1 lists ten tracks of four different genres for the classification task of identifying pop/rock songs among music pieces of other genres. Two training samples are used: a smaller sample with only four tracks (upper part of the table), and a larger one using all tracks.*

Table 13.1: Tracks for Training of Classification Models

Genre	Artist/Composer	Track
Pop/Rock	AC/DC	What Do You Do For Money Honey
Pop/Rock	Nirvana	Drain You
Classic	Beethoven, Ludwig van	Sonata No.17 in D minor Op.31 No.2
Classic	Händel, Georg Friedrich	Organ Concerto Op.4 No.1
Pop/Rock	Grönemeyer, Herbert	Mambo
Pop/Rock	Madonna	Push
Jazz	Coltrane, John	Summertime
Jazz	Mann, Herbie	Gospel Truth
Electronic	Faithless	Take The Long Way Home
Electronic	Prodigy	Fuel My Fire

Example 13.2 (Classification Models).   *Using the two training samples from Example 13.1, four decision tree models (cf. Section 12.4.3) are trained using a set of 13 MFCCs (see Section 5.2.3) for which the mean values are estimated for classification windows of 4 s length and 2 s overlap. Figure 13.1 shows the trees created from the smaller set (left subfigures (a), (b)) and the larger one (right subfigures (c), (d)), where the depth of the tree is limited either to two levels (upper subfigures (a), (c)) or four levels (lower subfigures (b), (d)). The numbers of positive and negative instances (classification windows) are given in brackets, e.g., the tree in the subfigure (a) classifies 7 windows of pop/rock tracks as not belonging to this class.*

So, which of these four models is best? This will be evaluated in what follows by means of a larger test sample of songs of the same genres as in the training samples.

13.2   Resampling

The situation in the above example is typical for practical evaluation situations that there is only one data set, wherefrom training samples L are taken. With resampling, such training samples are drawn randomly. This will be systematically discussed below.

The values of the quality measures p_k of the models a_k, $k = 1, \ldots, K$, depend on the underlying data distribution F. Let the learning sample L consist of n independent observations z_j, $j = 1, \ldots, n$, distributed according to some unknown distribution F. This is denoted by $L \sim F_n$. The most frequent situation in practice is that only one learning sample $L \sim F_n$ is available and there is no possibility to easily generate new samples. In this situation F is imitated by the empirical distribution function on the learning sample $\hat{F}_n$. *Resampling* is sampling of B independent learning samples from $\hat{F}_n$. In order to distinguish the resampled new learning samples from the original

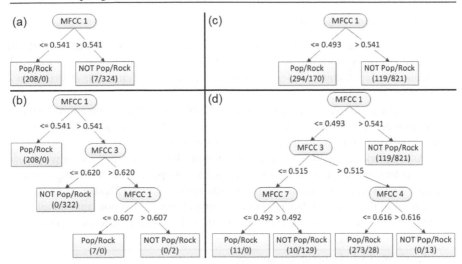

Figure 13.1: Decision trees trained with 13 MFCCs. Training samples: 4 tracks: left subfigures (a,b); 10 tracks: right subfigures (c,d). Maximal depth of the tree set to: 2: upper subfigures (a,c); 4: bottom subfigures (b,d).

learning sample, the new learning samples will be called *training samples* in the following.

$$L^1, \ldots, L^B \sim \hat{F}_n.$$

The quality of a model a_k might be assessed on the basis of the training samples L^i $(b = 1, \ldots, B)$. Therefore, by calculating

$$\hat{p}_{ki} = p(a_k, L^i), \quad b = 1, \ldots, B,$$

we get a random sample of B observations from the distribution of quality measures $p_k(\hat{F}_n)$ for each a_k.

In order to get real predictions, however, $p(a_k, \cdot)$ is normally not calculated on the training samples but on extra samples drawn from F, called *test samples* T^i.

Definition 13.2 (Test Sample Quality). Let a *test sample* $T \sim F_h$ with h independent observations be drawn from $\hat{F}_n$. Then, the *test sample quality* is defined by

$$\hat{p}_{ki} = \hat{p}(a_k, T^i).$$

In practice, the test samples are normally taken as the residue of L^i in L, i.e. $T^i = \bar{L}^i = L/L^i$. This is also assumed in the following subsection.

Note that in the above example we have another situation, where an extra, not too small, test sample is available apart from the learning sample. This means, though, that we deliberately restrict the learning sample to be small by artificially "holding out" the observations of the test sample. This procedure in a way mimics the practical situation that we can train ourselves typically only on small data sets.

13.2.1 *Resampling Methods*

In order to generate a random sample $\{p_{k1},\ldots,p_{kB}\}$ of B independent, identically distributed observations $p_{kb} = p(a_k, L^b)$ of the performance measure p for model a_k from B independent samples L^b drawn from the underlying data distribution F, resampling is used.

The term *resampling* indicates that new samples are drawn from an existing original sample. We will discuss variants of the three most well-known resampling methods *cross-validation*, *bootstrapping*, and *subsampling*.

As a generic resampling method we will consider Algorithm 13.1, in which each of the B learning samples is split into a training sample for model fitting and a test sample for model assessment. Note that the instruction *FITMODEL(L)* represents the fitting of the model a dependent on the model type. Also note that the elements of the set P of quality statistics are only the basis for equality tests or rankings.

Algorithm 13.1: Generic Resampling

Require: A learning sample L of n observations z_1 to z_n, the number of subsets
  B to generate and a loss function V.
  1: Generate B subsets of L named L^1 to L^B.
  2: $P \leftarrow \emptyset$
  3: **for** $i \leftarrow 1$ **to** B **do**
  4:   $\bar{L}^i \leftarrow L \setminus L^i$
  5:   $a \leftarrow FITMODEL(L^i)$
  6:   $p_i \leftarrow p(a, \bar{L}^i)$
  7:   $P \leftarrow P \cup \{p_i\}$
  8: **end for**
  9: **return** P

13.2.2 *Hold-Out*

Let us start with the special case of the above generic procedure with $B = 1$, i.e. where only one split into training and test sample is realized.

Definition 13.3 (Hold-Out or Train-and-Test Method). For large n the so-called *hold-out or train-and-test method* could be used, where the learning sample is divided into one training sample L' of smaller size and one test sample T: $L = L' \cup T$.

If n is so small that such an approach in infeasible, at first sight the following method appears to be most natural:

Definition 13.4 (Resubstitution Method). The *resubstitution method* uses for each model the original training sample also as the test sample.

Unfortunately, such an approach often leads to so-called *overfitting*, since the model was optimally fitted to the learning sample, and thus the error rate on this same sample will likely be better than on other, unseen, samples. With the help of the resampling methods described in the next subsections, such overfitting can be avoided.

13.2.3 Cross-Validation

Cross-Validation (CV) [26, 13] is probably one of the oldest resampling techniques. Like all other methods presented in this subsection, it uses the generic resampling strategy as described in Algorithm 13.1. The B subsets (line 1 of Algorithm 13.1) are generated according to Algorithm 13.2. Note that the instruction *SHUFFLE(L)* stands for a random permutation of the sample L. The idea is to divide the data set into B equally sized blocks and then use $B-1$ blocks to fit the model and validate it on the remaining block. This is done for all possible combinations of $B-1$ of the B blocks. The B blocks are usually called *folds* in the cross-validation literature. So a cross-validation with $B = 10$ would be called a *10-fold cross-validation*. Usual choices for B are 5, 10, and n.

Algorithm 13.2: Subsets for B-Fold CV

Require: A data set L of n observations z_1 to z_n and the number of subsets B to generate.

1: $L \leftarrow SHUFFLE(L)$
2: **for** $i \leftarrow 1$ **to** B **do**
3:    $L^i \leftarrow L$
4: **end for**
5: **for** $j \leftarrow 1$ **to** n **do**
6:    $i \leftarrow (j \bmod B) + 1$
7:    $L^i \leftarrow L^i \setminus \{z_j\}$
8: **end for**
9: **return** $\{L^1, \ldots, L^B\}$

The case $B = n$ is also referred to as *leave-one-out cross-validation* (LOOCV) because the model is fitted on the subsets of L, which arise if we leave out exactly one observation. With LOOCV, for a learning sample of size n a modeling method is applied to each subset of $n-1$ observations and tested on the n-th observation. This leads to n different models, tested on one observation each. This way, each observation of the learning sample is used exactly once as a test case for a model based on nearly the whole learning sample, neglecting nearly no information.

In classification, the error rate with respect to one resampled learning sample L^i is 1 or 0 for an incorrect or correct class prediction on the test case, respectively. As an overall quality criterion Ψ the error rate is calculated as "number of errors in test cases divided by n." For regression, the test case error is calculated as the individual quadratic loss, and the overall criterion as the mean quadratic loss, cf. Section 13.3.1.

Obviously, for large learning samples LOOCV is computer time intensive. In such cases, though, variants of the already mentioned train-and-test method, utilizing

Table 13.2: Variants of Cross-Validation: Number of Cases and Repetitions

	Leave-One-Out	*B*-Fold CV
training cases	$n-1$	$n-n/B$
test cases	1	n/B
repetitions	n	B

just one split of the original learning sample into a smaller new learning sample and a test sample, often produce a satisfying accuracy of the quality criterion.

Also in *B-fold cross-validation* with $B < n$, the cases are randomly partitioned in B mutually exclusive groups of (at least nearly) the same size. Each group is used exactly once as the test sample and the remaining groups as the new learning sample, i.e. as the training sample. In classification, the mean of the error rates in the B test samples is called *cross-validated error rate*. Table 13.2 gives an overview of the variants of cross-validation.

13.2.4 Bootstrap

The most important alternative resampling method to cross-validation is the *bootstrap*. We will only discuss the most classical variant here, called the *e0* bootstrap.

The development of the bootstrap resampling strategy [8] is ten years younger than the idea of cross-validation. Again, Algorithm 13.1 is the basis of the method, but the B subsets are generated using Algorithm 13.3. Note that the instruction *RANDOMELEMENT(L)* stands for drawing a random element from the sample L by means of uniformly distributed random number $\in \{1,\ldots,n\}$.

Algorithm 13.3: Subsets for the Bootstrap

Require: A data set L of n observations z_1 to z_n and the number of subsets B to generate.

1: **for** $i \leftarrow 1$ **to** B **do**
2:    $L^i \leftarrow \emptyset$
3:    **for** $j \leftarrow 1$ **to** n **do**
4:       $z \leftarrow RANDOMELEMENT(L)$
5:       $L^i \leftarrow L^i \cup \{z\}$
6:    **end for**
7: **end for**
8: **return** $\{L^1,\ldots,L^B\}$

The subset generation is based on the idea that instead of sampling from L without replacement, as in the CV case, we sample with replacement. This basic form of the bootstrap is often called the *e0 bootstrap*. One of the advantages of this approach is that the size of the training sample, in the bootstrap literature often also called the in-bag observations, is equal to the actual data set size. On the other hand,

Table 13.3: Bootstrap Method

	Bootstrap
training cases	n (j different)
test cases	$n - j$
repetitions	≥ 200

this means that some observations can and likely will be present multiple times in the training sample L^i. In fact, asymptotically only about 63.2% of the data points in the original learning sample L will be present in the training sample, since the probability not to be chosen n times is $(1 - 1/n)^n$, so the probability to be chosen is $1 - (1 - 1/n)^n \approx 1 - e^{-1} \approx 0.632$. The remaining 36.8% of observations are called out-of-bag and form the test sample as in CV.

Here the number of repetitions B is usually chosen much larger than in the CV case. Values of $B = 100$ up to $B = 1000$ are not uncommon. Do note, however, that there are n^n different bootstrap samples. So for very small n there are limits to the number of bootstrap samples you can generate. In general, $B \geq 200$ is considered to be necessary for good bootstrap estimation (cp. Table 13.3). This number of repetitions may be motivated by the fact that in many applications not only the bootstrap quality criterion is of interest, but the whole distribution, especially the 95% confidence interval for the true value of the criterion. For this, first the empirical distribution of the B quality measure values on the test samples is determined, and then the empirical 2.5% and 97.5% quantiles. With 200 repetitions, the 5th and the 195th element of the ordered list of the quality measures can be taken as limits for the 95% confidence interval, i.e. there are enough repetitions for an easy determination of even extreme quantiles. Note, however, that the bootstrap is much more expensive than LOOCV, at least for small learning samples.

The fact that with the bootstrap some observations are present multiple times in the training sample can be problematic for some modeling techniques. Several approaches have been proposed for dealing with this. Most add a small amount of random noise to the observations [8].

Please note that e0 can be approximated by *repeated 2-fold cross-validation*, i.e. by repeated 50/50 partition of the learning sample, or by repeated *2:1 train-and-test splitting*, because the e0 generates roughly 63.2% in-bag (train) observations and 36.8% out-of-bag (test) observations.

Another problem with adding some observations multiple times to the training sample is that we overemphasize their importance. This is called *oversampling*. This leads to an estimation bias for our quality measure. Instead of discussing variants of bootstrap which try to counter this, we will introduce another resampling method called subsampling which does not suffer from multiple observations.

13.2.5   Subsampling

Subsampling is very similar to the classical bootstrap. The only difference is that observations are drawn from L without replacement (see Algorithm 13.4). Therefore, the training sample has to be smaller than L or no observations would remain for the test sample. Usual choices for the subsampling rate $|L^i|/|L|$ are $4/5$ or $9/10$. This corresponds to the usual number of folds in cross-validation (5-fold or 10-fold). Like in bootstrapping, B has to be selected a priori by the user. Choices for B are also similar to bootstrapping, e.g., in the range of 200 to 1000.

Algorithm 13.4: Subsets for Subsampling

Require: A data set L of n observations z_1 to z_n, the number of subsets B to generate and the subsampling rate r.

1:   $m \leftarrow \lfloor r \cdot n \rfloor$
2:   **for** $i \leftarrow 1$ **to** B **do**
3:     $L' \leftarrow L$
4:     $L^i \leftarrow \emptyset$
5:     **for** $j \leftarrow 1$ **to** m **do**
6:       $d \leftarrow RANDOMELEMENT(L')$
7:       $L^i \leftarrow L^i \cup \{d\}$
8:       $L' \leftarrow L' \setminus \{d\}$
9:     **end for**
10: **end for**

13.2.6   Properties and Recommendations

Properties of Leave-One-Out and Cross-Validation   Leave-one-out cross-validation (LOOCV) has better properties for the squared loss in regression than for its 0-1 counterpart in classification and is an almost unbiased estimator for the mean loss [11]. Its near unbiasedness makes LOOCV an attractive candidate among the presented algorithms, especially when only few samples are available. But one should be aware of the following facts: LOOCV has a high variance [11, 32] as estimator of the mean loss, meaning quite different values may be produced if the data used for cross-validation slightly change. It also tends to select too complex models. In [23] theoretical reasons for this effect are presented, and subsampling and balanced leave-k-out CV are shown to be superior estimators in a simulation study. Reference [11] arrives at similar results regarding LOOCV and demonstrates empirically that 10-fold CV is often superior, suggesting a stratified version.

    For these reasons we recommend LOOCV mainly for efficient model selection, keeping in mind that this might lead to somewhat suboptimal choices.

Properties of the Bootstrap   The e0 bootstrap is pessimistically biased in the sense that it bases its performance values on models that use only about 63.2% of the data. Also, this estimator is known to have a low variance, and is especially good when the sample size is small and the error or noise in the data is high [32].

Independence, Confidence Intervals, and Testing In general, the generated training and test samples, and therefore the obtained performance statistics, will not be independent when sampling from a finite data set. This has negative consequences if confidence intervals for the performance measure should be calculated. The dependence structure is especially complicated for the commonly used cross-validation, where the split-up of the data in one iteration completely depends on all other split-ups. It can be shown that in this setting no unbiased estimator of the variance exists [2] and pathological examples can be constructed, where the performance of the variance estimator is arbitrarily bad. Reference [20] proposes a new variance estimator for CV that takes the dependence between sampled data sets into account and provides a much better foundation for interval estimators and subsequent statistical tests regarding location parameters.

13.3 Evaluation Measures

In this section, we will discuss several possibilities to measure classification performance. First, the measures based on loss between true and predicted response are introduced. Then, the confusion matrix is described, followed by common related measures and measures designed for imbalanced data sets. Afterwards, we show a simple way to aggregate evaluation measures for the calculation of prediction performance on larger entities (music pieces), when the models are applied to smaller entities (classification windows). Finally, we describe several groups of measures beyond classification performance which have a high practical relevance for music data analysis.

13.3.1 Loss-Based Performance

In supervised learning, the observations z have the form $z = \left[x^T y \right]^T$, where y is the response and x the vector of influential factors. Learning is aimed at the determination of predictions that deliver information about the unknown response exclusively on the basis of the influential factors. Therefore, for each of the K considered models, the constructed prediction function has the form $\hat{y} = a_k(x|L^i)$. The difference in the predicted response value $\hat{y}$ from the true response value y is typically represented by a scalar loss function $V(y, \hat{y})$.

Definition 13.5 (Loss in Classification and Regression). In *classification problems*, $\hat{y}$ typically is the predicted class of observations (or the vector of estimated conditional probabilities of class memberships for each class), and the *loss* is typically $V(y, \hat{y}) = 1$ if $y \neq \hat{y}$, and $V(y, \hat{y}) = 0$ otherwise. In *regression problems*, the *loss* of the predicted response value $\hat{y}$ relative to the true response value y is typically assumed to be quadratic, i.e. $V(y, \hat{y}) = (y - \hat{y})^2$.

Let us now define the corresponding quality measure p. Such a measure should combine the h individual losses of the different tested observations (examples) to one (real) number. This is realized by a function μ from $\mathbb{R}^h$ to $\mathbb{R}$.

In the case of a quadratic loss function $V(y, \hat{y}) = (y - \hat{y})^2$, the measure p is generally chosen as the mean value. This is also used in the classification case, i.e. for the 0-1 loss, leading to so-called *error rates*, i.e. (number of errors)/(number of observations). The quality measure is then called empirical risk. This risk is assumed to be evaluated on a test sample T^i independent of the training sample L^i.

Definition 13.6 (Empirical Risk). The *empirical risk* of the model a_k is defined as

$$\hat{p}_{ki} = \frac{1}{h} \sum_{j=1}^{h} (y_j - a_k(\boldsymbol{x}_j|L^i))^2, \tag{13.1}$$

where $\boldsymbol{z}_1, \ldots, \boldsymbol{z}_h$, $\boldsymbol{z}_i = \left[\boldsymbol{x}_i^T y_i\right]^T$ are the elements of a sample T^i independent of L^i.

In the regression case, $\hat{p}_{ki}$ is also called the *mean squared error* on T^i. In the classification case, $\hat{p}_{ki}$ is equal to the *misclassification (error) rate* on T^i:

Definition 13.7 (Misclassification Error Rate). The *misclassification error rate* of the model a_k is defined as

$$\hat{p}_{ki} = \frac{1}{h} \sum_{j=1}^{h} \mathbf{I}(y_j \neq a_k(\boldsymbol{x}_j|L^i)), \tag{13.2}$$

where $\boldsymbol{z}_1, \ldots, \boldsymbol{z}_h$ are the elements of a sample T^i independent of L^i and $\mathbf{I}$ is the indicator function being $= 1$ iff $y_j \neq a_k(\boldsymbol{x}_j|L^i)$.

13.3.2 Confusion Matrix

A common scheme for the evaluation of classification performance is the estimation of the *confusion matrix*. For all classes to predict, the rows in this matrix correspond to the predicted labels, the columns to the true labels, and the entries contain numbers of those classification instances corresponding to the predicted and true label of the cell.

Example 13.3 (Confusion Matrices). *Table 13.4 shows two confusion matrices after the application of models from Figure 13.1 (b,d) for a set of 120 audio recordings. For predictions on track level, the aggregation of classification results is applied as discussed later in Section 13.3.5. For instance, in the upper matrix 26 Pop/Rock songs are correctly classified, and 19 are wrongly predicted as not belonging to this class. No classical piece was incorrectly classified as Pop/Rock, etc.*

The extension of the smaller training sample (upper part of the Table 13.1) with Jazz and Electronic tracks significantly boosts the performance on music pieces which belong to these genres as can be seen in the bottom matrix of the Figure 13.4. Only two Electronic tracks are identified as belonging to Pop/Rock. The price for a better prediction of negative examples is paid here, however with an increased error for the identification of true Pop/Rock songs: only 13 instead of 26 are classified correctly. In the following we introduce how different aspects of performance can be measured.

Table 13.4: Two Confusion Matrices Using Training Sample with 4 Tracks (Upper Part) and 10 Tracks (Lower Part)

Predicted labels $\hat{y}$	True labels y					
	Pop/Rock	Classic	Jazz	Electronic	R&B	Rap
Pop/Rock	26	0	2	12	8	11
Other	19	15	13	3	7	4

Predicted labels $\hat{y}$	True labels y					
	Pop/Rock	Classic	Jazz	Electronic	R&B	Rap
Pop/Rock	13	0	0	2	1	3
Other	32	15	15	13	14	12

13.3.3 Common Performance Measures Based on the Confusion Matrix

Given W observations (i.e. classification windows in the above example), let y_w be the label of the w-th observation represented by its feature vector $\boldsymbol{x}_w$. Let $\hat{y}_w$ be the predicted label. We restrict ourselves to binary categorization, i.e. $y_w, \hat{y}_w \in \{0; 1\}$. Notice that the measures below are all possible empirical versions of quality measures p_{ki} for a classifier k, a test sample T^i with W observations, and a binary classification problem. In what follows, we ignore the indices k, i assuming that we want to estimate the quality measure for a fixed classifier on a fixed data set.

Definition 13.8 (Absolute Performance Measures). The number of *true positives* corresponds to the number of observations which belong to the positive class (i.e. class 1) and are correctly predicted:

$$m_{TP} = \sum_{w=1}^{W} y_w \cdot \hat{y}_w. \tag{13.3}$$

The number of *true negatives* corresponds to observations which do not belong to the positive class and are correctly predicted:

$$m_{TN} = \sum_{w=1}^{W} (1 - y_w) \cdot (1 - \hat{y}_w). \tag{13.4}$$

The number of *false positives* corresponds to the number of observations which do not belong to the positive class, but are misleadingly recognized as belonging to it:

$$m_{FP} = \sum_{w=1}^{W} (1 - y_w) \cdot \hat{y}_w. \tag{13.5}$$

The number of *false negatives* is the number of observations which belong to the positive class, but are recognized as not belonging to it:

$$m_{FN} = \sum_{w=1}^{W} y_w \cdot (1 - \hat{y}_w). \tag{13.6}$$

In contrast to absolute numbers of correct and wrong predictions in Definition 13.8, the following commonly estimated measures apply a normalization with respect to a particular class of observations.

Definition 13.9 (Relative Performance Measures).
Recall, or *sensitivity*, measures the number of true positives related to the overall number of positive observations:

$$m_{REC} = \frac{m_{TP}}{m_{TP} + m_{FN}}. \tag{13.7}$$

Precision measures the number of true positives related to the overall number of observations predicted as positives:

$$m_{PREC} = \frac{m_{TP}}{m_{TP} + m_{FP}}. \tag{13.8}$$

The two following measures characterize the performance on classification instances not belonging to the positive class or predicted as not belonging to it.
Specificity corresponds to the amount of correctly identified negative observations related to the overall number of negatives:

$$m_{SPEC} = \frac{m_{TN}}{m_{TN} + m_{FP}}. \tag{13.9}$$

Negative predictive value estimates the number of correctly identified negative observations relative to the overall number of observations predicted as not belonging to a class:

$$m_{NPR} = \frac{m_{TN}}{m_{TN} + m_{FN}}. \tag{13.10}$$

A very common performance evaluation measure is *accuracy*, which estimates the share of all correctly predicted observations:

$$m_{ACC} = \frac{m_{TP} + m_{TN}}{m_{TP} + m_{TN} + m_{FP} + m_{FN}} = \frac{m_{TP} + m_{TN}}{W}. \tag{13.11}$$

As a counterpart, the *relative error* or *misclassification error rate* measures the share of all wrongly predicted observations:

$$m_{RE} = \frac{m_{FP} + m_{FN}}{m_{TP} + m_{TN} + m_{FP} + m_{FN}} = \frac{m_{FP} + m_{FN}}{W} = 1 - m_{ACC}. \tag{13.12}$$

Note that precision and recall may have very different values. For instance, a high m_{PREC} together with a low m_{REC} indicates that the number of instances wrongly predicted as belonging to the positive class (m_{FP}) is significantly lower than the number of the positive instances predicted as not belonging to the positive class (m_{FN}):

$$m_{PREC} \gg m_{REC} \Leftrightarrow \frac{m_{TP}}{m_{TP} + m_{FP}} \gg \frac{m_{TP}}{m_{TP} + m_{FN}} \Leftrightarrow m_{FN} \gg m_{FP}. \tag{13.13}$$

Also note that the *relative error* corresponds to the above *misclassification error rate*.

Example 13.4 (Evaluation of Classification Models). *Let us now compare the models from Example 13.2 with respect to the evaluation measures discussed above. Recalling the example, models (a,b) are created from the smaller training sample of 4 songs, and models (c,d) are based on 10 training songs. The decision tree depth was limited to 2 levels for models (a,c) and 4 levels for (b,d). The measures are listed in Table 13.5.*

Table 13.5: Evaluation of Classification Models from Example 13.2

Mod.	m_{TP}	m_{TN}	m_{FP}	m_{FN}	m_{ACC}	m_{PREC}	m_{REC}	m_{SPEC}	m_{NPR}
(a)	25	42	33	20	0.56	0.43	0.56	0.56	0.40
(b)	26	41	34	19	0.56	0.43	0.58	0.55	0.68
(c)	16	61	14	29	0.64	0.48	0.36	0.81	0.68
(d)	13	69	6	32	0.68	0.68	0.29	0.92	0.68

If m_{ACC} is used as the only evaluation criterion, the first impression is that the larger training sample leads to a better classification performance (0.64/0.68 against 0.56/0.56). Larger trees do not significantly change m_{ACC} for the smaller training sample, but help to increase m_{ACC} from 0.64 to 0.68 when the models are trained on a larger training sample. The correct classification of 68% test tracks is quite appreciated, because in this example only MFCCs are used as features and the number of training tracks is very small.

However, if other measures are taken into account, the advantage of the larger training sample is not completely obvious. Lines (c,d) contain smaller m_{TP} values and m_{REC} decreases from 0.56/0.58 to 0.36/0.29. As discussed above in Expression 13.13, a high precision and a low recall mean that the number of songs wrongly recognized as belonging to Pop/Rock genre is lower than the number of true Pop/Rock songs recognized as not belonging to this genre.

Because the smaller training sample consists of only Pop/Rock and Classical pieces (cf. Table 13.1), it does not contain enough information to learn some other non-classical genres different from Pop/Rock. 12 of 15 Electronic pieces are recognized as Pop/Rock (see the upper confusion matrix in Figure 13.4). The larger training sample with more Jazz and Electronic examples helps to increase the recognition of non-Pop/Rock songs, but the number of correctly predicted Pop/Rock songs decreases.

Depending on the characteristics of training and test data, the correlation between evaluation measures may more or less vary, as investigated in [28]. An assessment of the classification performance with regard to several measures leads to multi-objective evaluation and optimization as discussed later in Section 13.6.

13.3.4 Measures for Imbalanced Sets

In many music classification scenarios the data sets are not balanced; consider the identification of a particular instrument in a large set of music pieces, or the classification of songs into specific music styles. If the share of positive examples in the test sample $(m_{TP}+m_{FN})/W \ll 1$, a model which simply classifies all observations as negatives would achieve a very high $m_{ACC} = 1 - (m_{TP}+m_{FN})/W$ and a very low m_{RE}.

For a credible evaluation and tuning of models created for the application on imbalanced sets, there exist several possibilities to measure aggregated performance for observations of both classes.

Definition 13.10 (Measures for Imbalanced Sets).
The *balanced relative error* is the average of relative errors for positive and negative observations and should be minimized:

$$m_{BRE} = \frac{1}{2}\left(\frac{m_{FN}}{m_{TP}+m_{FN}} + \frac{m_{FP}}{m_{TN}+m_{FP}}\right). \qquad (13.14)$$

The *F-measure* is a weighted combination of precision and recall which should be maximized:

$$m_F = \frac{(\alpha_F + 1)\cdot m_{PREC}\cdot m_{REC}}{\alpha_F \cdot m_{PREC} + m_{REC}}. \qquad (13.15)$$

α_F is the positive real number which controls the balance between m_{PREC} and m_{REC}. For an even balance, α_F is set to 1. Values higher than 1 increase the weight of the precision (e.g., for $\alpha_F = 2$ twice as recall in the denominator), and values below 1 favor the recall.

In [12], the *geometric mean* (the square root of the product) of recall and specificity was proposed which should be maximized:

$$m_{GEO} = \sqrt{m_{REC}\cdot m_{SPEC}}. \qquad (13.16)$$

The performance of a classification model can be also compared against the performance of a random classifier by means of the *Kappa statistic* [33, p. 163], which should be maximized and measures the difference between the number of correct predictions and the number of correct predictions of a random classifier R in relation to the difference between the number of observations and the number of correct predictions of a random classifier:

$$m_{KA} = \frac{(m_{TP}+m_{TN}) - (m_{TP}(R) + m_{TN}(R))}{W - (m_{TP}(R) + m_{TN}(R))}. \qquad (13.17)$$

For binary classification with a random unbiased classifier, expected values of correct predictions depend on the numbers of positive and negative observations: $E[m_{TP}(R)] = (m_{TP}+m_{FN})/2$ and $E[m_{TN}(R)] = (m_{TN}+m_{FP})/2$, so that $E[(m_{TP}(R)+m_{TN}(R))] = W/2$. The substitution of this term into Equation (13.17) leads to:

$$m_{KA} \approx \frac{(m_{TP}+m_{TN}) - W/2}{W - W/2} = \frac{2m_{TP}+2m_{TN}-W}{W}. \qquad (13.18)$$

Example 13.5 (Evaluation of Classification Models for Imbalanced Sets). *Table 13.6 lists $m_{BRE}, m_F, m_{GEO},$ and m_{KA} for the models of Example 13.2. Models (c) and (d)* *trained with the larger set perform worse w.r.t. m_F and m_{GEO} but are better when validated with m_{BRE} and m_{KA}. This discrepancy illustrates the complexity of a proper choice of an appropriate model and also data for training. Model (a) classifies correctly 56% of Pop/Rock songs and 56% of other tracks (cf. corresponding m_{REC} and m_{SPEC} values in Table 13.5). Model (d) classifies correctly 29% of Pop/Rock songs and 92% of other tracks. Here, a decision can be done according to the desired preference: a higher mean performance on tracks of both classes of model (d) or a low variance and a high minimum across performances on tracks of both classes of model (a). Generally, it is crucial to adapt the evaluation scheme to the requirements of the concrete application.*

Table 13.6: Evaluation of Balanced Performance for Classification Models from Example 13.2

Mod.	m_{TP}	m_{TN}	m_{FP}	m_{FN}	m_{BRE}	m_F	m_{GEO}	m_{KA}
(a)	25	42	33	20	0.44	0.49	0.56	0.12
(b)	26	41	34	19	0.44	0.50	0.56	0.12
(c)	16	61	14	29	0.41	0.43	0.54	0.28
(d)	13	69	6	32	0.40	0.41	0.52	0.37

Until now, most related studies use only a few evaluation measures. An interesting statistic is provided in [27]: from 467 analyzed works on genre recognition, accuracy is the most popular measure and is estimated in 82% of the studies. Recall is used for evaluation in 25% of the studies, precision in 10%, and the F-measure in 4%. This means that many models are tuned to the best performance with regard to a single or a few evaluation metrics, and may be of poor quality when other evaluation aspects play a role.

13.3.5 Evaluation of Aggregated Predictions

Sometimes in the classification of music data the predictions are done for individual observations, but the main task is to classify groups of these observations as entities. A typical example is the classification of music pieces into genres. Using available training data (tracks with given genres), it is often better to build models for smaller time frames (classification windows), but to categorize later music pieces as a whole because they usually contain parts with different instrumentation, harmonic and melodic properties, etc., so that feature values may have a strong variance between segments of the same song.

Let a model predict the label for some classification window w. If a music piece consists of W' classification windows (recall that W denotes the overall number of windows/observations in the training sample), the overall predicted relationship to a class can be estimated by majority voting, i.e. assigning the piece to the class which was predicted for the majority of classification windows contained in this piece as:

$$\hat{y}(\boldsymbol{x}_1,...,\boldsymbol{x}_{W'}) = \left\lceil \frac{\sum_{w=1}^{W'} \hat{y}_w}{W'} - \frac{1}{2} \right\rceil. \tag{13.19}$$

The advantage of this method is that it reduces the impact of outlier windows in a song: e.g., if a quiet intro and an intermediate part with string quartet in a rock song are recognized as belonging to the classical genre, the aggregated prediction may still be correct. A further discussion about reasonable sizes of classification windows is provided in Section 14.1.

Example 13.6 (Aggregation of Predictions for Genre Classification). *In the examples from the previous sections we observed that a larger training sample with more negative observations leads to a better identification of tracks which do not belong to the Pop/Rock genre. Figure 13.2 plots predictions for individual classification windows of 4 s with 2 s overlap. For 5 genres, 15 tracks each of the test sample are taken into account. A horizontal dash corresponds to a window which was wrongly predicted as belonging to Pop/Rock. The lengths of all songs were normalized, so that a broader dash may correspond to a single window for a shorter song. The upper subfigure plots classification results for the training sample with 4 tracks and the bottom subfigure for the training sample with 10 tracks.*

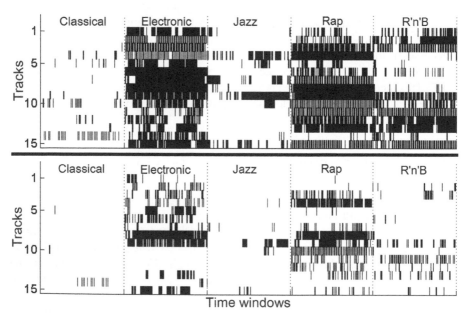

Figure 13.2: Recognition of the genre Pop/Rock, classification results for individual windows of 4 s with 2 s overlap for 75 tracks of 5 other genres. Top: small training sample; bottom: large training sample.

The balance between wrongly and correctly predicted windows can be easily recognized. For instance, classical music pieces were correctly predicted as not belonging to Pop/Rock for both training samples (cf. also Figure 13.4), but there exist some individual classification windows recognized as Pop/Rock, in particular for the 14-th piece (Adagio from Vivaldi's "The Four Seasons"). The number of misclassified windows is strongly reduced when the larger training sample is used (bottom subfigure). This holds for all genres (but there still remain 2 Electronic, 3 Rap, and 1 R'n'B tracks classified as Pop/Rock).

13.3.6 Measures beyond Classification Performance

The evaluation of music classification models is typically done with respect to classification performance. However, this may be problematic: for example, a classification model with a very low classification error may be extremely slow and very sensitive to algorithm parameters or the quality of music recordings. Or, an interactive learning system may require too much listener effort for acceptable performance. Even if still seldom taken into account, recently more attention is paid to measures beyond classification performance which are briefly discussed in this section. These measures are furthermore important for MIR systems, which are not restricted to classification and for which the classification quality cannot be directly estimated, e.g., similarity analysis, feature processing, or visualization of music collections.

We distinguish between the terms *evaluation focus* and *evaluation measure*. The targeted improvement of system/model properties can be adjusted by giving priority to three *evaluation focuses*: efficiency, generalization ability, and user satisfaction. The goal of high *efficiency* is to create a system, which requires the fewest resources possible. A system with a high *generalization ability* performs well for different data sets. A high *user satisfaction* is achieved if a system best matches specific user needs, e.g., providing highly interpretable models for a music scientist or a very short time required to understand the recommendation system from the listener perspective.

An *evaluation measure* is a function which outputs a numerical value for minimization/maximization. Below, we will give examples of such measures not characterizing classification performance like in Sections 13.3.1–13.3.5. The optimization of a single measure may help to increase system properties for one or several evaluation focuses. Examples are provided in Table 13.7. Four groups of measures (runtime, storage, stability, and user-related measures) will be briefly discussed in Sections 13.3.6.1–13.3.6.4. For example, a short online runtime characterizes not only an efficient, but also a user-friendly system. Other runtime measures may have less influence on user satisfaction: e.g., the time-consuming extraction of many audio features may be done in advance (offline) or on a server farm. Another example is the number of features: the selection of a few most relevant characteristics may not only decrease the demands on runtime and disc space, but also help to create more robust and generalizable models. A closely related task, the minimization of the training sample, may not only increase the efficiency but also reduce listener efforts for the definition of ground truth (labeled classes) for training. On the other side, a too small training sample has a negative impact on the generalization ability as well as on the classification performance.

Table 13.7: Evaluation Measures and the Impact of Their Optimization on Three Evaluation Focuses. +: Strong Impact; (+): Some Impact; -: No or almost No Impact. ↓: Measures to Minimize; ↑: Measures to Maximize

Evaluation measures		Evaluation focuses		
Group	Example measure	Efficiency	Generalization	User
Runtime	Online runtime ↓	+	-	+
Runtime	Offline runtime ↓	+	-	(+)
Storage	No. features ↓	+	+	-
Storage	training sample size ↓	+	+	+
Stability	Deviation of accuracy ↓	-	+	+
Stability	Model complexity ↓	(+)	+	(+)
User	Model interpretability ↑	-	-	+
User	Costs of active learning ↓	(+)	(+)	+

13.3.6.1   Runtime

If a classification model has a small error and achieves the best classification performance compared to other models, it may have a strong drawback being very slow for many possible reasons: the classification method may require transformations into higher dimensions, the search for optimal parameters may be costly, or data must be intensively preprocessed before the classification.

Runtime can be measured for different methods. For example, most music data analysis tasks rely on previously extracted features. The *runtime of feature extraction* depends on the source of the feature. Audio features often require several complex steps like the Fourier transform. It is possible to give priority to the extraction of features which may be extracted faster, on the other side increasing the danger that classification models trained with these features may perform worse. For instance, three audio features (autocorrelation, fundamental frequency, and power spectrum) required more than 65% of the overall extraction time of 25 features in [4].

For models themselves, one should distinguish between the *training runtime* and *classification runtime*. If new data instances are classified over and over again, for instance when new tracks are added to a music collection or an online music shop, the classification runtime becomes more important. In this situation, too high costs of the *optimization runtime* during the training stage, e.g., for the tuning of hyperparameters (see Section 13.4), may be less problematic.

It is also possible to shift the costs between individual steps to a certain extent: for example, the implementation and the extraction of complex high-level characteristics (e.g., recognition of instruments) may help to reduce runtime costs during classification so that rather simple approaches like k-NN and Naive Bayes would have similar or almost similar performance when compared with complex methods like SVMs. If the extraction of the features is done only once for a music instance, the user would be less influenced by a long "offline" runtime.

The biggest challenge of runtime-based evaluation is that it is very hard to achieve a reliable comparison between models. Implementations of algorithms may differ (cf. the difference between discrete and fast Fourier transforms, Section 4.4.1), and the properties of the environment (hardware, operating system load, dependency on external components) should be kept as constant as possible.

13.3.6.2 Storage Space

Another category of resources is the storage space which sometimes may have a negative correlation with runtime demands. For example, a data structure for a very fast and efficient search may require a significantly higher number of entries than another extreme solution, a strongly compressed archive.

If music data are characterized by features, the space necessary for their storage is referred to as the *indexing space*. A very simple possibility to measure the demands on indexing space is to count the number of saved values. Because many feature processing methods aim at the reduction of the indexing space, several related specific measures are discussed in Section 14.6.

Training samples for learning are matrices built with W observations for which F features and a label are stored. If these sets are created by users, the minimization of W keeps personal efforts for classification as small as possible. However, training samples with too few labeled observations may lead to a decrease of classification performance and an increase of the risk of overfitting.

Compared to music itself and extracted features, the space for the storage of classification models is usually less relevant. However, the storage size of the various classification models introduced in Section 12 may be very different. Models which measure distances in feature space like k-Nearest Neighbors require more space than compact decision trees with an integrated feature selection procedure. On the other hand, unpruned trees optimized for best classification performance may be very large and use the same feature in multiple nodes.

13.3.6.3 Stability

Stability measures whether an output of a classification system has a low variation for $B \gg 1$ experiments (i.e. training samples). For the measurement of stability w.r.t. an evaluation measure m, the standard deviation is estimated as:

$$s_m = \sqrt{\frac{1}{B-1} \cdot \sum_{i=1}^{B} (m_i - \bar{m})^2}, \tag{13.20}$$

where $\bar{m}$ is the mean value of m. Note that in principle the whole distribution of the evaluation measure is of interest. However, the stability is particularly important together with the mean performance. Also note that m does not necessarily need to be a classification performance measure. Stability can be estimated for most measures discussed in this chapter, such as runtime. There exist several possibilities to introduce variation for B experiments. We discuss here three cases: stability under test data variation, stochastic repetitions, and parameter variation.

To measure the *stability under test data variation*, the model is applied on B different and preferably non-overlapping test samples. Higher values of s_m mean that the model is more sensitive to data and it is harder to see in advance whether this model would be successful when applied to a new unlabeled data sample. If a test sample can be built from all available labeled data, B subsets can be built with the help of resampling discussed in Section 13.2. It is also thinkable, however, to create subsets with data instances which share a particular property, e.g., a genre (see example below), or belonging to the same cluster estimated by means of unsupervised learning (see Chapter 11).

Example 13.7 (Evaluation of Stability of Classification Performance for Negative Examples). *Let us measure the variance of classification performance across tracks of different* non-Pop/Rock genres *for models (b), (d) from the confusion matrices in Example 13.3. The mean and the standard deviation of classification performance measures for negative examples – estimated separately for the 5* non-Pop/Rock gen-res *Classical, Electronic, Jazz, Rap, and R'n'B tracks – are listed in Table 13.8. For each of these genres the corresponding 15 tracks were used as a test sample. Because* $m_{TN} + m_{FP} = 15$ *for both models,* $s_{m_{TN}} = s_{m_{FP}}$.

Table 13.8: Evaluation of Stability of $p = m_{TN}, m_{FP}$, and m_{SPEC}

Mod.	$\bar{m}_{TN}$	$s_{m_{TN}}$	$\bar{m}_{FP}$	$s_{m_{FP}}$	$\bar{m}_{SPEC}$	$s_{m_{SPEC}}$
(b)	8.4	5.4	0.6	5.4	0.56	0.36
(d)	13.8	1.3	1.2	1.3	0.92	0.09

As already discussed above, the extension of the training sample with Jazz and Electronic tracks significantly increases the performance for these genres, and the standard deviation of specificity decreases from 0.36 to 0.09. Even if it is not possible to reliably forecast the prediction quality of a model for tracks of other genres currently completely nonexistent in our database (say, African drum music), we can at least measure how the performance varies for genres nonexistent in the given training data. Such measure gives us a very rough estimator of the complexity of the classification task, in the example above, whether the identification of Pop/Rock songs is a rather simple task (because these songs have some unique properties compared to all possible other music genres) or a rather hard task (there exist other genres with very similar properties to Pop/Rock).

Stability under stochastic repetitions can be measured when some methods of a classification system provide output based on random decisions and results vary after repetitions for the same data. Because Random Forest selects random features for training of many trees during the training process (see Section 12.4.5), for such a classifier we may simply repeat the training B times and estimate the variation of performance of the created models as a measure of stability. Another candidate for the measurement of stability under stochastic repetitions is *evolutionary feature selection* (see Section 15.6.3). Even if this method is applied for a fixed training sample with a deterministic classifier (such as Naive Bayes), random selection of features may find several suboptimal feature sets after B repetitions of the evolutionary algorithm.

Finally, classification models may be sensitive to (hyper)parameters (cf. Definition 13.11), so that *stability under parameter variation* can be measured. Note that only a limited range of reasonable parameter values should be of interest. For instance, too small numbers of trees in Random Forest would lead to models with poor performance, and an increase of the number of trees above some threshold would only slow down the classification process. The search for a *reasonable range* of parameter values may be complex in practice and is beyond the scope of this chapter. To name a few possibilities, for a particular classifier it is possible to use settings recommended after theoretical investigations, the values can be selected from an interval of optimal parameter values found after the application for several classification tasks, and factorial or other statistical designs of experiments can be applied [1].

13.3.6.4 User-Related Measures

Most evaluation methodologies in studies on music data analysis are system-oriented and not user-oriented. One of the reasons is that it is much easier to measure success and improvement based on precise ground truth and efficiency of the system. However, such evaluation methods are not always sufficient. As stated in [31] with a reference to an earlier survey on MIR [7], "subjective musical experience varies not only between, but also within individuals, depending on affective and cultural context, associations between the music and events from episodic memory, and a host of other factors," i.e. the relationship between music data and music category may not be the same not only for different users, but may also change over time for the same person. In another study, only weak or no correlation was measured between system-based and user-based evaluation for the identification of similar songs [10].

A simple consequence of this discussion is that for many music applications it explicitly makes sense to apply multi-objective evaluation (cf. Section 13.6), estimating several less correlated measures relevant for a particular scenario, and not just reporting progress in terms of accuracy or classification error. A challenge of user-based evaluation is that it is typically expensive: asking for a feedback needs a lot of time and human effort. In particular, for a systematic optimization of a system or a classification model not only a single but many interactions are necessary.

In the following, we will list several examples of user-centered measures which can be estimated in addition to performance measures.

Listener satisfaction typically requires direct feedback from the user who may report the perceived quality filling in a questionnaire. Data to measure the listener satisfaction can be also automatically gathered without a direct interaction with a user: in a recommender system, the number of rejections and listening times of recommended tracks can be measured.

Measures of *listening context* make sense if music should be recommended for a certain purpose. Examples of such applications are the relation of beat times to runners step frequency [21] and measurements of heartbeats [16].

A high *interpretability* may lead to a higher personal satisfaction: for example, a systematic increase of the share of interpretable semantic features may help a music scientist to learn relevant properties of a composer's style, or a music listener to discover new music which shares some high-level characteristics of the previous personal preferences [30]. Not only features, but also classification models themselves can be measured in terms of interpretability. If a decision tree is constructed with semantic features, the depth of the tree can be minimized (too large trees are not interpretable anymore). In some studies, fuzzy classification was applied to music classification [9, 29]. Then, opposing goals can be the minimization of the number of fuzzy rules and the maximization of classification performance. It is worth mentioning that if the interpretability is important, each step in the algorithm chain should be verified. Consider the application of Principal Component Analysis (see Definition 9.48) on high-level features making their interpretation very hard at least when the number of components is high.

Particularly for music recommendation systems there are many user-related measures which may be completely uncorrelated with classification performance measures. For example, a "perfect" classifier in terms of accuracy would not take *novelty and surprise* into account, which may be important for some listeners. Then, the recommendation of music too similar to current user preferences would exclude the chance to discover new music or even change personal preferences. Another problematic issue is to use only performance measures if the *order* of the output is important. In a playlist, the order of mood and genre changes may be relevant, and it may be undesired to place tracks of the same artist too close after each other. Therefore, two playlists constructed with the same songs may have completely different impacts on listener satisfaction. More evaluation approaches for generated playlists are discussed in [5]. For a further discussion of measures related to music recommendation, we refer to Section 23.4.

13.4 Hyperparameter Tuning: Nested Resampling

Hyperparameters were already mentioned above as important. The optimization of such parameters is called hyperparameter tuning. For this, so-called nested resampling is applied. Let us first fix the meaning of these parameters as follows:

Definition 13.11 (Hyperparameters). Let *hyperparameters* be model or model estimator parameters that might influence model selection but have to be chosen prior to it.

An example of such a hyperparameter is the error weight C of an SVM. This parameter is typically chosen before model estimation and model selection, and does not restrict the model class of support vector machines, i.e. the model selection task is unchanged.

Since hyperparameters are to be fixed prior to model selection, they have to be varied in an extra sampling process, and one ends up with a *nested sampling* process. As an example, consider using subsampling with $B = 100$ in an outer loop for model evaluation and 5-fold cross-validation in an inner loop for hyperparameter selection. For each of the 100 training samples L^i from subsampling, a 5-fold cross-validation on the training sample is employed as an internal fitness evaluation to select the best setting for the hyperparameters of the model. The best obtained hyperparameters are used to fit the model on the complete training sample and calculate the quality measure on the test sample of the outer resampling strategy. Figure 13.3 shows this

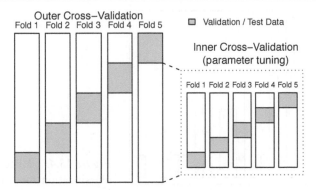

Figure 13.3: Nested resampling for (hyper)parameter tuning with two nested CVs.

process schematically for two nested cross-validations, namely 5-fold CV in both the inner and the outer resampling.

Example 13.8 (Hyperparameter Tuning). *In order to demonstrate nested resampling for hyperparameter tuning, we show results for the SVM with radial basis kernel in the piano-guitar distinction Example 11.2. For this SVM we vary the kernel width w and the error weight C. We will not only use MFCCs (as in Example 11.2) as tone characterization but four types of audio features which represent musically relevant properties of sound. Overall, this leads to 407 numeric non-constant features not including any missing values (cp. Section 14.2.3).*

Experiment for Hyperparameter Tuning in Piano-Guitar Classification

1. Select the SVM with radial basis kernel as a classifier.

2. Subsampling: Randomly select 600 observations from the full data set as a training sample for model selection. Retain the rest as a test sample.

3. Apply grid search on all powers of 2 in $[2^{-20}, 2^{20}]$ for both the kernel width w and the error weight C on these 600 observations with SVM. Performance is measured by 5-fold CV and MisClassification Error rate (MCE) (also denoted as relative error m_{RE}, cp. Equation (13.12)). Note: the data partitioning of the CV is held fixed, i.e. is the same for all grid points to reduce variance in comparisons.

4. Store the hyperparameters w and C with optimum MCE.

5. Train classifier with selected hyperparameters on all 600 instances of the training sample.

6. Predict classes in the test sample and store the test error.

7. Repeat steps (2)–(6) 50 times.

The results show that the chosen SVMs with tuned hyperparameters but without feature selection realize an empirical distribution of the MCEs with a range between 2% and 5% (see Figure 13.4).

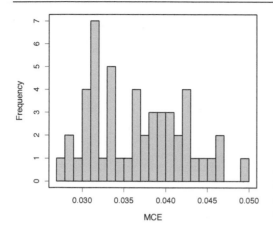

Figure 13.4: Histogram of misclassification rates from outer resampling for SVM hyperparameter tuning.

Bootstrapping or Subsampling? When combining model or hyperparameter selection with bootstrapped data sets in the outer loop of nested resampling, repeated observations can lead to a substantial bias toward more complex models. This stems from the fact that in the inner tuning loop measurements will occur both in the training sample and in the test sample with a high probability because some observations appear multiple times in the bootstrap sample so that more complex models "memorizing" the training data will seem preferable. In [3] subsampling was proposed and evaluated as a remedy since then there are no observations appearing multiple times in the sample.

13.5 Tests for Comparing Classifiers

Up to now we just described the differences between the realizations of the different evaluation measures. However, it is by no means clear whether the differences are "significant," i.e. whether one can give any "statistical guarantee" that one method is better than the other. In this section we will discuss this more formally and come back to the Quality Criterion Hypothesis in Section 13.1. We will give some exemplary statistical tests for such a hypothesis (cp. Section 9.7). However, let us state some warnings beforehand. Do not confuse "significance" with "relevance." One method may be significantly better than another, but not relevant in that the corresponding error rate is much too high to be acceptable. Relevance will not be discussed further since it is heavily problem dependent.

13.5.1 McNemar Test

Let us start with the comparison of two models or two classifiers. Here, we will consider tests on the equality of the whole distribution of evaluation measures for the two classifiers. The McNemar Test was designed for a variable with two nominal levels (success vs. non-success). As a basis, the so-called 2x2 table (contingency

Table 13.9: Contingency Table for McNemar Test

H_{FF} = no. of instances misclassified by both classifiers	H_{FT} = no. of instances misclassified by classifier 1 but not by classifier 2
H_{TF} = no. of instances misclassified by classifier 2 but not by classifier 1	H_{TT} = no. of instances correctly classified by both classifiers

table with 4 cells, cp. Definition 9.24) for dependent samples are used. Consider Table 13.9 for the comparison of two classifiers on the basis of a single test data set.

If the two classifiers were equally good, then the number of successes (correct classifications) should be as similar as possible. This leads to the equality $P_{TF} + P_{TT} = P_{FT} + P_{TT}$, i.e. $P_{TF} = P_{FT}$, where P stands for probability.

Therefore, the null hypothesis to be tested has the form

H$_0$: The probability of a success is equal for the two classifiers, i.e. $P_{FT} = P_{TF}$.

For fixed $q = H_{FT} + H_{TF}$, under **H$_0$** the frequency H_{FT} should be binomially distributed with success probability 0.5,

since there should be as many instances being successfully classified by method 1 but not by method 2 and vice versa. Thus, the test statistic

$$t = \frac{H_{FT} - q/2}{\sqrt{q/4}}$$

is approximately N(0,1) distributed, at least if q is big enough (then the binomial distribution is approximately normally distributed with $E(H_{FT}) = q/2$ and $var(H_{FT}) = q/4$). Therefore,

$$\chi^2 = \frac{(H_{FT} - H_{TF})^2}{H_{FT} + H_{TF}} = \frac{(H_{FT} - (H_{FT} + H_{TF})/2)^2}{(H_{FT} + H_{TF})/4} = t^2$$

is approximately χ^2-distributed with 1 degree of freedom (cp. Section 9.4.3).

For smaller sample sizes a corrected statistic is used:

$$\chi^2_{corr} = \frac{(|H_{FT} - H_{TF}| - 1)^2}{H_{FT} + H_{TF}}.$$

The McNemar Test rejects the equality of the goodness of the performance of the two classifiers, e.g., if the 95% quantile of the χ^2-distribution with 1 degree of freedom is exceeded, i.e.

$$\chi^2_{corr} > \chi^2_{0.95,1} = 3.84.$$

Example 13.9 (McNemar Test). *We will reconsider the models in the above Example 13.2 and test the equality of the goodness of model pairs. The only values we have to calculate are H_{FT} and H_{TF} on the test sample. Then, χ^2_{corr} can be derived. Table 13.10 presents the results. Note that if we reject* **H$_0$**, *we assume* **H$_1$**.

First, we compare models (a) against (b) and (c) against (d). For these pairs the size of the training sample is fixed, and the tree size varies. As we can observe, there is no significant difference between models, as $\chi^2_{corr} < 3.84$ for both cases.

Second, we compare models (a) against (c) and (b) against (d). Here, the tree size is fixed and the training sample varies. For smaller trees with depth 2 there is no significant difference between the two models, but for larger trees with depth 4 H_0 is rejected.

Table 13.10: Comparison of Classification Models from Example 13.2 by Means of McNemar Test

Pair of models	Fixed parameter	H_{TF}	H_{FT}	χ^2_{corr}	Hyp.
Impact of varying tree size					
(a), (b)	Smaller training samples (4 tracks)	1	1	0.50	H_0
(c), (d)	Larger training samples (10 tracks)	3	8	1.45	H_0
Impact of varying training sample size					
(a), (c)	Smaller trees (depth 2)	9	19	2.89	H_0
(b), (d)	Larger trees (depth 4)	14	29	4.56	H_1

13.5.2   Pairwise t-Test Based on B Independent Test Data Sets

Now we consider the case where the relative error rate is determined by two classifiers on the same B test data sets. This way, we get B error rates for the two classifiers, one for the B test data sets each. This is interpreted as B observations each of the error rate. In order to test whether the error rates E_1 and E_2 differ significantly, a t-Test is applied on the differences of the error rates of the two classifiers.

Definition 13.12 (One Sample t-Test). For a t-Test on expected value zero of the normal distribution of the differences $D := E_2 - E_1$ the null hypothesis is of the form $H_0 : \mu_D = 0$ with two-sided alternative hypothesis $H_1 : \mu_D \neq 0$.

For an unknown variance of the differences D and B test data sets, the test statistic has the form

$$ t = \frac{\bar{d}}{s_d / \sqrt{B}}, $$

where $\bar{d}$ is the mean and s_d the empirical standard deviation of the observed differences. This test statistic is t-distributed with $B - 1$ degrees of freedom (cp. Section 9.4.3).

The *t-Test* then has this form: If the absolute value of the test statistic t is greater than the $(1 - \alpha/2)$-quantile of the t_{B-1} distribution, α being the significance level of the test (see Section 9.7), then the null hypothesis is rejected, since then this hypothesis appears to be too improbable.

Example quantiles for $\alpha = 0.05$ and 4, 9, 100 degrees of freedom are: $t_{0.975,4} = 2.78$, $t_{0.975,9} = 2.26$, and $t_{0.975,100} = 1.98$.

Example 13.10 (t-Test). *We will again reconsider the models in the above Example 13.2 and test the equality of the mean goodness between all pairs of distinct models. This time, we generate $B = 100$ test samples with 4/5-subsampling from the*

whole test sample (each time, 4/5 of tracks with all corresponding classification windows are randomly selected for validation). The t-statistic is compared to the 97.5%
quantile $t_{0.975,99} = 1.98$.

The left part of Table 13.11 lists mean values of the relative classification error m_{RE} for the two tested models, the t-statistic, and the "assumed" hypothesis. The right part of the table contains the corresponding values for the balanced relative error m_{BRE}. We can observe that in all cases except for comparison of models (a) and (b), H_0 is rejected.

Table 13.11: Comparison of Classification Models from Example 13.2 by Means of t-Test

Pair	$\bar{m}_{RE}(1st, 2nd)$	t	Hyp.	$\bar{m}_{BRE}(1st, 2nd)$	t	Hyp.
(a), (b)	0.4465, 0.4451	0.51	H_0	0.4437, 0.4409	0.79	H_0
(a), (c)	0.4465, 0.3910	21.44	H_1	0.4437, 0.4159	8.13	H_1
(a), (d)	0.4465, 0.3732	28.36	H_1	0.4437, 0.3923	16.95	H_1
(b), (c)	0.4451, 0.3910	19.75	H_1	0.4409, 0.4159	9.52	H_1
(b), (d)	0.4451, 0.3732	27.98	H_1	0.4409, 0.3923	17.30	H_1
(c), (d)	0.3910, 0.3732	6.60	H_1	0.4159, 0.3923	9.88	H_1

13.5.3 Comparison of Many Classifiers

In the case of $K > 2$ classifiers and B independent test data sets, we have to compare K vectors of length B with estimated error rates. This is typically realized by means of a (Two-Way) Analysis of Variance.

Definition 13.13 ((Two-Way) Analysis of Variance). In the case of a nominally scaled independent variable (here classifier) and cardinally scaled dependent variable (here error rate) an analysis of variance can be applied. Here, we assume that the error rate E_{ij} of the jth classifier ($j = 1, \ldots, K$) on the i-th test data set ($i = 1, \ldots, B$) is additively composed of the overall mean μ, the data set effect α_i, the classifier effect β_j, as well as a random error ε_{ij}:

$$E_{ij} = \mu + \alpha_i + \beta_j + \varepsilon_{ij},$$

where the ε_{ij} are independently identically distributed and $\alpha_1 + \ldots + \alpha_B = 0, \beta_1 + \ldots + \beta_K = 0$, B = no. of test data sets, K = no. of classifiers.

Side conditions result from the fact that the overall mean error rate should be μ. By means of the data set effect α_i, the overall mean is only adapted to the actual data set. Interactions between data set and classifier are excluded here, i.e. the effect of the classifier does not depend on the data set. Such interactions could, though, be included in the model without problems.

Testing on significant differences in the classifier effects, i.e. checking the validity of the null hypothesis

$$H_0 : \mu_1 = \mu_2 = \ldots = \mu_K = \mu \quad \text{or} \quad \beta_1 = \beta_2 = \ldots = \beta_K = 0$$

with $\mu_j := \mu + \beta_j$ can be realized for normal model errors as follows:
First, the effect of classifier j is estimated by

$$b_j = \bar{e}_{\cdot j} - m = \frac{1}{B}\sum_{i=1}^{B} e_{ij} - \frac{1}{BK}\sum_{i=1}^{B}\sum_{j=1}^{K} e_{ij}, j = 1,\ldots,K.$$

Second, the mean contribution of all classifiers to the dependent variable E is estimated as the so-called Mean Squared Classifier effect:

$$MSC = \frac{SSC}{K-1} = \frac{B\sum_{j=1}^{K} b_j^2}{K-1}.$$

SSC is also called the Sum of Squared Classifier effects. Analogously, the data set effects $a_i, i = 1,\ldots,B$, are estimated by

$$a_i = \bar{e}_{i\cdot} - m = \frac{1}{K}\sum_{j=1}^{K} e_{ij} - \frac{1}{BK}\sum_{i=1}^{B}\sum_{j=1}^{K} e_{ij}, i = 1,\ldots,B.$$

The model error is then estimated by

$$me_{ij} = e_{ij} - m - a_i - b_j.$$

This leads to the Mean Squared Error

$$MSE = \frac{SSE}{(B-1)(K-1)} = \frac{\sum_{i=1}^{B}\sum_{j=1}^{K} me_{ij}^2}{BK - B - K + 1}.$$

SSE is also called the Sum of Squared Errors. Under the null hypothesis and the assumption that the model error term is normally distributed, the test statistic

$$F = \frac{MSC}{MSE}$$

is F-distributed with $(K-1)$ and $(B-1)(K-1)$ degrees of freedom (cp. Section 9.4.3).

Typical bounds for the rejection of the null hypothesis are the 95% quantiles of the F-distribution with different pairs of degrees of freedom, e.g., $(4,8),(4,16)$ for $K = 5, B = 3,5$ or $(9,18),(9,36)$ for $K = 10, B = 3,5$ leading to $F_{0.95,4,8} = 3.84$, $F_{0.95,4,16} = 3.01$, $F_{0.95,9,18} = 2.46$, $F_{0.95,9,36} = 2.15$.

However, when the null hypothesis is rejected, we only know that there are differences between the classifiers. In order to find out where these differences appear, we have to apply tests on subsets of the classifiers. For example, one can order the classifiers by their error rates and repeat the test without the best or the worst classifier. This is repeated until the null hypothesis is not rejected anymore. This way we get subsets of classifiers with non-significantly different error rates. For example, this could result in two groups of classifiers, e.g., classifiers with indexes $\{1,4,5\}$ significantly better than classifiers with indexes $\{2,3\}$. Please notice that this leads to the problem of multiple testing (cp. Section 9.7).

Example 13.11 (Two-Way Analysis of Variance).   *We will again reconsider the models in the above Example 13.2 and test on significant classifier effects. Again, we generate B = 100 test samples with 4/5-subsampling from the whole test sample. Then, we have B = 100 replicates and K = 4 classifiers. The F-statistic is compared to the 95% quantile $F_{0.95,3,297}$ = 2.64. For m_{RE}, F = 402.12, and for m_{BRE}, F = 128.97, indicating significant differences between the 4 models.*

13.6   Multi-Objective Evaluation

As discussed in Section 13.3, there exist some risks and pitfalls if the evaluation and optimization is done with respect to a single measure. A classification model with a very high accuracy can be extremely slow, too complex and not interpretable, perform worse on instances of less represented classes, have a poor generalization ability to classify new data, or simply not lead to a high listener satisfaction. That is why it is important to calculate several measures with a low correlation for the reliable evaluation.

Algorithms can be first optimized with regard to a single criterion, and for the validation of the final solution several other measures can be calculated. A more credible but also more time-consuming approach to find best compromise solutions is to apply multi-objective optimization as introduced in Section 10.4.

Let us end this chapter with some remarks on the design of an evaluation scenario. This should at least contain the three steps sketched in Figure 13.5.

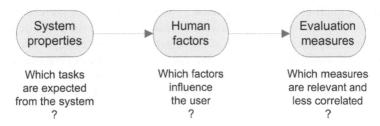

Figure 13.5: Three steps for the design of evaluation.

First, *system properties* for the evaluation should be checked. It is often necessary to think about details or less-obvious applications. For example, a music recommendation system aiming at a perfect recognition of current listener properties would fail if the discovery of new music and the evolution of the personal taste play a role.

Example 13.12 (Properties of genre recognition systems). *An extensive analysis of tasks related to genre recognition systems was applied in [27]. Here, ten experimental designs are listed, sorted by their application number: classify ("how well does the system predict genres?"), features ("at what is the system looking to identify genres?"), generalize ("how well does the system identify genre in varied data sets?"), robust ("to what extent is the system invariant to aspects inconsequential for identifying genre?"), eyeball ("how well do the parameters make sense with respect to identifying genre?"), cluster ("how well does the system group together music using the same genres?"), scale ("how well does the system identify the music genre with varying numbers of genres?"), retrieve ("how well does the system identify music using the same genres used by the query?"), rules ("what are the decisions the system is making to identify genres?"), compose ("what are the internal genre models of the system?").*

In the next step, *human factors* should be identified which influence the evaluation. Reference [22] distinguishes between four factors which influence human music perception: music content (properties of sound, e.g., timbre or rhythm), music context (metadata like lyrics or details of composition), user properties (musical experience, age, etc.), and user context (properties of listening situation like current activity or current mood, cp. Section 21.4.2). It is not easy to achieve a perfect match with regard to all these factors: for example, a recommender system may have to learn that a listener of classical music does not like organ (music content), does not prefer to listen to operas with lyrics based on fairy tales (music context), does not like Wagner because her/his parents listened Wagner's operas too often during her/his childhood (user properties), and does not like to listen to complex polyrhythmic pieces while driving because it may disturb her/his attention (user context).

Finally, *evaluation measures* should be selected which are relevant according to desired system properties and human factors. Another requirement is that these measures should be less correlated: if a data sample is well balanced, the maximization of accuracy may already lead to the minimization of balanced classification error and it is not necessary to evaluate the system using both measures or selecting both as criteria for multi-objective optimization. The measurement of correlation between measures is not always straightforward because there may be some dependencies which hold only for some regions of the search space. For example, increasing the number of features may lead to an increase of classification performance at first, but later lead to a decrease of the performance because too many irrelevant features would be identified as relevant. Also, the counter-strategy to decrease the number of features would not necessarily lead to a higher performance.

13.7 Further Reading

For a further discussion of evaluation measures and methods, we refer to literature on machine learning and classification. Visualization of classification performance with the help of the receiver operating characteristic (ROC), recall, precision, and cost curves is described in, e.g., [33, p. 168+]. Several measures for imbalanced sets are introduced in [24], among others from the field of medical diagnosis: Youden's index, likelihoods, and discriminant power; more measures for multi-labeled and hierarchical classification are also listed in [25]. Stability of classification models is analyzed in [14]. Other works measure complexity of models for a particular classifier, e.g., SVMs [19]. Several validation approaches besides the methods mentioned in this chapter with further references are provided in [18, p. 292+].

User-related evaluation is particularly relevant for music classification, and this subject receives more attention recently. To name a few studies, [15] outlines the history and statistics of user studies in MIR, [6] discusses related evaluation measures for music recommendation, and some general aspects of user-related evaluation are discussed in [17].

Bibliography

[1] T. Bartz-Beielstein. *Experimental Research in Evolutionary Computation.* Springer, 2006.

[2] Y. Bengio and Y. Grandvalet. No unbiased estimator of the variance of k-fold cross-validation. *Journal of Machine Learning Research*, 5:1089–1105, 2004.

[3] H. Binder and M. Schumacher. Adapting prediction error estimates for biased complexity selection in high-dimensional bootstrap samples. *Statistical Applications in Genetics and Molecular Biology*, 7(1):12, 2008.

[4] H. Blume, M. Haller, M. Botteck, and W. Theimer. Perceptual feature based music classification: A DSP perspective for a new type of application. In W. A. Najjar and H. Blume, eds., *Proc. of the 8th International Conference on Systems, Architectures, Modeling and Simulation (IC-SAMOS)*, pp. 92–99. IEEE, 2008.

[5] G. Bonnin and D. Jannach. Automated generation of music playlists: Survey and experiments. *ACM Computing Surveys*, 47(2):1–35, 2014.

[6] Ò. Celma. *Music Recommendation and Discovery: The Long Tail, Long Fail, and Long Play in the Digital Music Space.* Springer, 2010.

[7] J. S. Downie. Music information retrieval. *Annual Review of Information Science and Technology*, 37(1):295–340, 2003.

[8] B. Efron. Bootstrap methods: Another look at the jackknife. *The Annals of Statistics*, 7(1):1–26, 1979.

[9] F. Fernández and F. Chavez. Fuzzy rule based system ensemble for music genre classification. In *Proc. of the 1st International Conference on Evolutionary and Biologically Inspired Music, Sound, Art and Design (EvoMUSART)*, pp. 84–95. Springer, 2012.

[10] X. Hu and N. Kando. User-centered measures vs. system effectiveness in finding similar songs. In *Proc. of the 13th International Society for Music Information Retrieval Conference (ISMIR)*, pp. 331–336. FEUP Edições, 2012.

[11] R. Kohavi. A study of cross-validation and bootstrap for accuracy estimation and model selection. In *Proc. of the International Joint Conference on Artificial Intelligence (IJCAI)*, pp. 1137–1143, 1995.

[12] M. Kubat and S. Matwin. Addressing the curse of imbalanced training sets: One-sided selection. In *Proc. of the 14th International Conference on Machine Learning (ICML)*, pp. 179–186. Morgan Kaufmann, 1997.

[13] P. A. Lachenbruch and M. R. Mickey. Estimation of error rates in discriminant analysis. *Technometrics*, 10(1):1–11, 1968.

[14] T. Lange, M. L. Braun, V. Roth, and J. M. Buhmann. Stability-based model selection. In S. B. et al., ed., *Advances in Neural Information Processing Systems*

15 (NIPS), pp. 617–624. The MIT Press, 2002.

[15] J. H. Lee and S. J. Cunningham. Toward an understanding of the history and impact of user studies in music information retrieval. *Journal of Intelligent Information Systems*, 41(3):499–521, 2013.

[16] H. Liu, J. Hu, and M. Rauterberg. Music playlist recommendation based on user heartbeat and music preference. In *Proc. of the International Conference on Computer Technology and Development (ICCTD)*, volume 1, pp. 545–549. IEEE, 2009.

[17] J. Liu and X. Hu. User-centered music information retrieval evaluation. In *Proceedings of the Joint Conference on Digital Libraries (JCDL) Workshop: Music Information Retrieval for the Masses*. ACM, 2010.

[18] C. McKay. *Automatic Music Classification with jMIR*. PhD thesis, Department of Music Research, Schulich School of Music, McGill University, 2010.

[19] I. Mierswa. Controlling overfitting with multi-objective support vector machines. In H. Lipson, ed., *Proc. of the Genetic and Evolutionary Computation Conference (GECCO)*, pp. 1830–1837. ACM, 2007.

[20] C. Nadeau and Y. Bengio. Inference for the generalization error. *Machine Learning*, 52(3):239–281, 2003.

[21] M. Niitsuma, H. Takaesu, H. Demachi, M. Oono, and H. Saito. Development of an automatic music selection system based on runner's step frequency. In *Proc. of the 9th International Conference on Music Information Retrieval (ISMIR)*, pp. 193–198. Drexel University, 2008.

[22] M. Schedl, A. Flexer, and J. Urbano. The neglected user in music information retrieval research. *Journal of Intelligent Information Systems*, 41(3):523–539, 2013.

[23] J. Shao. Linear model selection by cross-validation. *Journal of the American Statistical Association*, 88(422):486–494, 1993.

[24] M. Sokolova, N. Japkowicz, and S. Szpakowicz. Beyond accuracy, F-score and ROC: A family of discriminant measures for performance evaluation. In A. Sattar and B. H. Kang, eds., *AI 2006: Advances in Artificial Intelligence – Proc. of the 19th Australian Joint Conference on Artificial Intelligence*, pp. 1015–1021. Springer, 2006.

[25] M. Sokolova and G. Lapalme. A systematic analysis of performance measures for classification tasks. *Information Processing and Management*, 45(4):427–437, 2009.

[26] M. Stone. Cross-validatory choice and assessment of statistical predictions. *Journal of the Royal Statistical Society, Series B*, 36(1):111–147, 1974.

[27] B. Sturm. A survey of evaluation in music genre recognition. In *Proc. of the 10th International Workshop on Adaptive Multimedia Retrieval (AMR)*. Springer, 2012.

[28] I. Vatolkin. Multi-objective evaluation of music classification. In W. A. G.

et al., ed., *Proc. of the 34th Annual Conference of the German Classification Society (GfKl), 2010*, pp. 401–410. Springer, Berlin Heidelberg, 2012.

[29] I. Vatolkin and G. Rudolph. Interpretable music categorisation based on fuzzy rules and high-level audio features. In B. Lausen, S. Krolak-Schwerdt, and M. Böhmer, eds., *Data Science, Learning by Latent Structures, and Knowledge Discovery*, pp. 423–432. Springer, Berlin Heidelberg, 2015.

[30] I. Vatolkin, G. Rudolph, and C. Weihs. Interpretability of music classification as a criterion for evolutionary multi-objective feature selection. In *Proc. of the 4th International Conference on Evolutionary and Biologically Inspired Music, Sound, Art and Design (EvoMUSART)*, pp. 236–248. Springer, 2015.

[31] D. Weigl and C. Guastavino. User studies in the music information retrieval literature. In *Proc. of the 12th International Society for Music Information Retrieval Conference (ISMIR)*, pp. 335–340. University of Miami, 2011.

[32] S. Weiss and C. Kulikowski. *Computer Systems that Learn*. Morgan Kaufmann, San Francisco, 1991.

[33] I. H. Witten and E. Frank. *Data Mining: Practical Machine Learning Tools and Techniques*. Elsevier, San Francisco, 2005.

Chapter 14

Feature Processing

IGOR VATOLKIN
Department of Computer Science, TU Dortmund, Germany

14.1 Introduction

After the extraction of features which represent music data, several steps may either be necessary, e.g., for subsequent classification (e.g., substitution of missing values) or may improve the classification performance (for example, removal of irrelevant features). The initial input of *feature processing* algorithms are previously extracted characteristics. The output are data instances for the training of classification or regression models.

Preprocessing consists of basic steps for the preparation of feature vectors. For a music piece, one of the goals is to create a matrix of F features for W time frames. Methods for preprocessing are discussed in Section 14.2.

After the feature matrix has been built from available observations, it may contain a very large number of entries because many features are estimated from short time frames: e.g., a time series of the spectral centroid extracted from 23-ms frames for a 4-min song contains more than 10,000 values. Primary tasks after the preprocessing are the reduction of the amount of data and the increase of quality of a subsequent classification or regression.

One group of methods operate on the feature dimension, keeping the number of matrix columns (time windows) unchanged. The number of rows (processed features) may remain the same, be reduced (e.g., after feature selection), or increased (after feature construction). Examples of methods for the *processing of feature dimension* are presented in Section 14.3. Another option is to focus on the time dimension (usually for individual features), for example by means of time series analysis or the aggregation of feature values around relevant musical events. Methods for the *processing of the time dimension* are introduced in Section 14.4.

Figure 14.1 illustrates how various feature processing methods influence the dimensionality of the feature matrix. Parts of the feature matrix to be processed are indicated by bordered transparent rectangles. Dashed areas mark parts of feature matrix with changed values after feature processing. Some of the methods do not

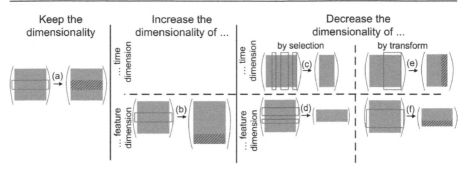

Figure 14.1: Impact of different feature processing methods on the dimensionality of the feature matrix.

change the dimensionality (Figure 14.1 (a)) and often belong to preprocessing techniques, e.g., the normalization of feature values. If new features are constructed from existing ones, the feature dimensionality increases (Figure 14.1 (b)). Generic frameworks for the creation of new features from existing ones are discussed in Section 14.5. As the number of time windows cannot be increased any more after the extraction of features, we do not consider methods which may increase the time dimensionality of the matrix (empty space above Figure 14.1 (b)). The reduction of dimensions can be achieved in two ways. Figure 14.1 (c,d) shows the selection of time intervals (e.g., verse of a song) or the selection of features (most relevant for classification). Another option is to apply transforms to certain time intervals (estimation of mean feature values around beat events) or certain features (principal component analysis, cp. Definition 9.48) (Figure 14.1 (e,f)).

Each feature processing step requires its own computing costs, and an improper application may even decrease the classification quality. The evaluation of feature processing is briefly discussed in Section 14.6.

If a feature matrix is optimized, e.g. for later classification of music data, data instances to classify may be constructed from the whole matrix or its parts. For example, when the feature matrix represents a single tone for the identification of an instrument, a complete column of the matrix might be summarized by one statistic (see Section 14.4.2). For other tasks, like classification into musical genres or listener preferences, the calculation of these statistics should be done separately for a set of *classification windows*: too many different musical segments with varying properties (harmonic, instrumental, rhythmic, etc.) would be mixed together if aggregated for the whole music piece. A constant length of several seconds (longer than a single note but shorter than a phrase or a segment) may be considered. The optimal length can, however, depend on the classification task, as investigated in our study on the recognition of personal preferences [33] where the length of classification window was optimized. For two complex classes, the optimal length was between 1 s and 5 s, and for a task very similar to the classification of classical against popular music the optimal estimated length was approximately 24 s. In another study, best results were reported for classification frames between 2 s and 5 s [4]. Another option is to aggregate feature statistics for classification windows with lengths adapted to time events of the musical structure (see Section 14.4.3). In [34], onset-based segmentation (cp. Section 16.2) outperformed other methods with windows of constant length and the aggregation of features for complete music tracks.

14.2 Preprocessing

In this section several groups of preprocessing methods are discussed: transforms of feature domains, normalization, handling of missing values, and harmonization of the feature matrix. Not all these methods are always necessary: some classification methods handle missing values themselves and/or do not require normalization. Moreover, an improper application of feature processing may even harm the ongoing classification. Therefore, for each classification scenario a careful choice of (pre)processing algorithms should be made. The impact of these methods can also be measured experimentally as discussed later in Section 14.6.

In the literature, there exists no clearly defined boundary between processing and preprocessing. For instance, [9] lists cleaning, integration, transformation, and reduction as preprocessing methods, hence counting data reduction among preprocessing.

14.2.1 Transforms of Feature Domains

Features can be represented either by numerical values (quantitative features) or categorical values (qualitative features, cf. Definition 9.9). Categorical values which cannot be directly compared on a numerical scale are called nominal. For example, a music time series could be labeled with regard to tonality as major or minor, or as one of typical structural parts of a song (intro, verse, bridge, refrain). If an ordering is possible, categorical features are referred to as ordinal. An example of an ordinal feature is the classification of songs into emotions after the valence-arousal model [26], see also Section 21.2.4. Emotions with a positive valence can be sorted based on their level of arousal: sleepy, calm, pleased, happy, excited.

Because many of the classification methods operate on quantitative features, qualitative features can be simply mapped to whole numbers which enumerate categories ("sleepy" to 1, "calm" to 2, etc. for the aforementioned example). Such mappings are called *random variables* (cp. Definition 9.4). Caution is necessary for ordinal features where the relation between values plays a role or several of such relations exist.

Example 14.1 (Numerating chord degrees). *Consider the mapping of a chord to the first five degrees of the key: tonic, supertonic, mediant, subdominant, or dominant. Handling this feature as nominal, alphabetical ordering may be natural:*

- *1 - dominant, 2 - mediant, 3 - subdominant, 4 - supertonic, 5 - tonic.*

If the feature is treated as ordinal, the position in the scale may be a promising order:

- *1 - tonic, 2 - supertonic, 3 - mediant, 4 - subdominant, 5 - dominant.*

Now consider that for a large collection of classical music pieces the appearances of chord degrees were measured. Suppose that tonic chords represent 25% of all chords, dominant to 12%, subdominant to 8%, mediant to 4%, and supertonic to 3%. Now we can order degrees by their relevance:

- *1 - tonic, 2 - dominant, 3 - subdominant, 4 - mediant, 5 - supertonic.*

For the classification of music styles or the identification of the composing period, the model may benefit from such ordering. A simple linear model could identify pieces for which generally less relevant degrees play a more important role than expected. Based on the exemplary measurements above, one can also consider the mapping of categories to real numbers:

- *0.25 - tonic, 0.12 - dominant, 0.08 - subdominant, 0.04 - mediant, 0.03 - super-tonic.*

Continuous features can also be transformed to categorical ones, e.g., if required for a classifier. Another example is the *discretization* of continuous values, e.g., to save indexing space (see Definition 9.4 for the difference between discrete and continuous variables and Section 12.4.1 for another discussion of discretization). A simple option is to limit the number of positions after the decimal place. Another common approach is to use histograms (an example was provided in Figure 9.1). Because some features may contain larger intervals with a very sparse number of values, the histograms can be constructed based on equal frequency (the number of observed feature values is equal in each histogram bin) and not based on equal width (each bin has the same length). Feature values may also be grouped by clustering (for related algorithms see Chapter 11), and the original continuous values mapped to the number of the corresponding cluster. A supervised classification scheme can be also integrated, for example, if feature intervals are divided using an entropy criterion as applied in decision trees; see Section 12.4.3.

Discretization may not only save space but also help to construct more meaningful, high-level interpretations of features. For example, the spectral centroid of an audio signal can be measured in Hz. On the other side, it is possible to map frequency ranges of each octave to one bin. Then, some harmonic and instrumental properties can be identified easier.

14.2.2 Normalization

Original ranges of feature values may be very different. In particular, distance-based classifiers, such as the k-nearest neighbors method (cp. Section 12.4.2), may overemphasize the impact of features with larger ranges. Many neural networks (cp. Section 12.4.6) expect input values between zero and one. But also for other classification methods the mapping of original feature values to the same interval may help to improve the performance. The task of *normalization* is to map original values to an interval of given range $[N_{min}, N_{max}]$.

The original values x_u^T of feature $u = 1, ..., F$ can be normalized with respect to the difference between the maximum and the minimum values of the feature (*min-max normalization*):

$$X'_{u,w} = \frac{X_{u,w} - min(\mathbf{x}_u^T)}{max(\mathbf{x}_u^T) - min(\mathbf{x}_u^T)} \cdot (N_{max} - N_{min}) + N_{min}. \tag{14.1}$$

$X_{u,w}$ denotes here the scalar value of u-th feature $\mathbf{x}_u^T$ from the w-th extraction window as element of the feature matrix $\mathbf{X}$ constructed from F features (rows) and W extraction windows (columns). Often, the target interval is $[0,1]$ (*zero-one normalization*).

Another option is to normalize to mean 0 and standard deviation 1 (*zero-mean or z-score normalization*) independent of maximum and minimum values:

$$X'_{u,w} = \frac{X_{u,w} - \bar{\mathbf{x}}_u^T}{s_{x_u}}, \tag{14.2}$$

where $\bar{\mathbf{x}}_u^T$ is the mean and s_{x_u} the (empirical) standard deviation of $\mathbf{x}_u^T$. The target range is here not equal to $[0,1]$ anymore, and this method may not work with all classifiers.

The problem of Equation (14.1) is that the maximum and the minimum estimated from some set of instances are not necessarily the same for features extracted from other music data. In that case normalization may lead to values below 0 or above 1 (out-of-range problem). Zero-mean normalization also does not guarantee that all values will be normalized to the interval $[0,1]$.

A procedure with several advantages is the *softmax normalization* [25], which is defined as follows

$$X'_{u,w} = \frac{X_{u,w} - \bar{\mathbf{x}}_u^T}{\lambda_S \cdot (s_{x_u}/2\pi)}, \tag{14.3}$$

and is plugged into the logistic function

$$X''_{u,w} = \frac{1}{1 + e^{-X'_{u,w}}}, \tag{14.4}$$

where λ_S is the control parameter for the linear response in standard deviations and should be set w.r.t. a desired level of confidence (cf. Definition 9.25). Feature values from the confidence interval are approximately linearly normalized. The default value of $\lambda_S = 2$ corresponds to the level of confidence $\approx 95.5\%$.

Softmax normalization has several advantages. For example, values from a middle region of the original range are normalized almost in a linear way and the values are always between zero and one. Although outliers are mapped to a short interval (with respect to their distance from most expected values), even for very large or small outliers with different original values the corresponding normalized values are not the same and the order is kept, in contrast to methods which map outliers to a single value.

If normalization is applied each time before the classification with a particular model, the normalization function should be the same independent of current data to classify. This means that parameters such as maximum and minimum in Equation (14.1), and mean and standard deviation in Equations (14.2) and (14.3) have to be

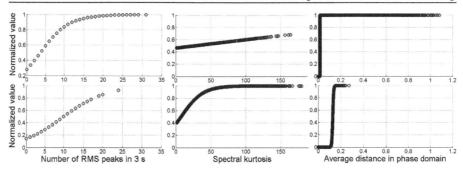

Figure 14.2: Normalization of three features with softmax. Upper row: estimation of mean and variance from classical and normalization for popular music pieces. Bottom row: estimation of mean and variance from popular and normalization for classical music pieces.

estimated only once. If new music pieces will have completely different distributions of features, the normalized values may be outside of the required range.

Figure 14.2 plots the normalizations of three features with softmax. The upper row contains instances where the estimation of mean and variance was done using a set of 15 classical music pieces, and normalization was applied to 15 popular songs. For the bottom row, mean and variance were estimated from popular songs and normalization was applied to classical music.

Original values are normalized differently in both rows. The average distance in phase domain (subfigures on the right) was a relevant feature for the distinction between classic and pop in [19], cf. Equation (5.38). After the estimation of mean and variance for classical music, a large share of popular music has normalized values very close to one, and the training of classification models would not sufficiently capture differences in popular music. Therefore, representative data samples should be analyzed not only for the training of classification models, but also for the definition of a normalization function.

An application of the same normalization method to all available features is not always necessary and may even lead to undesired effects. Some features do not require normalization because they are already scaled between zero and one (pitch class profiles, linear prediction coefficients). Sometimes it is known which values may be theoretically achieved as minimum or maximum: with a sampling rate of 44,100 Hz it is possible to analyze the frequencies up to the maximum limit of 22,050 Hz. So for frequency-related features (spectral centroid, spread, etc.) it is not necessary to normalize using softmax. Tempo in beats per minute, or duration of a music piece in seconds have no theoretical but practical limits to their ranges. Ratios (e.g., between the amplitudes of the first and the second periodicity peaks) can achieve very high values – up to infinity – if the second peak does not exist. Such extreme values can be replaced with the help of methods discussed in the next section.

14.2.3 Missing Values

Another relevant task of feature preprocessing is the handling of missing values. If some values of features are not available or not defined, this is problematic for many classification methods. However, for many real-world data mining applications, the appearance of missing values is rather common. Even if this happens not so often for music data, several sources of missing values exist:

Example 14.2 (Sources of missing values in music data).

- *Silence and very quiet signals. Some audio features cannot be extracted from time frames with too weak signal energy. Such frames may appear at the beginning or at the end of audio recordings. For instance, it is impossible to extract many frequency-domain characteristics if the signal contains only zeros after applying a DFT, cf. definitions in Section 5.2.2.*

- *Non-harmonic and noisy signals. When less harmonic events are part of a music piece (driving cars, noise of a helicopter), or strong digital effects like distortion are applied, it is sometimes not possible to extract the fundamental frequency. Then, further features based on the fundamental frequency (e.g., tristimulus or the ratio of harmonic components, cf. Section 5.3.4) cannot be calculated. Other examples are features based on the analysis of peaks. Depending on the algorithm for peak detection, it may not be possible to estimate characteristics of peaks. Then, a feature like "the width of the strongest spectral peak" is not defined.*

- *Frames where a feature is undefined. Some features require a certain number of frames before the extraction frame. Low-energy may be defined not as a fraction of the energy in the extraction frame to the whole energy of the signal (Definition 5.3), but as a relation of the energy in the extraction frame to the energy of a given number of frames before the extraction frame. For example, in jAudio [15] the mean RMS of 100 frames before the extraction frame has to be estimated.*

- *Non-availability at a certain time point. For less popular or recently composed/re-leased music pieces, many characteristics of metadata may be not available (listener tags, lyrics, moods, etc.).*

- *Undefined values as output of feature processing methods. Some feature processing methods may output "not a number" values for certain extraction frames. Similar to the above discussion of frames with undefined features, the estimation of structural complexity is not possible for a set of "early" frames, cf. Equation (14.8). After the harmonization of the feature matrix as introduced in the next section, some entries in the feature matrix may be intentionally filled with non-defined values (cf. example in Figure 14.3).*

There are many possible solutions for how to deal with missing or undefined values. Probably the simplest procedure is to remove instances with missing data from a training set. If only a small part of the training data is affected, this may help. In particular the removal of silence at the beginning and the end of audio recordings is useful. However, depending on the reason for missing values, this may lead to an undesired bias. For example, removal of music tracks with missing metadata may overemphasize the impact of popular songs.

If data instances with missing values are kept, these values can be replaced by a location measure (cf. Definition 9.10), for example, the mean or the median of the corresponding feature series. Because features with the same mean or median may have very different distributions, the missing values can be also replaced in the way that the standard deviation remains the same. If it is expected that data instances with missing values are rather uncommon (e.g., very noisy frames where the fundamental frequency cannot be estimated) or the reason for a missing value may carry some important information to learn (low popularity of songs with missing metadata), the value can be set to an outlying number, e.g., to zero for a feature with a positive definition space. However, a drawback might be that dissimilar music pieces would be characterized with the same values of corresponding features.

The replacement of all missing values in a series by the same single value may introduce even more problems. In particular, for time series of features based on audio signals, the values of neighboring positions may then have a weaker or stronger relation to each other. Even different features might often correlate to a smaller or bigger extent. Alternatively, a linear regression (Definition 9.32) can be applied to approximate missing values from other positions of the same feature or a multiple regression (Definition 9.34) might be used based on other features. The application of regression is limited, though, to single or several consecutive positions and is less reasonable for longer blocks of missing values in feature series.

These and more approaches to handle missing data are introduced in [13].

14.2.4   Harmonization of the Feature Matrix

Many processing methods operate on vectors of several features with the same dimensionality. However, the dimension of the features may differ considerably between the number of frames from several milliseconds (in particular spectral characteristics) and the length of the complete music piece (structural information, duration, metadata, etc.). To solve this problem, *harmonization* of raw feature vectors can be applied as sketched in Figure 14.3.

In the first step, the shortest extraction frame across all features is identified (frames of F_1 in Figure 14.3). The length of such frames is called l_{min}. Then, the length of each feature vector is extended to the number of frames of length l_{min} which are contained in the music piece. For features with frame length $l > l_{min}$, the feature value at position w is selected from the original (longer) consecutive extraction frames which overlap with the new smaller frame. The feature value of the original window with the largest contribution is selected.

Definition 14.1 (Harmonized feature matrix). If two original frames overlap the new (shorter) frame, then the value of the new frame w is calculated as follows:

$$\text{If} \quad l \cdot \left(\left\lfloor \tfrac{l_{min}(w-1)}{l} \right\rfloor + 1 \right) - l_{min} \cdot (w-1) > l_{min} \cdot w - l \cdot \left(\left\lfloor \tfrac{l_{min}(w-1)}{l} \right\rfloor + 1 \right), \quad \text{then}$$

$$X_{u,w} := x_u\left[\left\lfloor \tfrac{l_{min}(w-1)}{l} \right\rfloor + 1 \right], \quad \text{else} \quad X_{u,w} := x_u\left[\left\lfloor \tfrac{l_{min}(w-1)}{l} \right\rfloor + 2 \right], \quad \text{where}$$

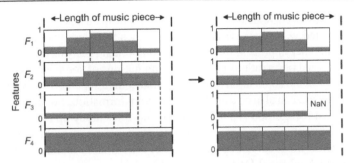

Figure 14.3: Harmonization of feature vectors with different dimensions for the estimation of the feature matrix.

u is the index of the current feature. $\left\lfloor \frac{l_{min}(w-1)}{l} \right\rfloor + 1$ and $\left\lfloor \frac{l_{min}(w-1)}{l} \right\rfloor + 2$ are the indices of larger original frames which overlap with the w-th shorter frame.

If only one longer frame covers the new shorter frame, its value is taken (cf. feature F_4 in Figure 14.3).

If no longer original frame with an overlap to the w-th shorter frame exists, the new value is set to "not a number" (cf. last value of feature F_3 in Figure 14.3).

There exist other options to construct a harmonized feature matrix, for instance, using interpolation between the values of the original frames. In that case, however, it should be guaranteed that the interpolated values are feasible and make sense with respect to the definition of the feature. Consider a C major cadence where the dominant triad (G) changes to a tonic (C), and the normalized strength of C in the chroma vector switches from a value near zero to a value near one. Then, an interpolation would blur the clear identification of this occurrence and make the recognition of a chord or a cadence more difficult.

14.3 Processing of Feature Dimension

We can roughly distinguish between two goals for processing methods which operate on feature dimension: to increase a number of features by the estimation of new characteristics, hopefully better suited for class separation, or to reduce a number of features saving storage requirements and computing costs.

Both options can be applied together or directly after each other. The first one, feature construction, is briefly introduced in Section 14.5.

For the purpose of reduction, some of the complete feature series (rows in the feature matrix) can be removed by means of feature selection, for example after the identification of highly correlated features or features which are less relevant for the corresponding task. Feature selection is discussed in Chapter 15.

Statistical transforms can be applied as a means for the reduction of the number of features. A prominent method is Principal Component Analysis (PCA, Definition 9.48). First, linear combinations of features (components) are estimated, with the goal to identify decorrelated dimensions with highest variance. Second, the components are sorted w.r.t. their variance, and the ones with less variance are finally removed. In another approach, Independent Component Analysis (ICA), it is assumed that the analyzed observations are linear combinations of some independent sources. In Section 11.6, the application of ICA to sound source separation is described. However, this method can also be used for feature transformation. For example, in a study on instrument recognition [7], an improvement of accuracy up to 9% is reported after the transformation of original features to more independent dimensions. Another option to select the most relevant dimensions after transformation is to apply Linear Discriminant Analysis (LDA), cf. Section 12.4.1.

Although these transformations can be applied efficiently, may reduce the complexity of classification models, and help to increase the classification quality, these algorithms also have disadvantages. If interpretable music features (moods, instruments, vocal characteristics, etc.) are used to train classification models, their meaning is lost after transformation. Later theoretical analysis of music categories becomes hard or impossible. Furthermore, the extraction of all original features is still required for new data instances even if a classification model is trained only on a few components.

14.4 Processing of Time Dimension

Methods which operate on the time dimension of the feature matrix often process features individually (an example of simultaneous processing of different features is multivariate regression discussed in Section 14.4.2). The number of values may remain the same (for example, after smoothing with a running average) or is reduced. In the following, we will discuss three groups of methods: sampling and order-independent statistics in Section 14.4.1, order-dependent statistics based on time series analysis in Section 14.4.2, and time processing based on musical structure in Section 14.4.3. The difference between order-independent and order-dependent statistics is that in the second case the temporal evolution of features is taken into account. Order-independent statistics produce the same values if the order of frames is permuted, i.e. the temporal development of features does not have an impact on these statistics.

Despite differences in operating methods, there are often no exact boundaries between these groups. A chain of feature processing algorithms may consist of techniques from all categories, e.g., the selection of time intervals from different structural parts of a music piece and a further estimation of time series characteristics.

14.4.1 Sampling and Order-Independent Statistics

Data sampling does not consider any high-level knowledge about music structure, and its main goal is to reduce the amount of data. One option is to select each k-th time frame and remove feature values of other frames. For the analysis of music pieces, a commonly applied method is to *select an interval* of a constant length from the music piece for further processing. In the literature, often an interval of 30 s is analyzed [28, 18]. The optimal length – depending on the application scenario – may be somewhere between hundreds of milliseconds (capturing a note or note sequence) up to several minutes for longer pieces with varying characteristics. Obviously, there is no statistical evidence to restrain the interval length to 30 s. One of the arguments for this particular number was based on legal issues: in some countries audio excerpts of 30-s length could be freely distributed.

Not only the length but also the starting point of the interval has an impact on the later analysis. In our previous study on the recognition of music genres and styles, the selection of 30 s from the middle of a music track performed better than 30 s taken from the beginning, and even better performance was achieved with the selection of 30 s after the first minute [32]. The latter method increases the probability to skip the song intro and capture vocal parts (verse or refrain) which are usually representative for popular music.

An interesting fact was observed in two studies on genre, artist, and style recognition, where very short time intervals of 250 ms and 400 ms were sufficient to recognize a class for human listeners [8, 10]. However, this does not mean that such intervals are optimal for automatic analysis of music pieces. The human brain may very quickly recognize previously learned patterns. Furthermore, the recognition of more complex music classes with differing structural segments may fail if based on too short segments because relevant "aspects of music such as rhythm, melody and melodic effects such as tremolo are found on larger time scales" [1, p. 31].

A commonly applied method for the aggregation of features is the estimation of a few statistics of each feature, including location measures (mean and median, Definition 9.10) and dispersion measures (standard deviation, quantiles, mode, Definition 9.11). To reduce the impact of outliers, the so-called *trimmed mean* can be estimated by sorting, removing a fixed percentage of extreme values, and then taking the mean. In [9], 2% and in [20] 2.5% are recommended. Skewness and kurtosis – also referred to as 3rd and 4th moments (see Definition 9.7) – describe the asymmetry of feature series and the flatness around its mean.

More information about the distribution of feature values can be saved using boundaries of confidence intervals (Definition 9.25) and histograms (Definition 9.6). For example, beat histograms are estimated for tempo prediction in [28] but may be useful to capture different levels of periodicity as well. As mentioned above in Section 14.2.1, histogram bins can be constructed either with an equal length or with an equal number of values. For both cases the optimal number of bins may not be known in advance.

Properties of features which are not normally distributed can be stored as parameters of a mixture of multiple (Gaussian) distributions [17].

14.4.2 Order-Dependent Statistics Based on Time Series Analysis

The intentional creation of sequences of musical events and the building of repetitive and similar patterns are an important part of music composition. The temporal progress of properties of sound can be characterized based on methods developed for time series analysis (cf. Definition 9.39). In contrast to the statistics from the previous section, here the temporal development of feature series is taken into account, and the estimated statistics are dependent on the order of extraction frames.

One of the methods to derive relevant properties of the original series without data reduction is to estimate first- and second-order derivatives, or to smooth the original series, e.g., with a running average. Also, order-independent statistics such as parameters of a Gaussian mixture model can be estimated for a sequence of larger *texture* windows. The transformation of a feature series to the phase domain with a subsequent estimation of characteristics such as average distance as done in [19, 20] can also be applied (the relevance of this feature to distinguish between classical and popular music is illustrated in Figures 14.2,15.2).

Another option is to save only a few characteristics of the series such as the parameters of an autoregressive model (Definition 9.40). Here, a feature value is predicted from P preceding values of the same feature. Such an application is introduced in [18] by means of P-th-order diagonal autoregressive (DAR) model:

$$X'_{u,w} = \sum_{p=1}^{P} A_{u,p} \cdot X_{u,w-p} + \varepsilon_u, \qquad (14.5)$$

so that each feature $u = 1, ..., F$ is independent of the other features but only dependent on its own past. ε is considered as white noise for each dimension.

Because of often existing correlations between different features, a more general multivariate autoregressive (MAR) model predicts the whole feature vector x_w based on the whole feature vector in past time periods:

$$x_w = \sum_{p=1}^{P} A_{u,p} \cdot x_{w-p} + \varepsilon_u, \qquad (14.6)$$

where ε_u is assumed multivariate noise with a full covariance matrix.

The estimation of optimal autoregressive coefficients can be achieved through the minimization of least square errors between original feature values and the model (see Definition 9.32).

Although MAR and DAR performed best for genre recognition in the original investigation [18], both models were outperformed by simpler statistics (mean, variance, and three quartiles) for the recognition of sounds in [22]. Another outcome of the latter study was that the recognition rate using DAR was closer to that of simple statistics if estimated on MFCCs, which were also used in [18]. For other feature sets, the difference was larger. This supports the suggestion that no "optimal" processing method can be recommended a priori for each possible feature set and classification task.

The estimation of correlation of time series with the same series shifted by different lags (autocorrelation) (see Definition 9.41) leads to *correlograms* which describe periodic properties of the underlying series. Then, various characteristics of a correlogram can be calculated, such as positions and amplitudes of strongest peaks or a decay of the correlation function estimated with a linear regression as proposed in [20].

A different strategy for feature aggregation is to perform a *modulation analysis* which treats a feature series as a downsampled time-domain signal. The modulation features are then obtained by computing a DFT along T_A frames of the series

$$X'_{u,w}[v] = \sum_{t=0}^{T_A-1} X_{u,w} \exp^{-i\frac{2\pi t v}{T_A}}, \tag{14.7}$$

where $v \in \{0,1,\ldots,T_A/2-1\}$ denotes the modulation frequency bin and i the imaginary unit. High energy at low or high modulation frequencies then corresponds to slow or fast changes of the feature values, respectively. Thus, by appropriately summarizing the squared magnitudes of the modulation spectrum the energies of the summarized bands serve as descriptors which model the temporal structure of the short-time features. McKinney and Breebaart [16] propose to subsume the modulation frequency bins to four bands which correspond to the frequency ranges of 0 Hz (average across observations), 1–2 Hz (musical beat rates), 3–15 Hz (speech syllabic rates), and 20–43 Hz (perceptual roughness).

Structural complexity is a method proposed in [14] to capture relevant structural changes of feature values in a larger analysis window. First, sets of F_z features are selected to represent the z-th of Z (if desired interpretable) properties. In the original contribution, chroma is represented by a 12-dimensional feature vector, and also rhythm and timbre are characterized. Later, [29] extended these properties to instrumentation, chords, harmony, and tempo/rhythm. For each short extraction frame w_a in the analysis window, a number T_A of frames before and after w_a (including w_a-th frame) are taken into account. The difference between vectors representing the summary of T_A preceding frames $\boldsymbol{wp}$ and T_A succeeding frames $\boldsymbol{ws}$ is measured by the Jensen–Shannon divergence:

$$d_{JS}(\boldsymbol{wp},\boldsymbol{ws}) = \frac{d_{KL}(\boldsymbol{wp},\frac{\boldsymbol{wp}+\boldsymbol{ws}}{2}) + d_{KL}(\boldsymbol{ws},\frac{\boldsymbol{wp}+\boldsymbol{ws}}{2})}{2}, \tag{14.8}$$

where $d_{KL}(\boldsymbol{wp},\boldsymbol{ws})$ is the Kullback–Leibler divergence:

$$d_{KL}(\boldsymbol{wp},\boldsymbol{ws}) = \sum_{k=1}^{F_z} \boldsymbol{wp}_k \cdot \log_2\left(\frac{\boldsymbol{wp}_k}{\boldsymbol{ws}_k}\right), \tag{14.9}$$

and summary vectors are built as follows,

$$\boldsymbol{wp}_k = \frac{1}{T_A} \sum_{w=w_a-T_A}^{w_a-1} X_{k',w}, \quad \boldsymbol{ws}_k = \frac{1}{T_A} \sum_{w=w_a}^{w_a+T_A-1} X_{k',w}, \tag{14.10}$$

where k' is the index in the complete feature matrix $\boldsymbol{X}$ corresponding to the k-th feature representing property z.

14.4.3 Frame Selection Based on Musical Structure

If ever possible, the knowledge of musical events should be integrated in frame selection for intelligent music data processing.

Table 14.1 lists several levels of structure with relation to score events. The most granular level is represented with notes of the score and onset events for audio. Another source of information at this level is the Attack-Decay-Sustain-Release (ADSR) envelope sketched in Figure 2.16. Each of the four intervals is characterized by its specific characteristics (inharmonic components in attack phase, stable energy in sustain phase, decreasing energy in release phase, etc.). Therefore, an improper aggregation of features from different intervals may complicate further analysis. Often, a simplified model of the envelope is calculated, the Attack-Onset-Release (AOR) envelope, where an onset corresponds to the time point with a maximum energy after the attack phase, and all remaining components of the sound are assigned to release phase. See also Section 16.3 for ADSR analysis.

Table 14.1: Levels of Structure for Feature Aggregation

Level	Score events	Audio events
Individual events	Notes	Attack-Decay-Sustain-Release intervals, onsets
Sequence of events	Motifs, phrases	Segments
Repetitive patterns	Measures, rhythmic accentuation	Tatums, beats

The grouping of musical events is explained in the theory of form as introduced in Section 3.7 and is the essential element of music composition. The variability of forms can be achieved by means of horizontal access (sequences of notes organized into motifs and phrases) and vertical access (harmonic structure of monody, heterophony, and polyphony, cf. Section 3.4). Changes on this level have a strong influence on the audio signal. The estimation of boundaries of longer segments is usually done with the help of self-similarity analysis (see Definition 16.1 and the example in Figure 14.4).

The third level in Table 14.1 characterizes patterns of strong and weak accents where a sequence of accents (measure) describes the rhythm, and the number of these repetitions in a given time interval corresponds to the tempo. Longer structural parts of music pieces are often characterized by a constant tempo and the same rhythmic pattern. For an audio signal, the corresponding structure can be described with tatum and beat events, where tatum corresponds to the shortest and beat to the strongest perceived entity of repetition. This structure has a certain abstract extent: beats and tatums do not necessarily coincide with onsets (played notes) because of breaks, syncopes, and varying length of notes. See also Section 20.2.3 for another discussion of metrical levels.

Example 14.3 (Extraction of Musical Events). *The knowledge of musical events can be used for data reduction based on the selection of particular time frames or segments. This is illustrated in Figure 14.4. The top left subfigure shows the first bars of Beethoven's "Für Elise." If the score is not available, notes and temporal structure have to be extracted from audio signal. The other subfigures on the left plot the magnitude of the spectrum, the root mean square of the signal, and estimated beat, tatum, and onset events. Then, only time frames with particular events can be selected for*

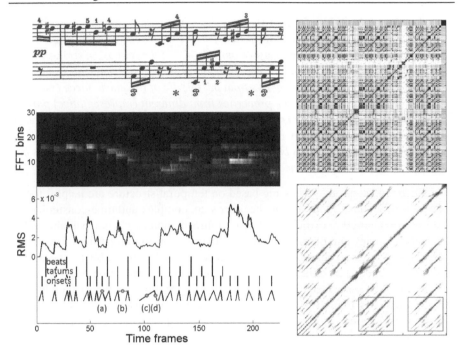

Figure 14.4: Analysis of music structure. Left subfigures (from top to down): the score of the first bars of Beethoven's "Für Elise," first 30 bins of the magnitude spectrum, the root mean square of the signal, the time events extracted from audio (beats and tatums after [6], onsets with MIR Toolbox [11]). Right subfigures: self-similarity matrix of the complete music piece, its variant enhanced by means of thresholding (both matrices are estimated with SM Toolbox [21]).

further processing. Examples of such frames are marked with small filled circles in the left bottom subfigure: (a): onset frame, (b): interonset frame, (c): middle of attack interval, (d): middle of release interval. The choice of the method depends on the application. For automatic analysis of harmony, frames between onsets with a stable sound may be preferred. For the identification of instruments, more relevant features may be extracted from the middle of the attack interval which contain instrument-specific inharmonic components such as stroke of a piano key or noise of a violin bow.

A general structure of a music piece can be estimated from Self-Similarity Matrices (SSM) which measure distances between feature vectors as introduced in Definition 16.1. The right top subfigure plots an SSM based on chroma features, and the right bottom subfigure an enhanced SSM variant. Segments with high similarity are visualized in the matrix as dark diagonal stripes. The information about the structure of a music piece can be used differently for data reduction. One option is to remove features from segments which are already contained in the feature matrix (or are very similar to such feature). Two examples of such segments are marked with rectangles on the plot of the enhanced SSM (bottom right subfigure). Another procedure is to select only a limited number of short time frames from each segment: here it is assumed that the relevant characteristics of the segment do not have a strong variation and may be captured by a sample, e.g., from the middle of the segment. Compared to the "blind" method of selecting 30 s from the middle of the music piece (see discussion in Section 14.4.1), here the properties from different (representative) parts of a music piece are maintained despite of strong data reduction. Because structures may have several and not always coincident layers (consider a segment with the same sequence of harmonic events but other instrumentation), the analysis of SSMs based on different features can be necessary.

Data reduction and processing based on temporal structure are helpful for classification and music analysis. On the other side, complex and time-consuming algorithms are required to extract the structural information if it is not available at hand. Because of many simultaneously playing sources, varying properties of instruments (such as progress of the envelope), and also applied digital effects, the accuracy of these methods has some limitations and an accurate resolution of musical structure cannot be guaranteed by state-of-the-art algorithms. Some of the challenges are discussed in Part III of the book. Identification of onset events is addressed in Section 16.2, tempo recognition in Chapter 20, and structure segmentation in Section 16.4.

14.5   Automatic Feature Construction

There exist plenty of methods with lots of parameters for the extraction of audio and other features from music data. However, even for related problems, very different features may be relevant, and the optimal parameters for their extraction cannot always be known in advance. For example, generic features which are optimal for the identification of all four groups of instruments in polyphonic mixtures could not outperform the features which are optimal for a classification task of an individual instrument in most cases [30].

The idea behind *automatic feature construction* (also referred to as feature synthesis or feature generation) is to provide a generic and flexible framework for the construction of features which are best suited for a particular classification task. In this section we only briefly discuss the motivation for automatic feature construction and describe several basic operating principles of the related algorithms.

Transforms of original feature dimensions and various linear and non-linear operators may allow better separability between classes. Figure 14.5 (a) provides an artificial example where no linear separation is possible between instances of two classes. After a nonlinear mapping to new feature dimensions, this separation becomes possible (Figure 14.5 (b)). Note that this separation is also possible for a projection onto the horizontal axis only (sum of squares) so that this improvement is achieved together with the reduction of dimensionality.

Figure 14.5 (c) presents distributions of the mean value of the 1st MFCC for a set of 62 cello (dashed line) and 30 flute (solid line) tones. This feature does not seem to be very relevant for the distinction between the two instruments. However,

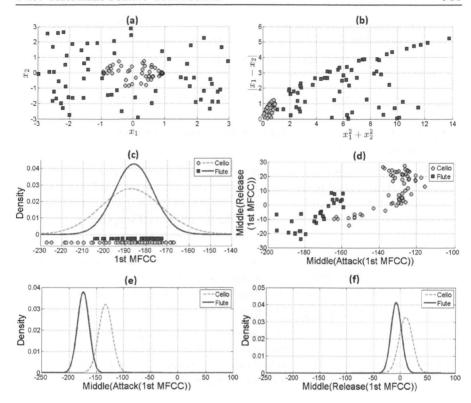

Figure 14.5: Examples of better separability between classes with new constructed features. (a,b): artificial example; (c–f): distinction of cello and flute tones.

we may integrate the knowledge of time events for the construction of new feature dimensions (see discussion in Section 14.4.3). In Figure 14.5 (d), two new features are generated from the original ones: it is distinguished between frames from the attack interval and the release interval, and only the middle of these intervals is taken into account. This leads to a better separation ability of new feature dimensions which produce more distinctive distributions for the two classes as shown on Figures 14.5 (e,f).

Pachet and Roy [23] introduce a general framework for the construction of so-called analytical features. The extraction of an individual feature is described by a sequence of operators allowing exactly one operation in each step. Mierswa and Morik [19] distinguish between several categories of operations and allow the construction of method trees capable of extracting multiple features. The following categories of operations are proposed (examples are provided in parentheses):

- *transforms* change the space of input series (DFT, phase space transform, auto-correlation),

- *filters* do not change the space and estimate some function for each value of input series (logarithmic scaling, band-pass filtering, moving average),

- *markups* assign elements of input series to some properties (audio segmentation, clustering),

- *generalized windowing* splits input series into windows of given size and overlap, and

- *functions* save single values for a complete series (characteristics of the strongest peak, location measures).

The implementation as a tree helps to save computing time if the same transform is used as input several times (see Example 14.4). On the other side, in [23, p. 5] it is argued that "the uniform view on windowing, signal operators, aggregation operators, and operator parameters" improves the flexibility and simplifies the generation process.

Example 14.4 (Feature Construction). *The following example of a feature generation chain after [23] estimates the maximum amplitude of the spectrum after FFT for frames of 1024 samples. Then, the minimum across all frames is saved.*

- *Min(Max(Sqrt(FFT(Split(x,1024))))).*

The following example of a feature generation tree after [19] in Figure 14.6 saves several features after the estimation of the spectrum (characteristics of three strongest spectral peaks, strongest chroma bin) as well as several time domain–based characteristics (zero-crossings, periods of two strongest peaks after autocorrelation).

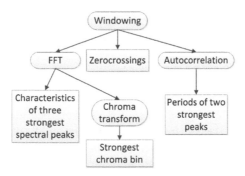

Figure 14.6: An example for a feature generation tree.

Two general problems have to be resolved for a successful automatic feature generation. First, a set of operations and transforms should be designed. For instance, [23] lists more than 70 operators from basic mathematical constructs (maximum, minimum, absolute value, etc.) to complex transforms (FFT, estimation of mel spectrum, various filters). The next challenge is to find a strategy for the exploration of a huge search space: l^k different chains of length l using k possible operators are possible. As discussed in Chapter 10, stochastic methods such as evolutionary algorithms are in particular suited for such complex optimization problems. Genetic programming is applied in [19, 23], and particle swarm optimization in [12]. During the evolutionary process, several operations for the variation of feature candidates can be applied:

Example 14.5 (Genetic Operators for Feature Construction). *Substitution of an operator with another, e.g.:*

- *Centroid(Low-Pass Filter(FFT(Split(x,1024)))) ↦*
  Centroid(High-Pass Filter(FFT(Split(x,1024))))

Removal of an operator:

- *Centroid(Low-Pass Filter(FFT(Split(x,1024)))) ↦*
  Centroid(FFT(Split(x,1024)))

Addition of a new operator:

- *Centroid(Low-Pass Filter(FFT(Split(x,1024)))) ↦*
  Centroid(Low-Pass Filter(Log(FFT(Split(x,1024)))))

Crossover between two chains:

- *{ Centroid(Low-Pass Filter(FFT(Split(x,1024)))) ,*
  Max(Peak Positions(Autocorrelation(Differentiation(Split(x,1024))))) } ↦
  Max(Peak Positions(Low-Pass Filter(FFT(Split(x,1024))))

14.6 A Note on the Evaluation of Feature Processing

The typical purpose of feature processing is to prepare classification instances and to improve the *quality* of classification models. Thus, the impact of processing algorithms can be validated with classification quality measures discussed in Chapter 13. The difference between outcomes of experiments with and without a particular processing method can be tested for significance by means of statistical tests for paired observations. Several tests were presented in Sections 9.7 and 13.5.

Another relevant evaluation issue is *computing costs*. The application of a simple normalization formula may require a marginal amount of time in relation to a complete feature extraction and classification chain. On the other side, methods like evolutionary feature construction may require hours and days of optimization time for a small improvement in quality. A thorough justification of each computing step should also be done with respect to the availability of resources in a concrete scenario: a complex algorithm which only slightly improves the quality may be meaningless for mobile devices but reasonable if it can be parallelized and applied on a server grid.

Computing costs can be measured *empirically* by the estimation of runtime (with the drawback of possible variance depending on hardware and a computing load) or *theoretically* by the estimation of computational complexity (with the drawback that in particular for complex methods, the optimal implementation cannot be always guaranteed and may depend on hardware, and for stochastic methods the precise estimation of costs is not always possible or may strongly depend on the problem).

Some of feature processing methods are explicitly designed to reduce the dimensionality of data. To evaluate the success of *data reduction*, we can measure the difference between the amount of data before and after feature processing. Considering feature processing in general as an intermediate step between the extraction of raw features and classification or regression (training), we can estimate the reduction rate of feature processing m_{FPRR} as the share of values in the feature matrix after feature processing in relation to the original number of values:

$$m_{FPRR} = \frac{F \cdot W}{\sum_{u=1}^{F^*} \left(W^{**}(u) \cdot F^{**}(u) \right)}. \tag{14.11}$$

F and W are dimensions of the feature matrix $\boldsymbol{X}$ after processing (number of features resp. number of classification windows). F^* is the number of original features, where the number of dimensions for feature $\boldsymbol{x}_u$ is characterized with $F^{**}(u)$ and the number of corresponding extraction frames with $W^{**}(u)$.

For each method that operates on only the feature or time dimension (cf. Sections 14.3,14.4), its specific data reduction performance can be measured. For the reduction of features the feature reduction rate m_{FRR} is:

$$m_{FRR} = \frac{F}{F^*} \tag{14.12}$$

(F is the number of features after and F^* before processing). Similarly, the time reduction rate m_{TRR} is defined as:

$$m_{TRR} = \frac{W}{W^*}, \tag{14.13}$$

where W is the number of time windows (number of values in feature series) after and W^* before the processing. Instead of measuring the number of time windows, the sum of the lengths of all extraction time intervals can be estimated and divided by the length of the complete music piece.

The interpretation of measures given above is not always straightforward; consider the two following examples. Even if the number of features is strongly reduced after PCA (e.g., only the two strongest components are selected), all the original features are still necessary for the determination of the components. Also, successful data reduction based on music structure may require algorithms for the estimation of boundaries between structural segments, which themselves require the extraction of underlying features from complete music pieces for the building of the self-similarity matrix (see Figure 14.4).

The outcome of a successful data reduction is, however, not only the reduced amount of data, but also an increase of generalization performance of a classification or regression model. Models built with fewer features and trained with fewer outlying instances may tend to be (but not necessarily are) more stable and may not overfit towards training sets as well. The stability of classification models after feature processing can be evaluated as discussed in Section 13.3.6.3.

The three groups of evaluation criteria (quality, resources, and data reduction rates) are often in conflict: algorithms which best improve classification quality or very efficiently reduce the number of features keeping high relevance of remaining ones often have to pay the price of high computing costs. Also, classification quality alone suffers from too strong data reduction. For an unbiased evaluation of methods, multi-objective optimization can be applied (see Section 10.4). In Section 15.7 we will discuss how the simultaneous optimization of two evaluation criteria can be integrated into feature selection.

14.7 Further Reading

Because feature processing itself was, until now, seldom a topic of main interest in studies on music data analysis, many of the following references describe different aspects of feature processing as a general task in data mining. However, these methods – in particular those which have been developed for the analysis of time series – are often a good choice for music data analysis. A comprehensive survey of feature processing algorithms with many practical examples is provided in [25].

The removal of noise is one of the preprocessing methods which may improve access to relevant information in (music) time series. Noise can be a part of the complete feature time series which represents music data (e.g., for audio recordings of poor quality). Several smoothing techniques which may help in this situation are presented in [25, chapter "Series Variables"]. On the other side, individual noisy classification windows (silent frames, spoken comments, wrongly applied digital effects) can be identified using unsupervised classification (see Chapter 11), among others with methods for outlier detection.

Unsupervised classification can also be applied for the selection of instances for the supervised classification. In a previous study we examined the suitability of instance selection with n-grams for music classification [31]. Also, n-grams were integrated into the search for repetitive patterns [24] thus enabling access to the recognition of structure and a following reduction of frames, as discussed in Section 14.4.3. An example of the application of evolutionary algorithms for instance selection is provided in [5].

Many classification methods produce results of best quality for certain distributions, and often a normal distribution is assumed. The performance may suffer from features with a high variance of density: linear models do not distinguish between sparsely occupied areas of the feature definition space and areas with a high density of observations. This problem can be addressed by means of a redistribution of feature values as discussed in [25, chapter "Normalizing and Redistributing Values"].

A general overview of methods for data reduction (unsupervised and unsupervised) is provided in [27]. Some of the multivariate methods (factor analysis, multidimensional scaling, linear discriminant analysis, etc.) are presented in detail in [2], chapter "Dimensionality Reduction." The application of the wavelet transform for data reduction is briefly discussed in [9]. Several non-linear techniques on dimension reduction are presented in [3].

Bibliography

[1] P. Ahrendt. *Music Genre Classification Systems - A Computational Approach.* PhD thesis, Informatics and Mathematical Modelling, Technical University of

Denmark, 2006.

[2] E. Alpaydin. *Introduction to Machine Learning.* The MIT Press, Cambridge London, 2010.

[3] Y. Bengio, O. Delalleau, N. L. Reoux, J.-F. Paiement, P. Vincent, and M. Ouimet. Spectral dimensionality reduction. In I. Guyon, M. Nikravesh, S. Gunn, and L. A. Zadeh, eds., *Feature Extraction. Foundations and Applications,* volume 207 of *Studies in Fuzziness and Soft Computing,* pp. 137–165. Springer, Berlin Heidelberg, 2006.

[4] J. Bergstra, N. Casagrande, D. Erhan, D. Eck, and B. Kégl. Aggregate features and ADABOOST for music classification. *Machine Learning,* 65(2-3):473–484, 2006.

[5] J. R. Cano, F. Herrera, and M. Lozano. Using evolutionary algorithms as instance selection for data reduction in KDD: An experimental study. *IEEE Transactions on Evolutionary Computation,* 7(6):561–575, 2003.

[6] A. J. Eronen, V. T. Peltonen, J. T. Tuomi, A. P. Klapuri, S. Fagerlund, T. Sorsa, G. Lorho, and J. Huopaniemi. Audio-based context recognition. *IEEE Transactions on Audio, Speech, and Language Processing,* 14(1):321–329, 2006.

[7] A. J. Eronen. Musical instrument recognition using ICA-based transform of features and discriminatively trained HMMs. In *Proc. of the 7th International Symposium on Signal Processing and Its Applications (ISSPA),* pp. 133–136. IEEE, 2003.

[8] R. O. Gjerdingen and D. Perrott. Scanning the dial: The rapid recognition of music genres. *Journal of New Music Research,* 37(2):93–100, 2008.

[9] J. Han, M. Kamber, and J. Pei. Data preprocessing. In *Data Mining: Concepts and Techniques,* Series in Data Management Systems, pp. 93–124. Morgan Kaufmann Publishers, 2012.

[10] C. L. Krumhansl. Plink: "Thin slices" of music. *Music Perception: An Interdisciplinary Journal,* 27(5):337–354, 2010.

[11] O. Lartillot and P. Toiviainen. MIR in MATLAB (II): A toolbox for musical feature extraction from audio. In S. Dixon, D. Bainbridge, and R. Typke, eds., *Proc. of the 8th International Conference on Music Information Retrieval (IS-MIR),* pp. 127–130. Austrian Computer Society, 2007.

[12] T. Mäkinen, S. Kiranyaz, J. Raitoharju, and M. Gabbouj. An evolutionary feature synthesis approach for content-based audio retrieval. *EURASIP Journal on Audio, Speech, and Music Processing,* 2012(23), 2012. `doi:10.1186/1687-4722-2012-23`.

[13] B. M. Marlin. *Missing Data Problems in Machine Learning.* PhD thesis, Department of Computer Science, University of Toronto, 2008.

[14] M. Mauch and M. Levy. Structural change on multiple time scales as a correlate of musical complexity. In *Proc. of the 12th International Society for Music Information Retrieval Conference (ISMIR),* pp. 489–494. University of Miami, 2011.

[15] D. McEnnis, C. McKay, and I. Fujinaga. jAudio: Additions and improvements. In *Proc. of the 7th International Conference on Music Information Retrieval (ISMIR)*, pp. 385–386. University of Victoria, 2006.

[16] M. F. McKinney and J. Breebaart. Features for audio and music classification. In *Proc. of the 4th International Conference on Music Information Retrieval (ISMIR)*. The Johns Hopkins University, 2003.

[17] G. J. McLachlan and D. Peel. *Finite Mixture Models*. Wiley Series in Probability and Statistics, New York, 2000.

[18] A. Meng, P. Ahrendt, J. Larsen, and L. K. Hansen. Temporal feature integration for music genre classification. *IEEE Transactions on Audio, Speech, and Language Processing*, 15(5):1654–1664, 2007.

[19] I. Mierswa and K. Morik. Automatic feature extraction for classifying audio data. *Machine Learning Journal*, 58(2-3):127–149, 2005.

[20] F. Mörchen, A. Ultsch, M. Thies, and I. Löhken. Modeling timbre distance with temporal statistics from polyphonic music. *IEEE Transactions on Audio, Speech, and Language Processing*, 14(1):81–90, 2006.

[21] M. Müller, N. Jiang, and H. G. Grohganz. Sm toolbox: MATLAB implementations for computing and enhancing similarity matrices. In *Proc. of 53rd Audio Engineering Society (AES)*. Audio Engineering Society, 2014.

[22] S. Ntalampiras, I. Potamitis, and N. Fakotakis. Exploiting temporal feature integration for generalized sound recognition. *EURASIP Journal on Advances in Signal Processing*, 2009. doi:10.1155/2009/807162.

[23] F. Pachet and P. Roy. Analytical features: A knowledge-based approach to audio feature generation. *EURASIP Journal on Audio, Speech, and Music Processing*, 2009. doi:10.1155/2009/153017.

[24] N. Patel and P. Mundur. An n-gram based approach for finding the repeating patterns in musical data. In M. H. Hamza, ed., *EuroIMSA*, pp. 407–412. IASTED/ACTA Press, 2005.

[25] D. Pyle. *Data Preparation for Data Mining*. Morgan Kaufmann, San Francisco, 1999.

[26] J. A. Russel. A circumplex model of affect. *Journal of Personality and Social Psychology*, 39(6):1161–1178, 1980.

[27] N. J. Salkind, ed. *Encyclopedia of Measurement and Statistics*. SAGE Publications, Inc., 2007.

[28] G. Tzanetakis and P. Cook. Musical genre classification of audio signals. *IEEE Transactions on Speech and Audio Processing*, 10(5):293–302, 2002.

[29] I. Vatolkin. *Improving Supervised Music Classification by Means of Multi-Objective Evolutionary Feature Selection*. PhD thesis, Department of Computer Science, TU Dortmund, 2013.

[30] I. Vatolkin, A. Nagathil, W. Theimer, and R. Martin. Performance of specific vs. generic feature sets in polyphonic music instrument recognition. In R. C. P.

et al., ed., *Proc. of the 7th International Conference on Evolutionary Multi-Criterion Optimization (EMO)*, volume 7811, pp. 587–599. Springer, 2013.

[31] I. Vatolkin, M. Preuss, and G. Rudolph. Training set reduction based on 2-gram feature statistics for music genre recognition. Technical report, TR13-2-001, Faculty of Computer Science, Technische Universität Dortmund. Presented at the 2012 Workshop on Knowledge Discovery, Data Mining and Machine Learning (KDML), 2013.

[32] I. Vatolkin, W. Theimer, and M. Botteck. Partition based feature processing for improved music classification. In W. A. G. et al., ed., *Proc. of the 34th Annual Conference of the German Classification Society (GfKl), 2010*, pp. 411–419. Springer, Berlin Heidelberg, 2012.

[33] I. Vatolkin, W. Theimer, and G. Rudolph. Design and comparison of different evolution strategies for feature selection and consolidation in music classification. In *Proc. of the IEEE Congress on Evolutionary Computation (CEC)*, pp. 174–181. IEEE, 2009.

[34] K. West and S. Cox. Finding an optimal segmentation for audio genre classification. In *Proc. of the 6th International Conference on Music Information Retrieval (ISMIR)*, pp. 680–685. University of London, 2005.

Chapter 15

Feature Selection

IGOR VATOLKIN
Department of Computer Science, TU Dortmund, Germany

15.1 Introduction

As we discussed in Chapters 11 and 12, the task of classification methods is to organize and categorize data based on features and their statistical characteristics. There exist many music classification scenarios, from classification into genres and styles to identification of instruments and cover songs, recognition of mood and tempo, and so on. Even similar tasks within the same application scenario may require different features. For example, for genre classification of classic against pop, the relative share of percussion may be a relevant feature. If popular music contains mostly rap songs, more relevant features may describe vocal characteristics, and for progressive rock, the important features may rather describe harmonic properties.

A manual solution is to select the best-suited features for each task with the help of a music expert, who would carefully analyze the data provided for each class. This approach requires high manual effort with no guarantee of the optimality of the selected features. Another method is to create a large feature set only once for different classification tasks. Then, some of these features would surely be relevant for a particular classification task. As we will discuss later, too many irrelevant features (which would be an unavoidable part of this large set) often lead to overfitting: the performance of classification models suffers because some of the irrelevant features are then identified as relevant.

In this chapter we address the search for relevant features by means of *automatic feature selection*, an approach to identify most relevant features and to remove redundant and irrelevant ones before the training of classification models. As the terms "relevance," "redundancy," and "irrelevance" are central for feature selection, we shall start with their definitions in Section 15.2, before the general scope of feature selection will be discussed in Section 15.3. Section 15.4 outlines the design steps for a feature selection algorithm. Several functions for the measurement of feature relevance are listed in Section 15.5, followed by three selected algorithms in Section 15.6. Multi-objective feature selection is introduced in Section 15.7.

15.2 Definitions

Definition 15.1 (Relevant Feature). In the context of classification, a feature x is called *relevant* if its removal from the full feature set $\mathscr{F}$ will lead to a decreased classification performance:

- $P(y = \hat{y}|\mathscr{F}) < P(y = \hat{y}|\mathscr{F} \setminus \{x\})$ and
- $P(y \neq \hat{y}|\mathscr{F}) > P(y \neq \hat{y}|\mathscr{F} \setminus \{x\})$,

where y describes the labeled (correct) class, $\hat{y}$ the predicted class, and $P(y = \hat{y}|\mathscr{F})$ is the probability that the predicted label $\hat{y}$ is the correct label y when all features $\mathscr{F}$ are used.

The first basic goal of feature selection is to keep relevant features. The second one is to remove non-relevant features which may be categorized as redundant or irrelevant. The following definitions describe the difference between the two latter kinds of features.

Definition 15.2 (Redundant Feature). If a feature subset $\mathscr{S}$ exists which does not contain x so that after the removal of $\mathscr{S}$ the non-relevant feature x would become relevant, this feature is called *redundant*:

- $P(y = \hat{y}|\mathscr{F}) = P(y = \hat{y}|\mathscr{F} \setminus \{x\})$ and
- $\exists \mathscr{S} \subseteq \mathscr{F}, \{x\} \cap \mathscr{S} = \emptyset : P(y = \hat{y}|\mathscr{F} \setminus \mathscr{S}) \neq P(y = \hat{y}|\mathscr{F} \setminus \{\mathscr{S} \cup \{x\}\})$.

This means that a redundant feature may be removed without the reduction of performance of a classifier, but may contain by itself some relevant information about the target class (e.g., have significantly different distributions for data of different classes). Often, strongly correlated features are redundant. However, this is not always true (see Example 15.1). Note that correlation is measured by the empirical correlation coefficient r_{XY} as introduced in Definition 9.20.

Example 15.1 (Redundancy and Correlation). *Figure 15.1 (a) plots the temporal evolution of two features which are aggregated in classification windows of 4 s and have a high negative correlation: $r_{XY} = -0.94$ for Schubert, "Andante con moto" (left subfigure) and $r_{XY} = -0.86$ for the Becker Brothers, "Scrunch" (middle subfigure). The distribution of these features is visualized in the right subfigure. The two features have a high grade of redundancy: linear separation of classes with either a vertical or a horizontal dashed line with individual projections of features leads to only few errors. Figure 15.1 (b) also shows two highly anti-correlated features: $r_{XY} = -0.98$ for Schubert, "Andante con moto" and $r_{XY} = -0.82$ for the Becker Brothers, "Scrunch." However, the combination of the two features leads to an almost perfect linear separation of the two classes in contrast to the individual projections of these features.*

Definition 15.3 (Irrelevant Feature).
In contrast to redundant features, the removal of an *irrelevant* feature x does not affect the performance of a classifier:

- $P(y = \hat{y}|\mathscr{F}) = P(y = \hat{y}|\mathscr{F} \setminus \{x\})$ and

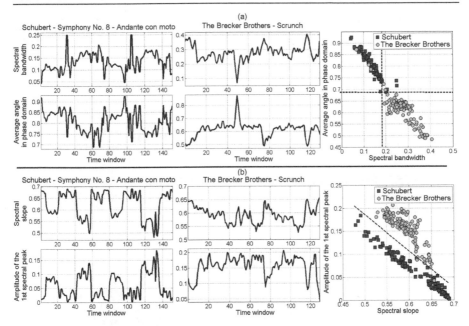

Figure 15.1: Examples of correlated features which have (a) high individual redundancy and (b) are relevant in their combination. The original values were normalized, and the mean value was estimated for each classification window of 4 s.

- $\forall \mathcal{S} \subseteq \mathcal{F}, \{x\} \cap \mathcal{S} = \emptyset : P(y = \hat{y} | \mathcal{F} \setminus \mathcal{S}) = P(y = \hat{y} | \mathcal{F} \setminus \{\mathcal{S} \cup \{x\}\}).$

Examples of a *rather irrelevant* and two *rather relevant* features for the classification between classic and rock genres are provided in Figure 15.2. Here, normal distributions (cf. Chapter 9) of three features are plotted for a classical music piece (Chopin) and a hard rock song (AC/DC). A *perfectly irrelevant* feature would have the same distribution for both songs. However, this may occur rather seldom in practice. The distribution of the chromagram tone with the maximum strength is very similar for both songs and the relevance for the related classification task is very low. On the other side, distributions of energy and distances in phase domain, Equation (5.38), are well discriminable and may be helpful for the classification of classic against rock.

Features which are individually irrelevant may become relevant together as illustrated in Figure 15.3. The left subplot presents a theoretical example ("2 x 2 chessboard"). The right subplot contains distributions of two features for two classical and two pop music pieces. The combination of the two features helps to improve the classification performance, although individual projections of these features are not very informative. Therefore it may be especially useful to estimate not only individual relevance of features, but also relevance of sets of features. To measure relevance, redundancy, and irrelevance of a *feature set* $\mathcal{X}$, $\{x\}$ should be replaced with $\mathcal{X}$ in the Definitions 15.1–15.3.

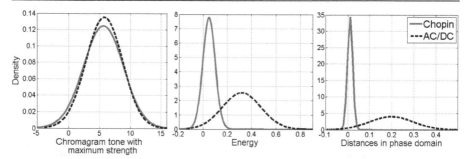

Figure 15.2: Examples of rather irrelevant and rather relevant features. Thick lines: normal distributions of features for Chopin, "Mazurka in e-Moll Op. 41 No. 2". Dashed lines: distributions for AC/DC, "Back in Black."

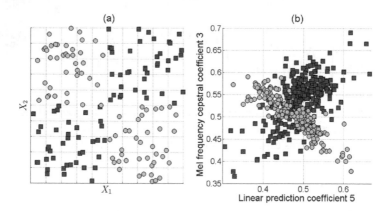

Figure 15.3: Examples of pairs of features which are individually irrelevant but relevant in their combination. Squares and circles represent instances of two different classes. (a): Theoretical example constructed after [8, p. 10] (features are uniformly distributed between 0 and 1). (b): Distribution of two audio signal features for two classical music pieces (squares, Jean Sibelius, "Symphony No. 5 Es-major op. 82" and Georg Friedrich Haendel, "Opus 4 No. 1") and two pop music pieces (circles, Herbert Grönemeyer, "Mambo" and Sonny Rollins, "No Moe").

Let m be a measure which describes the success of feature selection (see Section 13.3 for lists of measures for and beyond classification performance).

Definition 15.4 (Feature Selection).
The task of *feature selection* is to find the optimal set of features represented by $\boldsymbol{q}^*$:

$$\boldsymbol{q}^* = \arg\min_{\boldsymbol{q}} \left[m\left(\boldsymbol{y}, \hat{\boldsymbol{y}}, \Phi(\mathscr{F}, \boldsymbol{q})\right) \right],$$

where $\mathscr{F}$ is the full feature set, $\Phi(\mathscr{F}, \boldsymbol{q})$ is the set with selected features, and $\boldsymbol{q}$ is a binary vector of length F which indicates whether a feature u is selected ($q_u = 1$) or not ($q_u = 0$). Labeled class indicators **y** and predicted classes **ŷ** are required as inputs for the estimation of classification performance on selected data instances (but may be unnecessary for other evaluation measures).

Evaluation measures which have to be maximized, such as accuracy, can be easily adapted to Definition 15.4 as follows:

$$\max\{m(x)\} = -\min\{-m(x)\} \Rightarrow \arg\max\{m(x)\} = \arg\min\{-m(x)\}. \quad (15.1)$$

According to the classification of feature processing methods in Section 14.1, feature selection is a processing method which operates only on the feature dimension, cf. Figure 14.1 (d).

15.3 The Scope of Feature Selection

The basic task of feature selection is to remove non-relevant features from the full feature set. In particular, too many irrelevant features lead to a decrease of performance of classification methods. Classification models are typically created from a limited set of data instances (training set), and some irrelevant features may be by chance discriminative for this particular training set. The probability of this situation increases when a very large number of features are available in contrast to a rather small training set. A proper application of feature selection may decrease the danger of overfitting, i.e. to avoid the situation when a classification model performs well on some sets of music pieces, but the classification quality suffers for other sets (see also the discussion in Chapter 13).

Another important topic for any automatic classification approach are the corresponding computational costs. Feature selection helps to reduce them in two ways. First, the representation of data by feature vectors with fewer entries helps to reduce the required storage space. This also holds for classification models trained with fewer features. That advantage can be simply illustrated for music classification: a music piece of 4 minutes with a sampling rate of 44,100 Hz corresponds to a time series of 10,584,000 values. A feature extracted from non-overlapping time windows with 512 samples then has 20,671 values. Because many audio signal features were developed in recent years and also parameters of the extraction for an individual feature may vary, the representation of a single music piece may contain millions of entries. Scaled to music collections of thousands of music pieces, this leads to very large demands for disc space.

Secondly, the training of classification models and the application of a classification rule may be realized faster if the number of features is significantly reduced. This is important for large music collections, but also if new classes are created over and over again (after updates of listener preferences), or if new songs to be classified are frequently added to a collection.

It is less obvious that feature selection may also help to better understand the classification tasks. If features are interpretable and have a relation to music theory (harmony and melody characteristics, emotions, etc.), the selected set of relevant features will serve as the information source for a further (theoretical) analysis of the properties of the related class. This advantage is lost after feature processing methods which do not keep the interpretability, such as Principal Component Analysis (see Definition 9.48): even if original features have some musical meaning, these properties may be hidden in a new feature space after the transform of dimensions.

Finally, it is worth highlighting the role of feature selection when features are automatically constructed for a concrete classification task (see Section 14.5). Feature selection then helps to restrict the number of features in a typically very large search domain because of many available operators and transforms.

15.4 Design Steps and Categorization of Methods

Langley proposed in [11] four basic steps required for the design of a feature selection algorithm.

First, the *starting point* describes the initial set of features to evaluate: either an empty feature set, the full feature set, or some part of the full feature set.

Second, the *organization of search* manages the iterative update of feature sets for their evaluation. The most straightforward and costly approach is to evaluate all possible non-empty $2^F - 1$ feature subsets of the F features. Another option is the branch-and-bound search [13]. Two further and complementary methods are to apply forward feature selection adding features one by one starting with an empty feature set, or removing them starting with the full feature set (backward feature selection). The decisions as to which feature to add or to remove should be based on some relevance criterion, e.g., correlation with a target variable, information gain (see Section 15.5), or classification error. Both approaches may be combined as a floating search with alternating forward and backward steps [14]. As we have seen above, some less relevant features may become relevant by their combination. A heuristic search enables the addition and/or removal of several features at the same time based on some random component, as in evolutionary feature selection presented in Section 15.6.3. This may be especially helpful if many irrelevant features become relevant by combination (recall Figure 15.3).

The third design step for a feature selection algorithm is to choose the *evaluation strategy*. Reference [8] distinguishes between three general strategies: *filters*, *wrappers*, and *embedded methods*. Filters evaluate feature sets without the training of classification methods by means of a relevance criterion like correlation with the target. Wrappers rate the feature sets after the training of a classification model and measure the quality of classification. They often achieve better results but have a larger danger of overfitting and are typically significantly slower than filters. Embedded methods are directly integrated into a classification algorithm. They may best improve the performance of a particular classifier. Their disadvantage is that they cannot be simply applied to other classification methods (as wrappers). This is especially important if an ensemble of several different classifiers performs better than an individual classification method, as observed, e.g., for music instrument recognition [24].

The fourth and the last decision is to define the *stopping criterion*. For instance, the number of iteration steps can be simply limited, the process can be stopped if no significant improvements are achieved anymore, or if these improvements remain below a selected threshold.

15.5 Ways to Measure Relevance of Features

There exist a lot of different options to measure relevance of individual features and feature sets. In this section we provide a list of frequently applied relevance functions based on correlation, probability distributions, and information theory.

15.5.1 Correlation-Based Relevance

The ability of a feature x_u (represented as a row vector of the feature matrix X with feature values for a set of labeled data instances) to explain the target class y is often measured based on the correlation between x_u and y, e.g., by the empirical correlation coefficient (see Definition 9.20). In that case only linear relationships between the two variables (feature and class) are taken into account. If some interpretable high-level music descriptors belong to the set of features (shares of particular instruments in a chord, harmonic properties, etc.), it may be preferable to measure these relationships on ordinal scale. Then, Spearman's rank correlation coefficient can be estimated (Definition 9.23).

For a set $\mathscr{S}$ of k features, their inter-group correlation can be estimated together with the correlation to the target class for the measurement of the redundancy [9]:

$$c(\mathscr{S},y) = \frac{k \cdot \overline{\rho}(\mathscr{S},y)}{\sqrt{k + (k-1) \cdot \overline{\rho}(\mathscr{S})}}. \tag{15.2}$$

Here, $\overline{\rho}(\mathscr{S},y)$ denotes the mean correlation between all features in $\mathscr{S}$ and the label vector y, and $\overline{\rho}(\mathscr{S})$ the mean inter-correlation of all features in $\mathscr{S}$.

Another approach proposed in [10] is to calculate the distance between instances which are close to each other in the feature space but belong to different classes (the corresponding Relief algorithm is described below in Section 15.6.1). For a fixed number of iterations $t = 1, ..., I$, the weight of the u-th feature W_u is updated according to:

$$W_u(t) = W_u(t-1) - \left(X_{uw} - X_{(nearest\text{-}hit)w}\right)^2 + \left(X_{uw} - X_{(nearest\text{-}miss)w}\right)^2, \tag{15.3}$$

where w is the index of a randomly selected instance, *nearest-hit* corresponds to the instance which belongs to the same class as the instance w and is closest to the instance w according to the Euclidean distance in feature space, and *nearest-miss* is the instance which belongs to another class and is closest to the instance w. The target of this formula is to increase weights for features which have similar values for instances of the same class and different values for instances of different classes. After I iterations, the relevance is normalized to:

$$R_{RELIEF}(x_u, y) = \frac{W_u(I)}{I}. \tag{15.4}$$

The estimation of distances between k- instead of 1-nearest neighbors leads to another relevance function [8]:

$$R_{RELIEF_K}(\boldsymbol{x}_u, \boldsymbol{y}) = \frac{\sum_{j=1}^{W} \sum_{k=1}^{K} |X_{ij} - X_{(nearest\text{-}miss_k)j}|}{\sum_{j=1}^{W} \sum_{k=1}^{K} |X_{ij} - X_{(nearest\text{-}hit_k)j}|}. \tag{15.5}$$

Similar to $\overline{p}(\mathscr{S}, \boldsymbol{y})$, inter-group correlation between features can be estimated with Relief [9].

15.5.2 Comparison of Feature Distributions

Another group of relevance measures is based on the properties of feature distributions, whose differences can be validated with statistical tests. Recall the examples from Figure 15.2 with the distributions of the same feature for data of different classes. Let us assume that some feature has a Gaussian distribution. To distinguish between exactly two classes, the t-test statistic can be estimated according to Definition 9.28, where X should correspond to values of the feature for N instances belonging to the first class, and Y to values of this feature for M instances which belong to the second class. As the null hypothesis H_0 it is suggested that there is no significant difference between two distributions, i.e. the feature is irrelevant (Figure 15.2, left subfigure). If H_0 is rejected, the feature is reported to be relevant. The p-value can be used for the exact measurement of a relevance function.

Typically, we cannot assume that the extracted features are normally distributed. Then, the (less powerful) nonparametric Wilcoxon rank-sum statistic can be calculated; see Definition 9.29.

For a large number of features and the estimation of relevance by means of statistical tests, the probability to reject at least one of the hypotheses incorrectly can be very high. Then, the Bonferroni correction can be applied; see the discussion about multiple testing at the end of Section 9.7.

The so-called success of a Bayes classifier (cf. Section 12.4.1) can also be treated as a relevance function. For a feature x_u and classes $y^1, \ldots, y^G$, the output class is assigned to

$$\hat{y}(x_u) = \underset{c \in \{1, \ldots, G\}}{\arg\max} P(y^c(x_u)), \tag{15.6}$$

where $P(y^c)$ is the probability of class y^c. Let I_c denote the union of all intervals for which $\hat{y}(x_u) = c$. For the example in Figure 15.2, right subfigure, $I_1 \approx (-0.029, 0.04)$ for Chopin ($c = 1$) and $I_2 \approx [-0.1, -0.029] \cup [0.04, 0.5]$ for AC/DC ($c = 2$). Then, the success of the classifier can be measured as:

$$R_{NB} = \sum_{c=1}^{G} P(x_u \in I_c, y^c) = \sum_{c=1}^{G} P(x_u \in I_c | y^c) \cdot P(y^c). \tag{15.7}$$

$P(y^c)$ can be simply estimated as the share of the instances belonging to y^c in the training set, and $P(x_u \in I_c | y^c)$ corresponds to the sum of areas (integrals) below the distribution of x_u for the union of intervals I_c.

If a feature x_u and a class y^c are statistically independent of each other (i.e., the feature is irrelevant), $P(x_u, y^c) = P(x_u) \cdot P(y^c)$ (cf. Definition 9.2), and the difference between the joint and the independent distributions of a feature and a class is equal to zero. The Kolmogorov distance sums up these differences:

$$R_{KOL}(x_u) = \sum_u \sum_{c=1}^{G} |P(x_u, y^c) - P(x_u) \cdot P(y^c)| = \sum_u \sum_{c=1}^{G} |P(x_u|y^u) \cdot P(y^c) - P(x_u) \cdot P(y^c)|$$

(15.8)

so that $R_{KOL} = 0$ for a completely irrelevant feature.

15.5.3 Relevance Derived from Information Theory

Other commonly applied relevance functions are based on information theory investigated by Shannon in the 1950s [19]. The basic idea behind this theory is that if there are G independent, different possible messages of equal probability, $\log_2 G$ bits (information) are required to encode a message.

If training data contains W instances of G different classes, the general probability to draw an instance of the class c can be roughly estimated as the share of the corresponding instances:

$$P(c) = \frac{\sum_{\substack{w=1 \\ y_w=c}}^{W} 1}{G}.$$

(15.9)

Then, we can calculate the amount of information required to describe each class using its probability as a weight:

$$H(y) := -\sum_{c=1}^{G} P(c) \cdot \log_2 P(c).$$

(15.10)

$H(y)$ is referred as *entropy*, which measures the average amount of information required for the identification of a class for W instances.

For the feature x_u, a relevance function called *information gain* is measured as the difference between the general entropy and the entropy after the estimation of the distribution of this feature:

$$IG(y, x_u) = H(y) - H(y|x_u),$$

(15.11)

where

$$H(y|x_u) := -\sum_u P(x_u) \cdot \sum_{c=1}^{G} P(c|x_u) \cdot \log_2 P(c|x_u).$$

(15.12)

The decision tree classifier C4.5 uses the information gain ratio as a relevance function for the decision to select a feature in the split node [15] (see also Section 12.4.3):

$$IGR(y, x_u) = \frac{H(y) - H(y|x_u)}{H(x_u)}.$$

(15.13)

During the construction of a decision tree, in each node the information gain is maximized for both subtrees, using some threshold value of feature x_u for the optimal splitting.

Another variant of information gain is *symmetrical uncertainty*, which reduces the bias of features with more values and is normalized to $[0; 1]$:

$$SU(\boldsymbol{y}, \boldsymbol{x}_u) = 2 \frac{IG(\boldsymbol{y}, \boldsymbol{x}_u)}{H(\boldsymbol{y}) + H(\boldsymbol{x}_u)}. \tag{15.14}$$

Symmetrical uncertainty was applied in [9] for the measurement of correlation between features as well as for the correlation between a feature and a class.

A general framework for simultaneous reduction of redundancy between selected features and maximization of relevance to a class is discussed in [5]. The minimal redundancy condition W_H and the maximal relevance condition V_H are defined as:

$$W_H = \frac{1}{|\Phi(\mathscr{F}, \boldsymbol{q})|^2} \sum_{k=1}^{F} q_k \cdot \sum_{u=1}^{F} q_u \cdot H(\boldsymbol{x}_u, \boldsymbol{x}_k), \tag{15.15}$$

$$V_H = \frac{1}{|\Phi(\mathscr{F}, \boldsymbol{q})|} \sum_{u=1}^{F} q_u \cdot H(\boldsymbol{x}_u, \boldsymbol{y}), \tag{15.16}$$

where $\boldsymbol{q}$ is a binary vector indicating the selected features which build the set $\Phi(\mathscr{F}, \boldsymbol{q})$ according to Definition 15.4. In [5] it is proposed to maximize $V_H - W_H$ or V_H / W_H.

15.6 Examples for Feature Selection Algorithms

In this section, we provide implementation details of three feature selection algorithms. Relief is a fast filter approach which uses relevance measures defined in Equations (15.4) and (15.5). The floating search is a strategy which can be applied for filters and wrappers and is often superior to simpler forward or backward selection, because it allows both extension and reduction of a feature set with regard to a relevance function. An evolutionary search is even more flexible; the stochastic component helps to overcome local optima.

15.6.1 Relief

In Algorithm 15.1, the pseudocode of the Relief algorithm [10] is sketched. The original procedure is extended for K neighbors using Equation (15.5). The following inputs and parameters of the algorithm are required: $\boldsymbol{X} \in \mathbb{R}^{F \times W}$ is the matrix of F feature dimensions and W classification frames, $\boldsymbol{y} \in \mathbb{R}^{W}$ are the binary class relationships, I the number of iterations of Relief, and τ the threshold value to decide if a feature is relevant. The binary vector $\boldsymbol{q}$, which indicates the features to select, is reported as output.

The algorithm distinguishes between "nearest-hits" (classification instances which are closest to the selected instance and belong to the same class) and "nearest-misses" (instances which are closest to the selected instance but belong to another class).

First, the weights of nearest-hits and nearest-misses are initialized to zero (lines 1–5). Then, for I iterations, a random instance r is selected (line 7), and nearest-hits and nearest-misses are estimated (lines 8–16). The numerator and denominator

---

Algorithm 15.1: Relief for K Neighbors

---

input : X, y, I, K, τ
output: q

1   double$[F][2]$ W ; // for numerator and denominator of Equation (15.5)
2   **for** $u = 1$ **to** F **do**
3     $W[u][1] = 0;$
4     $W[u][2] = 0;$
5   **end**
6   **for** $t = 1$ **to** I **do**
7     $r = \text{random}(1, W);$
8   double$[K][F]$ $N_{pos} = \text{getPositiveNeighbors}(x_r, K);$
9   double$[K][F]$ $N_{neg} = \text{getNegativeNeighbors}(x_r, K);$
10   **if** $y_r == 1$ **then**
11   double$[K][F]$ $N_H = N_{pos}$; // nearest-hits
12   double$[K][F]$ $N_M = N_{neg}$; // nearest-misses
13   **else**
14   double$[K][F]$ $N_H = N_{neg}$; // nearest-hits
15   double$[K][F]$ $N_M = N_{pos}$; // nearest-misses
16   **end**
17   double$[F]$ $d_M = 0$; // distances to nearest-hits
18   double$[F]$ $d_H = 0$; // distances to nearest-misses
19   **for** $u = 1$ **to** F **do**
20   **for** $k = 1$ **to** K **do**
21         $d_M[u] = d_M[u] + \text{abs}(X[r][u] - N_M[k][u]);$
22         $d_H[u] = d_H[u] + \text{abs}(X[r][u] - N_H[k][u]);$
23   **end**
24       $W[u][1] = W[u][1] + d_M[u];$
25       $W[u][2] = W[u][2] + d_H[u];$
26   **end**
27   **end**
28   double$[F]$ w ; // for Equation (15.5)
29   double$[F]$ q ; // output
30   **for** $u = 1$ **to** F **do**
31     $w[u] = W[u][1]/W[u][2];$
32   **if** $w[u] \geq \tau$ **then**
33       $q[u] = 1$; // selected feature
34   **else**
35       $q[u] = 0$; // removed feature
36   **end**
37   **end**

---

of Equation (15.5) are calculated in lines 19–26 (the sum of distances between the selected instance and K neighbors for F feature dimensions). Finally, the weights of each feature w are estimated (line 31), and the decision if the feature should be selected is made based on the threshold τ (lines 32–36).

15.6.2 Floating Search

In the floating search, the starting point is usually the empty set. Features are added during the forward stage and removed during the backward stage. For both stages, a greedy search may be applied: first all features are sorted according to some relevance function and then are added one by one during the forward stage for a given number of iterations. The task of the following backward stage is to remove the redundant features. The actually selected feature set must be also evaluated by means of some relevance function, such as redundancy and relevance conditions from Equations (15.15), (15.16) (as a filter evaluation strategy), or a classification performance measure after the validation of a model created with selected features (as a wrapper evaluation strategy). It is also possible to switch randomly between forward and backward stages.

The algorithm terminates when a given number of iterations is achieved, or when no significant increase of the relevance is measured along the given number of iterations. In [21], it is proposed to use an archive which keeps best found feature sets of all previously examined set sizes.

15.6.3 Evolutionary Search

Evolutionary algorithms (EAs) are heuristics which are particularly useful for the optimization of complex problems where the functions to minimize are multi-modal, non-convex, and/or not differentiable (see Chapter 10). EAs were first applied for feature selection in [20].

Algorithm 15.2 provides a pseudocode for evolutionary feature selection. The solutions are stored in the population matrix $P \in \mathbb{B}^{(\mu+\lambda)\times F}$, where μ corresponds to the actual number of solutions, and λ is the number of new generated solutions in each iteration. Note that solutions $q_1, ..., q_{(\mu+\lambda)}$ are stored as columns in P. At the beginning, μ solutions are initialized randomly (lines 3–13). Then, from randomly selected pairs of (so-called parent) solutions λ, so-called offspring solutions are generated (lines 17–21). In general, the number of parents may vary. In order to generate the offsprings, two (so-called genetic) operators are applied: first, a new solution receives some properties from the first and other from the second parent. In lines 22–29, this is achieved by means of a (so-called) uniform crossover operator, where bits are "inherited" from both parents with equal probability. Afterwards, a (so-called) mutation operator is applied for the stochastic exploration of the search space (lines 30–35). Here, each bit is flipped with the probability $1/F$.

After the evaluation of the offspring population by means of the evaluation function m, the μ best solutions are kept for the next iteration step (s stores the indices of individuals after sorting, line 38). The evolutionary loop continues until the final

---

Algorithm 15.2: Evolutionary Feature Selection

---

input : X, y, μ, λ, I
output: P

```
1  boolean[μ+λ][F] P ; // population of μ parents and λ offsprings
2  double[μ+λ] m ; // for evaluation of parents and offsprings
   // initialization of the first parent population
3  for k = 1 to μ do
4      for u = 1 to F do
5          r = random(0,1);
6          if r= 0 then
7              P[k][u] = 0;
8          else
9              P[k][u] = 1;
10         end
11     end
12     m[k] = getEvaluationFunction(P[k]);
13 end
14 int t = 1;
   // evolutionary loop
15 while t < I do
       // generate offsprings
16     for k = 1 to λ do
17         int par₁ = random(1,μ);
18         int par₂ = random(1,μ − 1);
19         if par₂ ≥ par₁ then
20             par₂ = par₂ + 1 ; // the two parents should differ
21         end
           // crossover
22         for u = 1 to F do
23             r = random(0,1);
24             if r= 0 then
25                 P[μ+k][u] = P[par₁][u];
26             else
27                 P[μ+k][u] = P[par₂][u];
28             end
29         end
           // mutation
30         for u = 1 to F do
31             r = random(1,F);
32             if r= F then
33                 P[μ+k][u] = 1 − P[μ+k][u];
34             end
35         end
36         m[μ+k] = getEvaluationFunction(P[μ+k][k]);
37     end
       // selection
38     int[μ+λ] s = sort(m);
39     boolean[μ+λ][F] P_NEW ; // next population
40     for k = 1 to μ do
41         P_NEW[k] = s[k];
42     end
43     P = P_NEW;
44 end
```

---

number of iterations I is achieved. Another stopping criterion may be the limit of runtime or the stagnation in the improvement of best found solutions over several iterations.

Another option for the mutation is to prioritize the switching of "bits off" because one of the basic targets of feature selection is to reduce the number of used features. This can be done using an asymmetric bit flip with a probability $P_M(u)$ for each position $u \in \{1, ..., F\}$ as applied in [3]:

$$P_M(u) = \frac{\gamma}{F} \cdot (|q_u - P_{01}|). \qquad (15.17)$$

Here, γ is the general strength of mutation equal to the expected number of flips for a symmetric mutation, q_u is the binary selection value 0 or 1 for each position u, and P_{01} controls the balance between the switching of bits on and off and should be set to a value below 0.5, e.g., 0.05.

Further, we may increase the probability for switching "bits on" for features which stronger correlate with the classification class:

$$P_M(u) = \frac{\gamma}{F} \cdot (|q_u - P_{01}|) \cdot (|q_u - |\rho(\boldsymbol{x}_u^T, \boldsymbol{y})||), \qquad (15.18)$$

where $\boldsymbol{x}_u^T$ is the vector of values of feature u for the training set, $\boldsymbol{y}$ are the corresponding labels, and ρ is the correlation coefficient (Definition 9.17). This mutation operator was particularly successful in [3].

15.7   Multi-Objective Feature Selection

As in many optimization problems and real-world applications, the selection of feature sets according to a single criterion may lead to a decreased performance w.r.t. other criteria. For instance, a "perfect" feature set for the prediction of some class in a specific music collection may perform poorly on other music pieces and thus have a low generalization ability. Also, features which are particularly relevant may be very costly to extract, require more storage space than other characteristics, or may be less interpretable thus leading to classification models which can be hardly understood by music scientists, critics, and listeners. Addressing feature selection as a multi-objective problem (see Section 10.4), we may overcome such "over-optimization."

Definition 15.5 (Multi-Objective Feature Selection). The task of *Multi-Objective Feature Selection* (MO-FS) is to find the optimal set of features with regard to K evaluation functions $m_1, ..., m_K$, which should be minimized; cf. Definition 15.4:

$$\boldsymbol{q}^* = \underset{q}{\arg\min}[m_1(\mathbf{y}, \hat{\mathbf{y}}, \Phi(\mathscr{F}, \mathbf{p})), ..., m_K(\mathbf{y}, \hat{\mathbf{y}}, \Phi(\mathscr{F}, \boldsymbol{q}))].$$

Example 15.2 (Multi-Objective Feature Selection). *Figure 15.4 illustrates the optimization of four pairs of criteria; see Section 13.3 for definitions of the corresponding measures. The task was to identify the genre Electronic after [23]. Each point is associated with a feature set where the classification was done with a random forest or a linear support vector machine (see Chapter 12).*

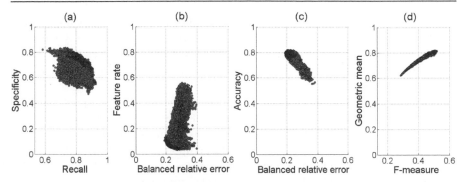

Figure 15.4: Four pairs of simultaneously optimized objectives. The feature selection was applied to the recognition of the genre Electronic. (a): maximization of both measures; (b): minimization of both measures; (c): minimization of error and maximization of accuracy; (d): maximization of both measures.

For the two left subfigures, the multi-objective feature selection leads to larger sets of compromise solutions. In Figure 15.4 (a), the goal was to maximize recall m_{REC} and specificity m_{SPEC}. The compromise non-dominated solutions are distributed between the points $[m_{REC} = 0.732; m_{SPEC} = 0.834]$ and $[m_{REC} = 0.927; m_{SPEC} = 0.557]$. In Figure 15.4 (b), the goal was to minimize the feature reduction rate m_{FRR}, Equation (14.12), and to maximize the balanced relative error m_{BRE}. The best compromise solutions are distributed between sets of 68 features ($m_{FRR} = 0.107$) and 21 features ($m_{FRR} = 0.033$), where m_{BRE} increases from 0.19 to 0.273. For mobile devices with limited resources, the reduction of requirements on storage space and runtime may be relevant (models with fewer features are trained faster and more important, classify new music faster) so that an increase of the classification error would be acceptable to a certain level.

In Figures 15.4 (c,d), the multi-objective optimization makes less sense because of stronger anti-correlation between balanced relative error and accuracy and the correlation between F-measure and geometric mean: the optimization of one of the two criteria may be sufficient.

The challenge is to identify those evaluation criteria which play the essential role in a concrete classification scenario. Such a decision may be supported by means of empirical validation using Equations (10.4) and (10.5) in Section 10.4.

The evaluation functions for MO-FS can be chosen from groups of measures presented in Chapter 13, e.g., for the simultaneous minimization of resource demands and maximization of classification quality and listener satisfaction. Even closely related measures may be relevant and only weakly correlated, such as classification performance on positive and negative examples. The estimation of a single combined measure such as balanced error rate or F-measure may not be sufficient here because of the fixed balance between the original measures. For some music-related classification and recommendation scenarios, the desired balance between *surprise* and *safety* cannot be always identified in advance. Surprise means that the rate of false positives is accepted below some threshold (the listener appreciates the identification of some negative music pieces as belonging to a class). Safety means that for a very low rate of false positives, a higher rate of false negatives is accepted to a certain level (it is desired to keep the number of recommended songs which do not belong to a class as small as possible).

Evolutionary algorithms are very well suited for MO-FS: the optimization of a population of solutions helps to search for not only one but for a set of compromise solutions. Feature selection w.r.t. several objectives becomes a very complex problem for large feature sets, and stochastic components (mutation, self-adaptation) are particularly valuable to overcome local optima. The first application of EA for MO-FS was introduced in [6] and for music classification in [26].

An evolutionary loop for FS as presented in Algorithm 15.2 can be simply extended to multi-objective FS through the estimation of several fitness functions for the selection of individuals. Then, a metric such as hypervolume (Definition 10.7) may measure the quality and the diversity of solutions in the search space. The comparison of single solutions (sets of features) can be done based on Pareto dominance (Definition 10.3) and algorithms with a fast non-dominated sorting, such as SMS-EMOA (Algorithm 10.9), may be useful for the efficient search for trade-off feature sets.

15.8   Further Reading

The relevance of individual features and feature sets can be measured by functions other than those mentioned in Section 15.5. Several statistical tests are discussed in [8] for that purpose and examples of their application are the selection of features for classification into genres [4] and the recognition of instruments [1]. Another interesting proposal is to generate features with a random statistical distribution as "probes" and to discard those features whose relevance would be below the estimated relevance of probe features [22].

It can also be distinguished between features particularly suited for the recognition of a specific class against features which are relevant for different tested classes ("allrounder" features against "specialists" for the recognition of genres [12] and the identification of "generic" and "specific" features for instrument recognition [25]). For multi-class problems such feature evaluation is addressed in [28].

Not only were many other algorithms for feature selection developed, many extensions to the methods discussed in Section 15.6 are available, albeit often rather simple or longer established methods are applied for feature selection in music classification. Further improvements and the analysis of Relief-based methods are discussed in [17]. Several variants of the floating search are described in [8]. One of the extensions to EAs for feature selection is their combination with a local search [27].

High-level, meaningful audio features may be derived from other characteristics, e.g., moods from signal features [18]. These features may themselves act as the source of information for the prediction of further descriptors, like the recognition of instruments based on low-level timbre descriptors and further prediction of moods and genres based on instrumentation. The application of machine learning for the subsequent extraction of music audio features on several levels and the optimization with multi-objective feature selection ("sliding feature selection") was proposed in [23] for a more interpretable music classification into genres and styles. This procedure is briefly discussed in Section 8.2.2.

Last but not least, it is important to mention the importance of a proper evaluation of feature sets. Reference [16] pointed out the danger of overfitting when cross-validation is applied to evaluate feature subsets. The re-evaluation of a previous study on music classification with an independent test set is described in [7]. A possible way to reduce the danger of over-optimization is to distinguish between inner and outer validation loops [2]. For an introduction into resampling methods and evaluation measures, see Chapter 13.

Bibliography

[1] B. Bischl, M. Eichhoff, and C. Weihs. Selecting groups of audio features by statistical tests and the group lasso. In *Proc. of the 9. ITG Fachtagung Sprachkommunikation*. VDE Verlag, Berlin, Offenbach, 2010.

[2] B. Bischl, O. Mersmann, H. Trautmann, and C. Weihs. Resampling methods for meta-model validation with recommendations for evolutionary computation. *Evolutionary Computation*, 20(2):249–275, 2012.

[3] B. Bischl, I. Vatolkin, and M. Preuß. Selecting small audio feature sets in music classification by means of asymmetric mutation. In R. Schaefer et al., eds., *Proc. of the 11th International Conference on Parallel Problem Solving From Nature (PPSN)*, pp. 314–323. Springer, 2010.

[4] H. Blume, M. Haller, M. Botteck, and W. Theimer. Perceptual feature based music classification - a DSP perspective for a new type of application. In W. A. Najjar and H. Blume, eds., *Proc. of the 8th International Conference on Systems, Architectures, Modeling and Simulation (IC-SAMOS)*, pp. 92–99. IEEE, 2008.

[5] C. H. Q. Ding and H. Peng. Minimum redundancy feature selection from microarray gene expression data. *Journal of Bioinformatics and Computational Biology*, 3(2):185–205, 2005.

[6] C. Emmanouilidis, A. Hunter, and J. MacIntyre. A multiobjective evolutionary setting for feature selection and a commonality-based crossover operator. In *Proc. of the IEEE Congress on Evolutionary Computation (CEC)*, volume 1, pp. 309–316. IEEE, 2000.

[7] R. Fiebrink and I. Fujinaga. Feature selection pitfalls and music classification. In *Proc. of the 7th International Conference on Music Information Retrieval (ISMIR)*, pp. 340–341. University of Victoria, 2006.

[8] I. Guyon, M. Nikravesh, S. Gunn, and L. A. Zadeh, eds. *Feature Extraction. Foundations and Applications*, volume 207 of *Studies in Fuzziness and Soft Computing*. Springer, Berlin Heidelberg, 2006.

[9] M. A. Hall. *Correlation-Based Feature Selection for Machine Learning*. PhD thesis, Department of Computer Science, The University of Waikato, 1999.

[10] K. Kira and L. A. Rendell. A practical approach to feature selection. In D. H. Sleeman and P. Edwards, eds., *Proc. of the 9th International Workshop on Machine Learning (ML)*, pp. 249–256. Morgan Kaufmann, 1992.

[11] P. Langley. Selection of relevant features in machine learning. In *Proc. of the AAAI Fall Symposium on Relevance*, pp. 140–144. AAAI Press, 1994.

[12] F. Mörchen, A. Ultsch, M. Thies, and I. Löhken. Modeling timbre distance with temporal statistics from polyphonic music. *IEEE Transactions on Audio, Speech, and Language Processing*, 14(1):81–90, 2006.

[13] P. M. Narendra and K. Fukunaga. A branch and bound algorithm for feature subset selection. *IEEE Transactions on Computers*, 26(9):917–922, 1977.

[14] P. Pudil, J. Novovičová, and J. Kittler. Floating search methods in feature selection. *Pattern Recognition Letters*, 15(10):1119–1125, 1994.

[15] J. R. Quinlan. *C4.5: Programs for Machine Learning*. Morgan Kaufmann, San Mateo, 1993.

[16] J. Reunanen. Overfitting in making comparisons between variable selection methods. *Journal of Machine Learning Research*, 3:1371–1382, 2003.

[17] M. Robnik-Sikonja and I. Kononenko. Theoretical and empirical analysis of ReliefF and RReliefF. *Machine Learning*, 53(1-2):23–69, 2003.

[18] P. Saari, T. Eerola, and O. Lartillot. Generalizability and simplicity as criteria in feature selection: Application to mood classification in music. *IEEE Transactions on Audio, Speech, and Language Processing*, 19(6):1802–1812, 2011.

[19] C. E. Shannon. A mathematical theory of communication. *Bell System Technical Journal*, 27(3):379–423, 1948.

[20] W. W. Siedlecki and J. Sklansky. A note on genetic algorithms for large-scale feature selection. *Pattern Recognition Letters*, 10(5):335–347, 1989.

[21] P. Somol, P. Pudil, J. Novovicová, and P. Paclík. Adaptive floating search methods in feature selection. *Pattern Recognition Letters*, 20(11-13):1157–1163, 1999.

[22] H. Stoppiglia, G. Dreyfus, R. Dubois, and Y. Oussar. Ranking a random feature for variable and feature selection. *Journal of Machine Learning Research*, 3:1399–1414, 2003.

[23] I. Vatolkin. *Improving Supervised Music Classification by Means of Multi-Objective Evolutionary Feature Selection*. PhD thesis, Department of Computer Science, TU Dortmund, 2013.

[24] I. Vatolkin, B. Bischl, G. Rudolph, and C. Weihs. Statistical comparison of classifiers for multi-objective feature selection in instrument recognition. In M. Spiliopoulou, L. Schmidt-Thieme, and R. Janning, eds., *Data Analysis, Machine Learning and Knowledge Discovery*, Studies in Classification, Data Analysis, and Knowledge Organization, pp. 171–178. Springer, 2014.

[25] I. Vatolkin, A. Nagathil, W. Theimer, and R. Martin. Performance of specific vs. generic feature sets in polyphonic music instrument recognition. In R. C. P. et al., ed., *Proc. of the 7th International Conference on Evolutionary Multi-Criterion Optimization (EMO)*, volume 7811 of *Lecture Notes in Computer Science*, pp. 587–599. Springer, 2013.

[26] I. Vatolkin, M. Preuß, and G. Rudolph. Multi-objective feature selection in music genre and style recognition tasks. In N. Krasnogor and P. L. Lanzi, eds., *Proc. of the 13th Annual Genetic and Evolutionary Computation Conference (GECCO)*, pp. 411–418. ACM Press, 2011.

[27] Z. Zhu, S. Jia, and Z. Ji. Towards a memetic feature selection paradigm. *IEEE Computational Intelligence Magazine*, 5(2):41–53, 2010.

[28] Z. Zhu, Y.-S. Ong, and J.-L. Kuo. Feature selection using single/multi-objective memetic frameworks. In C.-K. Goh, Y.-S. Ong, and K. C. Tan, eds., *Multi-Objective Memetic Algorithms*, volume 171 of *Studies in Computational Intelligence*, pp. 111–131. Springer, Berlin Heidelberg, 2009.

Part III

Applications

Chapter 16

Segmentation

NADJA BAUER, SEBASTIAN KREY, UWE LIGGES, CLAUS WEIHS
Department of Statistics, TU Dortmund, Germany

IGOR VATOLKIN
Department of Computer Science, TU Dortmund, Germany

16.1 Introduction

Segmentation is a task necessary for a variety of applications in music analysis. In Chapter 17 we will find it useful to segment a piece of music into small single parts corresponding to notes in sheet music to allow, e.g., for later transcription. This particular kind of segmentation is typically called onset detection and may be based, e.g., on time-domain or frequency-domain features indicating the fundamental frequency f_0 of a certain part of the tone (see Section 16.2).

An even finer segmentation splits a tone into parts such as attack, sustain, decay, and eventually noise (see Section 2.4.5). This is useful to allow for instrument classification from a piece of sound, for example. After finding relevant features for this low-level task, a clustering method (see Chapter 11), e.g., the k-means method, can be applied to yield a reduced number of features for subsequent classification (see Section 16.3).

We can, however, also aim at a segmentation that corresponds to larger parts of a piece of music like refrains, for example. In musicology, a typical first step of the analysis of compositions is to structure the piece into different phases. This is a time-intensive task, which is usually done by experts. For an analysis of large collections of music this is not feasible. Therefore, an automatic structuring method is desirable. As no ground truth is available, unsupervised learning methods like clustering are a sensible approach to solving this task (see Section 16.4).

Overall, the size of segments highly depends on the final application. Therefore, the methods used to achieve the goal also depend on the application. This chapter introduces basic concepts for constructing methods and hence algorithms that allow for segmenting music into smaller parts that are desirable for subsequent applications.

16.2   Onset Detection

Onset detection is an important step for music transcription and other applications like timbre, meter, or tempo analysis (see Chapter 20). It relates to a rather fine partitioning of a musical signal. In this section, we first discuss the definition of tone onsets and then introduce a common approach to onset detection.

16.2.1   Definition

For the tone onset definition, the concept of so-called transients (see Section 2.4.5) is essential. Transient signals are located in the attack phase of music tones (Figure 2.16). Transients are non-periodic and characterized by a quick change of frequency. They usually occur by interaction between the player and the musical instrument, which is necessary to produce a new tone. Reference [2] defines a tone onset to be located – in most cases – as the start of the transient phase.

The work of [24] summarizes three definitions of a tone onset: physical onset (first rising from zero), perceptual onset (time where an onset can first be perceived by a human listener) and perceptual attack onset (time where the rhythmic of a tone can first be perceived by a human listener). Reference [38, p. 334] conducted a study which found out that the perceptual onset "lies between about 6 and 15 dB below the maximum level of the tone." Reference [7] criticizes that the study did not consider complex musical signals.

Whether the physical or the perceptual tone onset definition is used very much depends on the data format. The MIDI file format (Section 7.2.3) contains all information about music notes, including onsets. Hence we can imagine these to be physical onsets. The perceptual definition is more suitable for the WAVE format (Section 7.3.2) that is typically used if real music pieces have to be annotated by human listeners.

There are two kinds of onset detectors: offline and online ones. For offline detectors, information of a whole music recording can by used for analysis. This case is well studied and there exist many algorithms. Many applications like hearing aids require, however, online (or real-time) approaches. Here, tone onsets should be detected in time or with minimal delay, also called *latency time*. The latency should not exceed a few tens of milliseconds, as human beings perceive – depending on the tempo of music pieces – two tone onsets separated by less than 20 to 30 ms as simultaneous [34].

Music instruments differ in the kind of tone "producing." There exist many different instrument types (see also Section 18.3): percussion instruments (like bass drum or timpani), string or bowed instruments (like guitar or violin), keyboard instruments (like piano or accordion), or wind instruments (like flute or trumpet).

Example 16.1 (Temporal visualization). *The basic onset detection procedure is illustrated in this chapter by means of two music pieces: monophonic recordings of the first strophe of the Hallelujah song[1] played by piano and flute, respectively. Fig-*

---

[1]Also known as the German song "Ihr seid das Volk, das der Herr sich ausersehn," http://www. gesangbuchlieder.de/gesangbuchlieder. Accessed 20 May 2015.

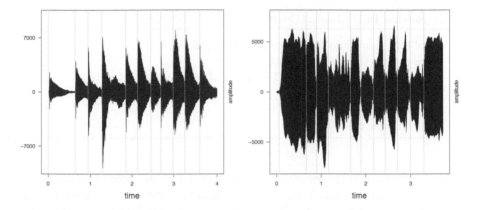

Figure 16.1: Example of a monophonic piano recording.

Figure 16.2: Example of a monophonic flute recording.

Figure 16.3: Sheet music for the first strophe of the Hallelujah song.

ures 16.1 and 16.2 present the amplitude envelope of the recordings, where the grey vertical lines mark the true onset times. The associated sheet music is presented in Figure 16.3. While for percussion or string instruments new tone onsets are necessarily marked with a major or minor amplitude increase, this is not always the case for wind instruments (especially for legato playing). This indicates the challenge of finding a universal approach for onset detection which suits for all kinds of music instruments.

16.2.2 Detection Strategies

In what follows, we provide an overview of strategies for tone onset detection in musical signals. The detailed and well-structured tutorial provided by [2] is summarized and extended by some newer approaches. The classical detection procedure is presented in Algorithm 16.1. We will explain the most important issues of each of the steps of this algorithm. Also, we will introduce one approach (transient peak classification, in Section 16.6) which slightly deviates from this structure.

16.2.2.1 Step 1: Splitting the Signal

As in previous chapters (e.g., Chapter 2), the ongoing audio signal is first split into (possibly overlapping) windows of M samples. For each window, the Short Time Fourier Transform (STFT, Section 4.4) is computed. In order to profit from the Fast

---

Algorithm 16.1: Classical Onset Detection Procedure

---

1 Split the signal into small (overlapping) windows.
2 Pre-process the data (optional).
3 Compute an Onset Detection Function (ODF) in each window.
4 Normalize the ODF.
5 Threshold the normalized ODF.
6 Localize the tone onsets.

Discrete Fourier Transformation (FFT, Section 4.4.3), M should be assigned only to specific numbers, e.g., powers of two. In the current onset detection literature, a window size of ca. 22 ms (2048 samples for the sampling rate of 44.1 kHz) is typically used. Note that small window sizes allow for a good time representation while large sizes provide a high spectral resolution. A further important issue is the hop size h, which is the distance in samples between neighboring window starting points. The lower h is, the more overlapping are the produced windows. In the case of $M = h$, the windows are disjunct (no overlap).

16.2.2.2   Step 2: Pre-Processing

A music signal is a complex time series containing important information. Pre-processing can be applied from many points of view: Separation of the ongoing signal into several frequency bands or attenuation of strong dynamically changing effects.

The motivation for the first approach can be found in the human auditory system, which consists of about 3000 auditory nerve fibers where each fiber responds to a special frequency (band) (see Chapter 6). The original signal can be separated either in terms of special filter banks (Section 4.6) or based on an auditory model (see Section 6.3). The number of frequency bands varies in the literature while five to six bands are usual. We, however, use Meddis' auditory model for illustration [19] based on forty channels (bands) which represent the frequencies between 250 Hz (channel 1) and 7500 Hz (channel 40). In Figures 16.4 and 16.5, the so-called auditory images of monophonic piano and trumpet tone sequences are presented. In the figures, frequency bands are located at the vertical axis, the horizontal axis specifies time progression, and grey shading indicates different channel activities. After dividing the signal into distinct frequency bands, each band can be analyzed separately concerning the tone onsets. However, the information of all bands have to be combined afterwards to a final vector of onset times. Reference [1] compares many possible solutions of this task.

As we see in Figures 16.4 and 16.5, depending on the musical instrument, different frequency bands are essential for tone onset recognition. While for piano pieces the higher frequencies seem to be especially appropriate, we observe a certain delay in these frequencies for wind instruments like trumpet. Please notice that for trumpet

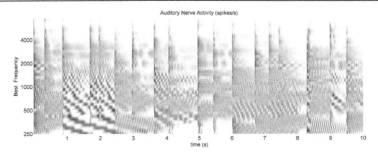

Figure 16.4: Auditory image of monophonic piano recording.

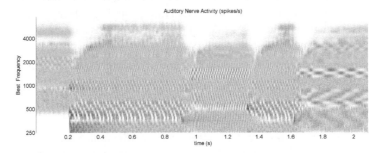

Figure 16.5: Auditory image of monophonic trumpet recording.

only the first 2 seconds are displayed in order to better demonstrate the delayed nerve activities for higher frequencies.

Reference [8] also divided the ongoing signal into several frequency bands for a hybrid approach. While in the upper bands, energy-based detector functions are applied in order to detect strong transients, frequency-based detectors are used for lower bands for exploring the soft onsets.

Another pre-processing approach is adaptive whitening [34]. The main idea is to re-weight the STFT in a data-dependent manner. STFT does not provide a good resolution in the low-frequency domain but contains excessive information for high frequencies. Adaptive whitening aims to bring "the magnitude of each frequency band into a similar dynamic range" [34, p. 315]. Define

$$q[\lambda,\mu] = \begin{cases} \max(|X_{\text{stft}}[\lambda,\mu]|, r, m \cdot q[\lambda-1,\mu]) & \text{if } \lambda > 0 \\ \max(|X_{\text{stft}}[\lambda,\mu]|, r) & \text{otherwise} \end{cases}$$

$$X_{\text{stft}}^{\text{aw}}[\lambda,\mu] \leftarrow \frac{X_{\text{stft}}[\lambda,\mu]}{q[\lambda,\mu]}.$$

(16.1)

The $X_{\text{stft}}[\lambda,\mu]$ denote Fourier coefficients (complex numbers) for the μ-th frequency bin of the λ-th window (cp. Section 4.5, Equation (4.30)).

$$|X_{\text{stft}}[\lambda,\mu]| = \sqrt{Re(X_{\text{stft}}[\lambda,\mu])^2 + Im(X_{\text{stft}}[\lambda,\mu])^2}$$

is the magnitude of these coefficients. The memory parameter m lies in the interval $[0,1]$ while an appropriate interval for the floor parameter r depends on the magnitude distribution. A value of $r > \max_{\forall \mu, \forall \lambda} (|X_{\text{stft}}[\lambda, \mu]|)$ eliminates the effect of adaptive whitening, while $r = 0$ and $m = 0$ cause absolute whitening (all magnitudes will be equal to 1). This simple and efficient approach shows a noticeable improvement for many online onset detectors.

16.2.2.3   Step 3: Onset Detection Functions

Applying an onset detection function to the original or pre-processed signal leads to the reduction of the signal to a feature vector. Several functions have been proposed in the literature: some of them use change in the spectral structure as an indicator for a tone onset, others consider phase deviation or just change in the amplitude envelop. There are also many model-based approaches.

Features-based Detection Functions   Let us first list the most important signal features regarding onset detection which were introduced in Chapter 5:

- Time-domain features: Zero-crossings (Section 5.1) and energy envelope (Section 5.2, Equation (5.3)).
- Frequency-domain features: Spectral centroid (Section 5.3), spectral skewness (Section 5.5), spectral kurtosis (Section 5.6), spectral flux (Section 5.10), and mel frequency cepstral coefficients (Section 5.2.3).
- Rhythmic features: High frequency content (Section 5.15), phase deviation (Section 5.16), and complex domain (Section 5.17).

Many modifications of phase deviation and complex domain features are described in [28]. Except for the energy envelop (in Equation (5.3)), all features can be computed online with the delay of one window, i.e. the delay is equal to hop size h.

Model-based Detection Functions   Many statistical model-based approaches consider onset detection as a (supervised) classification task as each window has to be assigned to one of two values: 1 (onset) or 0 (no onset) (cp. Chapter 12). In the simplest case, one could use the just listed signal features for the training of a user-defined classifier. However, more sophisticated, but also more time-consuming, solutions of the onset detection problem are also possible.

Example 16.2 (Neural Network Onset Detection). *As an example, we introduce the neural network modeling approach of [9] (cp. also Section 12.4.6) to illustrate such a procedure. The authors window the signal not only once – as usual – but twice: with a window size of 1024 samples (ca. 23 ms) but also with 2048 samples (ca. 46 ms). Then, STFTs are transformed to the mel spectrum (cp. Section 2.2.5) and additionally their positive first-order differences (between neighboring windows) are calculated for each of the two windowings. This results in four feature vectors, which provide the input for a neural network model. In this case, a relatively complex bidirectional long short-term memory neural network model with three hidden layers for each direction is considered. The output of the model is a probability for an onset in each window. The described algorithm performed well in the MIREX onset detection competitions in recent years.[2] The drawback is the time-consuming model training.*

16.2.2.4 Step 4: Normalization

Let us denote the feature vector obtained after applying an onset detection function to the signal with $\boldsymbol{odf} = (odf_1,\ldots,odf_W)^T$, where W is the number of windows. Depending on the used detection function, $\boldsymbol{odf}$ will cover different kinds of values. The aim of the normalization step is to bring this vector into a more uniform format. It is common to get rid of a possibly very fluctuating structure by means of smoothing and re-scaling the feature vector. The exponential smoothing operator is a rather popular one:

$$s(odf)_1 = odf_1$$
$$s(odf)_i = \alpha \cdot odf_i + (1-\alpha) \cdot s(odf)_{i-1}, \quad i = 2,\ldots,W. \tag{16.2}$$

The smoothing parameter α, $0 < \alpha < 1$, determines the influence of the past observations on the actual value: the smaller α is, the greater the smoothing effect.

Regarding re-scaling, many normalization methods have been proposed (see Section 14.2.2). Standardization, for example, is motivated by statistics:

$$\boldsymbol{n(odf)} = \frac{s(odf) - \overline{s(odf)}}{\sigma_{s(odf)}}. \tag{16.3}$$

The standardized vector $\boldsymbol{n(odf)}$ will then have mean 0 and standard deviation 1. However, $\min(\boldsymbol{n(odf)})$ and $\max(\boldsymbol{n(odf)})$ are unknown. A further method guarantees $\min(\boldsymbol{n(odf)}) = 0$ and $\max(\boldsymbol{n(odf)}) = 1$. Here

$$\boldsymbol{n(odf)} = \frac{s(odf) - \min(s(odf))}{\max(s(odf)) - \min(s(odf))}. \tag{16.4}$$

Note that the described kind of normalization just works offline. Reference [3] proposed and compared many online modifications of the normalization step.

Example 16.3 (Onset detection functions). *Figure 16.6 presents the features Spectral Flux (SF) and Energy Envelop (EE) as well as their normalized variants (n.SF and n.EE with $\alpha = 0.6$ and re-scaling to $[0,1]$) for a piano and a flute interpretation of the Hallelujah piece (Figures 16.1 and 16.2). While the first feature is computed in the spectral domain, for the second feature, just the amplitude envelope (time domain) is considered.*

Consider the first and the fourth piano tone: they last longer than the other tones and show a relevant change in spectral domain as well a minor amplitude increase toward their ends. This can be explained by the change of temporal and spectral characteristics of the signal during developing of long tones (possibly caused by overtones), which obviously results in at least one false detection. Interestingly, such patterns are not usual for synthetically produced tones. Unfortunately, differences

---

[2]http://www.music-ir.org/mirex/wiki/MIREX_HOME. Accessed 20 May 2015.

between real and synthetic music, with regard to onset detection, have not been extensively investigated yet.

As we see in Figure 16.6, the second tone onset of the flute is not reflected in the spectral flux feature. An appropriate explanation would be that the first and the second tone have the same pitch (cp. Figure 16.3) and the transient phase of the second tone was very short. The fourth flute tone contains relevant changes of spectral content so that many false detections would be expected here. This behavior is caused by the applied vibrato technique for this tone.

To summarize, SF and EE features appear to be suitable for the detection of piano tone onsets while EE does not appear to be a meaningful detection function for flute. Smoothing of the detection function can be seen as advisable in general.

16.2.2.5   Step 5: Thresholding

The (normalized) vector of detection features contains many small and intense peaks which provide an indication of a possible tone onset. A peak is, however, considered to indicate a tone onset only if its value exceeds a certain threshold. The simplest way is to define a fixed threshold whose value can be optimized on a training data set. As many detection functions reflect loudness variation, a fixed threshold would lead to a high error rate in case of pieces with lots of loudness variations. For this reason, dynamic thresholding approaches based on moving averages or moving medians became popular. For each window i, $i = 1, \ldots, W$, the threshold vector t is defined as

$$t_i = \delta + \beta \cdot ThreshFunction(|n(odf)_{i-l_T}|, \ldots, |n(odf)_{i+r_T}|), \qquad (16.5)$$

where *ThreshFunction* is either the *median* or the *mean* function and l_T and r_T are the numbers of windows left and right of the current frame to be considered. Note that frequently $l_T = r_T$ is chosen for offline detection, but for online detectors this distinction is of importance as r_T has to be small or even equal to zero. δ and β are additive and multiplicative threshold parameters, respectively. See Example 16.4 for a comparison of two choices for these parameters.

16.2.2.6   Step 6: Onset Localization

In the last step, tone onsets are localized according to t and $n(odf)$:

$$o_i = \begin{cases} 1, & \text{if } n(odf)_i > t_i \text{ and } n(odf)_i = \max(n(odf)_{i-l_O}, \ldots, n(odf)_{i+r_O}) \\ 0, & otherwise. \end{cases} \qquad (16.6)$$

o is the onset vector. l_O and r_O are additional parameters – number of windows left and right of the current window, respectively, for calculating the local maximum. $l_O = r_O = 0$ implies that every value with $n(odf)_i > t_i$ is a tone onset. For online applications, r_O should be chosen small or equal to zero.

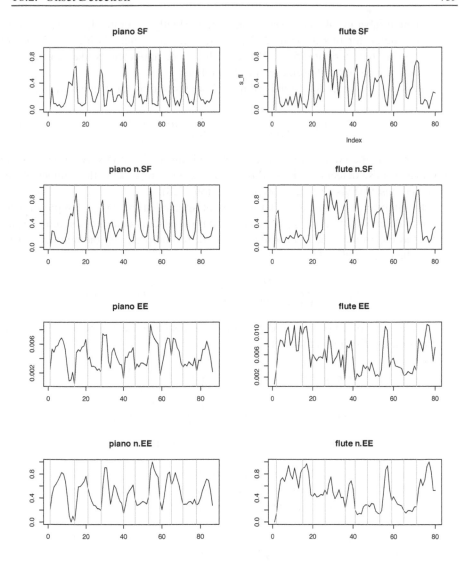

Figure 16.6: Spectral flux (SF), normalized SF (n.SF), energy envelop (EE), and normalized EE (n.EE) features for piano (left) and flute (right) interpretation of the Hallelujah piece.

16.2.3 Goodness of Onset Detection

Starting time points of windows with $o_i = 1$ compose a vector of onset times which is then compared to the time vector of true tone onsets. An onset is assumed to be correctly detected if it matches to one true onset within a certain tolerance interval. Such a tolerance is needed for various reasons, e.g., what we take as the ground truth, the manual labeling of true onsets, cannot be perfectly accurate. Usually, $\pm$ 50 ms are considered for this interval [2, 7], but also smaller intervals like $\pm$ 25 ms are employed in the literature [5, 3]. However, for some applications like hearing aids, this interval should be chosen even much smaller.

Goodness of onset detection is commonly measured by the F-measure [7]:

$$m_F = \frac{2m_{TP}}{2m_{TP} + m_{FP} + m_{FN}}, m_F \in [0,1], \tag{16.7}$$

where m_{TP} is the number of correctly detected onsets (True Positives), m_{FP} is the number of false detections (False Positives), and m_{FN} represents the number of undetected onsets (False Negatives)(cp. Definition 13.8).

To interpret the F-measure, let us look at special cases. $m_F = 1$ represents an optimal detection, whereas $m_F = 0$ means that no onset is detected correctly. Apart from these extremes, the F-measure is difficult to interpret. Let the number of true onsets be O_{true} leading to

$$m_F = \frac{2 \cdot (O_{true} - m_{FN})}{2 \cdot (O_{true} - m_{FN}) + m_{FP} + m_{FN}}, \quad m_F \in [0,1].$$

This relationship can be used to derive the dependence of the number of misclassifications on the F-value for three scenarios:

$$m_{FP} = 0 \quad \Longrightarrow \quad m_{FN} = \left(1 - \frac{m_F}{2 - m_F}\right) \cdot O_{true}$$

$$m_{FN} = 0 \quad \Longrightarrow \quad m_{FP} = \left(\frac{2}{m_F} - 2\right) \cdot O_{true}$$

$$m_{FP} = m_{FN} \quad \Longrightarrow \quad m_{FP} = m_{FN} = (1 - m_F) \cdot O_{true}$$

For example, if the number of false detections of onsets $m_{FP} = 0$, then $m_F = 0.8$ means that the number of undetected onsets $m_{FN} = \frac{1}{3}O_{true}$. $m_{FN} = 0$ corresponds to the case that all true onsets are detected, whereas $m_{FP} = m_{FN}$ corresponds to the case that the number of errors is the same for onsets and non-onsets.

Alternatively, the F-value can be defined using Recall (m_{REC}) and Precision (m_{PREC}) measures (cp. Definitions 13.9, 13.10):

$$m_F = \frac{2m_{PREC} \cdot m_{REC}}{m_{REC} + m_{PREC}}, \text{ where}$$

$$m_{PREC} = \frac{m_{TP}}{m_{TP} + m_{FP}} \quad \text{and} \quad m_{REC} = \frac{m_{TP}}{m_{TP} + m_{FN}}.$$

Note that there is a tradeoff between recall and precision, so that onset detection could be optimized in a multi-objective fashion (see Section 10.4).

In order to achieve good detection quality, a sophisticated optimization of algorithm parameters is essential. Of course, the tuned algorithm will work particularly well on music pieces close to the training set. This illustrates the importance of elaborating a training data set which considers many musical aspects. Reference [18]

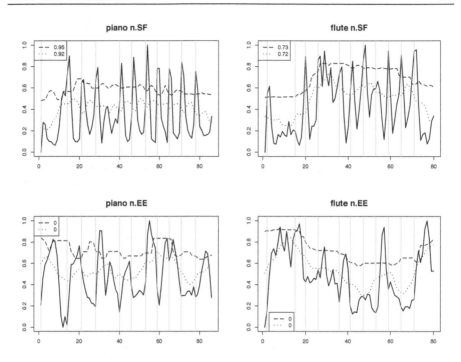

Figure 16.7: Thresholding for normalized SF (n.SF) and normalized EE (n.EE) for piano (left) and flute (right) interpretation of the Hallelujah piece. The dotted line corresponds to setting **SET1** and the dashed line represents **SET2**. The associated F-values are given in the legend.

proposed, for example, a way for building a representative corpus of classical music. Further research in this field is not only important for onset detection but also for many other music applications.

Example 16.4. *Figure 16.7 illustrates the thresholding and onset localizing procedure. We consider the above-mentioned normalized spectral flux (n.SF) and energy envelop (n.EE) features for piano and flute (as in Figure 16.6). Two possible parameter settings were exemplarily compared. The dotted line corresponds to the setting* **SET1**: $\delta = 0.1$, $\beta = 0.9$, *th.fun=mean and* $l_T = r_T = 5$ *while the dashed line represents* **SET2**: $\delta = 0.4$, $\beta = 0.7$, *th.fun=median and* $l_T = r_T = 10$. *The figure shows how important the correct choice of the thresholding parameters is. The dynamic threshold is especially sensible to even small variation of* δ *and* β. *The smaller* l_T *and* r_T *are, the more likely it is to detect smaller peaks of the detection function (dotted line). Please note the associated F-values in the legend.*

Figure 16.7 reflects some well-known facts: the simplest onset detection problem is given for monophonic string, keyboard, or percussive instruments. Furthermore, the spectral flux feature achieves remarkable results for almost all musical instruments.

16.3   Tone Phases

An even finer segmentation of a musical piece splits each tone into parts such as attack, sustain, decay and eventually noise (see Section 2.4.5). This is useful for instrument recognition from a piece of sound, for example. Features can again be derived from a pre-filtered time series divided into small windows. Perceptive Linear Prediction Coding (PLP) [13] and Mel Frequency Cepstral Coefficients (MFCCs; [6]) (see also Section 4.7) are widely used in the context of speech recognition. These features might also be used to model an instrument's tone. In order to characterize the development of a tone by a small number of aggregated features, we aim at clustering the small time windows (time frames) according to the different tone phases (at least attack, sustain, decay). For this, clustering methods, like k-means, are applied yielding aggregated features for a subsequent classification task like, e.g., instrument recognition (see [17]).

16.3.1   Reasons for Clustering

Especially during the attack phase, the sound of an instrument is often different from the rest of the time frames due to the mechanics which generate the sound (plucking, acceleration of the bow, etc.). The sound of instruments might also change over time for some other reason, e.g., because of vibrato or during fading at the end of a tone. Hence, for W time frames we consider a relatively large number of $p \cdot W$ features that should be clustered. Clustering allows the automatic partitioning of the time frames into groups, which differ from each other by some criterion. These groups will be interpreted as tone phases (attack, sustain, decay, and noise). Within the groups, the time frames are similar, resulting in cluster representatives for the individual sound phases of an instrument's sound.

16.3.2   The Clustering Process

After preprocessing and feature extraction we assume an F-dimensional feature vector $\boldsymbol{x}_w$ for every time frame $w = 1, \ldots, W$ that can be represented in form of a matrix $\boldsymbol{X} = (x_{uw})$ for features $u = 1, \ldots, F$. To these data, we apply clustering techniques to find representatives for the different tone phases *attack*, *sustain* and *decay* of an instrument's sound. An additional sound phase to be considered, as it is present in most recordings, represents the silent or noise-dominated parts at the beginning and the end of a recording. In our examples, we do not consider vibrato as an extra phase. As a result of these considerations, we assume four clusters representing the phases relevant in a typical recording of a tone of an instrument.

For each recorded note, after having clustered the feature vectors of all frames into the four mentioned clusters, we use their $k(=4)$ cluster centers (written as a matrix $\boldsymbol{X}^c = (x_{uk}^c); k = 1, \ldots, 4; u = 1, \ldots, F$) as representatives rather than the original W

Table 16.1: Overview of the Number of Observations and Features

Process step	Number of observations	Features
Raw data	1980 recordings	on average 180,000 samples
Feature extraction	800,000 time frames	p-dimensional sound features
Clustering	1980 recordings	4 clusters with p-dim. cluster centers
Clustering after noise removal	1980 recordings	3 clusters with p-dim. cluster centers

feature vectors for all frames. This greatly reduces complexity (see, e.g., Table 16.1 related to the example below), but still allows us to make use of the change in the instruments' sound. As the silence and noise cluster includes no useful information for the following classification task, this cluster center is completely dropped for a further complexity reduction. As results of the clustering process one might consider the clustered frames as well as the cluster centers useful for further classification.

Example 16.5 (Tone Phases of Different Instruments). *For clustering we use the k-means method (see Section 11.4.1) with 25 random starting points, which results in promising clustering results. In Figure 16.8 the almost perfect clustering of a piano note is exemplarily. Labels have been given to the different clusters according to their first occurrence in the sound. The time frames containing only silence and a bit of noise in the beginning and the end of the recording are grouped to a single cluster. The actual sound has a first phase of high energy and additional overtones of the hammer hitting the strings, followed by a phase where these additional overtones have subsided before the sound fades away. Strings or even wind instruments often also result in excellent clusterings, as can be seen in Figures 16.9 and 16.10. For the bowed viola sound in Figure 16.9, the cluster labeled as* attack *consists of two parts, where the bow accelerates (in the beginning) and decelerates after the* sustain *phase before the sound finally decays.*

The clustering of a contrabassoon (cp. Figure 16.11) shows crisp clusters only for the silence/noise part in the beginning and the end. A smaller number of clusters seems to be more sensible and the labeling of the phases attack, sustain, *and* decay

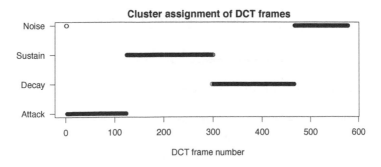

Figure 16.8: Clustering of the DCT frames of a piano note.

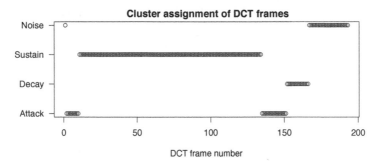

Figure 16.9: Clustering of the DCT frames of a viola note.

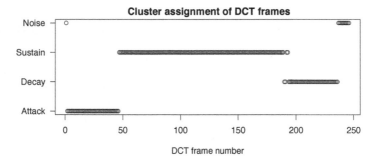

Figure 16.10: Clustering of the DCT frames of an alto saxophone note.

does not make much sense in this case. Therefore, we tried to apply an automated selection of the right number of clusters. Relative criteria to validate the number of clusters using the Dunn, Davies–Bouldin, *or* SD *indices as described in [12] suggest a minimum of 2 and a maximum of 5 clusters to be tried for all the instruments.*

Although usage of k-means with k = 4 clusters is not always optimal, we will use k = 4 later in our classification example.

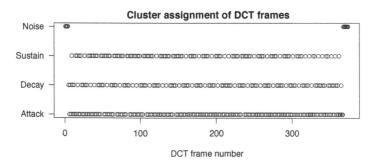

Figure 16.11: Clustering of the DCT frames of a contrabassoon note.

16.3.3   Refining the Clustering Process

The classical clustering methods like k-means or hierarchical agglomerative clustering (see Section 11.3) suffer from the same problem. They are all only distance based and ignore the temporal structure of music, which often results in unstable and uninterpretable clustering results. In [33] the authors suggest a method to introduce an order constraint in k-means clustering to reduce the risk of finding only locally optimal clustering solutions and thus increase the stability of the resulting clusters. Using this methodology we can stabilize the clustering of our musical sound features, and also get clusters which are a lot better to interpret and to comprehend.

Let us now apply this refined clustering of tone phases to instrument recognition.

Example 16.6 (Instrument Recognition Using Tone Phase Information). *We used sounds from the McGill Instrument Database, which consists of 1986 notes (3–5 seconds long) played on 38 different instruments with different playing techniques (with or without vibrato, pizzicato or bowed, clean or distorted, etc.) resulting in 60 different timbres. Between 6 and 88 recorded notes are available for each timbre, representing the tonal range of the instrument.*

Based on this data set, two classification tasks are discussed. One task is to discriminate between all instrument timbres. We drop the slapping and popping sounds of the electronic bass (only 6 examples available), resulting in 1980 notes in 59 classes (cp. Table 16.1). For the other task, the instruments are grouped in 25 instrument families (trumpets, flutes, bowed strings, etc.) resulting in an easier classification problem.

Using a Support Vector Machine (SVM) with the polynomial kernel the classification (see Section 12.4.4) gives convincing results. We used the implementation in the R [25] package `kernlab` *[16]. On the single note recordings of the McGill Instrument Database [21] we achieve a misclassification error of 10% for classifying the instruments in the 25 instrument families and 19% for discriminating between all 59 available instrument timbres in the database. Misclassification error is smaller for the problem with fewer classes, as expected.*

16.4   Musical Structure Analysis

In a more sophisticated setting we want to automatically extract the structure of a musical piece from a recording. There are many possible segmentations of one piece of music. For popular music, most of the time a segmentation into "verse," "bridge," and "chorus" is assumed [14, 32]. However, other so-called horizontal and vertical segmentations appear to be sensible also. Horizontally, we might want to distinguish segments like sequences of notes, motifs, or measures, and vertically one might look for different instruments, harmonics, rhythm, or dynamics.

A categorization of segmentation methods is presented in [22]. As principal targets the recognition of change (novelty), stability (homogeneity), and repetition are identified. These targets are also combined [22, 11, 15]. The methods mainly base on the extraction of audio features, followed by a similarity analysis of feature vectors for the construction of a self-similarity matrix [10] (see below). In some studies, segments or segment borders are predicted by classification models: supervised ([36], [37]), unsupervised ([17], [30]), or combined unsupervised and supervised [35].

In most cases, low-level features are used to identify changes in timbre or pitch. Though the usage of such features was often quite successful [14, 17, 30], such models are unable to throw light on the relevance of interpretable characteristics of a song or music style for segmentation. Interpretable high-level features like instrumentation, harmonics, melody, tempo, rhythm, and dynamics are suggested in [29] for the description of personal music categories and predicted by low-level features.

The evaluation of segmentation methods is often based on music pieces from individual genres or interpreters. For example, songs of the Beatles are often compared to other songs or genres, e.g., to piano music in [4] or Mazurkas in [26], [11]. The transition between segments may differ, e.g., cadences might be used in classical music and "turn-arounds" in jazz [23]. A big step towards more variability in the analyzed pieces of music offers the database of the project SALAMI [31] with 1400 music pieces of different genres and styles.

For the identification of segment boundaries and similar segments in a musical piece, the similarity between feature vectors is usually measured. The following definition is based on [20, pp. 178].

Definition 16.1 (Self-Similarity Matrix). Let $\boldsymbol{X} \in \mathbb{R}^{F \times W}$ be a feature matrix with $u = 1, ..., F$ individual feature dimensions over $w = 1, ..., W$ time frames, and let s be a similarity measure. The *Self-Similarity Matrix (SSM)* $\boldsymbol{S} \in \mathbb{R}^{W \times W}$ is defined as:

$$S_{m,n} = s(\boldsymbol{x}_m, \boldsymbol{x}_n) \; \forall m, n \in \{1, ..., W\}, \tag{16.8}$$

where $\boldsymbol{x}_w$ is the vector of feature values in time frame w.

After normalization, the similarity measure s has a value of 1 for a high similarity between vectors and a value of 0 for indicating no similarity. Similarity measures are counterparts $1 - d$ to (normalized) distance measures d discussed in Section 11.2. For an example of a similarity measure and the corresponding distance measure, see below.

Let $[a, b]$ denote a segment starting in a time frame a and ending in a time frame b. To identify a homogeneous segment like verse, bridge or chorus, we expect that some related features selected for the building of the SSM (instrumentation, tempo, etc.) have similar values in this segment. This means that the corresponding *diagonal block* of the original matrix $\boldsymbol{S}$, i.e. a matrix $\boldsymbol{B} \in \mathbb{R}^{(b-a+1) \times (b-a+1)}$ with $B_{m,n} = S_{a+m-1,b+n-1} \; \forall m, n \in \{1, ..., b-a+1\}$, is characterized by high values. An example of a block with a very high similarity of underlying features is the percussion intro in Figure 16.12 (dark rectangle in the bottom left corner of the matrix). The score of a block is defined as:

$$s(\boldsymbol{B}) = \sum_{m=a}^{b} \sum_{n=a}^{b} S_{m,n}. \tag{16.9}$$

The mean homogeneity in a block, independent of its size, can be measured by normalization to the number of entries in $\boldsymbol{B}$ as $\frac{1}{(b-a+1)^2} \cdot s(\boldsymbol{B})$. If similar and homogeneous segments are repeated in a music piece, the corresponding SSM would contain blocks with high similarity values which may appear in different parts of the matrix and not only around the main diagonal; for a more general definition of a block, see [20].

For the search of similar segments in S, the *path* of length L is defined as an ordered set of cells $\mathscr{P} = \{S_{m_1,n_1}, ..., S_{m_L,n_L}\}$ with $\varepsilon_1 \leq m_{l+1} - m_l \leq \varepsilon_2$ and $\varepsilon_1 \leq n_{l+1} - n_l \leq \varepsilon_2$ (ε_1 and ε_2 define the permitted step sizes; the path is parallel to the main diagonal if $\varepsilon_1 = \varepsilon_2 = 1$). The score of a path is defined as:

$$s(\mathscr{P}) = \sum_{l=1}^{L} S_{m_l,n_l}. \tag{16.10}$$

Because of possible tempo changes across similar segments, a path with a high score does not have to be necessarily parallel to the main diagonal of S. Examples of two paths with high scores are visible as dark stripes enclosed in marked rectangles in Figure 14.4, right bottom subfigure.

Visualization of Features   For easier judgement of the quality of a clustering result, a visual representation of the similarity between the feature vectors of different time frames is necessary. The most common method is a heatmap of the pairwise distances between feature vectors of different time frames. The darker the color is, the more similar the time frames are. As a *similarity measure*, the *cosine similarity*

$$s(\boldsymbol{x}_m, \boldsymbol{x}_n) = \frac{1}{2}\left(1 + \frac{\boldsymbol{x}_m^T \boldsymbol{x}_n}{||\boldsymbol{x}_m||\,||\boldsymbol{x}_n||}\right)$$

is usually best suited for visualizing the distances between features of frames m and n [22]. $d(\boldsymbol{x}_m, \boldsymbol{x}_n) = \frac{1}{2}\left(1 - \frac{\boldsymbol{x}_m^T \boldsymbol{x}_n}{||\boldsymbol{x}_m||\,||\boldsymbol{x}_n||}\right)$ is the corresponding *cosine distance*. In the heatmap, clusters are represented by white dots with the same vertical location (presented at the last time frame in the cluster). Hence, a direct comparison to the similarity of the time frames is possible, since areas with very high similarity are represented by dark colored squares.

Let us now give an example for the segmentation of a musical piece by using order-constrained solutions in k-means clustering (see [17]).

Example 16.7 (Structure Segmentation by Clustering).   *For this task we use longer time frames than in the instrument recognition setting in Example 16.6. Non-overlapping time frames of 3 seconds duration give sufficient temporal resolution. For a visual representation of the results, the plots described above will be used.*

Let us now consider our results for two different recordings of popular music. One is Depeche Mode's song "Stripped," which has a very clear structure. The other one is Queen's "Bohemian Rhapsody," a longer piece of music with a very diversified composition.

In Figure 16.12 the structure of Depeche Mode's "Stripped" is shown. The dark squares are easily visible and the clusters represent this structure very well. The number of clusters is estimated as 10.

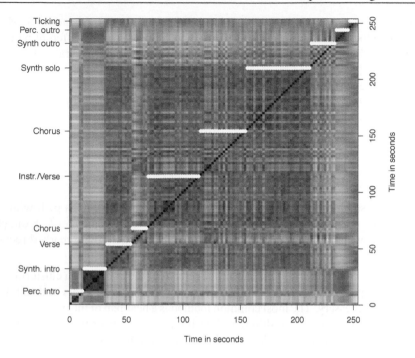

Figure 16.12: Musical structure of Depeche Mode's "Stripped."

For Queen's "Bohemian Rhapsody" the picture is a bit more difficult, see Figure 16.13. The squares are less dark but it is still possible to see the structure of the song. Even in this situation the cluster results correspond very well with the squares in the plot. This song does not follow the simple structure of most popular music. Hence a short annotation of the clusters is not possible. Listening to the music allows us to verify the result. The clusters end when the dominant instruments or voices change. Because of the more complex structure, more clusters are necessary.

16.5   Concluding Remarks

Segmentation in music analysis is still a hot research topic that is heavily under development. We gave some ideas how segmentation algorithms can be constructed by finding the right features, putting together well-known and newly generated methods.

This chapter gives three examples of segmentation procedures in music. Onset detection is mainly a pre-step to, e.g., pitch estimation and instrument recognition. Tone phase analysis by clustering into attack/sustain/decay phases can be, e.g., used for instrument recognition.

Automatic music structure analysis is realized by clustering the sound features of a whole song. The proposed order-constrained solutions in k-means clustering are very easy to interpret and stable. The method works for a complex musical piece

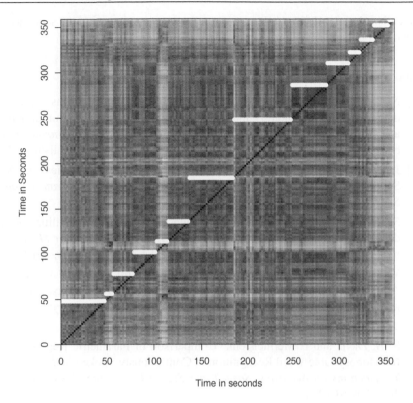

Figure 16.13: Musical Structure of Queen's "Bohemian Rhapsody."

like Queen's "Bohemian Rhapsody" as well as for simpler songs, with a very easy recognizable structure like Depeche Mode's "Stripped." The resulting cluster centers can be used for further tasks like music genre classification, where each part of a song is labeled separately instead of labeling the whole song in order to improve accuracy.

16.6 Further Reading

For onset detection, we highly recommend [2] and [28] for further reading. Other important publications that may be very helpful are [9] and [27]. Reference [27] proposes an online onset detection algorithm by means of transient peak classification – an approach which gained, in its several modifications, remarkable results in many MIREX competitions. The main idea is that pitches of the detection function caused by the transient phase of a tone onset should differ from pitches caused by random transients like noise. The pitch classification is then done by means of a statistical model. The proposal of [27] has a maximum delay of the length of the 8th part of an analysis window and is therefore recommendable for many online applications.

In musical structure analysis the extracted borders and segments could be hierarchically related ([4],[26],[14]), e.g. if smaller groups of certain note sequences are parts of a rougher division into verses and bridges.

Bibliography

[1] N. Bauer, K. Friedrichs, D. Kirchhoff, J. Schiffner, and C. Weihs. Tone onset detection using an auditory model. In M. Spiliopoulou, L. Schmidt-Thieme, and R. Janning, eds., *Data Analysis, Machine Learning and Knowledge Discovery*, volume Part VI, pp. 315–324. Springer International Publishing, 2014.

[2] J. P. Bello, L. Daudet, S. A. Abdallah, C. Duxbury, M. E. Davies, and M. B. Sandler. A tutorial on onset detection in music signals. *IEEE Transactions on Speech and Audio Processing*, 13(5):1035–1047, 2005.

[3] S. Böck, F. Krebs, and M. Schedl. Evaluating the online capabilities of onset detection methods. In *Proc. of the 13th International Society for Music Information Retrieval Conference (ISMIR)*, pp. 49–54. FEUP Edições, 2012.

[4] W. Chai. *Automated Analysis of Musical Structure*. PhD thesis, School of Architecture and Planning, Massachusetts Institute of Technology, 2005.

[5] N. Collins. Using a pitch detector for onset detection. In *MIREX Online Proceedings (ISMIR 2005)*, pp. 100–106, 2005.

[6] S. B. Davis and P. Mermelstein. Comparison of Parametric Representations for Monosyllabic Word Recognition in Continuously Spoken Sentences. *IEEE Transactions on Acoustics, Speech and Signal Processing*, ASSP-28(4):357–366, August 1980.

[7] S. Dixon. Onset detection revisited. In *Proc. of the 9th International Conference on Digital Audio Effects (DAFx)*, pp. 133–137. McGill University, 2006.

[8] C. Duxbury, M. Sandler, and M. Davies. A hybrid approach to musical note onset detection. In *Proc. of the 5th International Conference on Digital Audio Effects*, pp. 33–38, 2002.

[9] F. Eyben, S. Böck, B. Schuller, and A. Graves. Universal onset detection with bidirectional long short-term memory neural networks. In J. S. Downie and R. C. Veltkamp, eds., *Proc. of the 11th International Society for Music Information Retrieval Conference (ISMIR)*, pp. 589–594. International Society for Music Information Retrieval, 2010.

[10] J. Foote. Visualizing music and audio using self-similarity. In *Proc. ACM Multimedia*, pp. 77–80. ACM, 1999.

[11] H. Grohganz, M. Clausen, N. Jiang, and M. Müller. Converting path structures into block structures using eigenvalue decompositions of self-similarity matrices. In *Proc. of the 14th International Society for Music Information Retrieval Conference (ISMIR)*, pp. 209–215. International Society for Music Information Retrieval, 2013.

[12] M. Halkidi, Y. Batistakis, and M. Vazirgiannis. On clustering validation techniques. *Journal of Intelligent Information Systems*, 17(2-3):107–145, Decem-

ber 2001.

[13] H. Hermansky. Perceptual linear predictive (PLP) analysis of speech. *Journal of Acoustical Society of America*, 87(4):1738–1752, April 1990.

[14] K. Jensen. Multiple scale music segmentation using rhythm, timbre, and harmony. *EURASIP Journal on Advances in Signal Processing*, 2007. doi: 10.1155/2007/73205.

[15] F. Kaiser and G. Peeters. A simple fusion method of state and sequence segmentation for music structure discovery. In *Proc. of the 14th International Society for Music Information Retrieval Conference (ISMIR)*, pp. 257–262. International Society for Music Information Retrieval, 2013.

[16] A. Karatzoglou, A. Smola, K. Hornik, and A. Zeileis. kernlab - An S4 package for kernel methods in R. *Journal of Statistical Software*, 11(9):1–20, 2004.

[17] S. Krey, U. Ligges, and F. Leisch. Music and timbre segmentation by recursive constrained K-means clustering. *Computational Statistics*, 29(1–2):37–50, 2014.

[18] J. London. Building a representative corpus of classical music. *Music Perception*, 31(1):68–90, 2013.

[19] R. Meddis. Auditory-nerve first-spike latency and auditory absolute threshold: A computer model. *Journal of the Acoustical Society of America*, 119(1):406–417, 2006.

[20] M. Müller. *Fundamentals of Music Processing - Audio, Analysis, Algorithms, Applications*. Springer International Publishing, 2015.

[21] F. Opolko and J. Wapnick. McGill University master samples (CDs), 1987.

[22] J. Paulus, M. Müller, and A. Klapuri. Audio-based music structure analysis. In *Proc. of the 11th International Society on Music Information Retrieval Conference (ISMIR)*, pp. 625–636. International Society for Music Information Retrieval, 2010.

[23] J. Pauwels, F. Kaiser, and G. Peeters. Combining harmony-based and novelty-based approaches for structural segmentation. In *Proc. of the 14th International Society for Music Information Retrieval Conference (ISMIR)*, pp. 601–606. International Society for Music Information Retrieval, 2013.

[24] R. Polfreman. Comparing onset detection and perceptual attack time. In *Proc. of the 14th International Society for Music Information Retrieval Conference (ISMIR)*, pp. 523–528. International Society for Music Information Retrieval, 2013.

[25] R Core Team. *R: A Language and Environment for Statistical Computing*. R Foundation for Statistical Computing, Vienna, Austria, 2014.

[26] C. Rhodes and M. Casey. Algorithms for determining and labelling approximate hierarchical self-similarity. In *Proc. of the 8th International Conference on Music Information Retrieval (ISMIR)*, pp. 41–46. Austrian Computer Society, 2007.

[27] A. Roebel. Onset detection in polyphonic signals by means of transient peak classification. In *Proc. of the 6th International Society for Music Information Retrieval Conference (ISMIR)*, London, Great Britain, 2005. University of London.

[28] C. Rosão, R. Ribeiro, and D. Martins De Matos. Influence of peak selection methods on onset detection. In *Proc. of the 13th International Society for Music Information Retrieval Conference (ISMIR)*, pp. 517–522. FEUP Edições, 2012.

[29] G. Rötter, I. Vatolkin, and C. Weihs. Computational prediction of high-level descriptors of music personal categories. In B. Lausen, D. Van den Poel, and A. Ultsch, eds., *Algorithms from and for Nature and Life*, pp. 529–537. Springer, 2013.

[30] J. Serrà, M. Müller, P. Grosche, and J. Arcos. Unsupervised music structure annotation by time series structure features and segment similarity. *IEEE Transactions on Multimedia*, 16(5):1229–1240, 2014.

[31] J. B. L. Smith, J. A. Burgoyne, I. Fujinaga, D. D. Roure, and J. S. Downie. Design and creation of a large-scale database of structural annotations. In *Proc. of the 12th International Society for Music Information Retrieval Conference (ISMIR)*, pp. 555–560. University of Miami, 2011.

[32] J. B. L. Smith, C.-H. Chuan, and E. Chew. Audio properties of perceived boundaries in music. *IEEE Transactions on Multimedia*, 16(5):1219–1228, 2014.

[33] D. Steinley and L. Hubert. Order-constrained solutions in K-means clustering: Even better than being globally optimal. *Psychometrika*, 73:647–664, 2008.

[34] D. Stowell and M. Plumbley. Adaptive whitening for improved real-time audio onset detection. In *Proc. of the International Computer Music Conference (ICMC'07)*, pp. 312–319. Michigan Publishing, 2007.

[35] M.-Y. Su, Y.-H. Yang, Y.-C. Lin, and H. Chen. An integrated approach to music boundary detection. In *Proc. of the 10th International Society for Music Information Retrieval Conference (ISMIR)*, pp. 705–710. International Society for Music Information Retrieval, 2009.

[36] D. Turnbull, G. Lanckriet, E. Pampalk, and M. Goto. A supervised approach for detecting boundaries in music using difference features and boosting. In *Proc. of the 8th International Conference on Music Information Retrieval (ISMIR)*, pp. 51–54. Austrian Computer Society, 2007.

[37] K. Ullrich, J. Schlüter, and T. Grill. Boundary detection in music structure analysis using convolutional neural networks. In *Proc. of the 15th International Society for Music Information Retrieval Conference (ISMIR)*, pp. 417–422. International Society for Music Information Retrieval, 2014.

[38] J. Vos and R. Rasch. The perceptual onset of musical tones. *Perception & Psychophysics*, 29(4):323–335, 1981.

Chapter 17

Transcription

UWE LIGGES, CLAUS WEIHS
Department of Statistics, TU Dortmund, Germany

17.1 Introduction

In this chapter we describe methods for automatic transcription based on audio features. Transcription is transforming audio signals into sheet music, and it is in some sense the opposite of playing music from sheet music. The statistical core of transcription is classification of notes into classes of pitch (e.g. c, d, ...) and lengths (e.g. dotted eight note, quarter note, ...). A typical transcription algorithm includes at least some of the following steps:

1. Separation of the relevant part of music to be transcribed (e.g. human voice) from other sounds (e.g. piano accompaniment)

2. Estimation of fundamental frequencies

3. Classification of notes, silence and noise

4. Estimation of the relative length of notes and meter

5. Estimation of the key

6. Final transcription into sheet music

Note that step 1 is related to a pre-processing of the time series of the original musical audio signal. In step 2, time series modeling is used to estimate fundamental frequencies (cp. also Sections 4.8 and 6.4) which are to be classified into notes in step 3. In steps 4 and 5 these notes are fitted into meter and key. Finally, sheet music is produced in step 6.

Section 17.5 below will be organized along this list of steps and will present more details. In Sections 17.2, 17.3, and 17.4 we will comment on the analyzed audio data and describe the musical and statistical challenges of the transcription task. Transcription software is discussed in Section 17.6. For more information on transcription methods see, e.g., [20, 53].

17.2 Data

Most existing transcription systems have been invented for the transcription of *MIDI* data (see Section 7.2); both onset times and pitch are already exactly encoded in the data or for instruments such as piano and other plucked string or percussion instruments.

The transcription of MIDI data is not very difficult, because information related to pitch as well as the beginning and end of tones is already explicitly available within the data in digital form. Therefore, this information has not to be estimated from the sound signal for MIDI data. Transcription of WAVE data (see Section 7.3.2) or other types of audio data is harder. For WAVE data, transcribing plucked and stroked instruments (piano, guitar, etc.) is still simpler than, e.g., the transcription of melodies sung by a highly flexible human voice. Moreover, some properties of the data may have to be differently interpreted for different instruments. For example, sudden increases of the signal's amplitude may indicate new tones for some instruments like piano, but this may not be the case for other types of instruments like flute, violin, or the human voice.

For the following part of this chapter, the sound that has to be transcribed is given in form of a WAVE file, typically in CD quality with sampling rate 44,100 Hz and in 16 bit format (i.e. 2^{16} possible values).

Example 17.1 (Transcription). *For this chapter, as an example, we use the German Christmas song "Tochter Zion" (G.F. Händel) performed by a professional soprano singer. The singer is recorded in one channel and the piano accompaniment in the other channel of the stereo WAVE file.*

17.3 Musical Challenges: Partials, Vibrato, and Noise

If a tone is played or sung, it commonly does not only produce a single (co)sine wave oscillating with the fundamental frequency but also waves oscillating with integer multiples of the fundamental frequency. These waves are called partials of the whole tone (see Section 2.2). One challenge for transcription algorithms is the possible (almost) absence of the fundamental while some of the other partials are well observable.

It is particularly interesting to automatically transcribe one of the most complex musical instruments: the human voice. The human voice can adjust loudness and many other properties of the sound like vibrato and tremolo very easily within one single tone. Indeed, the sound characterization of the human voice has many more facets than for instruments because the sound varyies depending on technical and emotional expression [50, 22]. Hence robustness against such variations is very important for the design of transcription systems.

Another challenge for transcription algorithms is the presence of vibrato, some kind of intended or unintended adornment. The loudness of a singer's vibrato varies about 2–3 decibels while the pitch varies around one semitone [41] up to two semitones [27] around the desired pitch of the tone. The vibrato frequency is roughly 5–7

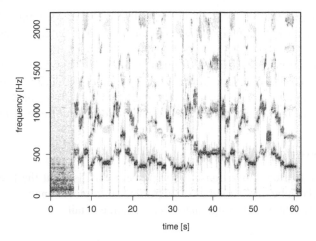

Figure 17.1: Spectrum, strong vibrato in sound performed by a professional singer. The vertical line indicates the start of the last 8 bars as shown in Figure 17.8.

Hertz. Models and detection methods for vibrato have been described, for example, by [39] and [31].

Example 17.2 (Transcription cont.). *The strong vibrato of the professional soprano singer (see Example 17.1) is shown by the nervously changing line of fundamental frequencies (the lower dark curve) in the spectrum given in Figure 17.1.*

A third challenge is the presence of noise in the signal. Noise might be caused by the environment of the music, but also by other instruments in a polyphonic performance if only one (say) instrument is of interest (predominant instrument recognition). For a more detailed discussion, see Section 17.5.2.

17.4 Statistical Challenge: Piecewise Local Stationarity

For most methods in time series analysis, both in the time and in frequency domains, at least some weak stationarity assumptions of the underlying process have to be valid (cp. Section 9.8.2). Unfortunately, even if processes of musical time series might be stationary in the mean, they are not stationary with respect to covariance (see Definition 9.40), because the tones (and hence the covariances) change quite frequently.

In [1] an algorithm for the segmentation of time series is developed and *piecewise local stationary processes* are defined as finite series of locally stationary processes. This definition is very useful for music time series: for n tones (corresponding to a series of n locally stationary processes), we expect to find at least $n - 1$ change points in the time series where some characteristic of the series changes. Unfortunately, changes from vowels to consonants (e.g. for a voice) or from one kind to another kind of tone generation (e.g. for a violin) within the same tone might lead to change points as well, which might prevent correct identification of, e.g., onsets by means of change points.

Most algorithms used in transcription apply Short Time Fourier Transformation (STFT), i.e. calculate periodograms of very small pieces (e.g. 23–46 ms, see Section 6.4) corresponding to windows (mostly overlapping by 50%) of the time series in order to detect the change points and estimate fundamental frequencies.

17.5 Transcription Scheme

A sequence of steps for a transcription process was listed at the beginning of this chapter and can be understood as steps from local to global analysis of a music time series. We will now go through these steps in some detail.

17.5.1 Separation of the Relevant Part of Music

As a first step of the transcription algorithm, the relevant part of music to be transcribed (e.g. human voice) has to be separated from other sounds (e.g. piano accompaniment). The outcome of such a separation is a time series of one relevant part of the music. To solve this sound source separation task, one of the commonly used standard methods is Independent Component Analysis (ICA) as proposed by [19] (see Section 11.6). It can separate as many sound sources as channels are available in a recoding, i.e. two channels can be separated for a typical CD quality stereo recording. Some disadvantages of ICA have been shown by [48], e.g., we cannot assume that the signals are really independent (as there are generated by performers playing the same piece of music).

Example 17.3 (Transcription cont.). *The two channels explained in Example 17.1 are mixed by a linear combination with equal weights (0.5). Hence we get two identical channels as shown in Figure 17.2. Applying ICA to the corresponding data matrix shows perfect results. In Figure 17.3 we see that the left channel starts with the piano accompaniment and the right channel contains the part of the soprano singer.*

17.5.2 Estimation of Fundamental Frequency

In the following sections, we assume the sound got well separated, e.g. by ICA, and we will only deal with one channel of monophonic sound now. Afterwards, we have to determine the fundamental frequency f_0 (cp. also Sections 4.8 and 6.4 for autocorrelation-based methods and the improved YIN algorithm). This is also called pitch estimation or f_0 estimation in the following.

References [26] and [54] propose a model for fundamental frequency estimation that combines the models of [12] and [39]. The first model [12] includes parameters for phase displacement, frequency displacement of partials, and trigonometric basis functions that model changes in amplitude. The second model [39] covers vibrato

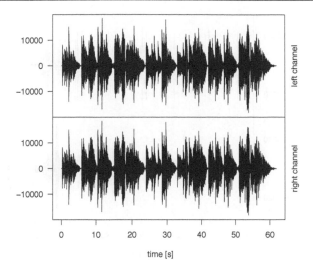

Figure 17.2: Two channels of the wave before unmixing via ICA.

using a sine wave around the "average audible" frequencies and their partials. The aim is to model well-known physical characteristics of the sound in order to estimate f_0 independently of other relevant factors that might influence estimation. Proposed methods to estimate the model are non-linear optimization of an error criterion such as the Mean Squared Error (MSE) between the real signal and the signal generated from the model after a transformation of the signals to the frequency domain.

The fundamental frequencies can, however, be estimated much faster than by the above modeling when using a heuristic approach as proposed in, e.g., [52]. In this approach several thresholds are applied to values of the periodograms $I_x[f_\mu] = |F_x[f_\mu]|^2$ (cp. Definition 9.47) derived from the complex DFT with coefficients $F_x[f_\mu]$ for Fourier frequencies f_μ on a window of size T from the original musical time series $x[t]$ in order to identify the peak representing the fundamental frequency. This is done using the following steps:

1. Restrict the frequencies f to a sensible region R defined by:

$$(I_x[f] > \text{threshold}_{noise}) \wedge (\text{lowerbound} < f < \text{upperbound}).$$

2. Identify the Fourier frequency f_v of the maximal peak:

$$I_x[f_v] \in [\text{lowerbound}, \min_R(f) \cdot \text{threshold}_{overtone}].$$

3. If there is a relevant frequency in $[l_2 \cdot f_v, u_2 \cdot f_v]$ at roughly $1.5 \cdot f_v$, we assume we found a higher partial and restart the algorithm at step 1 with a decreased threshold$_{noise}$.

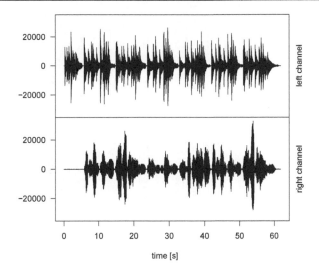

Figure 17.3: Two channels of the wave after unmixing via ICA.

Possible values are threshold$_{noise}$ = 0.1 (ignore noise), lowerbound = 80 Hz, upperbound= 5000 Hz (sensible frequency region), threshold$_{overtone}$ < 2 (keep below overtone 1), $l_2 = 1.3$, $u_2 = 1.7$ (search for overtone 2).

Unfortunately, just choosing the relevant peak is not sufficiently accurate given the resolution of the Fourier frequencies. Therefore, we have to estimate the fundamental frequency f_0 more precisely, e.g. by weighting the frequencies f^* and f^{**} of the two strongest Fourier frequencies' values $I_x[f^*]$ (strongest, see Figure 17.4) and $I_x[f^{**}]$ (second strongest) of that peak:

$$\hat{f}_0 := f^* + \frac{f^{**} - f^*}{2} \cdot \sqrt{\frac{I_x[f^{**}]}{I_x[f^*]}}. \qquad (17.1)$$

Alternatively, Quinn [37] uses an estimator, which in a similar way interpolates three Fourier coefficients, although he works directly on the complex DFT coefficients $F_x[f_\mu]$. The estimates are calculated in the following way:

1. Let μ^* be the maximizing index of $|F_x[f_\mu]^2|$ (see Figure 17.4). Note that $f^* = f_{\mu^*}$.
2. Let $\alpha_1 = Re(F_x[f_{\mu^*-1}]/F_x[f_{\mu^*}])$, $\alpha_2 = Re(F_x[f_{\mu^*+1}]/F_x[f_{\mu^*}])$, $\delta_1 = \alpha_1/(1-\alpha_1)$, and $\delta_2 = \alpha_2/(1-\alpha_2)$.
3. If both $\delta_1, \delta_2 > 0$, then $\delta := \delta_2$, else $\delta := \delta_1$.

Then, the estimated frequency of the peak is

$$\hat{f}_{0,\text{Quinn}} := (\mu^* + \delta)f_1, \qquad (17.2)$$

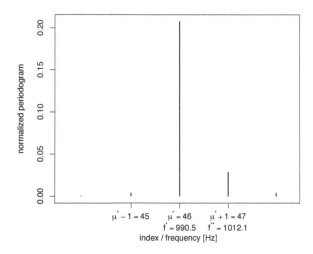

Figure 17.4: Part around the relevant peak in a periodogram showing which frequencies are used for Equations (17.1) and (17.2).

where f_1 is the first Fourier frequency. Note that $f_\mu = \mu \cdot f_1$. Hence Quinn proposes to shift away from f_{μ^*} by δ Fourier frequencies with $|\delta| < 1$.

Example 17.4 (Simulation of Frequency Estimation Methods). *Following [3] we generated time series $x_f(t) = \sin(2\pi f \cdot t/44100 + \phi) + \varepsilon_t, t = 1, \ldots, T$, where we used frequencies $f \in \{80, 81, \ldots, 1000\}$ Hz, while the noise variance σ^2 was varied from 0 to 1, and the phase ϕ was selected randomly from $[0, 2\pi]$ for the resulting sinusoids. Every signal was sampled $T = 2048$ (as a typical size of a window) times. Figure 17.5 shows the error distributions for the two estimators from Equations (17.1) and (17.2). It is clearly visible that the simple interpolation after Equation (17.1) results in the worst accuracy. It exhibits a much larger variance than the method of Quinn. Also the main mass of the distribution in Equation (17.1) is bimodal around zero.*

Example 17.5 (Peak Picking). *In some cases it turns out that finding the right peak representing the fundamental frequency is difficult. In such cases the estimation algorithms fail to estimate the correct fundamental frequency if the overtone sequence is not taken into account. An example of a periodogram showing a series of extremely strong overtones compared to the strength of the fundamental frequency is given in Figure 17.6. Here we see that the strongest overtone is the sixth one and 20 overtones are visible. The underlying signal was produced by a professional bass singer. The method based on Equation (17.1) estimates the fundamental frequency of the very first relevant peak of the tone, namely 141.35 Hz.*

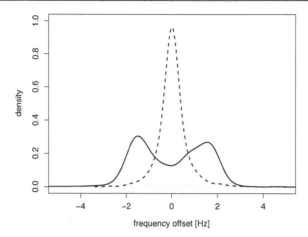

Figure 17.5: Empirical distributions of estimation errors of Equations (17.1, solid line), (17.2, dashed line).

17.5.3   *Classification of Notes, Silence, and Noise*

While it seems to be plausible to segment tones at first and to assign them to notes afterwards, this was found to be less useful in real applications with singing performances where the precessing steps are already rather error prone. Instead, a joint procedure could be used. Such a procedure has been proposed by [26], where first the classification into notes takes place by classifying an estimated frequency to the note with minimal (Euclidean) distance in cents of halftones. This is the same as the k-NN classification method (see Section 12.4.2) for $k = 1$ with the given distance learned on all possible halftones. In case of low energy in the signal, we can assume that only irrelevant noise is present, hence this is classified as silence.

Afterwards, a running median is applied to the time series of notes in order to smooth it. A running median is a median calculated in intervals moving from left to right in the time series. Finally, the segments are defined as the constant parts of the time series of smoothed notes, i.e. each change in the pitch of the smoothed notes implies a new segment.

Example 17.6 (Transcription cont.). *Applying the simple frequency estimation method from Equation (17.1) to the singer's data from Example 17.1, the estimated frequencies classified to the note with minimal (Euclidean) distance in cents of halftones can be found in Figure 17.7. The "real" sheet music has been translated to the grey shading, the black line indicates the estimated notes, and at the bottom an energy bar indicates the loudness. For the periodogram values that are used by the frequency estimation method, see Figure 17.1.*

Example 17.7 (Simulation of Frequency Estimation Methods cont.). *In Example*

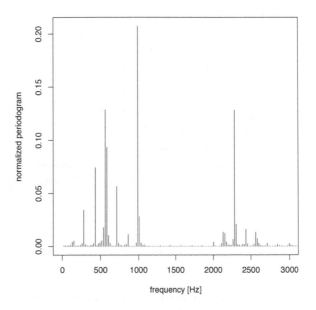

Figure 17.6: Periodogram showing an overtone series of a professional bass singer.

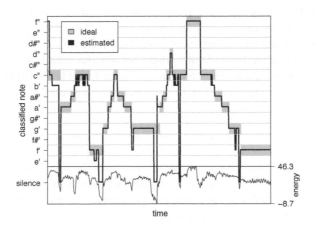

Figure 17.7: Visualization of classified notes and energy.

17.4 the minimum frequency difference (realized for the lowest tone of 80 Hz) that corresponds to a difference of 50 cents in halftones, is 2.38 Hz. Figure 17.5 shows that both estimators produce deviations mainly lower than this threshold.

As alternative methods, in Section 17.4 we already mentioned the SLEX [30] procedure and a segmentation algorithm for speech in [1]. Also, the segmentation of

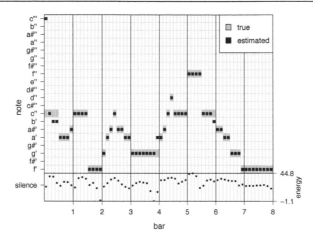

Figure 17.8: Collected information during a transcription procedure.

sound or notes has been discussed in Chapter 16. Segmentation of sound related to transcription has also been examined by [40].

17.5.4   Estimation of Relative Length of Notes and Meter

After the segmentation of notes, we have to quantize the notes, i.e. to estimate relative lengths of notes. In [28] *quantized melodies* are defined as "[...] melodies where the durations are integer multiples of a smallest time unit". For now, we assume that the tempo is fixed throughout a song. An obvious idea is to look for the least common multiple of the divisors of all note lengths to get the smallest time unit. For example, if there are quavers (length $\frac{1}{8}$), punctuated quavers $\left(\frac{1}{8} + \frac{1}{16}\right)$, quarters and half notes, the searched divisor is $\frac{1}{16}$.

Example 17.8 (Transcription cont.). *After pitch estimation, note classification and quantization, the information that has been derived is presented visually in Figure 17.8, which again shows the outcome of analyzing the last 8 bars of the German Christmas song "Tochter Zion" (G.F. Händel) performed by a professional soprano singer. Note that each segment corresponds to an eighth note. Obviously, the estimated pitches are correct most of the time.*

Unfortunately, quite large inaccuracies have to be expected in real data, because humans tend to start with notes too late, intentionally or not, and finish the notes too early, e.g., in order to breathe when singing. Hence quantization has to be very robust against such inaccuracies. Reference [2] has analyzed (intentional) variations of the tempo by famous pianists. Obviously, besides inexact length of notes, changes of tempo have to be expected as well.

Most published methods use sudden changes of the amplitude in order to track the tempo, segment the music and perform the quantization. One of these methods has been described by [10] and was extended later by [9] in order to take care of dynamic changes of the tempo during time. Alternatively, [8] proposes some Monte Carlo methods for tempo tracking and [55] uses Bayesian models of temporal structures. Reference [11] try to adapt the quantization to dynamic tempo changes. The "perceptual smoothness" of tempo in expressively performed music is analyzed by [14]. For more general findings on extracting tempo and other semantic features from audio data with signal processing techniques, see Chapter 5.

After a successful quantization, the meter has to be estimated. This is a rather difficult task, because even humans cannot always distinguish between, for example, $\frac{2}{4}$, $\frac{4}{4}$ and $\frac{4}{8}$ meters. Most of the time, it is, thus, assumed that the meter is externally given by the user of the algorithm. A detailed discussion about tempo and meter (metrical level) estimation is given in Chapter 20 and in Section 20.2.3 in particular. A rough distinction between $\frac{4}{4}$, and $\frac{3}{4}$ meters was proposed, e.g., in [51] by means of the number of quarters between so-called accentuation events.

17.5.5 Estimation of the Key

The basic idea for key estimation [7] is as follows. All notes from a piece of music can be tabulated. Depending on the frequencies (i.e. the number of occurrences) of the twelve different notes (including halftones), the most probable key can be estimated. See Chapter 19 for chord and hence implicitly also key recognition.

Bayesian modeling can also be used for key estimation. Temperley [43, 44] proposes such a model for a given piece or segment of music (cp. Chapter 16). Here, the probability computation is based not only on the relative frequency with which the twelve scale degrees appear in a key, but also on the probability of a segment (as defined in Section 16.4) being in the same key as the previous segment in the same piece (probability of modulation). Other more sophisticated approaches would also analyze the sequence of tones and chords. More references on estimating the key, e.g. by analysis of the pitch distribution using histograms, are given in Section 19.7.

17.5.6 Final Transcription into Sheet Music

In the preceding sections we have described how to estimate properties of the sound that are required for the transcription of sound to sheet music. The final part of producing the sheet music is a matter of music notation and score printing.

17.6 Software

The freely available R [38] package *tuneR* [26] is a framework for statistical analysis and transcription of music time series which provides many tools (e.g. for reading WAVE files, estimating fundamental frequencies, etc.) in the form of R functions. Therefore, it is highly flexible, extendable and allows experimenting and playing around with various methods and algorithms for the different steps of the transcription procedure. A drawback is that knowledge of the statistical programming language R is required, because it does not provide transcription on a single key press nor any graphical user interface – as opposed to commercial products.

A free and powerful software for music notation is LilyPond [29] which uses LATEX[24], the well-known enhancement of TEX[23]. Beside sheet music, LilyPond is also capable of generating MIDI files. Therefore it is possible to examine the results of transcription both visually and acoustically. The R package *tuneR* contains a function which implements an interface from the statistical programming language to LilyPond.

Finally, we discuss commercial software products. We have reviewed more than 50 software products of which only 7 provide the basic capabilities of transcription we ask for, which means taking a WAVE file and converting it to some format of Midi or sheet music like representation. Those we found are *AKoff Music Composer*,[1] *AmazingMIDI*,[2] *AudioScore*,[3] *Intelliscore*,[4] *Melodyne*,[5] *Tartini*,[6] and the *WIDI Recognition System*.[7]

From our point of view, the well known *Melodyne* is currently the best commercial transcription software we tried out for the singer's transcription. It performs all the steps required by a full featured transcription software, including key and tempo estimation. Its recognition performance is quite good (see Example 17.9), even with default settings on sound that has been produced by human voices. Some parameters can be tuned in order to improve recognition performance.

Example 17.9 (Transcription cont.). *The outcome of the example that has been continued throughout this chapter is the transcription of 8 bars of "Tocher Zion" given in Figure 17.10 for tuneR. Figure 17.11 shows the transcription for Melodyne. For comparison, the original notes of that part of "Tocher Zion" are shown in Figure 17.9.*

For both Melodyne *and* tuneR *we have optimized the quantization by specifying the number of bars and the speed. The quality of the final transcriptions is comparable. The software* tuneR *produces more "nervous" results. At some places, additional notes have been inserted where the singer slides smoothly from one note to another. The first note is estimated one octave too high due to an immensely strong second partial almost in absence of any other partials.* Melodyne *omits some notes. Here we guess that* Melodyne *smooths the results too much and detects smooth transitions of the singer even if the singer intended to sing a separate note.*

17.7   Concluding Remarks

A typical transcription algorithm includes steps related to pre-processing of the original musical audio signal, time series modeling to estimate fundamental frequencies,

---

[1]http://www.akoff.com/music-composer.html. Accessed 18 December 2015.

[2]http://www.pluto.dti.ne.jp/araki/amazingmidi/. Accessed 18 December 2015.

[3]http://www.sibelius.com/products/audioscore/ultimate.html. Accessed 18 December 2015.

[4]http://www.intelliscore.net/. Accessed 18 December 2015.

[5]http://www.celemony.com/. Accessed 18 December 2015.

[6]http://miracle.otago.ac.nz/tartini/. Accessed 18 December 2015.

[7]http://www.widisoft.com/. Accessed 18 December 2015.

Figure 17.9: Original sheet music of "Tochter Zion."

Figure 17.10: Transcription by *tuneR*.

Figure 17.11: Transcription by *Melodyne*.

classification into notes, fitting into meter and key, and sheet music production. This chapter gave an overview over some methods for all these steps. Note that in most steps there are noise and uncertainties involved and we have to make rather strong assumptions in order to get results which are still much worse than the original sheet music.

17.8 Further Reading

Reference [21] uses the spectral smoothness method for both sound source separation and polyphonic fundamental frequency estimation. Another method for sound source separation has been proposed by Viste and Evangelista [46, 47]. Their idea is to estimate the delays between the signals from different sources and put constraints on the deconvolution coefficients. They aim at audio coding and compression for formats like MPEG 3 [5], or the integration into hearing aids.

The SLEX (Smooth Localized Complex Exponential) transformation by [30] can segment bivariate non-stationary time series into almost stationary segments and it can be flexibly adapted to different time and frequency resolutions. For other related time series methods in the frequency domain, see also [4], [6], and particularly for signal analysis, see [45].

Many approaches for the estimation of the fundamental frequency, for both monophonic and polyphonic sound, have been published. Reference [18] proposes a method called *PreFEst* for the "predominant f_0 estimation" of melody and bass lines without requiring assumptions about the number of sound sources. Reference [13] describes a heuristic method for the identification of notes and [21] describes some method for polyphonic estimation of fundamental frequencies. Reference [42] extends the Fast Fourier Transformation (FFT) by "Non-Negative Matrix Factorization" (cp. Section 23.2.1.1) for polyphonic transcription. Bayes methods for the f_0 estimation of monophonic and polyphonic sound have been proposed by [49], [12], and again [16]. A rather theoretical work by [56] introduces Bayesian variable selection for spectrum estimation. In the MAMI project (Musical Audio-Mining, see [25]), software for the fundamental frequency estimation has been developed.

Reference [32] proposes "Algorithms for Nonnegative Independent Component Analysis" (N-ICA) in order to extract features of polyphonic sound, but applies it only to sound generated by MIDI instruments. Moreover, [33] suggests optimization using Fourier expansion for N-ICA and expresses his hope to extend the method to perform well for regular ICA. In another work, [34] proposes to use dictionaries of sounds, i.e., databases that contain many tones of different instruments played in different pitches. Using such dictionaries might overcome the problem that different tones containing a lot of partials may not be identifiable for polyphonic problems.

Under some circumstances the frequency of partials is slightly shifted from the expected value. This is a problem for the polyphonic case, if a partial's frequency cannot be assigned to a corresponding fundamental frequency. Hence this phenomenon has to be modeled as done in some recent work by [17].

Reference [35] modeled phenomena like pink noise (noise decreasing with frequency; also known as $1/f$ noise) using wavelet techniques in order to get a more appropriate model and hence better estimates. Later on, [36] also modeled other special kinds of unwanted noise or the sound of consonants that do not sound with a well-defined fundamental frequency. A more general article about wavelet analysis of music time series can be found in [15].

A general overview of music transcription methods can be found, e.g., in [20].

Bibliography

[1] S. Adak. Time-dependent spectral analysis of nonstationary time series. *Journal of the American Statistical Association*, 93:1488–1501, 1998.

[2] J. Beran. *Statistics in Musicology*. Chapman & Hall/CRC, Boca Raton, 2004.

[3] B. Bischl, U. Ligges, and C. Weihs. Frequency estimation by DFT interpolation: A comparison of methods. Technical Report 06/09, SFB 475, Department of Statistics, TU Dortmund, Germany, 2009. http://www.statistik.tu-dortmund.de/fileadmin/user_upload/Lehrstuehle/MSind/SFB_475/2009/tr06-09.pdf.

[4] P. Bloomfield. *Fourier Analyis of Time Series: An Introduction*. John Wiley and Sons, 2nd edition, 2000.

[5] K. Brandenburg and H. Popp. An introduction to MPEG Layer 3. *EBU Technical Review*, 2000.

[6] D. Brillinger. *Time Series: Data Analysis and Theory*. Holt, Rinehart & Winston Inc., NY, 1975.

[7] H. Brown, D. Butler, and M. Jones. Musical and temporal influences on key discovery. *Music Perception*, 11(4):371–407, 1994.

[8] A. Cemgil and B. Kappen. Monte Carlo methods for tempo tracking and rhythm quantization. *Journal of Artificial Intelligence Research*, 18:45–81, 2003.

[9] A. Cemgil, B. Kappen, P. Desain, and H. Honing. On tempo tracking: Tempogram representation and Kalman filtering. *Journal of New Music Research*, 29(4):259–273, 2001.

[10] T. Cemgil, P. Desain, and B. Kappen. Rhythm quantization for transcription. *Computer Music Journal*, 24(2):60–76, 2000.

[11] M. Davies and M. Plumbley. Causal tempo tracking of audio. In *Proceedings of the International Conference on Music Information Retrieval*, Audiovisual Institute, Universitat Pompeu Fabra, Barcelona, Spain, 2004.

[12] M. Davy and S. Godsill. Bayesian harmonic models for musical pitch estimation and Analysis. Technical Report 431, Cambridge University Engineering Department, Cambridge, 2002.

[13] S. Dixon. Multiphonic note identification. *Australian Computer Science Communications*, 17(1):318–323, 1996.

[14] S. Dixon, W. Goebl, and E. Cambouropoulos. Perceptual smoothness of tempo in expressively performed music. *Music Perception*, 23(3):195–214, 2006.

[15] G. Evangelista. Flexible wavelets for music signal processing. *Journal of New Music Research*, 30(1):13–22, 2001.

[16] S. Godsill and M. Davy. Bayesian modelling of music audio signals. In *Bulletin of the International Statistical Institute, 54th Session*, volume LX, book 2, pp. 504–506, Berlin, 2003.

[17] S. Godsill and M. Davy. Bayesian computational models for inharmonicity in musical instruments. In *IEEE Workshop on Applications of Signal Processing to Audio and Acoustics*, New Paltz, NY, October 16–19 2005.

[18] M. Goto. A predominant-F0 estimation method for polyphonic musical audio signals. In *Proceedings of the 18th International Congress on Acoustics (ICA'04)*, pp. 1085–1088, Kyoto, Japan, 2004. Acoustical Society of Japan.

[19] A. Hyvärinen, J. Karhunen, and E. Oja. *Independent Component Analysis*. John Wiley and Sons, NY, 2001.

[20] A. Klapuri and M. Davy, eds. *Signal Processing Methods for Music Transcription*. Springer, 2006.

[21] A. Klapuri. Multipitch estimation and sound separation by the spectral smoothness principle. In *IEEE International Conference on Acoustics, Speech and Signal Processing (ICASSP)*, 2001.

[22] B. Kleber. Evaluation von Stimmqualität in westlichem, klassischen Gesang. Diploma Thesis, Fachbereich Psychologie, Universität Konstanz, Germany, 2002.

[23] D. Knuth. *The TEXbook*. Addison-Wesley, 1984.

[24] L. Lamport. *LATEX, a Document Preparation System*. Addison-Wesley, 2nd edition, 1994.

[25] M. Lesaffre, K. Tanghe, G. Martens, D. Moelants, M. Leman, B. De Baets, H. De Meyer, and J.-P. Martens. The MAMI query-by-voice experiment: Col-

lecting and annotating vocal queries for music information retrieval. In _Proceedings of the International Conference on Music Information Retrieval_, 2003.

[26] U. Ligges. _Transkription monophoner Gesangszeitreihen_. Dissertation, Fachbereich Statistik, Universität Dortmund, Dortmund, Germany, 2006.

[27] J. Meyer. _Akustik und musikalische Aufführungspraxis_. Bochinsky, Frankfurt am Main, 1995.

[28] D. Müllensiefen and K. Frieler. Optimizing measures of melodic similarity for the exploration of a large folk song database. In _5th International Conference on Music Information Retrieval_, Barcelona, Spain, 2004.

[29] H.-W. Nienhuys, J. Nieuwenhuizen, et al. _GNU LilyPond: The Music Typesetter_. Free Software Foundation, 2005. Version 2.6.5.

[30] H. Ombao, J. Raz, R. von Sachs, and B. Malow. Automatic statistical analysis of bivariate nonstationary time series. _JASA_, 96(454):543–560, 2001.

[31] H. Pang and D. Yoon. Automatic detection of vibrato in monophonic music. _Pattern Recognition_, 38(7):1135–1138, 2005.

[32] M. Plumbley. Algorithms for nonnegative independent component analysis. _IEEE Transactions on Neural Networks_, 14(3):534–543, 2003.

[33] M. Plumbley. Optimization using Fourier expansion over a geodesic for non-negative ICA. In _Proceedings of the International Conference on Independent Component Analysis and Blind Signal Separation (ICA 2004)_, pp. 49–56, Granada, Spain, 2004.

[34] M. Plumbley, S. Abdallah, T. Blumensath, M. Jafari, A. Nesbit, E. Vincent, and B. Wang. Musical audio analysis using sparse representations. In _COMPSTAT 2006 — Proceedings in Computational Statistics_, pp. 104–117, Heidelberg, 2006. Physica Verlag.

[35] P. Polotti and G. Evangelista. Harmonic-band wavelet coefficient modeling for pseudo-periodic sound processing. In _Proceedings of the COST G-6 Conference on Digital Audio Effects (DAFX-00)_, Verona, Italy, December 7–9 2000.

[36] P. Polotti and G. Evangelista. Multiresolution sinusoidal/stochastic model for voiced-sounds. In _Proceedings of the COST G-6 Conference on Digital Audio Effects (DAFX-01)_, Limerick, Ireland, December 6–8 2001.

[37] B. G. Quinn. Estimating frequency by interpolation using Fourier coefficients. _IEEE Transactions on Signal Processing_, 42(5):1264–1268, 1994.

[38] R Core Team. _R: A language and environment for statistical computing_. R Foundation for Statistical Computing, Vienna, Austria, 2015.

[39] S. Rossignol, P. Depalle, J. Soumagne, X. Rodet, and J.-L. Collette. Vibrato: Detection, estimation, extraction, modification. In _COST-G6 Workshop on Digital Audio Effects_, 1999.

[40] S. Rossignol, X. Rodet, J. Soumagne, J.-L. Collette, and P. Depalle. Automatic characterisation of musical signals: Feature extraction and temporal segmentation. _Journal of New Music Research_, 28(4):281–295, 1999.

[41] W. Seidner and J. Wendler. *Die Sängerstimme*. Henschel, Berlin, 1997.

[42] P. Smaragdis and J. Brown. Non-negative matrix factorization for polyphonic music transcription. In *IEEE Workshop on Applications of Signal Processing to Audio and Acoustics*, pp. 177–180, October 2003.

[43] D. Temperley. Bayesian models of musical structure and cognition. *Musicae Scientiae*, 8(2):175–205, Fall 2004.

[44] D. Temperley. A probabilistic model of melody perception. In *Proceeding of the 7th International Conference on Music Information Retrieval*, pp. 276–279, 2006.

[45] H. Van Trees. *Detection, Estimation, and Modulation Theory, Part I*. Wiley-Interscience, Melbourne, FL, USA, reprint edition, 2001.

[46] H. Viste and G. Evangelista. Sounds source separation: Preprocessing for hearing aids and structured audio coding. In *Proceedings of the COST G-6 Conference on Digital Audio Effects (DAFX-01)*, Limerick, Ireland, December 6–8 2001.

[47] H. Viste and G. Evangelista. An extension for source separation techniques avoiding beats. In *Proceedings of the 5th Int. Conference on Digital Audio Effects (DAFx-02)*, Hamburg, Germany, September 26–28 2002.

[48] F. von Ameln. Blind source separation in der Praxis. Diplomarbeit, Fachbereich Statistik, Universität Dortmund, Dortmund, Germany, 2001.

[49] P. Walmsley, S. Godsill, and P. Rayner. Polyphonic pitch tracking using joint Bayesian estimation of multiple frame parameters. In *IEEE Workshop on Applications of Signal Processing to Audio and Acoustics*, New Paltz, NY, October 17–20 1999.

[50] J. Wapnick and E. Ekholm. Expert consensus in solo voice performance evaluation. *Journal of Voice*, 11(4):429–436, 1997.

[51] C. Weihs and U. Ligges. From local to global analysis of music time series. In K. Morik, J.-F. Boulicaut, and A. Siebes, eds., *Local Pattern Detection*, Lecture Notes in Artificial Intelligence 3539, pp. 217–231, Berlin, 2005. Springer-Verlag.

[52] C. Weihs and U. Ligges. Parameter optimization in automatic transcription of music. In M. Spiliopoulou, R. Kruse, A. Nürnberger, C. Borgelt, and W. Gaul, eds., *From Data and Information Analysis to Knowledge Engineering*, pp. 740–747, Berlin, 2006. Springer-Verlag.

[53] C. Weihs, U. Ligges, F. Mörchen, and D. Müllensiefen. Classification in music research. *Advances in Data Analysis and Classification*, 1:255–291, 2007.

[54] C. Weihs, U. Ligges, and K. Sommer. Analysis of music time series. In A. Rizzi and M. Vichi, eds., *COMPSTAT 2006 – Proceedings in Computational Statistics*, pp. 147–159, Heidelberg, 2006. Physica Verlag.

[55] N. Whiteley, A. Cemgil, and S. Godsill. Bayesian modelling of temporal structure in musical audio. In *7th International Conference on Music Information*

Retrieval, pp. 29–34, Victoria, Canada, 2006.

[56] P. Wolfe, S. Godsill, and W.-J. Ng. Bayesian variable selection and regularization for time-frequency surface estimation. *Journal of the Royal Statistical Society: Series B (Statistical Methodology)*, 66(3):575–589, 2004.

Chapter 18

Instrument Recognition

CLAUS WEIHS, KLAUS FRIEDRICHS, KERSTIN WINTERSOHL
Department of Statistics, TU Dortmund, Germany

18.1 Introduction

The goal of instrument recognition is the automatic distinction of the sounds of musical instruments playing in a given piece of music. Under most circumstances it is a difficult task since different musical instruments have different compositions of partial tones (cp. Definition 2.3), e.g., in the sound of a clarinet only odd partials occur. This composition of partials is, however, also dependent on other factors like the pitch, the played instrument, the room acoustics, and the performer [14]. Additionally, there are temporal changes within one tone like vibrato. Also, different non-harmonic properties, e.g. noise in the attack phase of a tone (cp. Section 2.4.7), are typical for many families of instruments. For a plucked string, e.g., the attack is the very short period between the initial contact of the plectrum or finger and the scraping of the string. For a hammered string, the attack is the period between the initial contact and the rebounding of the hammer (or mallet). Both, the plectrum and the hammer produce typical noise. Hence, expert knowledge for distinguishing the instruments is very specific and complex, and instead of expert rules, supervised classification (see Chapter 12) is usually applied.

The typical processing flow of instrument recognition is illustrated in Figure 18.1. It starts with an appropriate data set of labeled observations, labels corresponding to instruments or families of instruments. Dependent on the concrete application, the kind of data can be very different. For example, observations can be derived from single tones, but also from complete pieces of music. There are at least four dimensions which define the complexity of a specific instrument recognition task. These dimensions are described in detail in the next section.

The next processing step is the taxonomy applied to the data. The obvious one is a flat taxonomy, where each observation is directly assigned to an instrument label. However, due to the different degree of similarity between different pairs of instruments a hierarchical taxonomy makes also sense, which will be discussed in Section 18.3.

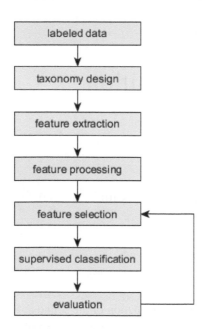

Figure 18.1: Instrument recognition.

For building a classifier the meaningful information of each observation has to be extracted. Therefore, appropriate features have to be extracted, which are discussed in Chapter 5. Naturally, for instrument recognition timbre features are the most important ones. However, even if we restrict ourselves to timbre-related features, the dimensionality of the feature-space is by far too high for further processing since there are lots of timbre features and most of them are generated framewise, which can yield hundreds of values of the same feature for one single tone. Hence, the next step is the aggregation of the values at different time points, e.g., by simple statistics like mean or variance. Particularly in the case of single-tone classification an additional temporal aggregation makes sense, since many instruments can be more easily distinguished by their evolution in time. For example a guitar and a piano usually differ much during the attack-phase but can sound very similar during the sustain-phase. This topic and additional feature processing techniques are described in detail in Chapter 14 (Feature Processing).

Even after the feature processing step, the dimensionality of the feature space might still be very high, impeding efficient classification. At the same time some meaningless features can even negatively affect the classification due to possible overfitting. Additionally, some classification methods, e.g. "linear discriminant analysis" (cp. Section 12.4.1), have problems dealing with redundant features. To further reduce the amount of features, a selection process can be applied. There exist two main methods: Filter methods select the features before any classification method is applied while wrapper methods select the features with respect to their classification performance (cp. Section 15.4).

For the actual classification all methods described in Chapter 12 can be applied. As mentioned there, a classifier is a map $f : X \rightarrow Y$. Here, X is a (reduced) set of features and Y is a set of labels of musical instruments or instrument families. Since it cannot be decided a priori which classification method is best and which are its best parameter settings, an appropriate evaluation step is needed. The most popular evaluation method is 10-fold cross-validation (see Chapter 13).

18.2 Types of Instrument Recognition

In this section we will discuss five aspects related to the complexity of instrument recognition.

Aspect 1: Tones   Classification can be carried out for single tones, tone intervals and tone chords (e.g. a piano can produce several tones at the same time), where the two latter tasks have a higher complexity due to possible overlapping of partials. Moreover, classification can be applied to entire tone sequences or to short tone segments. While meaningful segmentation, e.g., into single tones is a big challenge in itself (see Chapter 16), classification of entire tone sequences has the advantage of providing more information.

Aspect 2: Polyphony   Monophonic instrument recognition where only one instrument is playing at each point of time is much easier than the polyphonic variant where more than one instrument is playing at the same time. Most recent studies deal with the polyphonic variant. Here, the main challenge originates from overlapping partials of different tones which has a nonlinear effect on the features' values. For example, louder instruments can mask softer ones. In the easiest variant of polyphonic instrument recognition only the predominant instrument has to be classified. This task can be solved similar to the monophonic variant, but is much harder due to the additional "noise" from the accompanying instruments [18]. The complexity even increases if all accompanying instruments have to be detected. Here, the problem arises how to classify multiple correlated events which occur at the same time. The naive approach is generating one class for each possible combination of instruments which obviously results in too many classes. An alternative is starting with sound source separation (cp. Section 11.6) in order to apply monophonic instrument recognition afterwards. Naturally this concept fails if the sources are not separated well, a task which itself is still a hard challenge. The third possibility is multi-label classification [14].

Aspect 3: Instrument Types   Additionally, the complexity of the classification problem is influenced by the set of instruments to be distinguished. Many similar instruments make the problem more difficult and recognizing only the instrument family is obviously easier than recognizing the exact instrument. Naturally, also the pure amount of considered instruments has a big impact on the results. While some studies just deal with two instruments, others deal with dozens.

Aspect 4: Individual Instruments   In some applications only one specific representative of each instrument class has to be recognized. But for most applications the goal is getting a universally valid model which can recognize classes of instruments even for unseen representatives. This is not trivial since the timbre of musical instruments can be relatively different depending on their construction type. Therefore, on the one hand it is crucial to include as many representatives of the different instrument classes into the training data set as possible. On the other hand, for a realistic evaluation of a universally valid model, instruments which occur in the training set should not occur in the test set. Unfortunately, in practice it is difficult getting that much data and so in many studies the considered music pieces are only based on, say, three different representatives of each instrument class at most.

Aspect 5: Databases   There exist three *databases* commonly used for (monophonic) single tone classification: the McGill University Master Samples (MUMS) database [13], the University of Iowa Musical Instrument Samples [8] and the Real World Computing (RWC) Database [9]. However, the various other types of instrument recognition discussed in this section are lacking clear reference data sets. Hence, it is often difficult to compare results of different studies and the evaluation of accuracy results should take this point into account.

18.3   Taxonomy Design

The easiest taxonomy for instrument recognition is a flat taxonomy where each sample is directly assigned to one instrument. An example is shown in Figure 18.2. Since the problem complexity gets higher if the considered instruments have similar timbres like horn and trumpet, many approaches apply a hierarchical taxonomy, either a natural one using instrument families or an automatic one, e.g. from clustering groups of instruments. Hierarchical taxonomies can be seen as a tree or a directed graph with a classification task at each node. One advantage of hierarchical classification is that at each node, other features can be selected and other classification methods can be chosen, since each classification model is trained independently. In particular, the best set of features to distinguish a set of instruments can be very diverse for different tasks [16]. In principle, feature generation and feature processing could also be differently designed at each node. Nevertheless, this effort is rather unusual in practice.

Hierarchical classification works as follows. First, a classification model is determined for the top level of the taxonomy on the basis of all samples in the training set. Then, classification models are determined for the second level of the taxonomy, but only on those samples which were classified into the corresponding class on the second level, etc. Therefore, the drawback of hierarchical classification is that errors made at higher levels are propagated down to all levels below.

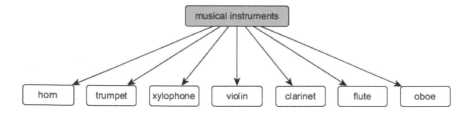

Figure 18.2: Flat taxonomy.

A typical natural taxonomy is to group the instruments into brass, percussion, strings and woodwinds (see Figure 18.3). This taxonomy has the disadvantage that

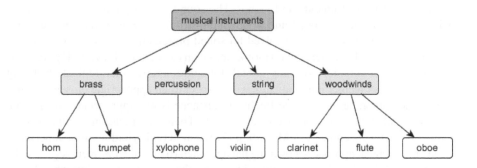

Figure 18.3: Hierarchical taxonomy.

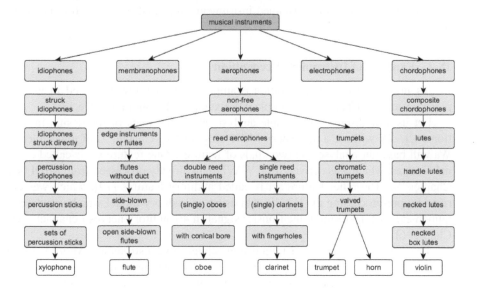

Figure 18.4: Hornbostel–Sachs system (extract).

not all instruments fit well into one (and only one) group. For example, the piano could be classified into string and percussion [14]. Another popular taxonomy
is the Hornbostel–Sachs system which considers the sound production source of
the instruments [17]. It consists of over 300 categories, ordered on several levels.
On the first level, instruments are classified into five main categories: idiophones,
membranophones, chordophones, aerophones, and electrophones. Idiophones are
instruments where the instrument body is the sound source itself without requiring stretched membranes or strings. This includes all percussion instruments except
drums. Membranophones are all instruments where the sound is produced by tightly
stretched membranes which includes most types of drums. Chordophones are all
instruments where one or more strings are stretched between fixed points, which in-

cludes all string instruments and piano. The sound of aerophones is produced by vibrating air like in most brass and woodwind instruments. Electrophones are all instruments where electricity is involved for sound producing, such as synthesizers or theremins. A small extract of the Hornbostel–Sachs system is shown in Figure 18.4.

In [6] an automatic taxonomy is built by agglomerative hierarchical clustering putting classes together automatically with respect to an appropriate closeness criterion. The authors argue that the Euclidean distance is not appropriate and instead, two probabilistic distance measures are tested. Their classification results, using a "support vector machine" (SVM) with a Gaussian kernel, yield a slight superiority of the hierarchical approaches over the flat approach (64% vs. 61% accuracy). On the other hand, following [5] there is no evidence for the superiority of hierarchical classification of single instruments in comparison to flat classification. However, in both studies, pizzicati and sustained instruments are quite well distinguished, whereas the classification of individual sustained instruments appeared to be much more error-prone.

18.4 Example of Instrument Recognition

In this book, instrument recognition examples were already discussed in various places. See, e.g., Chapter 6, Example 9.9, Example 11.2, and Example 13.4.

Let us now discuss an example analysis along the lines of the design in Section 18.1. We will use a data set of MIDI versions (see Chapter 7) of music pieces. We will base our analyses on two kinds of features. One feature set is called original features, which was derived from WAVE data directly processed from MIDI (see Chapter 7). The other feature set is called auditory features [19], which was derived by pre-processing the WAVE data using an auditory model which simulates a human ear by transforming the (music) signal in 40 auditory nerves / channels (see Chapter 6).

The classification task is distinguishing between five instruments. Multiple MIDI versions of each music piece are produced – one for each instrument – so that each of these instruments is playing the main voice in one version. Various features of these music pieces are extracted both from the original signals and from the corresponding auditory signals. The prediction quality of the classification rules based on these two kinds of features is compared. Other targets are to identify the most important features in order to reduce computation time and to identify particularly suitable classification methods.

18.4.1 Labeled Data

We analyze a set of 16 phrases of chamber music pieces recorded in MIDI which include a specific melody instrument (main voice) and one or more accompanying instruments. For the main voice, five melody instruments are compared, flute, clarinet, oboe, trumpet, and violin. The idea is to classify the data according to the predominant instrument (main voice). All phrases together consist of 170 predominant tones. Each piece was replicated 5 times by changing the melody instrument to one of the five instruments resulting in 80 phrases, respectively 850 tones, with 5 different class labels. The accompanying instruments may be a piano or strings and are not changed. The ISP toolbox[1] in MATLAB® is applied to convert the phrases into WAVE files with a sampling rate of 44,100 Hz.

The features derived from these data will be introduced in Section 18.4.3. The class variable (label) is an instrument indicator specifying the instrument of the main voice.

18.4.2 Taxonomy Design

For simplification, a flat taxonomy is applied. However, for comparison two hierarchical taxonomies are also tested. The first one is created based on the classical taxonomy which was shown in Figure 18.3. For our five instruments, this approach yields the tree which can be seen in Figure 18.5. The classification task on the first level is distinguishing between woodwind, trumpet and violin. When the observation is classified as woodwind the next step is distinguishing between clarinet, flute and oboe. This means that two classification models have to be trained. On the first level, this is simply achieved by changing the label of the specific woodwind instruments into woodwind and the second level is trained using only the woodwind observations.

The second hierarchical taxonomy is based on the Hornbostel–Sachs system (see Figure 18.4). This results in the tree shown in Figure 18.6. On the first level, all observations are classified into aerophone or violin, on the second level, aerophones are classified into reed aerophone, flute or trumpet, and on the last level, reed aerophones are classified into oboe or clarinet. We will apply these hierarchical taxonomies just in Section 18.4.5.4. All other evaluations in this section correspond to the flat taxonomy.

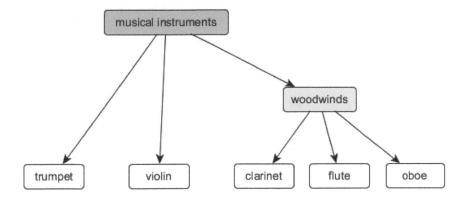

Figure 18.5: Hierarchical taxonomy for our example.

---

[1] http://kom.aau.dk/project/isound/

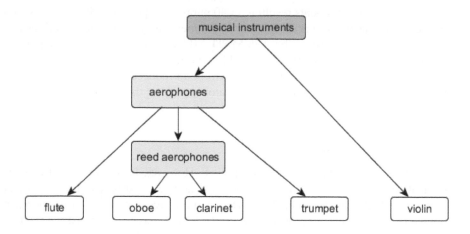

Figure 18.6: Hornbostel–Sachs system for our example.

18.4.3 Feature Extraction and Processing

The first data set we analyze is based on the WAVE signal directly and consists of 21 features (see below). Tones are separated by their known onsets and offsets. The second data set is based on the auditory model (see Chapter 6) and consists of 840 corresponding auditory features, i.e. the 21 features from 40 channels. Here, each music phrase is processed completely through the auditory model and separation is done afterwards.

In the feature processing step, just simple statistics like mean and variance are used to aggregate feature values of different frames.

The 21 characterizing features will be introduced in the following. Please note that in contrast to the definitions in Section 5.2, here, features are defined on complete tones and not frame-wise. Hence, the definitions can be simplified as they are independent of the frame index λ.

- "rms": "root mean square energy": Global energy of signal x: $e_{rms} = \sqrt{\frac{1}{n}\sum_{k=0}^{n-1}x[k]^2}$, where $x[k]$ is the amplitude at time (sampling instance) k (cp. Definition 5.2).

- "lowenergy": Percentage of frames with energy lower than mean (cp. Definition 5.2, Equation (5.4)).

- "mean spectral flux": Mean of spectral flux of successive frames (cp. Definition 5.10).

- "standard deviation of spectral flux": Standard deviation of spectral flux.

- "spectral rolloff": Smallest frequency index μ_{sr} below which at least 85% of the cumulated spectral magnitudes are concentrated (cp. Definition 5.8).

- "spectral brightness": Share of cumulated spectral magnitudes in frequencies greater equal 1500 Hz (cp. Definition 5.9). Note that the values of "brightness" and "rolloff" are less dependent on the instrument than on the tone pitch.

- "irregularity": Degree of variation of the amplitudes of successive partial tones (cp. Definition 5.11). This measures whether successive partial tones have similar energy. For some instruments, such partial tones have very different energies. For example, for clarinets the odd partial tones exhibit nearly the whole energy of a tone.

- "entropy": Shannon Entropy: $H(X) = -\sum_{\mu=0}^{M-1} p(|X[\mu]|) \log_2 p(|X[\mu]|)$, where X is the DFT of the time signal and $p(|X[\mu]|) = \frac{|X[\mu]|}{\sum_{\nu=0}^{M-1} |X[\nu]|}$ is the share of the μth frequency bin with respect to the cumulated spectral magnitudes of all bins. $H(X)$ measures the degree of spectral dispersion of an acoustic signal and is taken as a measure for tone complexity. The entropy is minimal for pure tones (only one frequency) and maximal for white noise, i.e. for signals where all frequency bins have identical spectral magnitudes.

- "mfcc" 1-13: First 13 "Mel Frequency Cepstral Coefficients" describing the spectral form of acoustic signals (cp. Section 5.2.3).

The features are computed by means of the "MIR Toolbox" [11] in the software MATLAB®.

18.4.4 Feature Selection and Supervised Classification

In this subsection, we describe how a study can be designed to derive meaningful results. On the one hand, we compare the classification performance of features derived from original and auditory signals over all five instruments; on the other hand, we look at "One-vs-All" classifications (cp. Section 12.4.4) in order to assess how good one instrument can be separated from all others. Last but not least, we study which features are particularly important for class separation.

Overall, we discuss whether there are classification methods particularly appropriate for instrument classification. We compare the following classification methods (cp. Chapter 12): linear discriminant analysis (LDA), quadratic discriminant analysis (QDA), support vector machine (SVM) (linear (SVML), polynomial kernel (SVMP), radial basis kernel (SVMR)), decision trees (CART), random forests (RF), and k-nearest-neighbor (k-NN).

Forward and backward feature selection (see Section 15.4) are applied to the original feature set in order to identify the most important features which eventually might even improve classification results. Here, forward selection means that at each iteration, the feature is added which maximally decreases the misclassification error measured by 10-fold cross-validation. As the stopping criterion we define that the improvement of the classification error between two consecutive iterations is < 0.01. In an analogous manner, features are removed by backward selection. Here the minimum improvement is set to -0.001, which means that also a slight increase of the classification error is allowed in order to get a less complex model.

Applying these methods to the bigger auditory feature set leads to huge time complexities, which means at least backward feature selection is almost impracticable. However, a way to reduce the complexity is grouping the features into feature groups and handling each group as one single feature for forward and backward selection,

Table 18.1: Error Rates Using All Instruments and All Features

Data	CART	LDA	QDA	SVMR	SVML	SVMP	RF	k-NN
original	0.36	0.25	0.24	0.21	0.21	**0.19**	0.22	0.27
auditory	0.32	0.78	—	0.15	0.15	**0.14**	0.16	0.27

respectively. There are two natural grouping mechanisms since the features can be categorized by two dimensions: the channel number and the feature name. The first approach is to combine the same features over all channels into one group and the second approach is to combine all features generated within the same channel into one group. This results in 21 feature groups for the first approach and 40 groups for the second approach. Channel-based grouping has the additional advantage of neglecting entire channels, which not only reduces computing time for feature generation but also the computing time for the auditory modeling process.

Some of the classification methods have free parameters. These parameters are tuned by means of a grid search, i.e. we test all parameter combinations on a grid and take that combination with the lowest classification error rate. For SVML we tune the "cost" parameter C, for SVMR C and the kernel width γ, and for SVMP C, γ, the increment q, and the polynomial degree d (see Section 12.4.4). For k-NN the parameters k and λ in the Minkowski distance are tuned (see Section 12.4.2). For "One-vs-All"-classifications and for feature selection, the default parameters are used due to runtime restrictions.

18.4.5   Evaluation

For error estimation, generally the mean of 10 random repetitions of 10-fold cross-validation is taken. In tuning, only 3-fold cross-validation is carried out.

18.4.5.1   Multiple Classes

In Table 18.1 the error rates for the different classification models are shown when all classes are included in the analysis. To compare the quality of these results, let us consider the result of a random classifier. For our five-class problem with uniform distributed class labels we can obviously expect an error rate of 80% which is much worse than the results of the best classification methods. For the original data, the polynomial SVM is best (18.62% error rate) and CART is worst (35.67% error rate). For the auditory data, the linear SVM is best (15.38% error rate) and the error rates for radial and polynomial SVM and for random forests are nearly equally good (all not higher than 17%). The LDA method is distinctly worst (77.66% error rate). This is due to collinear features (cp. Section 9.8.1), which could be removed by a prior feature selection. (Note that we have 840 auditory features.) However, the collinear features are no problem for the other classification methods. Except for LDA and QDA, the errors were always lower for auditory data. Note that QDA was not applicable to auditory data because of the big number of features involved.

Table 18.2: Error Rates One vs. Rest; Single Class Instrument Specified

Data	CART	LDA	QDA	SVMR	SVML	SVMP	RF	k-NN
flute original	0.14	0.16	0.16	0.08	0.14	0.16	0.09	0.10
flute auditory	0.13	0.46	—	**0.05**	0.07	0.15	0.07	0.09
clarinet original	0.09	0.07	0.06	0.05	0.05	0.09	0.06	0.06
clarinet auditory	0.09	0.47	—	0.05	0.05	0.12	**0.04**	0.07
oboe original	0.13	0.13	0.17	0.10	0.12	0.16	0.09	0.14
oboe auditory	0.14	0.47	—	**0.07**	0.08	0.18	**0.07**	0.11
trumpet original	0.15	0.16	0.17	0.13	0.15	0.19	0.11	0.13
trumpet auditory	0.11	0.48	—	0.10	**0.07**	0.19	0.09	0.13
violin original	0.09	0.09	0.13	**0.05**	0.08	0.14	**0.05**	0.07
violin auditory	0.09	0.47	—	0.06	0.06	0.15	**0.05**	0.09

Table 18.3: Error Rates for Feature Selection in the Multi-Class Case (All Instruments)

Method	CART	LDA	QDA	SVMR	SVML	SVMP	RF	k-NN
no selection orig.	0.36	0.25	0.24	0.21	0.21	0.19	0.22	0.27
forward orig. (fo)	0.38	0.27	0.26	0.23	0.27	0.22	0.23	0.21
backward orig. (bo)	0.34	0.24	0.22	0.24	0.20	0.18	0.21	0.22
no select. auditory	0.32	0.78	—	0.15	0.15	0.14	0.16	0.27
forward audit. (fa)	0.30	0.21	0.22	0.21	0.20	0.22	0.20	0.21
channel groups fa	0.29	0.17	0.23	0.16	0.18	0.17	0.17	0.20
channel groups ba	0.28	0.75	—	0.13	0.14	**0.12**	0.15	0.18
feature groups fa	0.32	0.18	0.28	0.15	0.18	0.16	0.17	0.20
feature groups ba	0.30	0.18	—	0.14	0.15	0.14	0.15	0.21

18.4.5.2 *"One-vs-All" Classification*

Table 18.2 shows the estimated error rates for the "One-vs-All" classifications. For example "clarinet auditory, RF" = 0.04 means that we get an error rate of 4% if we apply random forest on the auditory features for distinguishing clarinet tones from the rest. For the auditory data, the LDA method delivers by far the highest error rates, which again is due to collinear features. For flute, SVMR is best for both the original and the auditory data. For clarinet, SVML and SVMR as well as RF are appropriate methods. For oboe, RF appears to be most appropriate. For trumpet, RF for the original data and SVML for the auditory data appear to be best suited. For violin, SVMR and RF appear to be best for both the original and the auditory data. Overall, in comparison SVML, SVMR, and RF are better than the other methods. Also, the auditory data gives better error rates than the original data. Clarinet and violin can be best separated from the other instruments, clarinet even better than violin.

18.4.5.3   Feature Selection

In order to select the most important features for class separation, we apply forward
and backward feature selection to both feature sets and all considered classification methods as well as all classes (instruments). Due to the huge time complexity
backward selection is not performed on the auditory model-based features. Instead,
grouping-based feature selection is additionally applied. The results are shown in
Table 18.3. Note that also here the large auditory data set with redundant features is
problematic for LDA and QDA, which leads to problems for the backward selection
variants.

Let us have a closer look at the classification method performing best in this
study. For the feature set based on the original signal, as mentioned above the best
method is SVMP with the parameters identified in parameter tuning leading to an
error rate of 18.62%. Then, by forward selection eight features are included in the
model (brightness, entropy, mfcc1, mfcc2, mfcc3, mfcc4, mfcc5, mfcc11) leading to
an error rate of 22.00%. By backward selection, only the features irregularity and
mfcc8 are eliminated. Retuning the parameters of SVMP then leads to the improved
error rate of 17.65%.

For the feature set based on the auditory model, group-based forward selection
is not only faster but also leads to better results for most classification methods than
standard forward selection. Also here, the best method on the full feature set is
SVMP with an error rate of 14.27%. By standard forward selection nine features
are picked (rms of channel 30, mean spectral flux of channel 39, mfcc1 of channel
8, mfcc2 of channels 32 and 37, mfcc3 of channel 38, mfcc5 of channel 28, mfcc6
of channel 13 and mfcc13 of channel 26) leading to an error rate of 21.76%. By
channel-based grouping and forward selection, four channels are included (15, 29,
34 and 39) which means the new feature set consists of $4 \cdot 21 = 84$ features. This
feature set leads to the error rate 17.18%. By applying the backward variant five
channels are neglected (1, 2, 4, 11 and 14), which leads to an error rate of 12.35%,
the overall best result. To sum up these results, higher channels – which have higher
best frequencies – seem to be the more important for instrument recognition than
lower ones. By feature-based grouping and forward selection, three features are
selected (mfcc2, mfcc5 and mfcc7) leading to a model with $3 \cdot 40 = 120$ features
and an error rate of 16.03%. By the backward variant just two features are removed
(irregularity and mfcc6) leading to an error rate of 14.00%.

A sequential combination of channel-based grouping and feature-based grouping
might even further improve the results. Additionally, a second parameter tuning for
the polynomial SVM on the reduced feature sets might lead to another enhancement.

18.4.5.4   Evaluation of Hierarchical Taxonomies

Let us now compare the hierarchical taxonomies shown in Figure 18.5 and Figure 18.6 to the flat taxonomy. Again, the experiments are separately applied to the
original and the auditory features. Linear SVM and Random Forest are applied on
each node and the best model is chosen on each node individually. The linear SVM
is again tuned like in the previous experiments. No feature selection is applied. The

Table 18.4: Error Rates for the 2 Nodes of the Classical Variant of Hierarchical Classification

Classification Task	SVML	RF
level1 original	0.14	0.14
level2 original	0.18	0.20
level1 auditory	**0.08**	0.10
level2 auditory	**0.14**	0.15

Table 18.5: Error Rates for the 3 Nodes of Hornbostel–Sachs Taxonomy

Classification Task	SVML	RF
level1 original	0.08	0.06
level2 original	0.23	0.18
level3 original	**0.08**	0.14
level1 auditory	**0.04**	0.05
level2 auditory	**0.12**	0.14
level3 auditory	0.11	0.13

individual classification performance of each node is again measured by 10 random repetitions of 10-fold cross-validation. An observation is classified correctly if it is correctly classified in all corresponding nodes of the hierarchy. The results for the individual nodes are shown in Table 18.4 and Table 18.5. As can be seen, the best classification method is not the same for all nodes. Also, the optimal value for the "cost" parameter of the linear SVM varies. The overall classification errors for all taxonomies by applying the individual best classification method on each node are listed in Table 18.6. In this simple example, a hierarchical taxonomy seems not to be better than the simpler flat taxonomy. However, more individual decisions at each node by applying feature selection and more classification methods might improve the results of hierarchical taxonomy. While in this example misclassification costs are set constant, a hierarchical taxonomy might benefit in classification scenarios

Table 18.6: Overall Error Rates for the 3 Taxonomies

Taxonomy	best combined model
flat taxonomy original	0.21
classical hierarchical taxonomy original	0.23
Hornbostel–Sachs system original	0.21
flat taxonomy auditory	**0.15**
classical hierarchical taxonomy auditory	**0.15**
Hornbostel–Sachs system auditory	0.16

with class dependent costs. For example it could be argued that it is worse to classify an observation that is labeled as oboe as violin than to classify it as clarinet.

18.4.6   Summary of Example

Using all instruments (classes) and all features the data of the auditory features lead to lower error rates than the original data. This tends to be true also for "One-vs-All" classifications, except for LDA. Particularly for clarinet and violin, the error rates are low, i.e. these instruments can be best distinguished from the other instruments. From feature selection, one can see that the entropy feature and the mfcc features appear to be most important for instrument discrimination. In this example, hierarchical and flat taxonomies lead to nearly the same results. Comparing the different classification methods, the SVM variants and random forests lead to the best discriminations.

18.5   Concluding Remarks

Instrument recognition is a typical problem for supervised classification. The level of difficulty is dependent on the actual application, which can be characterized by five aspects: tones, polyphony, instrument types, individual instruments, and data base. However, for all types of instrument recognition the main challenge is an appropriate choice of features. Therefore, after usually hundreds of features are extracted, this large feature set is appropriately reduced in order to simplify the problem for the classification methods. The whole procedure of instrument recognition is illustrated by means of an example.

18.6   Further Reading

There are many methods applied to instrument recognition in the literature, but not discussed in this chapter. For example, taking the temporal development into account instead of just computing the mean and the variance of each feature, can lead to improved results [10]. In [12] it is shown that especially the beginning of a tone, the so-called attack transient, contains much information about the specific instrument. By taking only features from the attack transient they achieve nearly as good results as with additional features from the whole tone. Instead of just dealing with the harmonic partials of a tone, in [20] also the inharmonic attack of each note is considered. In their experiments, this approach outperformed other state-of-the-art algorithms.

In [3], single piano and guitar tones are classified by means of various music features. In a first study, four different kinds of mid-level features are taken into account for classification. Three spectral features (mfcc, pitchless periodogram, and simplified spectral envelope) and one temporal feature (absolute amplitude envelope) are used for the classification task. The spectral features characterize the distribution of overtones, the temporal feature the energy of a tone over time. In a second study a very large number of low-level features proposed in the literature and the mid-level features are used for the classification task after feature selection.

In [4], intervals and chords played by instruments of four families (strings, wind, piano, plucked strings) are used to build classification rules for the recognition of the musical instruments on the basis of the same groups of mid-level features, again by means of feature selection.

In [1], again the same groups of mid-level features and common statistical classification algorithms are used to evaluate by statistical tests whether the discriminating power of certain subsets of feature groups dominates other group subsets. The authors examine if it is possible to directly select a useful set of groups by applying logistic regression regularized by a group lasso penalty structure. Specifically, the methods are applied to a data set of single piano and guitar tones.

In [16], multi-objective feature selection is applied on data sets which are based on intervals and chords. The first objective is the classification error and the second one is the number of features. The authors argue that a smaller number of features yield better classification models since the danger of overfitting is reduced. Additionally, smaller feature sets also need less storage and computing time. Their experimental results show decreased error rates by applying feature selection. Furthermore, it is shown that the best set of features might be very diverse for different kinds of instruments. In [15], this study is extended by comparing the results of the best specific feature sets for concrete instruments to a generic feature set, which is the best compromise for classifying several instruments. By applying their experiments to four different classification tasks, they conclude that it is possible to get a generic feature set which is almost as good as the specific ones.

In [2], solo instruments accompanied by a keyboard or an orchestra are distinguished. Instead of classification on a note-by-note level, they classify on entire sound files. The authors argue that most of the features used for monophonic instrument recognition do not work well in the context of predominant instrument recognition and only use features based on partials. First, they estimate the most dominant fundamental frequencies for all frames. Afterwards, 6 features are generated on each of the lowest 15 partials, which yields 90 features altogether. One drawback of this approach is that it depends strongly on the goodness of predominant F0 estimation, a problem which itself is not solved, yet. Using a Gaussian classifier they get an accuracy of 86% for 5 instruments.

In [14], several strategies for multi-label classification of polyphonic music are explained and compared. Additionally, specific characteristics for multi-label feature selection are discussed. In [7], hierarchical classification is applied to multi-label classification. This means, e.g., first classify the dominant instrument, then the next one, etc.

Bibliography

[1] B. Bischl, M. Eichhoff, and C. Weihs. Selecting groups of audio features by statistical tests and the group lasso. In *9. ITG Fachtagung Sprachkommunikation*, Berlin, Offenbach, 2010. VDE Verlag.

[2] J. Eggink and G. Brown. Instrument recognition in accompanied sonatas and concertos. In *Proceedings of the IEEE International Conference on Acoustics,*

Speech and Signal Processing, pp. IV–217–IV–220. IEEE, 2004.

[3] M. Eichhoff, I. Vatolkin, and C. Weihs. Piano and guitar tone distinction based on extended feature analysis. In A. Giusti, G. Ritter, and M. Vichi, eds., *Classification and Data Mining*, pp. 215–224. Springer, 2013.

[4] M. Eichhoff and C. Weihs. Recognition of musical instruments in intervals and chords. In M. Spiliopoulou, L. Schmidt-Thieme, and R. Jannings, eds., *Data Analysis, Machine Learning and Knowledge Discovery*, pp. 333–341. Springer, 2013.

[5] A. Eronen and A. Klapuri. Musical instrument recognition using cepstral coefficients and temporal features. In *Proceedings of the IEEE International Conference on Acoustics, Speech and Signal Processing (ICASSP)*, pp. II–753–II–756. IEEE, 2000.

[6] S. Essid, G. Richard, and B. David. Hierarchical classification of musical instruments on solo recordings. In *Proceedings of the IEEE International Conference on Acoustics, Speech and Signal Processing*, pp. V–817–V–820. IEEE, 2006.

[7] S. Essid, G. Richard, and B. David. Instrument recognition in polyphonic music based on automatic taxonomies. *IEEE Transactions on Audio, Speech, and Language Processing*, 14(1):68–80, 2006.

[8] L. Fritts. University of Iowa musical instrument samples. url: http://theremin.music.uiowa.edu/MIS.html, 1997.

[9] M. Goto. Development of the RWC music database. In *Proceedings of the 18th International Congress on Acoustics (ICA 2004)*, volume 1, pp. 553–556. International Commission for Acoustics, 2004.

[10] C. Joder, S. Essid, and G. Richard. Temporal integration for audio classification with application to musical instrument classification. *IEEE Transactions on Audio, Speech, and Language Processing*, 17(1):174–186, 2009.

[11] O. Lartillot and P. Toiviainen. A MATLAB toolbox for musical feature extraction from audio. In *International Conference on Digital Audio Effects*, pp. 237–244. LaBRI, Université Bordeaux, 2007.

[12] M. Newton and L. Smith. A neurally inspired musical instrument classification system based upon the sound onset. *The Journal of the Acoustical Society of America*, 131(6):4785–4798, 2012.

[13] F. Opolko and J. Wapnick. *McGill University master samples collection on DVD*. McGill [University], 2006.

[14] T. Sandrock. *Multi-Label Feature Selection with Application to Musical Instrument Recognition*. PhD thesis, Stellenbosch University, 2013.

[15] I. Vatolkin, A. Nagathil, W. Theimer, and R. Martin. Performance of specific vs. generic feature sets in polyphonic music instrument recognition. In *Evolutionary Multi-Criterion Optimization*, pp. 587–599. Springer, 2013.

[16] I. Vatolkin, M. Preuß, G. Rudolph, M. Eichhoff, and C. Weihs. Multi-objective

evolutionary feature selection for instrument recognition in polyphonic audio mixtures. *Soft Computing*, 16(12):2027–2047, 2012.

[17] E. von Hornbostel and C. Sachs. Classification of musical instruments: Translated from the original German by Anthony Baines and Klaus P. Wachsmann. *The Galpin Society Journal*, pp. 3–29, 1961.

[18] A. Wieczorkowska, E. Kubera, and A. Kubik-Komar. Analysis of recognition of a musical instrument in sound mixes using support vector machines. *Fundamenta Informaticae*, 107(1):85–104, 2011.

[19] K. Wintersohl. *Instrumenten Klassifikation mit Hilfe eines auditorischen Modells*. Bachelor Thesis, Department of Statistics, TU Dortmund University, 2014.

[20] J. Wu, E. Vincent, S. Raczynski, T. Nishimoto, N. Ono, and S. Sagayama. Polyphonic pitch estimation and instrument identification by joint modelling of sustained and attack sounds. *IEEE Journal of Selected Topics in Signal Processing*, 5(6):1124–1132, 2011.

Chapter 19

Chord Recognition

GEOFFROY PEETERS
Sound Analysis and Synthesis Team, IRCAM, France

JOHAN PAUWELS
School of Electronic Engineering and Computer Science, Queen Mary University of London, England

19.1 Introduction

Chords are abstract representations of a set of musical pitches (notes) played (almost) simultaneously (see also Section 3.5.4). Chords, along with the main melody, are often predominant characteristics of a music track. Well-known examples of chord reductions are the "chord sheets" where the background harmony of a music track is reduced to a succession of symbols over time (C major, C7, ...) to be played on a guitar or a piano.

In this chapter, we describe how we can automatically estimate such chord successions from the analysis of the audio signal. The general scheme of a chord recognition system is represented in Figure 19.1. It is made of the following blocks that will be described in the next sections:

1. A block that defines a set of chords that will form the dictionary over which the music will be projected (see Section 19.2),

2. A block that extracts meaningful observations from the audio signal: chroma or Pitch Class Profile (PCP) features extracted at each time frame (see Section 19.3),

3. A block that creates a representation (knowledge-driven see Section 19.4.1) or a model (data-driven see Section 19.4.2) of the chords that will be used to map the chords to the audio observation,

4. The mapping of the extracted audio observations to the models that represent the various chords. This can be achieved on a frame basis (see Section 19.5) but leads to a strongly fragmented chord sequence. We show that simple temporal smoothing methods can improve the recognition. In Section 19.6 we show how chord

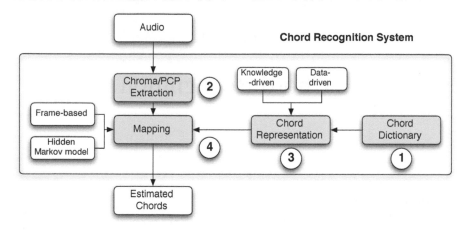

Figure 19.1: General scheme of an automatic system for chord recognition from audio.

pattern recognition and temporal smoothing can be performed simultaneously by training and then Viterbi-decoding a hidden Markov model.

Interdependency between chords and keys are then included in the hidden Markov model to perform joint chord and key recognition (Section 19.7). Section 19.8 introduces some measures for key and chord recognition performance and compares results between the two approaches. We close the chapter by showing what other methods have been used to tweak chroma-based chord estimation and give a summary of existing chord recognition tools.

19.2   Chord Dictionary

When developing a chord recognition system, the first step is to choose the dictionary of chords over which the harmonic content of the music track will be projected. Examples of possible chord dictionaries are given in Table 19.1 (see also Section 3.5.4). One can reduce the harmonic content to the main 24 major and minor chord triads or include chord tetrads (major 7, minor 7, dominant 7) or pentads (major 9, dominant 9). Depending on this choice, the observation of the notes {c,e,g,b} in the audio can either be mapped to a C-M or a C-M7 label. The larger the chord dictionary, the more precise the harmonic reduction to chords, but also the more difficult the task. This difficulty comes not only from the increase in classes (chord labels can be considered as classes in a machine learning sense), but also from the equivalence between some chords. For example C-M6: {c,e,g,a} has the same notes as A-m7: {a,c,e,g} although not in the same order.

From a musical point of view, a single chord can be played in various ways. C-M can be played in root position {c,e,g}, in first inversion {e,g,c} or second inversion {g,c,e} (see also Figure 3.18). Since most current chord estimation methods rely on mapping the audio content to the twelve semitone pitch classes independently of

Table 19.1: Dictionary for the Root Notes and Three Possible Dictionaries for the Type of Chords

Root-note	Type of the chord
c, c#, d, d# ... b	**Triads**: major (C-M: c,e,g) , minor (C-m: c, e♭ , g), suspended (C-sus2: c, d, g / C-sus4: c, f, g), augmented (C-aug: c, e, g#), diminished (C-dim: c, e♭, g♭)
c, c#, d, d# ... b	**Tetrads**: major 7 (C-M7: c, e, g, b), minor 7 (C-m7: c, e♭, g, b♭), dominant 7 (C-7: c, e, g, b♭), major 6 (C-M6: c, e, g, a), minor 6 (C-m6: c, e♭, g, a) ...
c, c#, d, d# ... b	**Pentads**: major 9 (C-M9: c, e g, b, d), dominant 9 (C-9: c, e, g, b♭, d) ...

their octave positions, it is not possible to distinguish whether the chord has been inverted or not. For this last reason, chords are often estimated jointly with the local key. The root of a chord is then expressed as a specific degree in a specific key. The choice of C-M6 will be favored in a C-Major key while A-m7 will be favored in a A-minor key.

When estimating chords from the audio signal, we will also rely on enharmonic equivalence, i.e. we consider the note c# to be equivalent to d♭, and also consider the chord F#-M to be equivalent to G♭-M.

19.3 Chroma or Pitch Class Profile Extraction

Chords represent a set of notes played almost simultaneously. It therefore seems natural to estimate the chords of a music track from a previous estimation of the existing pitches in its audio signal (multiple-pitch estimation). However, multiple-pitch estimation is still a difficult and a very computer-time-consuming task. For this reason, most algorithms that estimate chords from an audio signal use another approach: the extraction of chroma [34], also known as Pitch Class Profile (PCP) [6] (see also Section 5.3.1).

The notion of chroma/PCP is derived from Shepard [30], who proposes to factor the pitch of a signal into values of chroma (denoted here by $p \in [1, 12]$) and tone height or octave (denoted here by $o \in [1, O]$).[1] For example, if one chooses the reference $p = 1$ for the note c, then a4 (440 Hz) is factored as the chroma $p = 10$ at the octave $o = 4$ (octaves in scientific notation, see Section 2.2.4).

The chroma/PCP representation is obtained by mapping the energy content of the spectrum of an audio signal to the 12 semitone pitch-classes (c, c#, d, d#, e, ...). More precisely, to compute the value at the chroma p, we add the energy existing at all frequencies corresponding to the possible pitches of this chroma. For example, to obtain the chroma $p=10$, we add the energy at the frequencies corresponding to all

---

[1]It should be noted that the two-component theory of pitch was originally proposed by Hornbostel [33].

possible octaves of the a: a0, a1, a2, a3 .... The representation is computed at each time frame λ. In the following we denote by $t_{chroma}[p,\lambda]$ the set of chroma vectors over time, also known as a chromagram.

Unlike multiple-pitch estimation, chroma/PCP is a mapping and not a an estimation. Therefore it is not prone to errors.

19.3.1 Computation Using the Short-Time Fourier Transform

Chroma/PCP can be computed starting from the Short-Time Fourier Transform $X_{stft}[\lambda,\mu]$ (see Section 2.4.2 or Section 4.4), where $\mu \in [0,\dots,N-1]$ denotes the frequency and λ the time frame. For each pitch class $p \in [1,12]$, the value of $t_{chroma}[\lambda,p]$ is computed simply by adding the energy of $X_{stft}[\lambda,\mu]$ at the frequencies μ corresponding to the pitch class p:

$$t_{chroma}[\lambda,p] = \sum_{\mu \in p} X_{stft}[\lambda,\mu]^2 . \tag{19.1}$$

To know which frequencies μ correspond to a specific p, we first convert the frequencies μ of the Discrete Fourier Transform (DFT) to the Hz scale: $f_\mu = f_s \frac{\mu}{N}$ where f_s is the sampling rate and N the number of points of the DFT. We then convert f_μ to the MIDI scale: $m_\mu = 12 \log_2 \frac{f_\mu}{440} + 69$ (for a tuning of a4 $= 440$ Hz), $m_\mu \in \mathbb{R}$. For example, $m_{450Hz} = 69.3891$.

Hard-Mapping The value of the chroma at $p \in \mathbb{N}$ is then found by summing the energy of the spectrum at all frequencies μ that correspond to the chroma p, i.e. such that $rem([m_\mu], 12) + 1 = p$ (where $[x]$ is the "round to nearest integer" function and rem is the "remainder after division" operation). This method provides a "hard" mapping. For example, the energy at $m_{452Hz} = 69.4658$ will be entirely assigned to m=69 (p=10) while $m_{453Hz} = 69.5041$ to m=70 (p=11).

Soft-Mapping In order to avoid this "hard" mapping, a "soft" mapping is often used. In this, the energy at m_μ is assigned to different chroma with a weight inversely proportional to the distance between m_μ and the closest pitches. In the previous example, $m_{453Hz} = 69.5041$ will equally contribute to m=69 ($p = 10$) and m=70 ($p = 11$). For this, the summation is done through a windowing operation $g\left(\frac{1}{2}|m - m_\mu|\right)$ where $m \in \mathbb{N}$ is one of the MIDI notes, $m_\mu \in \mathbb{R}$ is the value of the frequency μ converted to the MIDI scale, and $|x|$ denotes the absolute value. $g(x)$ is designed such that $g(0) = 1$ and is zero outside the interval $x \in [-1/2, 1/2]$ (see Table 2.4 Section 2.4.2). Therefore $g(x)$ takes its maximum value for $m_\mu = m$ and is equal to zero outside the interval $[m-1, m+1]$. Common choice of $g(x)$ are the triangular, Hanning, or tanh functions.

We illustrate this process in Figure 19.2.

19.3.2 Computation Using the Constant-Q Transform

Limitations of the DFT The ability of a spectral transform to separate adjacent frequencies is defined by its "frequency resolution". More precisely, when using an

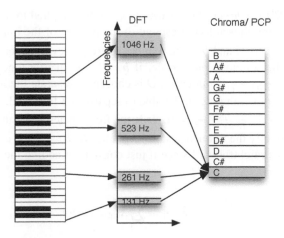

Figure 19.2: Chroma computation from the DFT.

analysis window $w[k]$ of duration L, the frequency resolution obtained is proportional to the width at -6dB of the main lobe of its DFT: $Bw = \frac{Cw}{L}$ where Cw is a constant specific to each window type (for example $Cw = 1.81$ for a Hamming window). Given that the DFT uses the same analysis window $w[k]$ for all frequencies μ, it has a constant resolution over frequencies μ. This resolution can be too large to separate the frequencies of the lowest adjacent pitches in the spectrum. For example, the notes c3 and c#3 are separated by only 7.8 Hz, which is smaller than the frequency resolution Bw provided by a Hamming window of $L=80$ ms which is $Bw=22.62$ Hz.

Constant-Q Transform   Because of this, the Constant-Q Transform (CQT), which has a variable resolution over frequencies, is often used to compute the chroma/PCP representation. Q is the quality factor defined as $Q = \frac{f_\mu}{f_{\mu+1}-f_\mu}$. In the CQT, we impose Q to be constant over the frequencies f_μ. Since we want to be able to "resolve" adjacent frequencies, we impose $f_{\mu+1} - f_\mu \geq Bw$ or $f_{\mu+1} - f_\mu = \alpha Bw$ with $\alpha \geq 1$. We therefore have

$$Q = \frac{f_\mu}{f_{\mu+1} - f_\mu} = \frac{f_\mu}{\alpha Bw} = \frac{f_\mu N_\mu}{\alpha Cw f_s}, \tag{19.2}$$

where N_μ is the duration of the analysis window in samples and f_s is the sampling rate. In practice, αCw is often omitted and the rule $Q = \frac{f_\mu N_\mu}{f_s}$ is used. In order to have Q constant over frequencies the duration N_μ of the analysis window, $w[k]$ is set inversely proportional to f_μ. It is therefore denoted by $w_\mu[k]$.

The Constant-Q transform (see also Section 4.5) is then defined as

$$X_{\text{cqt}}[\lambda, \mu] = \frac{1}{N_\mu} \sum_{k=0}^{N_\mu-1} w_\mu[k] x[k + R\lambda] \exp\left(-i2\pi \frac{Q}{N_\mu} k\right). \tag{19.3}$$

Use of the CQT for Chroma/PCP Extraction   When applied to music analysis, the
frequencies f_μ are chosen to correspond to the pitches of the musical scale: $f_\mu =$
$f_{min} 2^{\frac{\mu}{12b}}$ where $\mu \in \mathbb{N}$ and b is the number of frequency bins per semitone (if $b =$
1, we obtain the semitone pitch-scale, if $b = 2$, the quarter-tone pitch-scale). In
this case, $Q = \frac{1}{2^{1/(12b)}-1}$. In practice, Q is chosen to separate the lowest considered
pitches. For example, in order to be able to separate c3 from c#3, Q is chosen such
that $Q = \frac{f_\mu}{f_{\mu+1}-f_\mu} = \frac{130.8}{138.6-130.8} = 16.7$. If the frequencies f_μ of the Constant-Q are
chosen to be exactly the frequencies of the pitches (if $b = 1$), the computation of the
chroma/PCP is straightforward since it just consists of adding the values for which
$rem(m_\mu, 12) + 1 = p$:

$$t_{chroma}[\lambda, p] = \sum_{o=1}^{O} X_{cqt}[\lambda, p + 12o] \tag{19.4}$$

where $o \in [1, O]$ denotes the octave number.

19.3.3   Influence of Timbre on the Chroma/PCP

Ideally, when a musical instrument plays a pitch of c (whatever octave), we would
like the chroma/PCP representation to have a single non-zero at $p = 1$: $t_{chroma}[\lambda, p] =$
$[1, 0, 0, 0, 0, 0, 0, 0, 0, 0, 0, 0]$. However, a musical instrument does not produce a single frequency at the pitch f_0, but a harmonic series at $f_0, 2f_0, 3f_0, 4f_0, 5f_0 \ldots$ The
corresponding amplitudes $(a_1, a_2, a_3, a_4, a_5 \ldots)$ define the "timbre" of the instrument.
These harmonics will create components in the chroma/PCP that do not necessarily represent the pitch (see Table 19.2). For example, for a pitch c3 (130.81 Hz),
its third harmonic $(3f_0 = 392.43$ Hz) has the same frequency as the note g4, its
fifth harmonic $(5f_0 = 654.06$ Hz) is close to the note e5 ... When we consider the
first five harmonics, the corresponding chroma/PCP vector will be $t_{chroma}[\lambda, p] =$
$[a_1 + a_2 + a_4, 0, 0, 0, a_5, 0, 0, a_3, 0, 0, 0, 0]$. It is clear that the chroma/PCP representation depends on the timbre of the musical instrument, so it is very likely that the
same note played by two different instruments results in two different chroma/PCP
representations. Chroma is therefore said to be timbre sensitive. This is illustrated
on Figure 19.3.

To deal with this problem, three different strategies can be used:

1. *Whiten the spectrum* (i.e., give the spectrum a flat spectral shape: $a_1 = a_2 =$
   $a_3 = a_4 = a_5 \ldots$) before computing the chroma/PCP. Doing so will emphasize the
   presence of the higher harmonics in the chroma/PCP but will make it equal for
   all musical instruments, and hence will make chroma/PCP timbre independent.
   This effect can be achieved by binarizing spectral peaks that are harmonically
   related [36], by frequency-dependent compression [28] or by applying cepstral
   liftering [15].

2. *Remove as much as possible the presence of higher harmonics* in the spectrum
   before computing the chroma/PCP representations. Some suitable techniques include attenuating harmonics through the calculation of a harmonic power spec-

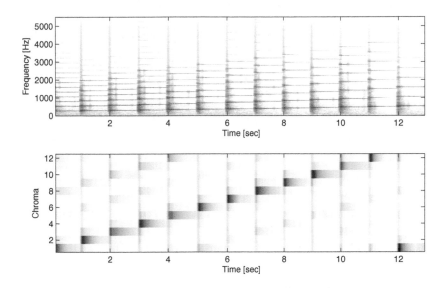

(a) Chromatic scale starting from c played by a piano.

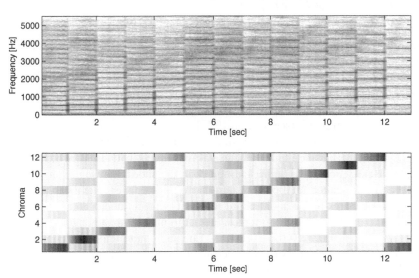

(b) Chromatic scale starting from c played by a violin.

Figure 19.3: On each figure, the top part represents the spectrogram $X_{\text{stft}}[\lambda, \mu]$, the bottom the corresponding chromagram $t_{\text{chroma}}[\lambda, p]$. The influence of the higher harmonics of the notes are clearly visible in the form of extra values in the chromagram.

Table 19.2: Harmonic Series of the Pitch c3, Corresponding Frequencies f_μ, Conversion into MIDI Scale m_μ, Conversion to Chroma/PCP p

Pitch	Harmonic	Frequency f_μ	MIDI-scale m_μ	Chroma/PCP p
c3	f_0	130.81	48.00	1 (= c)
	$2 f_0$	261.62	60.00	1 (= c)
	$3 f_0$	392.43	67.01	8.01 ($\simeq$ g)
	$4 f_0$	523.25	72.00	1 (= c)
	$5 f_0$	654.06	75.86	4.86 ($\simeq$ e)
	...	...	...	...

trum [15] or using a pitch salience spectrum that takes the energy of higher harmonics into consideration too [28].

3. *Keep the chroma/PCP vector as it is* and consider the existence of the higher harmonics in the chords representation (see Section 19.4).

It is also possible to combine the first approach with one of the two others.

19.4 Chord Representation

19.4.1 Knowledge-Driven Approach

The most obvious way to represent a chord in a computer is to create a vector representing the existence of the 12 semitone pitch classes. Such a vector is often named a "chord template": $T_c[p]$ where c is the index of the chord. The chord templates corresponding to the chords C-M, C-m and C-dim are displayed in Figure 19.4. Since these chord templates have the same description space as the chroma/PCP vector, it is possible to compute a distance between $T_c[p]$ and a chroma/PCP vector $t_{chroma}[\lambda, p]$ at a given time frame λ (see Section 19.5.1).

As mentioned before, in order to deal with the problem generated by the existence of the higher harmonics of musical instrument sounds, it is possible to consider the existence of the higher harmonics in the chord representation. This is done by creating so-called "audio chord templates." For this, each existing pitch p in the chord contributes to the audio chord template with an audio note template. If we consider only the first five harmonics, the audio note template of $p = 1$ is defined by $[a_1 + a_2 + a_4, 0, 0, 0, a_5, 0, 0, a_3, 0, 0, 0, 0]$. The audio note-templates corresponding to the other p are obtained by circular permutation. To get the values of the amplitudes a_h, Izmirli [10] studied the average values of the a_h on a piano data set. Gomez [7] proposes a theoretical model of the amplitudes in the form $a_h = 0.6^{h-1}$.

19.4.2 Data-Driven Approach

Another possibility is to learn the chord representations from a collection of audio files annotated over time into chord labels. Two machine-learning approaches can be used for that.

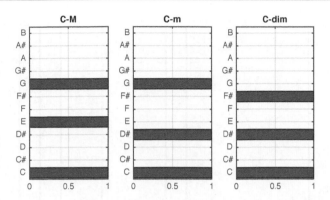

Figure 19.4: Chord templates $T_c[p]$ corresponding to the chords C-M, C-m and C-dim.

In the "generative" approach, all the extracted chroma/PCP vectors corresponding to a given annotated chord c are used to train a "generative" model of this chord. For each chord c, we represent the probability of observing a specific chroma/PCP vector given the chord: $P(t_{chroma}[\lambda, p] | c)$. This is known as the "likelihood." Common choices for the "generative model" are the multivariate normal/Gaussian distribution model or the multivariate Gaussian mixture model [19]. In the case of multivariate normal/Gaussian distribution mode, each chord c is represented by a model $\mathcal{N}(\mu_c, \Sigma_c)$ where μ_c and Σ_c are the mean vector and covariance matrix learned from the set of $t_{chroma}[\lambda, p]$ that belong to the specific chord c.

In the "discriminant approach," a single model is trained to best "discriminate" (i.e. to find the best decision boundaries between) the values of chroma/PCP vectors corresponding to the various chords c. Examples of such models are Support Vector Machines [5] or Convolutional Neural Networks [9] (see also Chapter 12).

19.5 Frame-Based System for Chord Recognition

The task of chord recognition consists of finding the chord succession $c(\lambda)$ with $c \in [1, C]$ (where C is the size of the chord dictionary) that "best explains" the succession of observations $t_{chroma}[\lambda, p]$. Depending on the method (see below), "best explains" can mean "that minimizes the distance" or "with the highest likelihood."

Depending on the choice of the chord representation (chord templates or generative model) and the amount of musical knowledge we want to introduce in the algorithm, various systems can be used.

19.5.1 Knowledge-Driven Approach

The simplest system consists of finding at each frame λ, the chord label c that minimizes a distance $d(x = T_c[p], y = t_{chroma}[\lambda, p])$. Such a distance can be a simple

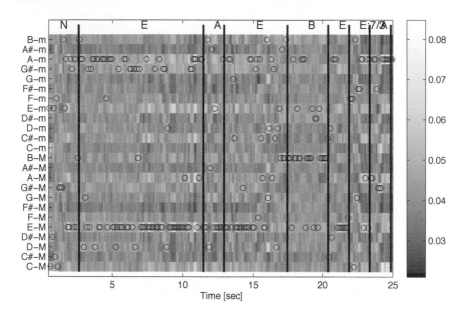

Figure 19.5: Example of frame-based chord recognition on the track "I Saw Her Standing There" by The Beatles.

Euclidean distance

$$d(\underline{x},\underline{y}) = \sqrt{\sum_{p=1}^{12}(x_p - y_p)^2},\tag{19.5}$$

or to be independent from the norm of the Chroma/PCP vector (hence independent of the local loudness of the signal), a one-minus-cosine distance:

$$d(\underline{x},\underline{y}) = 1 - \frac{\underline{x}\cdot\underline{y}}{||\underline{x}||^2||\underline{y}||^2}.\tag{19.6}$$

Divergences, such as Itakura–Saito or Kullback–Leibler [17] can also be used instead of distances in order to highlight additions or deletions of components p in the vectors.

Example 19.1. *In Figure 19.5, we illustrate the chord recognition results obtained by using the one-minus-cosine distance. In this example, we use a dictionary of size $C = 24$ consisting of all major and minor chords. At each frame λ, we have chosen the chord $c \in C$ that has the minimal distance to $t_{\text{chroma}}[\lambda, p]$. In the figure, we display the distances between the vectors over time $t_{\text{chroma}}[\lambda, p]$ and all the chords $c \in C$ in a matrix. The chord that corresponds to the minimal distance at each frame is indicated by a circle. The top row of the figure shows the ground-truth chord label. The audio signal corresponds to the track "I Saw Her Standing There" by The Beatles.*

19.5.2 Data-Driven Approach

When the chords are represented by a generative model (such as the Gaussian mixture models) instead of templates, the distance is replaced by the computation of the probability of observing a chord c given the observed chroma/PCP vector: $P(c \mid t_{\text{chroma}}[\lambda, p])$. This probability is known as the "posterior probability" and the method that consists of choosing the c that maximizes this probability, known as the "Maximum-a-Posteriori" (MAP) method. $P(c \mid t_{\text{chroma}}[\lambda, p])$ is derived from the "likelihood" model $P(t_{\text{chroma}}[\lambda, p] \mid c)$ using Bayes' rule (see Section 12.4.1). As described above, the "likelihood" model $P(t_{\text{chroma}}[\lambda, p] \mid c)$ has been trained on a collection of audio files representing each possible chord of the dictionary.

$P(c \mid t_{\text{chroma}}[\lambda, p])$ is computed for each possible chord c and the chord c leading to the MAP is chosen as the chord $c(\lambda)$ for time frame λ.

19.5.3 Chord Fragmentation

Whatever knowledge-driven or data-driven technique is used to map chroma observations onto chord representations, the resulting estimation of chords over time is usually quite fragmented (visible as an unrealistically high number of jumps between chords). This is because the decision of which chord best matches the observation is taken independently at each time frame λ, without considering adjacent frames. We however know that the frame rate of the system is higher than a realistic rate for changing chords. Therefore, we expect the chord output to appear clustered over time and to have clear changes, without oscillating back and forth between the new and the old chord.

To reduce this fragmentation problem, two simple processes can be used.

The first consists of applying a low-pass filter (see Chapter 4) over time λ to each of the C distance functions $d(c, t_{\text{chroma}}[\lambda, p])$ or to each of the posterior probability functions $P(c \mid t_{\text{chroma}}[\lambda, p])$. The decision at time λ (choice of the minimal distance or the MAP) is then taken on the smoothed in time function.

The second consists of directly applying a median filtering over time to the output of the decision function $c(\lambda)$.

19.6 Hidden Markov Model-Based System for Chord Recognition

Another, potentially more powerful, way to achieve stability over time is the use of a hidden Markov model (HMM) [26] (see also Example 9.16). In an HMM, the transition probabilities between states allow us to constrain the sequence of decoded states over time. In the case of chord estimation, the chord labels c_i, $i \in [1, C]$ form the hidden states from which we observe the Chroma/PCP features.

The hidden Markov model for chord recognition is defined by

- the initial probability of the state/chord c_i: $P_{init}(c_i)$,
- the emission probability of each state/chord c_i: $P(t_{\text{chroma}}[\lambda, p] \mid c_i)$, and
- the transition probabilities between states/chords, i.e. the probability to transit from chord c_i at time λ to chord c_j at time $\lambda + 1$: $P_{trans}(c_j \mid c_i)$.

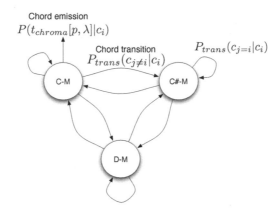

Figure 19.6: Hidden Markov model for chord estimation.

A visual representation of a simple HMM consisting of only three chords is depicted in Figure 19.6.

The *initial probabilities* are usually considered uniformly distributed.

For the *emission probabilities*, we can reuse and apply with little or no change the same mapping strategies between chord labels and audio signals as we used in the frame-based system (by converting the distances $d(x = T_c[p], y = t_{\text{chroma}}[\lambda, p])$ to probabilities or by computing the likelihood $P(t_{\text{chroma}}[\lambda, p] | c_i)$ given the trained statistical models $\mathcal{N}(\mu_c, \Sigma_c)$).

The *self-transition probabilities* $P_{trans}(c_{j=i} | c_i)$ regulate how easy it is to change between chords and can therefore prevent fragmentation. They take on the same role as the low-pass or median filtering of the chord output. In contrast to the latter, the Viterbi algorithm [26] used to decode the HMM does not just consider the chosen output of each frame for the temporal smoothing. It rather considers all options for each frame, also the locally suboptimal, to find the sequence of labels that optimally combines the observations and the requirement of temporal stability.

The *change-transition probabilities* $P_{trans}(c_{j\neq i} | c_i)$ allow us to encode musical rules concerning the transitions between chords. They allow us to take into account the fact that music is not just a "bag of chords" but a specific temporal succession of them, i.e., certain transitions are more likely to appear than others. These combinations form the rules that are taught in music theory. Much like the mapping between chroma observations and chord representations, these transition probabilities can be set manually based on a musicological knowledge or can be optimized on an annotated data set using machine learning techniques [18]. These transition probabilities will of course depend on the considered musical style and epoch as do the musical rules.

19.6.1 Knowledge-Driven Transition Probabilities

A simple theoretical model, used in [1], that can be used to derive change probabilities, is the doubly nested circle of fifths. It expresses the distance between all major and minor chords. A visual representation can be seen in Figure 19.7(a). It is formed by arranging major and minor chords on two concentric circles. When moving clockwise, the distance is a rising perfect fifth between successive roots, when moving counter-clockwise, a falling perfect fifth.

The distance between two chords is then calculated by counting the shortest number of steps along the circle that has to be taken to go from one triad to the other. Changing from the "major" circle to the "minor" circle counts as one step. A drawback of this model is that it does not extend to other chord types, such as seventh chords. A number of alternative chord distances that are more extensive in this regard are compared in [27]. The premise that all models that are based on chord distances share, is that close chords are likely to follow each other in sequence. These distances are then transformed into transition probabilities, where a small distance leads to a high probability of transition. The matrix of transition probabilities that corresponds to the doubly nested circle of fifths can be seen in Figure 19.7(b).

Example 19.2. *In Figure 19.8, we illustrate the chord recognition results obtained using a hidden Markov model for the same track as Figure 19.5. The observation probabilities of the HMM have been taken as the one-minus-cosine distance (normalized such that $\sum_i P(t_{\text{chroma}}[\lambda, p]|c_i) = 1$). The transition probabilities of the HMM have been taken as $P_{trans}(c_j|c_i) = 1 - d(c_i, c_j)/6$ where $d(c_i, c_j)$ is the distance between two chords c_i and c_j in the doubly nested circle of fifths as explained before. It is then normalized such that $\sum_j P_{trans}(c_j|c_i) = 1 \ \forall i$. By circles, we represent the most likely path over time as decoded by the Viterbi algorithm given the hidden Markov model parameters and the observations over time $t_{\text{chroma}}[\lambda, p]$. Compared to the results illustrated in Figure 19.5, the path obtained by the HMM-based system is much less fragmented over time.*

19.6.2 Data-Driven Transition Probabilities

Instead of explicitly relying on musicological theory to derive the transition probabilities, those can also be determined by the statistical analysis of a corpus of symbolic music. In its most basic form, this amounts to counting the proportion of occurrence of each chord pair. If the set comes with synchronized audio, the HMM can also be trained such that the optimal combination of observation and transition probabilities can be found that maximizes recognition performance on that set. These approaches are not always exclusive either. Since a data-driven approach can lead to a local optimum which is musically meaningless, one can use a theoretical model to initialize the training.

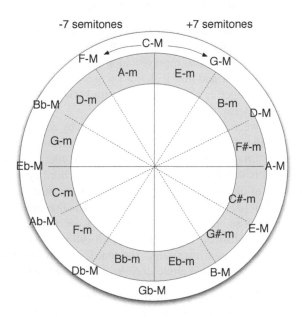

(a) Doubly nested circle of fifths for major and minor chords.

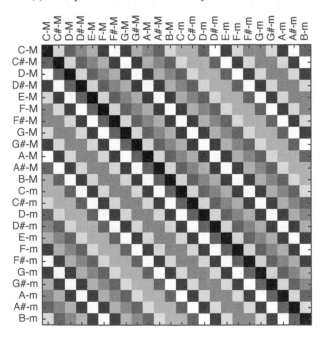

(b) Corresponding transition matrix $P_{trans}(c_j \mid c_i)$.

Figure 19.7: Deriving a transition matrix from a theoretic model of chord distance.

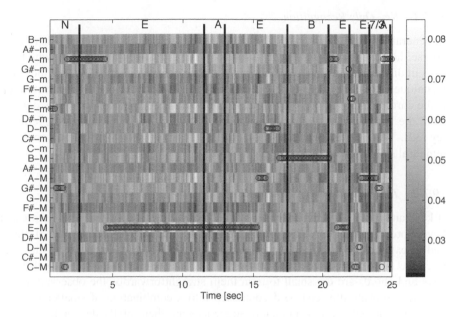

Figure 19.8: Example of HMM-based chord recognition on the track "I Saw Her Standing There" by The Beatles.

19.7 Joint Chord and Key Recognition

In order to define change probabilities that go beyond the most elementary chord transitions, we need to express chords in terms of their key. In music theory, harmony is mostly analyzed as the movement of chord degrees in a key.[2] Therefore we need the key to be available at the point where we try to recognize the chords. There are two kinds of approaches to accomplish this: (1) we build a sequential system where the key is recognized first, and is then followed by a chord recognition step; or (2) the key and chord are recognized simultaneously.

When discussing key recognition, there is one additional distinction that needs to be made which does not exist in chord recognition. We need to decide if we want to find the *global* key or the *local* key. The former assigns a single label that applies to the whole music piece, whereas the latter segments the song into segments of a constant key and labels them. Global estimation has the advantage that it is easier, as there are more observations to base the decision on. Its disadvantage is that in pieces with key changes, it will inevitably lead to a loss of information, namely the secondary keys and the location of the changes. In our use case, this will cause the interpretation of chords as degrees relative to the key to be wrong.

---

[2] It should be noted that, in this approach, chords that do not belong to the scale of the key cannot be determined.

19.7.1   Key-Only Recognition

In general, key recognition is conceptually similar to chord recognition. There is of course the difference in time scale, especially for global key recognition, which results in differences in analysis window sizes and/or duration modeling, but the same techniques can easily be reused. Here too, the most popular signal representation is the chroma/PCP vector. They are commonly matched with key templates derived from the perceptual experiments of Krumhansl [12], or the variations of Temperley [32]. When estimating local keys, an HMM can be used to provide the required temporal stability, as in [25].

19.7.2   Joint Chord and Key Recognition

If keys and chords are recognized simultaneously, the states of the HMM represent key–chord combinations. This leads to a strong increase in the number of states and an exponential growth of the transition matrix. To keep the number of variables tractable, and also because the current size of data sets annotated with both local keys and chords are too small to train them straightforwardly, the observation and transition probabilities can be decomposed into a combination of smaller parts as in [22]. The optimal key and chord sequences are then jointly decoded. Another option is to consider only the global key, which causes the majority of the values in the transition matrix to be set to zero. This can equivalently be seen as constructing a separate key-dependent HMM for each key, as in [13], instead of one large HMM. Finally, when the key is determined beforehand, the state space stays the same, but the optimal key path becomes deterministic instead of probabilistic, which brings the number of transitions that should be investigated at each step back down to the same number as for a chord-only HMM.

In combined key and chord recognition systems, it is common to express chords as a relative degree to a key (see also Section 3.5.4). For example in the key of C-major, the chord C-M is the first degree (denoted by I-M), D-m is the second (II-m), G-M the fifth (V-M) and so on. It is then common to tie transition probabilities together such that they are invariant under transposition. This means that a II-m to V-M change is as likely in C-major as in G-major, but not in C-minor however. The main point of expressing chords relative to a key, is that we can then reason about chord transitions in the same way as in the study of music theory, but there is also evidence that this representation of harmony reduces confusion about the expected chords when compared to an absolute sequence of chords without the context of a key [29]. On the other hand, according to [22], the improvement in recognition performance due to the better modeling of chord changes is only modest compared to the improvements brought by a better modeling of chord duration.

A drawback of the usage of a standard HMM, is that it can only take into account pairwise transitions, whereas we know from music theory that looking at a wider context can be even more enlightening. For instance, a D-m / G-M / C-M sequence is more representative of a C major key than either D-m / G-M or G-M / C-M sequences. Therefore multiple attempts have been made to include higher-order models of musical context, ranging from frame-based approaches [4], over lattice rescoring of HMM output [11], to a full search over all key and chord trigram sequences [23].

19.8 Evaluating the Performances of Chord and Key Estimation

In order to compare the performance of the different methods for chord and/or key recognition we discussed in the previous section, we need a way to quantify the correctness of their estimated outputs. We therefore need an evaluation procedure that takes a sequence of timed key or chord labels and numerically expresses the extent to which it resembles a certain reference sequence, preferably obtained through manual annotation. This resemblance can be measured according to a number of different aspects.

19.8.1 Evaluating Segmentation Quality

A first way to compare sequences is their degree of segmentation. We only need to take into account the positions of the key or chord changes for this, not their exact labels. Consequently, the same procedures as used to evaluate other segmentation tasks can be used here (see Chapter 13). A commonly used pair of measures is based on the directional Hamming divergence [14] (see Section 11.2). It is calculated by matching each segment in the tested sequence to the segment in the reference (the manually annotated piece of music) that overlaps it most, and then adding the durations of the non-covered parts in the tested sequence. Normalized by the sequence length and subtracted from one to make an increase in value correspond to an increase in performance, this gives a measure for over-segmentation. A corresponding measure for under-segmentation can be achieved by swapping the reference sequence and the sequence under test.

19.8.2 Evaluating Labeling Quality

A second aspect of evaluating key or chord sequences is the correspondence in harmonic content. This is complementary to the segmentation evaluation, because here the positions of the changes are not important, only the proportion of time both sequences match. What exactly constitutes a match between the two sequences will be made clear by the following steps.

Sequence Alignment   First of all, we line up the two sequences we want to compare and take the union of all key or chord change positions. This brings the two sequences onto a common time scale, which we call evaluation segments. We end up with a list of label pairs and associated durations. An example of this procedure for two chord sequences can be seen in Figure 19.9.

Label Selection   Then we decide which label pairs we want to include in the evaluation. For a first overall score, all segments should be included of course, but in some cases it makes sense to evaluate only on a subset of them. If the dictionary of the algorithm is such that no output can be generated that suitably resembles the reference, that segment can be dropped such that the score can span its full range. For

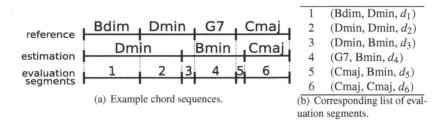

(a) Example chord sequences.

1	(Bdim, Dmin, d_1)
2	(Dmin, Dmin, d_2)
3	(Dmin, Bmin, d_3)
4	(G7, Bmin, d_4)
5	(Cmaj, Bmin, d_5)
6	(Cmaj, Cmaj, d_6)

(b) Corresponding list of evaluation segments.

Figure 19.9: Creating a list of evaluation segments from two sequences.

example, it makes sense to drop segments annotated with diminished chords from the evaluation when the algorithm can only output major and minor chords, such as segment 1 in Figure 19.9. Another possible use case is to limit the evaluation only to certain categories of labels, for instance, to compare the performance on triads with the one on tetrads (segment 4 versus segments 2,3,4,6). In all cases, this decision of inclusion should be based on the reference label only.

Harmonic Content Correspondence   Finally, the retained label pairs are compared to each other and a score is assigned to the evaluation segment, which then gets weighted by the segment's duration. All segment scores of the whole data set are summed together and divided by the total duration of retained segments to arrive at the final result. The pairwise score itself can be calculated according to a number of methods.

Obviously, when both labels are the same (taking into account enharmonic variants and the transformation to the previously defined chord and key vocabularies), the score is 1. The remaining question is if, and how, a difference is made between erroneous estimations that are close and those that are completely off.

The case of related keys is reasonably well defined. In the Music Information Retrieval Evaluation Exchange (MIREX)[3] audio key detection contest, a part of the score is assigned to keys that are a perfect fifth away from, relative or parallel to the reference key. For C major, these are F and G major (perfect fifth away), A minor (relative minor), and C minor (parallel minor). They get 0.5, 0.3, and 0.2 points, respectively. The best score obtained in MIREX-2014 was 0.8683.

For chords, there is less consensus about how to account for related chords, because the chord dictionary is typically more complex, so many times, almost-correct estimations do not contribute at all. One option is to consider chords as sets of chromas and to take the precision and the recall of these sets (see Section 13.3.3). This is useful to detect over- or under-estimation of the chord cardinality when mixing triads and tetrads, but cannot measure if the root has been estimated correctly. Therefore it can be complemented by a score that only looks at a match between the roots.

Just like for designing a recognition algorithm, the evaluation procedure requires a dictionary on which the music will be projected, and the transformation rules to achieve this. This is used to bring the reference and tested sequence into the same

---

[3]http://www.music-ir.org/mirex/wiki/MIREX_HOME. Accessed 22 June 2016.

space. So far we have assumed that this evaluation dictionary is the same as the algorithmic dictionary, which is the easiest and most recommended option, but this is not always possible. A notable example is when we want to compare multiple algorithms with different vocabularies to the same ground truth, as is done for the MIREX audio chord estimation task.

To handle these differences, extra rules need to be formulated about which segments should be included in the evaluation and which evaluation dictionary should be used. To this end, a framework for the rigorous definition of evaluation measures is explained in [24]. The accompanying software, as used in MIREX too, is freely available online.[4] Naturally, it can also be used for the evaluation of a single algorithm on its own.

When multiple algorithms are compared to each other, it is also important to know to what degree their differences are statistically significant. The method used for MIREX is described in [2], and is also freely available as an R package.[5]

19.9 Concluding Remarks

In this chapter, we have described the basic techniques for building the blocks of a chord recognition system: the audio signal representations, the representation of the chord labels, and the joint estimation of several musically related parameters.

While the performances of each of these blocks can be improved, one should not forget that the aim of a chord estimation system is to reduce the harmonic content of a music piece to a set of chord labels over time. Considering this, the performances of such a system are not only limited by the performances of the system itself but also by the possibility to efficiently reduce the harmonic content of a music piece to a set of chord labels. While multi-pitch estimation can be applied to any music piece that contains harmonic sounds, chord estimation implies some specific temporal and vertical organization of those pitches. For example, chord reduction for modal, rap or electronic music is questionable since these music genres usually do not rely on a vertical organization of pitches. A chord estimation algorithm should therefore ideally rely on the analysis of previously estimated pitches (as does a musicologist for music analysis) that would allow deciding whether this chord reduction is applicable or not. However, given the current performances of multi-pitch estimation systems, current chord recognition systems directly relate the chord labels to the audio signal without using any pitch transcription. The results obtained using this direct approach are yet impressive when a chord reduction is applicable.

19.10 Further Reading

We finish this chapter by providing short descriptions and further readings on possible variations around the techniques used for each of the blocks of a chord recognition system.

---

[4]https://github.com/jpauwels/MusOOEvaluator
[5]https://bitbucket.org/jaburgoyne/mirexace

19.10.1 Alternative Audio Signal Representations

Beat-synchronous chroma/PCP. One common variation is the computation of the chroma/PCP vector in a beat-synchronous way [20]. In this, a beat-tracking algorithm is first used (see Chapter 20) to estimate the beat positions b_i (where $i \in \mathbb{N}$ is the beat number). Chroma/PCP are then extracted to represent every beat of a track. This is done by centering the frame positions on the beats b_i. This method allows us to make the chroma/PCP tempo-invariant. Beat-synchronous chroma/PCP are especially useful for popular music in which chords often last an integer number of beats (usually 2 or 4). Having chroma/PCP attached to beat duration therefore allows us to more easily estimate the duration of the chords.

Multi-band chroma/PCP. Another common variation is to compute separate chroma/PCP vectors for the lower and higher parts of the spectrum [14]. Each part is then considered to bring different observations related to chords.

Tonal centroid. Despite their obvious appeal due to the proximity to the theoretical definition of chords, chroma/PCP vectors are not the only type of signal representations that have been tried. Harte et al. [8] proposed a six-dimensional vector named the *tonal centroid*, which emphasizes the intervals of a third and a fifth because these are the most discriminative for chord recognition. This representation is used in the key-dependent HMMs of [13].

Full spectra. Another option is to keep the full spectrum, instead of reducing it to a single octave chroma/PCP vector. The redundancy present in the different octaves then needs to be dealt with by the mapping to the internal chord representation itself. This is more complicated, but potentially more powerful. Such an approach has been tried in pioneering work [16], before the establishment of the chroma/PCP vector as de facto standard, and has recently seen renewed interest with the advent of deep-learning techniques for neural networks [9].

19.10.2 Alternative Representations of the Chord Labels

Other than changing the signal representation, there are also alternatives for the chord representations. In addition to generative models such as Gaussian distributions and mixtures thereof, discriminative methods can be used as well. A simple frame-based classifier can be implemented as a support vector machine, as in [35] (see also Chapter 12). A discriminative counterpart for a system that can take into account the broader musical context, can be achieved by using a linear-chain conditional random field [3].

19.10.3 Taking into Account Other Musical Concepts

Beside the joint estimation with keys, other musical concepts can be included in the modelling of musicological context. Just like with the inclusion of the key, the idea is that there is a dependency between chords and the other notion that can help to narrow down chords to more specific positions and combinations.

Joint chord/meter recognition. A first example is the co-recognition with metric position. Here the premise is that a chord change is more likely to happen at some positions in the measure than at others. An intuitive example would be that it is more likely to change chords on the first beat of a measure than on the second. This dependency can be taken into account both with chords directly [20], as well as in a larger context of chords in a key [14].

Joint chord/bass line recognition. Another musical concept that is intertwined with chords is the bass line. The bass note that is played together with a chord often gives an indication of the chord itself. Because there are fewer interfering harmonics of other notes in the lower part of the spectrum, the bass note can also be estimated comparatively easily. The combination of these two qualities makes the bass line a valuable addition to a chord context model, as demonstrated in [31].

Joint chord/key/structure recognition. A final case of including other concepts, is the co-recognition of chords and keys with musical structure. It is based on the idea that certain chord combinations, especially when expressed in a key, are indicative of structural endings. Examples are cadences in classical music or typical turnarounds in jazz and blues. The effect on the chord and key recognition output seems negligible in this case, but it provides an alternative method of structure estimation [21].

Bibliography

[1] J. P. Bello and J. Pickens. A robust mid-level representation for harmonic content in music signals. In *Proceedings of the 6th International Conference on Music Information Retrieval (ISMIR)*, pp. 304–311, 2005.

[2] J. A. Burgoyne, W. B. de Haas, and J. Pauwels. On comparative statistics for labelling tasks: What can we learn from Mirex Ace 2013? In *Proceedings of the 15th Conference of the International Society for Music Information Retrieval (ISMIR)*, pp. 525–530, 2014.

[3] J. A. Burgoyne, L. Pugin, C. Kereliuk, and I. Fujinaga. A cross-validated study of modelling strategies for automatic chord recognition in audio. In *Proceedings of the 8th International Conference on Music Information Retrieval (ISMIR)*, pp. 251–254, 2007.

[4] H.-T. Cheng, Y.-H. Yang, Y.-C. Lin, I.-B. Liao, and H. H. Chen. Automatic chord recognition for music classification and retrieval. In *Proceedings of the IEEE International Conference on Multimedia and Expo (ICME)*, pp. 1505–1508. IEEE Press, 2008.

[5] D. P. Ellis and A. V. Weller. The 2010 Labrosa chord recognition system. In *MIREX 2010 Extended Abstract (14th Conference of the International Society for Music Information Retrieval)*, 2010.

[6] T. Fujishima. Realtime chord recognition of musical sound: A system using Common Lisp music. In *Proceedings of the International Computer Music Conference (ICMC)*, pp. 464–467. International Computer Music Association, 1999.

[7] E. Gómez. Tonal description of polyphonic audio for music content processing. *INFORMS Journal on Computing, Special Cluster on Computation in Music*, 18(3):294–304, 2006.

[8] C. Harte, M. Sandler, and M. Gasser. Detecting harmonic change in musical au-
     dio. In *Proceedings of the 1st ACM Workshop on Audio and Music Computing
     Multimedia*, pp. 21–26, New York, NY, USA, 2006. ACM.

[9] E. J. Humphrey and J. P. Bello. Rethinking automatic chord recognition with
     Convolutional Neural Networks. In *Proceedings of the IEEE International
     Conference on Machine Learning and Applications (ICMLA)*, pp. 357–362.
     IEEE Press, 2012.

[10] O. Izmirli. Template based key finding from audio. In *Proceedings of the
     International Computer Music Conference (ICMC)*, pp. 211–214. International
     Computer Music Association, 2005.

[11] M. Khadkevich and M. Omologo. Use of hidden Markov models and factored
     language models for automatic chord recognition. In *Proceedings of the 10th
     International Conference on Music Information Retrieval (ISMIR)*, pp. 561–
     566, 2009.

[12] C. L. Krumhansl and E. J. Kessler. Tracing the dynamic changes in perceived
     tonal organization in a spatial representation of musical keys. *Psychological
     Review*, 89(4):334–368, July 1982.

[13] K. Lee and M. Slaney. Acoustic chord transcription and key extraction from
     audio using key-dependent HMMs trained on synthesized audio. *IEEE Trans-
     actions on Audio, Speech and Language Processing*, 16(2):291–301, February
     2008.

[14] M. Mauch and S. Dixon. Simultaneous estimation of chords and musical con-
     text from audio. *IEEE Transactions on Audio, Speech and Language Process-
     ing*, 18(6):1280–1289, August 2010.

[15] J. Morman and L. Rabiner. A system for the automatic segmentation and classi-
     fication of chord sequences. In *Proceedings of the 1st ACM Workshop on Audio
     and Music Computing Multimedia*, pp. 1–10. ACM, 27 October 2006.

[16] S. H. Nawab, S. Abu Ayyash, and R. Wotiz. Identification of musical chords
     using constant-Q spectra. In *Proceedings of the IEEE International Conference
     on Acoustics, Speech and Signal Processing (ICASSP)*, volume 5, pp. 3373–
     3376. IEEE Press, 2001.

[17] L. Oudre, Y. Grenier, and C. Févotte. Chord recognition using measures of fit,
     chord templates and filtering methods. In *Proceedings of the IEEE Workshop
     on Applications of Signal Processing to Audio and Acoustics (WASPAA)*. IEEE
     Press, 2009.

[18] H. Papadopoulos. *Joint Estimation of Musical Content Information*. PhD thesis,
     Université Paris VI, 2010.

[19] H. Papadopoulos and G. Peeters. Large-scale study of chord estimation al-
     gorithms based on chroma representation. In *Proceedings of the IEEE Fifth
     International Workshop on Content-Based Multimedia Indexing (CBMI)*. IEEE
     Press, 2007.

[20] H. Papadopoulos and G. Peeters. Joint estimation of chords and downbeats

from an audio signal. *IEEE Transactions on Audio, Speech and Language Processing*, 19(1):138–152, January 2011.

[21] J. Pauwels, F. Kaiser, and G. Peeters. Combining harmony-based and novelty-based approaches for structural segmentation. In *Proceedings of the 14th Conference of the International Society for Music Information Retrieval (ISMIR)*, pp. 138–143, 2013.

[22] J. Pauwels and J.-P. Martens. Combining musicological knowledge about chords and keys in a simultaneous chord and local key estimation system. *Journal of New Music Research*, 43(3):318–330, 2014.

[23] J. Pauwels, J.-P. Martens, and M. Leman. Modeling musicological information as trigrams in a system for simultaneous chord and local key extraction. In *Proceedings of the IEEE International Workshop on Machine Learning for Signal Processing (MLSP)*. IEEE Press, 2011.

[24] J. Pauwels and G. Peeters. Evaluating automatically estimated chord sequences. In *Proceedings of the IEEE International Conference on Acoustics, Speech and Signal Processing (ICASSP)*. IEEE Press, 2013.

[25] G. Peeters. Musical key estimation of audio signal based on hidden Markov modelling of chroma vectors. In *Proceedings of the International Conference on Digital Audio Effects (DAFx)*, pp. 127–131. McGill University Montreal, September 18–20 2006.

[26] L. Rabiner. A tutorial on hidden Markov model and selected applications in speech. *Proceedings of the IEEE*, 77(2):257–285, 1989.

[27] T. Rocher, M. Robine, P. Hanna, and M. Desainte-Catherine. A survey of chord distances with comparison for chord analysis. In *Proceedings of the International Computer Music Conference (ICMC)*, pp. 187–190. International Computer Music Association, 2010.

[28] M. P. Ryynänen and A. P. Klapuri. Automatic transcription of melody, bass line, and chords in polyphonic music. *Computer Music Journal*, 32(3):72–86, Fall 2008.

[29] R. Scholz, E. Vincent, and F. Bimbot. Robust modeling of musical chord sequences using probabilistic n-grams. In *Proceedings of the IEEE International Conference on Acoustics, Speech and Signal Processing (ICASSP)*, pp. 53–56. IEEE Press, 2009.

[30] R. N. Shepard. Circularity in judgements of relative pitch. *Journal of the Acoustical Society of America*, 36:2346–2353, 1964.

[31] K. Sumi, K. Itoyama, K. Yoshii, K. Komatani, T. Ogata, and H. G. Okuno. Automatic chord recognition based on probabilistic integration of chord transition and bass pitch estimation. In *Proceedings of the 9th International Conference on Music Information Retrieval (ISMIR)*, pp. 39–44, 2008.

[32] D. Temperley. What's key for key? The Krumhansl-Schmuckler key-finding algorithm reconsidered. *Music Perception*, 17(1):65–100, 1999.

[33] E. M. von Hornbostel. Psychologie der Gehörerscheinungen. In A. Bethe et al.,

eds., *Handbuch der normalen und pathologischen Physiologie*, chapter 24, pp. 701–730. J.F. Bergmann-Verlag, 1926.

[34] G. H. Wakefield. Mathematical representation of joint time-chroma distributions. In *Proc. of SPIE conference on Advanced Signal Processing Algorithms, Architectures and Implementations*, volume 3807, pp. 637–645. SPIE, 1999.

[35] A. Weller, D. Ellis, and T. Jebara. Structured prediction models for chord transcription of music audio. In *Proceedings of the IEEE International Conference on Machine Learning and Applications (ICMLA)*, pp. 590–595. IEEE Press, 2009.

[36] Y. Zhu and M. S. Kankanhalli. Precise pitch profile feature extraction from musical audio for key detection. *IEEE Transactions on Multimedia*, 8(3):575–584, June 2006.

Chapter 20

Tempo Estimation

JOSÉ R. ZAPATA
Department of TIC, Universidad Pontificia Bolivariana, Colombia

20.1 Introduction

Rhythm, along with harmony, melody and timbre, is one of the most fundamental aspects of music; sound by its very nature is temporal. Rhythm in its most generic sense is used to refer to all of the temporal aspects of music, whether it is represented in a score, measured from a performance, or existing only in the perception of the listener. In order to build a computer system capable of intelligently processing music, it is essential to design representation formats and processing algorithms for the estimation of rhythmic content of music [24]. Tempo estimation (also referred to as tempo induction or tempo detection) is the computational approach to estimate the rate of the perceived musical pulses and normally referred to as the foot tapping rate.

Content analysis of musical audio signals has received increasing attention from the research community, specifically in the field of music information retrieval (MIR) [42]. MIR aims to retrieve musical pieces by processing not only text information, such as artist name, song title or music genre, but also by processing musical content directly in order to retrieve a piece based on its rhythm or melody [54]. Since the earliest automatic audio rhythm estimation systems proposed in [21, 50, 13] in the mid to late 1990s, there has been a steady growth in the variety of approaches developed and the applications to which these automatic systems have been applied. The use of automatic rhythm estimation has become a standard tool for solving other MIR problems, e.g. structural segmentation [38], chord detection [40], music similarity [31], cover song detection [47], automatic remixing [28], and interactive music systems [48]; by enabling "beat-synchronous" analysis of music.

While many different tempo estimation and beat tracking techniques have been proposed in recent years, e.g. for beat tracking, see [18, 9, 43, 15, 5, 11, 56] and for tempo estimation [20, 19, 44], recent comparative studies of rhythm estimation systems [57] suggest that there has been little improvement in the state of the art in recent years [41] and the method by Klapuri [33] is still widely considered to represent the state of the art for both tasks.

Current approaches for tempo estimation focus on the analysis of mainstream popular music with clear and stable rhythm and percussion instruments, which facilitates this task. These approaches mainly consider the periodicity of intensity descriptors (principally onset detection functions) to locate the beats, and then to estimate the tempo. Nevertheless, they usually fail when they are analyzing other music genres like classical music, because this type of music often exhibits tempo variations; in other words, it does not include clear percussive and repetitive events. The same problem appears with a capella or choral music, acoustic music, different jazz styles and pop music [24].

The goal of this chapter is to provide basic knowledge about automatic music tempo estimation. The remainder of the chapter is structured as follows: Section 20.2 gives an overview of the principal definitions and relations of the rhythm elements. The system steps to estimate the tempo are presented in Section 20.3, emphasizing the computation of the onset detection functions for tempo estimation. Section 20.4 presents the evaluation methods for tempo estimation approaches. Then, in Section 20.5, a simple implementation of a tempo estimation system is provided, and finally in Section 20.6, some applications of automatic rhythm estimation are described.

20.2 Definitions

Musical rhythm is used to refer to the temporal aspects of a musical work and the pulse (which is a regular sequence of events); its components are beat, tempo, meter, timing and grouping, and are presented in Figure 20.1. For the sake of understanding the computational approaches of automatic rhythm estimation methods in Western music, we assumed that the musical pulse which can be felt by a human being is related to the beat. This chapter is focused on estimating the time regularity of musical beats in audio signals (tempo estimation) related to the beats per minute in a song.

20.2.1   Beat

Beat is any of the events or accents in the music and is characterized by [27, p. 391] as what listeners typically entrain to as they tap their foot or dance along with a piece of music.

The beat perception is an active area of research in music cognition, in which there has long been an interest in the cues listeners use to extract a beat. Reference [53] lists six factors that most researchers agree are important in beat finding (i.e. in inferring the beat from a piece of music). These factors can be expressed as preferences:

1. for beats to coincide with note onsets,
2. for beats to coincide with longer notes,
3. for regularity of beats,
4. for beats to align with the beginning of musical phrases,

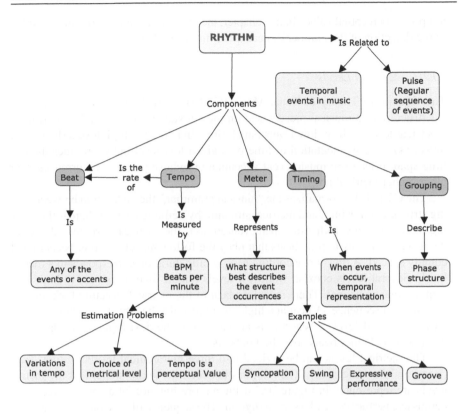

Figure 20.1: Rhythm components.

5. for beats to align with points of harmonic change, and

6. for beats to align with the onsets of repeating melodic patterns.

20.2.2   *Tempo*

Tempo is defined as the number of beats in a time unit (usually the minute). There is usually a preferred regularity, which corresponds to the rate at which most people would tap or clap in time with the music. However, the perception of tempo exhibits a degree of variability. Differences in human perception of tempo depend on age, musical training, music preferences and the general listening context [34]. They are, nevertheless, far from random and most often correspond to a focus on a different metrical level and are quantifiable as simple ratios (e.g. 2, 3, 1/2 or 1/3) [45]. The computation of tempo is based on the periodicities extracted from the music signal and the time differences between the beats. Therefore, the absolute beat positions are not necessarily required for estimating the tempo. The principal computational tempo estimation problems are related to tempo variations in the same musical piece, the correct choice of the metrical level (see Section 20.2.3), and mainly because the tempo is a perceptual value. In this chapter, the automatic tempo estimation is related to the detection of the number of beats per minute (BPM).

20.2.3   Metrical Levels

The Generative Theory of Tonal Music (GTTM) [37] defines meter as the metrical structure of a musical piece based on the coexistence of several regularities (or "metrical levels"), from low levels (small time divisions) to high levels (longer time divisions). The segmentation of time by a given low-level pulse provides the basic time span to measure music event accentuation whose periodic recurrences define other higher metrical levels.

The GTTM also formalizes the "musical grammar," the distinction between grouping structure (phrasing), and metrical structure by defining rules. Whereas the grouping structure deals with time spans (durations), the metrical structure deals with duration-less points in time beats that obey the following rules. First, beats must be equally spaced. A division according to a specific duration corresponds to a metrical level. Several levels coexist, from low levels (small time divisions) to high levels (longer time divisions). There must be a beat of the metrical structure for every note in a musical sequence. A beat at a high level must also be a beat at each lower level. At any metrical level, a beat that is also a beat at the next higher level is called a downbeat, and other beats are called upbeats.

The metrical levels can be divided into three hierarchical levels: *Tatum*, *Tactus* (Beat) and *Bar* (measure). The relations between the audio signal and the metrical levels are represented in Figure 20.2 using a representation of an audio excerpt of a percussive performance of a samba rhythm. The sequence of note onsets, related with each drum hit of the audio, is shown in Figure 20.2(b). The *tatum*, the lowest metrical level, is defined by [3] as the shortest commonly time interval, Figure 20.2(c). The *tactus* or beat, Figure 20.2(d), is defined by [37, p.21] as the preferred human tapping tempo and the computational approach of this task is called *beat tracking*. Not all the beats are aligned with note onsets or audio signal changes, which is due to the existence of rhythm deviations like syncopation, swing, groove, and expressive performance among others. The bar, Figure 20.2(e), is the highest metrical level and is typically related to the harmonic change rate or to the length of a rhythmic pattern.

20.2.4   Automatic Rhythm Estimation

The aim of automatic rhythm estimation is parsing acoustic events that occur in time into more abstract notions of tempo, timing and meter. Algorithms described in the literature differ in their goals; some of them derive beats and tempo of a single metrical level, others try to derive the complete transcription (i.e. musical scores, see Chapter 17), others aim to determine some timing features from musical performances (such as tempo changes, event shifts or swing factors), others focus on the classification of music signals by their overall rhythmic similarities, while others look for rhythm patterns. Nevertheless, most of the algorithm approaches share some

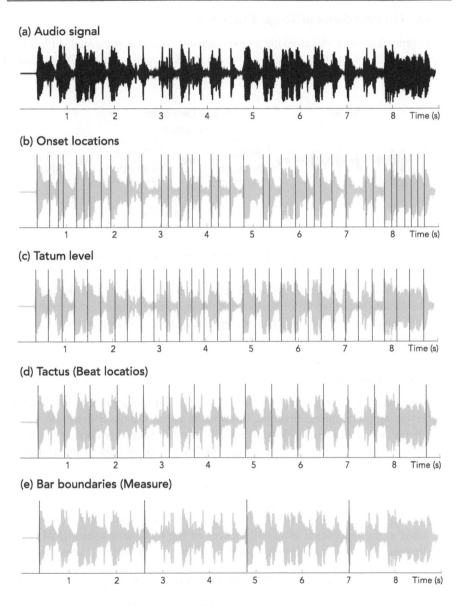

Figure 20.2: Metrical structure for the first seconds of the song *Kuku-cha Ku-cha* by "Charanga 76". (a) Audio signal. (b) Note onset locations. (c) Lowest metrical level: the Tatum. (d) Tactus (Beat locations). (e) Bar boundaries (Measure).

functional aspects (feature list creation, tempo induction, Figure 20.3), as pointed out by [24].

20.3   Overall Scheme of Tempo Estimation

The general scheme for automatic tempo estimation methods presented in Figure 20.3, represents the main steps to estimate the tempo of an audio signal. This scheme includes a feature list creation block and a tempo induction block.

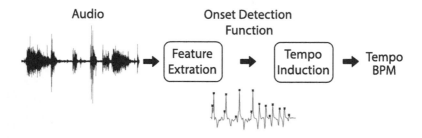

Figure 20.3: General tempo estimation scheme.

Following the recent results in [51, ch. 4] and [56] which show the importance of the feature list creation for rhythm estimation through input features rather than tempo induction models, this chapter emphasizes understanding of the onset detection functions for tempo estimation.

20.3.1   Feature List Creation

Feature list creation transforms the audio waveform into a temporal series of features representing predominant rhythmic information called the onset detection function (ODF) or novelty function. For descriptions of onset-related features, refer to Sections 5.4 and 16.2. The following list complements these lists of most important features.

- *Energy flux* [35]. Like the low-energy feature presented in Section 5.2, Equation (20.1) is calculated by segmenting the signal into short time Fourier transform frames ($x[\lambda, k]$). From these frames, each input feature sample $EF(\lambda)$ is calculated as the magnitude of the differences of the root mean square (RMS) value between the current short time Fourier transform frame and its predecessor:

$$EF(\lambda) = |RMS(x[\lambda, k]) - RMS(x[\lambda, k-1])|. \qquad (20.1)$$

The performance of the energy flux ODF is higher if the music audio signal presents percussive instruments or clear note onsets, but the performance drops with other kind of music.

- *Spectral Flux*. This onset detection function, proposed in [39] and presented in Definition 5.10 and Equation (20.2), describes the temporal evolution of the magnitude spectrogram calculated by computing the short time Fourier transform (STFT) $X[\lambda, \mu]$ in the frames. From these frames, each spectral flux sample $SFX(\lambda)$ is calculated as the sum of the positive differences in magnitude between each frequency bin of the current short time Fourier transform frame and its predecessor:

$$SFX(\lambda) = \sum_{\mu=0}^{M/2} H(|X[\lambda,\mu]| - |X[\lambda-1,\mu]|), \qquad (20.2)$$

where $H(x) = \frac{x-|x|}{2}$ is the half-wave rectifier function. This equation differs from Equation (5.12) because the square absolute difference is not calculated as presented in Definition 5.10.

- *Spectral flux log filtered* is an onset detection function introduced by Böck et al. [4] based on spectral flux, but the linear magnitude spectrogram is filtered with a pseudo Constant-Q filter bank (see Section 4.5) as follows:

$$X_\lambda^{logfilt}(\mu) = \log(\gamma \cdot (|X_\lambda(\mu)| \cdot F(\lambda,\mu)) + 1). \qquad (20.3)$$

where the frequencies are aligned according to the frequencies of the semitones of the Western music scale over the frequency range from 27.5 Hz to 16 kHz, using a fixed window length for the STFT. The resulting filter bank, $F(\lambda,\mu)$, has B = 82 frequency bins with λ denoting the bin number of the filter and μ the bin number of the linear spectrogram. The filters have not been normalized, resulting in an emphasis of the higher frequencies, similar to the high frequency content (*HFC*) method presented in Definition 5.15. From these frames, in Equation (20.4) each input feature sample is calculated as the sum of the positive differences in logarithmic magnitude (using γ as a compression parameter, e.g. $\gamma = 20$) between each frequency bin of the current STFT frame and its predecessor:

$$SFLF(\lambda) = \sum_{\mu=1}^{B=82} H\left(\left|X_\lambda^{logfilt}(\mu)\right| - \left|X_{\lambda-1}^{logfilt}(\mu)\right|\right). \qquad (20.4)$$

Nowadays the automatic rhythm description algorithms with better performance use the spectral flux log filtered as the onset detection function in their systems.

- *Complex spectral difference* [17]. This feature, presented in Definition 5.17 and Equation (20.5), describes the temporal evolution of the magnitude and phase spectrogram calculated from the short time Fourier transform. The feature has a large value if there is a significant change in magnitude or deviation from expected phase values, different from the spectral flux that only computes magnitude changes in frequency. $X_T[\lambda,\mu]$ is the expected target amplitude and phase for the current frame and is estimated based on the values of the two previous frames assuming constant amplitude and rate of phase change.

$$CSD(\lambda) = \sum_{\mu=0}^{M/2} |X[\lambda,\mu] - X_T[\lambda,\mu]|. \qquad (20.5)$$

- *Beat Emphasis Function* [10]. This ODF emphasizes the periodic structure in musical excerpts with a steady tempo. The beat emphasis function is defined as a weighted combination of the sub-band complex spectral difference functions, Equation (20.5), which emphasize the periodic structure of the signal by deriving a weighted linear combination of 20 sub-band ($S_\mu(\lambda)$) onset detection functions, where the weighting function $w(\mu)$ favors sub-bands with prominent periodic structure.

$$BEF(\lambda) = \sum_{\mu=1}^{B=20} w(\mu) \cdot S_\mu(\lambda). \qquad (20.6)$$

- *Harmonic Feature* [26]. This is a method for harmonic change detection and is calculated in Equation (20.7) by computing a short time Fourier transform. *HF* uses a modified Kullback–Leibler distance measure, see Equation (14.9), to detect spectral changes between frequency ranges of consecutive frames $X[\lambda, \mu]$. The modified measure is thus tailored to accentuate positive energy change.

$$HF(\lambda) = \sum_{\mu=1}^{B} \log_2 \left(\frac{|X[\lambda, \mu]|}{|X[\lambda - 1, \mu]|} \right). \qquad (20.7)$$

- *Mel Auditory Feature* [18]. This feature is calculated from a short time Fourier transform magnitude spectrogram and is based on the MFCC (Section 5.2.3). In Equation (20.8) each frame is then converted to an approximate "auditory" representation in 40 bands on the mel frequency scale and converted to dB, $X_{mel}(\mu)$. Then the first-order difference in time is taken and the result is half-wave rectified. The result is summed across frequency bands before some smoothing is performed to create the final feature.

$$MAF(\lambda) = \sum_{\mu=1}^{B=40} H\left(|X_{\mathrm{mel}}[\lambda, \mu]| - |X_{\mathrm{mel}}[\lambda - 1, \mu]|\right). \qquad (20.8)$$

The auditory frequency scale is used to balance the periodicities in each perceptual frequency band.

- *Phase Slope Function* [30]. This feature is based on the group delay, which is used to determine instants of significant excitation in audio signals and is computed as the derivative of phase over frequency $\tau(\lambda)$ (presented before in Section 5.16), as can be seen in Equation (20.9). Reference [30] uses this concept as an onset detection function: using a large overlap, an analysis window is shifted over the signal and for each window position the average group delay is computed. The obtained sequence of average group delays is referred to as the phase slope function (PSF). To avoid the problems of unwrapping, the phase spectrum of the signal for the computation of group delay can be computed as

$$\tau(\lambda) = \frac{X_{\mathbb{R}e}(\lambda) \cdot Y_{\mathbb{R}e}(\lambda) + X_{\mathbb{I}m}(\lambda) \cdot Y_{\mathbb{I}m}(\lambda)}{|X(\lambda)|^2}, \qquad (20.9)$$

where $X(\lambda)$ and $Y(\lambda)$ are the Fourier transforms of $x[k]$ and $kx[k]$, respectively. The phase slope function is then computed as the negative of the average of the group delay function. The performance of the phase slope function is higher in musical signals with simple rhythmic structure and little or no percussive content.

- *Bandwise Accent Signals* [33]. This feature estimates the degree of musical accent as a function of time at four different frequency ranges. This ODF is calculated from a short time Fourier transform and used to calculate power envelopes at 36 sub-bands on a critical-band scale. Each sub-band is up-sampled by a factor of two, smoothed using a low-pass filter with a 10-Hz cutoff frequency, and half-wave rectified. A weighted average of each band and its first-order differential is taken, $E_\mu(\lambda)$. In [33] each group of 9 adjacent bands (i.e. bands 1–9, 10–18, 19–27 and 28–36) are summed up to create a four-channel input feature.

$$BAS(\lambda) = \sum_{\mu=1}^{36} E_\mu(\lambda). \tag{20.10}$$

20.3.2 Tempo Induction

Tempo induction uses the onset detection function result to estimate periodicities in the signal; for computational simplicity, a fundamental assumption is made: the tempo is stable in the music audio signal. The most used methods are:

- *Autocorrelation.* The autocorrelation function is a common signal processing technique for periodicity computation as presented in Sections 2.2.7 and 4.8, which is applied to the onset detection function and the tempo can be selected from the local maximum peaks of the resulting signal; see Figure 20.5. The position of the peaks represents the time lag, which is converted to BPM as presented in Equation (20.11). This the most used method to estimate the tempo; some systems that use autocorrelation are [9, 11, 56].

$$BPM = 60/TimeLag \tag{20.11}$$

- *Comb Filterbank.* The comb filterbank uses a bank of resonator filters, each tuned to a possible periodicity, where the output of the resonator indicates the strength of that particular periodicity. This method also "implicitly encodes aspects of the rhythmic hierarchy" [49, p. 594]. More information about filter banks can be seen in Section 4.6. Examples of systems that use comb filterbanks are [33, 49].

- *Time Interval Histogram.* The information of the onset events can be extracted from the onset detection function and the time interval between the onsets is used to calculate a histogram, whose maximum peak gives us the tempo value. To estimate the tempo, the time difference between the onsets is more important than the specific time position of each one. This method is called the IOI (inter-onset interval) histogram; an example of a system that uses this method is [14].

20.4 Evaluation of Tempo Estimation

While the efficacy of automatic rhythm estimation systems can be evaluated in terms of their success with these end applications, e.g. by measuring chord detection accuracy, considerable attention has been given to the tempo estimation evaluation through the use of annotated test databases, in particular the MIR community has made a considerable effort to standardize evaluations of MIR systems.

The evaluation databases are made up of songs with the annotated ground truth consisting of a single tempo value (e.g., from the score). Accordingly, the output of the tempo estimation systems is the overall BPM value of a piece of music.

The evaluation measures that are mainly used to evaluate tempo estimation algorithms return a state of the task accomplishment given by two separated metrics:

- Metric 1: The tempo estimation value is within 4% (the precision window) of the ground-truth tempo. This measure is used to evaluate the accuracy of the algorithm to detect the general BPM of the song.

- Metric 2: The tempo estimation value is within 4% (the precision window) of 1, 2, $\frac{1}{2}$, 3, $\frac{1}{3}$ times the ground-truth tempo. This measure is used to take into account problems of double or triple deviation of the tempo estimation.

The algorithm with the best average score of *Metric 1* and *Metric 2* will achieve the highest rank.

Complementary to this evaluation method, there is a specific task in the Music Information Retrieval Evaluation eXchange (MIREX)[16] initiative to evaluate the estimation of the perceptual tempo, given by two tempo values, because a piece of music can be perceived faster or slower than its notated tempo. Perceptual tempo estimation algorithms estimate two tempo values in BPM (T1 and T2, where T1 is the slower of the two tempo values). For a given algorithm, the performance, P, for each audio excerpt will be given by the following equation:

$$P = ST1 * TT1 + (1 - ST1) * TT2, \qquad (20.12)$$

where ST1 is the relative perceptual strength of T1 (given by ground truth data, varies from 0 to 1.0). TT1 is the ability of the algorithm to identify T1 using *Metric 1* to within 8% (the precision window), and TT2 is the ability of the algorithm to identify T2 using *Metric 1* to within 8%. The algorithm with the best average P-score will achieve the highest rank in the task.

Many approaches to tempo estimation have been proposed in the literature, and some efforts have been devoted to their quantitative comparison. The first public evaluation of tempo extraction methods was carried out in 2004 by [25] evaluating the accuracy of 11 methods at the ISMIR audio estimation contest; an updated tempo evaluation comparison is presented in [57]. In 2005, 2006, and 2010 to the present, the MIREX initiative[1] continued the evaluation comparison of tempo estimation systems.

20.5   A Simple Tempo Estimation System

Previous research [57] suggests that the best tempo estimator systems use frequency decomposition and periodicity detection prior to multi-band integration. Based on this and the work of Dixon (2003), we now sketch an implementation of a simple

---

[1]http://www.music-ir.org/. Accessed 22 June 2016.

tempo estimation MATLAB® algorithm computing the energy correlation for 8 different frequency bands.

1. Load the audiofile and compute STFT using a FFT of 2048 points and a hop size of 512.

```
[x,fs] = audioread(filename); %read audio file
hopsize = 512; % STFT hopsize
nfft = 2048; % FFT size
noverlap = winsize - hopsize;
% Short-time fourier transform calculation
sp = spectrogram(x,winsize,noverlap,nfft, fs);
nfr = size(sp,1); % sp variable length
```

2. Compute the energy for 8 different frequency bands. The first band should go up to 100 Hz and the remaining bands should be equally spaced (logarithmically) and one octave wide to cover the full frequency range of the signal (cp. Figure 20.4).

```
nb = 8; % number of frequency bands
lowB = 100; % lowband in Hz
% frequency axis in bins
fco=[0,lowB*(fs/2/lowB).^((0:nb-1)/(nb-1))]/fs*nfft;
fco = round(fco); % Round each element to nearest integer
energy = zeros(nfr,nb); %memory allocation
for fr = 1:nfr
  for i = 1:nb
    lower_bound = 1+fco(i);
    upper_bound = min(1+fco(i+1), size(sp,2));
    % energy calculation
    energy(i, fr) = sum(abs(sp(lower_bound:upper_bound, fr)).^2);
  end
end
energy = 10*log10(energy); % energy in Log-scale
```

3. For each band, compute the autocorrelation of the energy values for the first 6 seconds of the song (cp. Figure 20.5 (upper part)).

```
corrtime = 6; % number of seconds of the analyzed window
nfr_corr = round(corrtime*fs/hopsize)
corr_matrix = zeros(nfr_corr,bands); %memory allocation
for nband = 1:nb
    e = energy(nband, 1:nfr_corr);
    x = xcorr(e-mean(e)); %Cross-correlation
    x = x / x(nfr_corr); %normalize
    %Correlation signal
    corr_matrix(nband,:) = x(nfr_corr:2*nfr_corr-1);
    kk(nband) = mean(e);
end
```

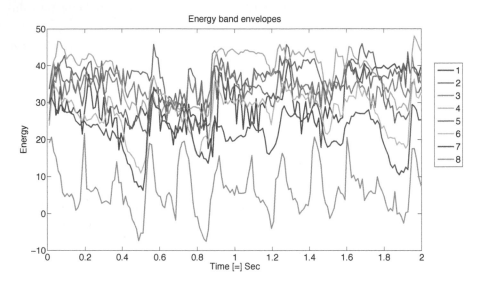

Figure 20.4: Energy in 8 bands.

```
% Band weighting
weight = sum(kk);
kk = kk /weight;
for o = 1:nb;
    corr_matrix(o,:) = corr_matrix(o,:).*kk(o);
end
```

4. Sum of all the correlation signals, normalize the amplitude and plot the result
   signal with the horizontal axe in BPM values in a range between 30 Bpm and 240
   Bpm. (cp. Figure 20.5 (lower part)).

```
sum_corr_matrix = sum(corr_matrix, 1)/nb;
tt = [0:hopsize:(nfr_corr-1)*hopsize]/fs;
bpm = 60./tt;
```

5. Finally locate the maximum values as the most prominent tempo values, if you
   have a ground truth verify that one of the values correspond to the annotated
   tempo.

20.6   Applications of Automatic Rhythm Estimation

There are several areas of research for which automatic rhythm estimation is relevant:

- Estimation of tempo and variations in tempo for performance analysis consider
  the interpretation of musical works, for example, the performer's choice of tempo

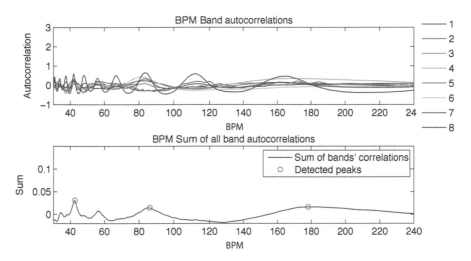

Figure 20.5: Autocorrelation of the 8 bands in BPM.

and expressive timing. These parameters are important in conveying structural and emotional information to the listener [6].

- Rhythm estimation is necessary for automatic score transcription from musical signals, like music transcription [2], chord detection [40], and structural segmentation [38]. See also Chapter 17.

- Rhythm data is used in audio content analysis for automatic indexing and content-based retrieval of audio data, such as in multimedia databases and libraries, like music similarity [31] and cover-song detection [47].

- Automatic audio synchronization with devices such as lights, electronic musical instruments, recording equipment, computer animation, and video with musical data. Such synchronization might be necessary for multimedia or interactive performances or studio post-production work. The increasingly large amounts of data processed in this way lead to a demand for automation, which requires that the software involved operate in a "musically intelligent" way, and the interpretation of beat is one of the most fundamental aspects of musical intelligence [14].

Other applications are source separation [46], interactive music accompaniment [48], automatic remixing [28], real-time beat-synchronous audio effects [52], and biorhythms detection [1] among others.

20.7 Concluding Remarks

Automatic estimation of musical rhythm is not a simple task. It seems to involve two processes: (1) a bottom-up process that enables faster perception of beats from scratch, and (2) a top-down process, due by a persistent mental framework, that induced a perceptual guide of the organization of incoming events [12]. Implementing both reactivity to the music interpretation and persistence of internal representation in a computer program, is a challenge. It is important to say that rhythm estimation does not solely call for handling timing features. Moreover, despite the somewhat automatic inclusion of rhythm estimation systems as temporal processing components in different applications, tempo estimation and beat tracking itself is not considered a solved problem.

In the small number of comparative studies of automatic beat tracking algorithms with human tappers [29, 49, 9, 41, 7] musically trained individuals are generally shown to be more adept at tapping the beat than the best computational systems. Given this gap between human performance and computational beat trackers, automatic rhythm estimation is not yet a solved problem.

20.8 Further Reading

Extensive information about automatic rhythm description can be found in the PhD theses [22, 8, 55] and the books [36, ch. 6], [32, 23] and a general scheme of automatic rhythm estimation methods is provided in [24]. An updated year-by-year comparison of different tempo estimation systems and beat trackers can be found at http://www.music-ir.org/mirex/.

Furthermore, beat tracking systems can be considered one of the fundamental problems in music information retrieval (MIR) research and the tempo can be computed using the beat positions. Thereby, it is recommended to study beat tracking systems to have a better approximation to automatic rhythm description systems. Some beat tracker algorithms are [33, 56, 11, 14, 18], whose common aim is to "tap along" with musical signals.

Bibliography

[1] C. Barabasa, M. Jafari, and M. D. Plumbley. A robust method for S1/S2 heart sounds detection without ECG reference based on music beat tracking. In *10th International Symposium on Electronics and Telecommunications*, pp. 307–310. IEEE, 2012.

[2] J. P. Bello. *Towards the Automated Analysis of Simple Polyphonic Music: A Knowledge-Based Approach.* PhD thesis, University of London, London, 2003.

[3] J. Bilmes. *Timing Is of the Essence: Perceptual and Computational Techniques for Representing, Learning, and Reproducing Expressive Timing in Percussive Rhythm.* PhD thesis, Massachusetts Institute of Technology, 1993.

[4] S. Böck, F. Krebs, and M. Schedl. Evaluating the online capabilities of onset detection methods. In *Proceedings of the 13th International Society for Music Information Retrieval Conference (ISMIR)*, pp. 49–54, Porto, 2012.

[5] S. Böck and M. Schedl. Enhanced beat tracking with context-aware neural networks. In *Proceedings of the 14th International Conference on Digital Audio Effects (DAFx-11)*, pp. 135–139, 2011.

[6] E. Clarke. Rhythm and timing in music. In *The Psychology of Music*, pp. 473–500. Academic Press, San Diego, 2nd edition, 1999.

[7] N. Collins. Towards a style-specific basis for computational beat tracking. In *Proceedings of the 9th International Conference on Music Perception and Cognition*, pp. 461–467, 2006.

[8] M. E. P. Davies. *Towards Automatic Rhythmic Accompaniment*. PhD thesis, Queen Mary University of London, 2007.

[9] M. E. P. Davies and M. D. Plumbley. Context-dependent beat tracking of musical audio. *IEEE Transactions on Audio, Speech, and Language Processing*, 15(3):1009–1020, 2007.

[10] M. E. P. Davies, M. M. D. Plumbley, and D. Eck. Towards a musical beat emphasis function. In *IEEE Workshop on Applications of Signal Processing to Audio and Acoustics (WASPAA)*, pp. 61–64, New Paltz, NY, 2009. IEEE.

[11] N. Degara, E. Rua, A. Pena, S. Torres-Guijarro, M. E. P. Davies, M. D. Plumbley, and E. Argones. Reliability-informed beat tracking of musical signals. *IEEE Transactions on Audio, Speech and Language Processing*, 20(1):290–301, 2012.

[12] P. Desain and H. Honing. Computational models of beat induction: The rule-based approach. *Journal of New Music Research*, 28(1):29–42, 1999.

[13] S. Dixon. Beat induction and rhythm recognition. In *The Australian Joint Conference on Artificial Intelligence*, pp. 311–320, 1997.

[14] S. Dixon. Automatic extraction of tempo and beat from expressive performances. *Journal of New Music Research*, 30(1):39–58, 2001.

[15] S. Dixon. Evaluation of the audio beat tracking system BeatRoot. *Journal of New Music Research*, 36(1):39–50, 2007.

[16] J. S. Downie. The music information retrieval evaluation exchange (2005–2007): A window into music information retrieval research. *Acoustical Science and Technology*, 29(4):247–255, 2008.

[17] C. Duxbury, J. Bello, M. E. P. Davies, and M. D. Sandler. Complex domain onset detection for musical signals. In *Proceedings of the 6th Conference on Digital Audio Effects (DAFx)*, volume 1, London, UK, 2003.

[18] D. Ellis. Beat tracking by dynamic programming. *Journal of New Music Research*, 36(1):51,60, 2007.

[19] M. Gainza and E. Coyle. Tempo detection using a hybrid multiband approach. *IEEE Transactions on Audio, Speech, and Language Processing*, 19(1):57–68, 2011.

[20] A. Gkiokas, V. Katsouros, and G. Carayannis. Tempo induction using filterbank analysis and tonal features. In *Proceedings of the 11th International Society on Music Information Retrieval Conference (ISMIR)*, pp. 555–558, 2010.

[21] M. Goto and Y. Muraoka. A beat tracking system for acoustic signals of music. In *the Second ACM Intl. Conf. on Multimedia*, pp. 365–372, 1994.

[22] F. Gouyon. *A Computational Approach to Rhythm Description Audio Features for the Computation of Rhythm Periodicity Functions and Their Use in Tempo Induction and Music Content Processing.* PhD thesis, Universitat Pompeu Fabra, 2005.

[23] F. Gouyon. *Computational Rhythm Description: A Review and Novel Approach.* VDM Verlag Dr. Müller, Saarbrücken, Germany, 2008.

[24] F. Gouyon and S. Dixon. A review of automatic rhythm description systems. *Computer Music Journal,* 29(1):34–54, 2005.

[25] F. Gouyon, A. P. Klapuri, S. Dixon, M. Alonso, G. Tzanetakis, C. Uhle, and P. Cano. An experimental comparison of audio tempo induction algorithms. *IEEE Transactions on Audio, Speech, and Language Processing,* 14(5):1832–1844, 2006.

[26] S. Hainsworth and M. Macleod. Onset detection in musical audio signals. In *Proceedings of the International Computer Music Conference (ICMC),* pp. 136–166, Singapore, 2003.

[27] S. Handel. *Listening: An Introduction to the Perception of Auditory Events.* MIT Press, Cambridge MA, 1989.

[28] J. A. Hockman, J. P. Bello, M. E. P. Davies, and M. D. Plumbley. Automated rhythmic transformation of musical audio. In *Proceedings of the 11th International Conference on Digital Audio Effects (DAFx-08),* pp. 177–180, Espoo, Finland, 2008.

[29] A. Holzapfel, M. E. P. Davies, J. R. Zapata, J. L. Oliveira, and F. Gouyon. Selective sampling for beat tracking evaluation. *IEEE Transactions on Audio, Speech and Language Processing,* 20(9):2539–2548, 2012.

[30] A. Holzapfel and Y. Stylianou. Beat tracking using group delay based onset detection. In *Proc. of ISMIR—International Conference on Music Information Retrieval,* pp. 653–658, Philadelphia, 2008.

[31] A. Holzapfel and Y. Stylianou. Parataxis: Morphological similarity in traditional music. In *Proceedings of the 11th International Society for Music Information Retrieval Conference (ISMIR),* pp. 453–458, 2010.

[32] A. Klapuri and M. Davy. *Signal Processing Methods for Music Transcription.* Springer, 2006.

[33] A. P. Klapuri, A. J. Eronen, and J. Astola. Analysis of the meter of acoustic musical signals. *IEEE Transactions on Audio, Speech, and Language Processing,* 14(1):342–355, 2006.

[34] E. Lapidaki. *Consistency of Tempo Judgments as a Measure of Time Experience in Music Listening.* PhD thesis, Northwestern University, 1996.

[35] J. Laroche. Efficient tempo and beat tracking in audio recordings. *Journal of the Audio Engineering Society,* 51(4):226–233, 2003.

[36] A. Lerch. *An Introduction to Audio Content Analysis: Applications in Signal Processing and Music Informatics.* IEEE Press, Berlin, 2012.

[37] F. Lerdahl and R. Jackendoff. *A Generative Theory of Tonal Music.* MIT Press, Cambridge, MA, 1983.

[38] M. Levy and M. B. Sandler. Structural segmentation of musical audio by constrained clustering. *IEEE Transactions on Audio, Speech and Language Processing,* 16(2):318–326, 2008.

[39] P. Masri. *Computer Modelling of Sound for Transformation and Synthesis of Musical Signal.* PhD thesis, University of Bristol, Bristol, UK, 1996.

[40] M. Mauch, K. Noland, and S. Dixon. Using musical structure to enhance automatic chord transcription. In *Proceedings of the 10th International Society for Music Information Retrieval Conference (ISMIR),* pp. 231–236, 2009.

[41] M. F. McKinney, D. Moelants, M. E. P. Davies, and A. Klapuri. Evaluation of audio beat tracking and music tempo extraction algorithms. *Journal of New Music Research,* 36(1):1–16, 2007.

[42] E. Pampalk. *Computational Models of Music Similarity and Their Application to Music Information Retrieval.* PhD thesis, Vienna University of Technology, 2006.

[43] G. Peeters. Beat-tracking using a probabilistic framework and linear discriminant analysis. In *Proceedings of the 12th International Conference on Digital Audio Effect, (DAFx),* pp. 313–320, 2009.

[44] G. Peeters. Template-based estimation of tempo: Using unsupervised or supervised learning to create better spectral templates. In *Proceedings of the International Conference on Digital Audio Effect (DAFx),* pp. 6–9, 2010.

[45] P. Polotti and D. Rocchesso, eds. *Sound to Sense—Sense to Sound: A State of the Art in Sound and Music Computing.* Logos Verlag Berlin GmbH, 2008.

[46] Z. Rafii and B. Pardo. REpeating Pattern Extraction Technique (REPET): A simple method for music/voice separation. *IEEE Transactions on Audio, Speech, and Language Processing,* 21(1):71–82, 2013.

[47] S. Ravuri and D. P. W. Ellis. Cover song detection: From high scores to general classification. In *IEEE International Conference on Acoustics, Speech and Signal Processing (ICASSP),* pp. 65–68. IEEE, 2010.

[48] A. Robertson and M. D. Plumbley. B-Keeper: A beat-tracker for live performance. In *International Conference on New Interfaces for musical expression (NIME),* pp. 234–237, New York, 2007.

[49] E. Scheirer. Tempo and beat analysis of acoustic musical signals. *The Journal of the Acoustical Society of America,* 103(1):588–601, 1998.

[50] E. D. Scheirer. Pulse tracking with a pitch tracker. In *IEEE Workshop on Applications of Signal Processing to Audio and Acoustics,* Mohonk, NY, 1997.

[51] A. M. Stark. *Musicians and Machines: Bridging the Semantic Gap in Live Performance.* PhD thesis, Queen Mary, U. of London, 2011.

[52] A. M. Stark, M. D. Plumbley, and M. E. P. Davies. Real-time beat-synchronous audio effects. In *International Conference on New Interfaces for Musical Ex-*

pression, pp. 344–345, 2007.

[53] D. Temperley and C. Bartlette. Parallelism as a factor in metrical analysis. *Music Perception*, 20(2):117–149, 2002.

[54] R. Typke, F. Wiering, and R. C. Veltkamp. A survey of music information retrieval systems. In *Proceedings of the International Symposium on Music Information Retrieval, (ISMIR)*, pp. 153–160, 2005.

[55] J. R. Zapata. *Comparative Evaluation and Combination of Automatic Rhythm Description Systems*. PhD thesis, Universitat Pompeu Fabra, 2013.

[56] J. R. Zapata, M. E. P. Davies, and E. Gómez. Multi-feature beat tracking. *IEEE/ACM Transactions on Audio, Speech, and Language Processing*, 22(4):816–825, 2014.

[57] J. R. Zapata and E. Gómez. Comparative evaluation and combination of audio tempo estimation approaches. In *Audio Engineering Society Conference: 42nd International Conference: Semantic Audio*, pp. 198 – 207, Ilmenau, 2011. Audio Engineering Society.

Chapter 21

Emotions

GÜNTHER RÖTTER
Institute for Music and Music Science, TU Dortmund, Germany

IGOR VATOLKIN
Department of Computer Science, TU Dortmund, Germany

21.1 Introduction

Six basic emotions are known: fear, joy, sadness, disgust, anger, and surprise [17]. These basic emotions are objects of emotion theories. Each emotion has three components: the personal experience of an emotion, activation (which is a higher activity of the sympathetic nervous system), and action. When talking about emotions in music, action means the anticipation of a structure in music. There is a connection between the emotional expression in language and in music. The melodic contour, the range, the change rate of tones and the tempo of a spoken sentence in a specific emotion are very similar to music expressing the same emotional state. Still, music does not cause real emotions; it rather works like a pointer that elicits stored emotional experiences. Music not only deals with basic emotions, but also with moods and emotional episodes.

21.1.1 What Are Emotions?

Emotions are mainly used to steer behavior in important situations of life and to support the organism with an adequate amount of energy. Emotions have an evolutionary significance for adaption and selection. As emotions have a high demand of energy, they are only used for exceptional situations in life. When emotions appear they are often accompanied with bodily changes. Emotions always have a cause that can be clearly described. They may not be confused with moods like relaxation or with drives like hunger or sexuality. The term "emotion" should furthermore be closed off from personality traits like aggression and introversion as well as attitudes like tolerance and hospitality.

  The facial expressions of the six basic emotions can be recognized by most ethnic groups from all over the world. This means that emotional behavior is not the result of a learning process, but is an anthropological constant. In the following, various emotion theories shall be explained but first it has to be remarked that these theories mostly deal with basic emotions and not with "aesthetic" ones.

21.1.2 Difference between Basic Emotions, Moods, and Emotional Episodes

In spoken language, one normally does not distinguish between emotion, emotional episode, and mood. However, there are slight differences. Emotions last for several seconds only up to a few minutes at most. An emotion that lasts for several hours is called an emotional episode. Moods (cf. Section 21.4.2) instead are long lasting and can endure a couple of days. Apart from the most obvious distinctive features, there are further characteristics in which emotions and moods differ. If one is in a specific mood, one is more likely to show corresponding emotions. "It is as if the person is seeking an opportunity to indulge the emotion relevant to the mood" [18, p. 57]. In a depressed mood, for instance, it is likely to be sad, too. One furthermore has difficulties in regulating and controlling the emotion supported by the mood. In addition, moods in comparison to emotions do not have a certain facial expression that is typical of it.[1]

21.1.3 Personality Differences and Emotion Perception

Besides the general process of remembering under emotional influence, there are differences in gender and age. This should be kept in mind for the evaluation of automatic music emotion recognition systems, which will be discussed in Section 21.6. If, for instance, the labels for the training of supervised classification models are provided by student males, they may not correspond to emotional categories perceived by other human groups. Women remember emotional events more easily than males but often forget about other information because emotions are encoded in different parts of the brain. Elderly people instead tend to forget about negative emotions fast in comparison to younger ones. The amygdala (a pea-sized part of the temporal lobe in the brain that is responsible for an emotion's evaluation) in an elderly person reacts to positive and negative events to the same extent, whereas the amygdala in a younger person mainly reacts to negative events.

The perception and the reactions on stimuli also change with increasing age. This has been proven by the study by Neiss et al. [44]. They investigated the response to positive, neutral and negative stimuli in two subject groups; the first group consisted of younger adults aged 24–40 and the second one of older adults aged 65–85. In the end, they found out that older adults tend to rate positive stimuli more positively than younger adults. Instead, younger adults are more prone to negative stimuli. Again, this is due to the older adults' amygdala activity, which is lower for negative stimuli than for positive stimuli. However, female subjects from the older group usually rated more extreme than any other group.

---

[1]http://www.paulekman.com/wp-content/uploads/2013/07/Moods-Emotions-And-Traits.pdf. Accessed 24 February 2016.

In the following, we will start with an overview of emotional theories and models (Section 21.2). Section 21.3 briefly introduces the relationship of speech and emotion. Emotions in music are discussed in Section 21.4. Groups of features helpful for automatic emotion prediction are introduced in Section 21.5. Section 21.5.7 shows how individual feature relevance can be measured for a categorical and a dimensional music emotion recognition system. In Section 21.6, the history and targets of automatic music emotion recognition systems are outlined, with a list of databases with freely available annotations and a discussion of classification and regression methods applied in recent studies.

21.2 Theories of Emotions and Models

There are two ways of arranging emotion [10]. On the one hand, the classification of basic emotions is regarded as a categorical approach. On the other hand, there is a dimensional classification where an emotion can be arranged in a orthogonal co-ordinate system. One axis defines the level of arousal (low arousal to high arousal) whereas the other axis defines valence (positive to negative). Every emotion can be mapped within this system. Wundt even created a three-dimensional system, reaching from tension to relief, lust to reluctance, arousal to sedation [68, pp. 208–219].

Most of the emotion theories contain three components: activation or arousal, the individual perception of emotions, and the action that results from it.

Mandler's emotion theory (1975) [41] can be transferred to emotions in music, cf. Figure 21.1. He states that activation is a consequence of an action being interrupted. In this theory the environment delivers information about the quality of an emotion, too. Consequently, a specific action of the subject follows.

What does this mean according to the perception of music? Action in the case of listening to music means anticipation of a musical structure. If a specific, unexpected, element occurs in music, the anticipation is interrupted which leads to arousal. Hans Werbik [65] has shown that there is a reversed u-shaped relation between the number of interruptions and the preference of music.

A low number of interruptions means that the anticipation of a musical structure works almost perfectly and, thus, causes boredom. In contrast, a high number of interruptions nearly terminates the whole process of anticipating. Anger and insecurity are the consequences. Listeners prefer a medium number of interruptions, which cause positive emotions. There is only a statistical problem with the reversed u-shaped curve: each subject has a personal opinion depending on musical experience and knowledge of music.

21.2.1 Hevner Clusters of Affective Terms

Beneath bodily processes, the subjective experience plays an important role concerning emotional processes. If bodily changes are objectively determinable, the experience can be conveyed by language. Hence, descriptions of impressions are used in music psychology, which have to be interpreted. Additionally, lists of adjectives can offer some characteristics. Excerpts from pieces of music as well as examples of

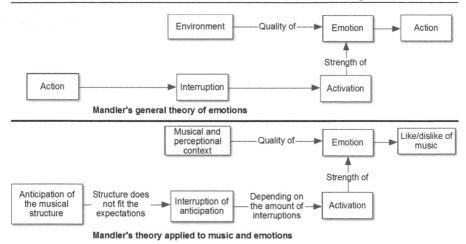

Figure 21.1: Mandler's emotion theory and its adaption to music.

original and arranged music (variation of condition) and musical elements (intervals, rhythms)[2] are evaluated. Overall, numerous investigations were developed between the late 19th century and the 1970s, which can only be presented partially.

Hevner tried to determine the emotional expressions of musical parameters by changing pieces of music [24]. She either altered the melody or changed the compositions' mode from major to minor. The musical examples have to be assigned to different adjectives that were circularly arranged according to their relation. In these adjective circles, bars were drawn with their length depending on how much the adjective fit the given example. Simple harmonies appeared happy, graceful and detached, whereas dissonant and more complex melodies appeared more powerful, exciting and desolate. As expected, major seemed more happy and graceful, minor instead dignified, desolate and dreamy. A moving rhythm appeared happy and graceful contrastingly, the firm rhythm appeared powerful and dignified. Still, it was not always possible to make explicit judgments and it was not always simple to reasonably change the original composition.

A recent meta-study was conducted by Gabrielsson and Lindström who used the Hevner adjective circle [22]. They therefore examined every study about musical expression since the 1930s and asserted that 18 musical parameters have already been investigated on its emotional expression.[3] A rising melody, for example, can either be connected with dignity, seriousness and tension or with fear, surprise, anger, and power. Falling melodies can on the one hand be regarded as graceful, powerful and detached or on the other hand be linked to boredom and gusto. Another methodical problem is that only one musical parameter can be investigated at a time.

---

[2]This type of description has already been investigated by Huber [25].

[3]These parameters are amplitude, articulation, harmonics, intervals, volume, alterations of volume, ambit, melody, sound sequence, melody movement, key, pitch, alteration of notes, rhythm, tempo, audio quality, tonality, and musical form.

De la Motte-Haber states:

> Psychologists' regular procedure of isolating and varying the conditions to gather the effect of an alteration of conditions is only partly applicable when it comes to music examples. Isolated alterations like the change of tempo for example, can have caricaturing effects on the subjects or cause uncertainties at least of the subtext [43, p. 28].

21.2.2 Semantic Differential

Another method of describing musical expression is by means of the semantic differential. Semantic differentials are lists of contrasting adjectives in which is marked how much one or other adjective is applicable. The connected points reveal a characteristic profile of evaluation. In this process, adjectives can either be judgmental (e.g., nice, ugly, interesting, boring), associative like angular, round, pale and colorful, or adjectives that describe an emotional expression like happy, sorrowful, dreamy or pedestrian. De la Motte-Haber used this technique for a study examining ten rhythms, which differed in measure, frequency of events, and homogeneity and played them in three different tempi to subject groups [43].

Fast rhythms were described as happier than slower ones, whereby the intensity of happiness was not dependent on the metronome but the subjective perception of tempo as well as the frequency of events and some other characteristics.

21.2.3 Schubert Clusters

Schubert enhanced the Hevner adjective circle in 2003, see Table 21.1. A total of 133 musically experienced people were asked to rate 90 adjectives, consisting of the 67 adjectives Hevner had used and 23 nonmusical, additional ones, according to their ability of describing any kind of music properly. In a forced-choice response, the subjects had to rate each word on a scale from 0 (= totally unsuitable) to 7 (= very suitable). Every word with a mean value below 4 was deleted from the list. The remaining 46 adjectives were placed into nine different clusters.

Table 21.1: Emotion Clusters by Schubert [71]

Cluster A	Cluster B	Cluster C	Cluster D	Cluster E	Cluster F	Cluster G	Cluster H	Cluster I
Bright	Humorous	Calm	Dreamy	Tragic	Dark	Heavy	Dramatic	Agitated
Cheerful	Light	Delicate	Sentimental	Yearning	Depressing	Majestic	Excited	Angry
Happy	Lyrical	Graceful			Gloomy	Sacred	Exhilarated	Restless
Joyous	Merry	Quiet			Melancholy	Serious	Passionate	Tense
	Playful	Relaxed			Mournful	Spiritual	Sensational	
		Serene			Sad	Vigorous	Soaring	
		Soothing			Solemn		Triumphant	
		Tender						
		Tranquil						

21.2.4 Circumplex Word Mapping by Russell

In the two-dimensional circumplex emotion word mapping by Russell there are two axes, cf. Figure 21.2. The vertical axis illustrates the level of arousal whereas the horizontal axis shows whether the affect is positive or negative. Various emotions are arranged on this diagram. Angry, for example, is a rather negative affect with an upper-medium level of arousal.

Thayer's model instead allows a more accurate way of placing the emotions since there are more axes than in Russell's mapping. Thayer created a circular diagram that is divided into eight parts. The horizontal axis reaches from tension to calm and the vertical axis from tiredness to energy. Up to this point the two diagrams are almost similar, but Thayer additionally inserted two diagonal axes that measure from calm-tiredness to tension-energy and from tension-tiredness to calm-energy.

The third model has been created by Barrett and Russell. In the two-dimensional model the horizontal axis reaches from unpleasant to pleasant and the vertical one reaches from deactivation to activation. Sixteen emotions are circularly arranged around the axes.

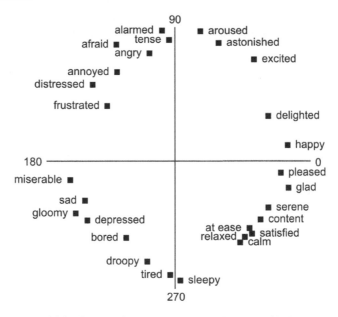

Figure 21.2: Circumplex word mapping by Russell after [47].

21.2.5 Watson–Tellegen Diagram

In the emotion diagram by Watson and Tellegen (Figure 21.3), four axes split the circular diagram into eights. These axes reach from a low positive affect to a high positive affect, from unpleasantness to pleasantness, from disengagement to strong

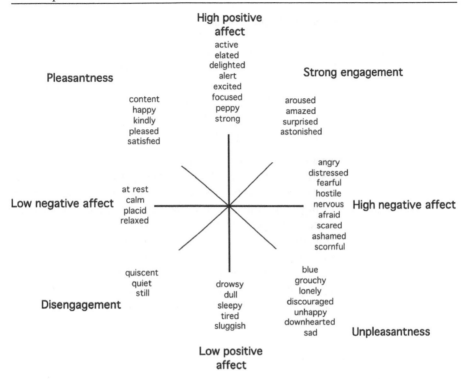

Figure 21.3: Tellegen and Watson diagram after [56].

engagement, and from low negative affect to high negative affect. On these axes, several emotions are listed, one below the other.

21.3 Speech and Emotion

There is a close connection between a musical melody and the speaking voice. Even the melody of the speaking voice is able to deliver emotions. The melody can be investigated for certain parameters, namely frequency, range, variability, loudness, and tempo. These parameters of the speaking voice have huge similarities to a musical melody, which is empirically proven nowadays [28]. Music is, thus, obviously based on the melody of the speaking voice. Here might be the origin of music making. Playful dealing and testing with the emotional speaking voice turned into an autonomous discipline known today as "singing".

Another way of analyzing speech phonetically is by using the software Praat.[4] By means of this program, it is possible to analyze speech parameters such as energy intensity, pitch, standard deviation, jitter, shimmer, autocorrelation, noise-to-harmonics ratio, and harmonics-to-noise ratio.

A practical example for this phenomenon is the polices' effort of analyzing a

---

[4]http://www.fon.hum.uva.nl/praat. Accessed 24 February 2016.

Table 21.2: Emotional Expression in Speech and Music [50, p. 300]

Emotions	F_0 level	F_0 range	F_0 variability	Loudness	Tempo
Joy	High	?	Large	High	Fast
Anger	High	Large	Large	High	Fast
Fear	High	Large	Small	?	Fast
Sadness	Low	Small	Small	Low	Slow

hostage taker's speech parameters to prevent escalating situations. These exceptional circumstances are often stress related, which causes the speaking voice to change. The police analyze and categorize the hostage taker's emotional and affective conditions with the help of linguistic features. Nowadays they also focus on analyzing the voice in terms of voice level, speaking rate, volume, speech melody, and inarticulate noises. By analyzing these characteristics it is possible to prove or disprove the authenticity of the hostage taker's message. Especially variations of voice frequency give reliable evidence of the hostage taker's state of arousal. Obviously, investigation of emotion recognition in speech seems to be advanced compared to emotion recognition in music. This is due to the less complex situation of emotion recognition in music because of fewer parameters.

21.4   Music and Emotion

21.4.1   Basic Emotions

Many models focus on describing the relation between intention and comprehension of an expression as well as the expression's emotional effect. Balkwill and Thompson [2] claim that due to psychological processes, culture-specific and also non-culture-specific components have an effect on music. This is proven by studies where people of the Western world first had to listen to Indian Raggas and later on describe their emotions while listening. The described emotions were similar to the emotions intended by the music.

Important factors of the "intercultural comprehensibility" of music are tempo and timbre. The authors base their conclusion on the study by Mandler, who claims that musical cognition results from an interaction of fulfilled and unfulfilled listening expectations, which lead to a direct activation [2]. Kreutz critically notes[5]:

> The confines of the models are reached with the hearers' relation between understood musical expression and observed emotions. Processes as induction, empathy and infection in the course of communication [35, p. 556].

This becomes even more complicated, taking into account that a biographical

---

[5]Kreutz in this context mentions a non-checked model of six components by Huron. In this model occurs an emotional meaning analysis through six different systems (reflexive, denotative, connotative, associative, emphatic, critical). This model ranges from a reflexive, unconscious reaction to a critical examination of the authenticity of emotional expressions.

connection and the current (emotional) situation influence the hearers' listening experience. In this context it is necessary to mention two studies by Knobloch et al., who found out that men listen to sad music in case of lovesickness only, whereas women also listen to sad music when they're in love [32, 31].

Some emotions can be found in music. Despite their distinct acoustic structure, rock and popular music both imitate 18th- and 19th-century musical expression because they convey the same basic emotions (especially the positive ones). Disgust, prudence, and interest are usually not expressed in music in opposition to joy, anger, fear, and sadness.

In the following, some examples are given:

Example 21.1 (Joy). *(1) "Maniac" from the film Flashdancer by Michael Sembello. Similar to the speaking melody, in "Maniac" joy is expressed by a high frequency, large intervals. (2) "All My Loving" by The Beatles. (3) "Happy" by Pharrell Williams. (4) "Don't Worry Be Happy" by Bobby McFerrin. (5) "Lollipop" by The Chordettes. When expressing joy in music, large intervals and irregularity of instrumentation are used to convey surprise, which is a part of joy.*

Example 21.2 (Fear). *(1) "Bring Me to Life" by Evanescence. (2) "Gott" from Beethoven's "Fidelio." (3) "This Is Halloween" by Marilyn Manson. (4) "Psycho Theme" by Bernard Herrmann. One can find strangeness and intransparency cues as well as sudden exclamations in fearful music. According to de la Motte-Haber [10], fear has a dual nature: "stiffening in horror" on the one hand and "simultaneous desire to run away" on the other.*

Example 21.3 (Sadness). *(1) "Marche Funèbre" by Frederick Chopin. (2) "Sonata in G Minor" by Albioni. (3) "Goodbye My Lover" by James Blunt. (4) "Der Weg" by Herbert Grönemeyer. In music, sadness can be characterized by small intervals, a falling melody, a low sound level (small activation), and a slow rhythm. Sometimes, sad songs are ambivalent because they might include encouraging and soothing elements.*

Example 21.4 (Anger). *(1) "Enter Sandman" by Metallica. (2) "Lose Yourself" by Eminem. (3) "Line and Sinker" by Billy Talent. (4) "Königin der Nacht" by Mozart. The expression of anger is often highly similar in music and spoken language. Audio frequency, range, variability, and tempo are high, which are indicative of an increased pulse rate. Due to their related expression of joy and anger, some pieces might be "emotionally confused." An example is the first movement of Johann Sebastian Bach's "Brandenburg Concerto no. 3" (BWV 1048). One can either interpret the musical expression as joyful or angry (measures 116–130).*

Mathematically, there would be over a thousand shadings of emotions if there were only six grades of intensity of an emotion and four different emotions expressed. Sometimes there are, however, less basic emotions within a song. As a result, the musical expression of the basic emotions is intensified through language and thus, easier to understand.

21.4.2   Moods and Other Affective States

Sometimes it is not about the expression of basic emotions at all: the first movement of Maurice Ravel's "Gaspard de la Nuit" expresses an atmosphere or mood rather than a basic emotion. But the situation is even more complicated: Scherer und Zentner tried to reveal deficiencies in research in this area first and then created conceptual clarity [51].

In Table 21.3, they distinguish between six affective states: preferences, emotions, moods, interpersonal stances, attitudes, and personality traits. These conditions are related to certain characteristics such as intensity, duration, synchronization, result relatedness, willingness to evaluate, tempo of change, and willingness to act. Anger for example is symbolized by a high intensity, a short duration, and a high synchronization.

Furthermore, there is a high tempo of change as well as a strong tendency to act. This means that psychological and physical changes occur simultaneously. A mood is very long lasting, has a medium intensity and a low synchronization. Result relatedness and the willingness to evaluate are rather small, and the tempo of change and willingness to act instead are high, though not as high as compared to emotions.

Scherer and Zentner doubt that music can provoke real emotions: "It is rather unlikely, of course, that one will be able to find as intense and highly synchronized response patterns as found in the case of violent range leading to fighting, for instance" [51, p. 384]. Equally, it is hard to tell whether music causes the same level of willingness to act as an emotion would [20]. The author claims that research should stop thinking that these reactions were conventional, emotional processes. It is rather about emotional episodes[6] where every component of the table is involved. Scherer and Zentner state:

> We believe that progress in this area will be difficult as long as researchers remain committed to the assumption that real intense emotions must be traditional basic emotions, such as fear or anger, for which one can identify relatively straightforward action tendencies such as fight or flight. Progress is more likely to occur if we are prepared to identify emotion episodes where all of the components shown in the table are in fact synchronized without there being a concrete action tendency or a traditional, readily accessible verbal label. In order to study these phenomena, we need to free ourselves from the tendency of wanting to assign traditional categorical labels to emotion processes [51, p. 384].

According to Scherer and Zentner, the term "affection" cannot be described through conventional literature about emotions. This emotion occurs quite often during the reception of music, accompanied with goose bumps, teary eyes, as well as hot and cool shivers, even though one cannot really describe one's feeling and their trigger. It might be hard to prove, but Konecni assumes:

---

[6]Konecni called them "aesthetic mini-episodes" that are embedded in everyday life, cf. [34].

Table 21.3: Design Feature Delimitation of Different Affective States after [51]. VL: Very Low; L: Low; M: Medium; H: High; VH: Very High

Type of affective state: brief definition (*examples*)	Design feature						
	Intensity	Duration	Synchronization	Event focus	Appraisal elicitation	Rapidity of change	Behavioural impact
Preferences: evaluative judgements of stimuli in the sense of liking or disliking, or preferring or not over another stimulus (*like, dislike, positive, negative*)	L	M	VL	VH	H	VL	M
Emotions: relatively brief episodes of synchronized response of all or most organismic subsystems in response to the evaluation of an external or internal event as being of major significance (*angry, sad, joyful, fearful, ashamed, proud, elated, desperate*)	H	L	VH	VH	VH	VH	VH
Mood: diffuse affect states, most pronounced as change in subjective feeling, of low intensity but relatively long duration, often without apparent cause (*cheerful, gloomy, irritable, listless, depressed, buoyant*)	M	H	L	L	L	H	H
Interpersonal stances: affective stance taken toward another person in a specific interaction, colouring the interpersonal exchange in that situation (*distant, cold, warm, supportive, contemptuous*)	M	M	L	H	L	VH	H
Attitudes: relatively enduring, affectively coloured beliefs and predispositions towards objects or persons (*liking, loving, hating, valuing, desiring*)	M	H	VL	VL	L	L	L
Personality traits: emotionally laden, stable personality dispositions and behavior tendencies, typical for a person (*nervous, anxious, reckless, morose, hostile, envious, jealous*)	L	VH	VL	VL	VL	VL	L

It is perhaps the ultimate humanistic moment, and it may well include an elitist element: of feeling privileged to regard Mozart as a brother, of sensing the larger truth hidden in the pinnacles of human achievement, and yet realizing, with some resignation, their minuscule role in the universe [33, p. 339].

Due to the lack of appropriate terminology, researchers have mostly avoided the phenomenon of affection. Additionally, it is possible that some reactions might be provoked through music that aren't emotional at all and thus are not definable with the help of the table. This might be an additional field of study for further research.

The previous explanations have illustrated the complexity of the relation between music and emotions. Recognition of emotions is one of the most challenging tasks in music data handling for a computer scientist. Automatic prediction of emotions from music data is often treated as a classification or a regression problem [71]. Methods from Chapters 5, 8, 12, 14, and 15 may be integrated into a music emotion recognition (MER) system. The goal of classification-based MER systems is to predict emotions from categorical (discrete) emotion models and regression-based MER systems to estimate numerical characteristics from dimensional emotion models (for examples of models, see Section 21.2). In the following we will discuss groups of music characteristics which are helpful in recognizing emotions.

21.5 Factors of Influence and Features

As outlined in Sections 21.2–21.4, there are diverse theoretical explanations of coherence between certain musical characteristics and perceived emotions. Features related to music theory that can be extracted from audio or score are discussed in Sections 21.5.1–21.5.5. If the score is not available, transcription of audio signal to symbolic representation may be applied; cf. Chapter 17. Other feature sources are mentioned in Section 21.5.6. Note that individual relevance of features may strongly depend on data and prediction goals, so that "recommended" features are not always the same in different studies.

21.5.1 Harmony and Pitch

The basics of harmony are discussed in Chapter 3. One of the most prominent examples of related characteristics is the *mode*. Music in major is often perceived as happy, joyful or majestic, music in minor as sad, melancholic, or desperate. A significant share of Western popular music is composed either in major or minor. Other modes may also produce certain emotional affects. For example, Locrian is a mode with a tritone instead of a fifth, causing the music to be perceived as more distressed or mystical. An algorithm for mode extraction from the MIR Toolbox [36] was the best individual contributor to the prediction of valence in [11].

Music intervals can be grouped according to *consonance*: perfect consonant, imperfect consonant, and dissonant intervals; cf. Figure 3.7. Consonant intervals sound more pleasing to the ear, so that the balance between consonant and dissonant intervals may help to distinguish between positive and negative emotions or valence.

In polyphonic music, chords based upon consonant and dissonant intervals have an unquestioned emotional impact.

The *temporal progression* of harmonies is also relevant. For a satisfying perception of enclosed music sequences, dissonant intervals should be resolved. A typical example is a cadence, consider Figures 3.10, 3.24–3.27. A temporary change to a secondary dominant or a modulation (cf. Section 3.5.5) may evoke a surprising stimulus, whereas multiple repetitions of the same cadence may be associated with boredom or frustration. Properties of harmonic sequences may be characterized by probabilities of transitions between chords by means of Markov models (cf. Example 9.16) or generalized coincidence function (see Figure 3.8).

The first step of the extraction of harmonic features from audio signal is the estimation of the semitone spectrum (Definition 2.7) or chroma (Section 5.3.1). Simple statistics may help to recognize emotions, e.g., a high average pitch tends to correlate with a higher arousal and a high deviation of pitch with a higher valence [71, p. 51]. Several harmonic characteristics are available in the MIR Toolbox [36]: Harmonic Change Detection Function (HCDF), key and its clarity, alignment between major and minor, strengths of major/minor keys, tonal novelty, and tonal centroid. Based on chroma vector, individual strengths of consonant and dissonant intervals can be measured [60, p. 31]. Statistics of chord progression include numbers of different chords, their changes, and most frequent chords [60, p. 32], or longest common chord subsequence as well as chord histograms [71, p. 191]. For methods to extract chords from audio, refer to Chapter 19.

Harmony-related features are often integrated in MER systems [4]. Including these features increased the performance of a system for the prediction of induced emotions in [1]. The share of minor chords was relevant for the recognition of "sadness," the share of tritones for "tension," and the share of seconds for "joyful activation." In another study [14], the MIR Toolbox features key clarity and tonal novelty were among the most relevant features for regression-based modeling of anger and tenderness. HCDF and tonal centroid were among the best features selected by Relief (cf. Section 15.6.1) for the recognition of emotion clusters in [45].

Example 21.5 (Relevant Harmony and Pitch Features). *Figure 21.4 shows distributions of best individually relevant harmony and pitch features for the recognition of emotions after Examples 21.1–21.4. A set of audio descriptors was extracted for 17 music pieces with AMUSE [62]. Feature values from extraction frames between estimated onset events were selected, normalized and aggregated for windows of 4 s with 2 s overlap as mean and standard deviation. The most individually relevant features were identified by means of the Wilcoxon test (see Definition 9.29). For visualization, probability density was calculated using Gaussian kernel (function* KSDENSITY *in MATLAB ®). The prediction of anger and sadness seems to be easier than fear and joy, but the error is high in all cases when only the distribution of the most relevant feature is taken into account alone. For the list of best features across all tested groups (harmony and pitch, timbre, dynamics, tempo and rhythm), see Example 21.9 in Section 21.5.7.*

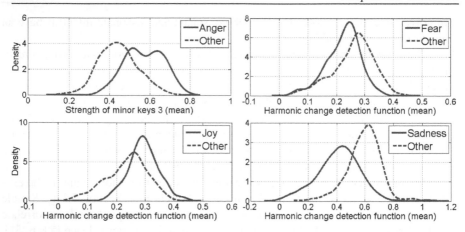

Figure 21.4: Probability densities for the most individually relevant harmony and pitch features for the recognition of emotions from Examples 21.1–21.4.

21.5.2 *Melody*

In music with vocals, melody has a large impact on listeners' attention: a general high melodiousness may correlate with invoked tenderness and low melodiousness with tension [1]. As discussed in the previous section, the grade of consonance or dissonance may be relevant. In contrast to harmonic properties, for the measurement of *consonance in a melody* interval strengths should be extracted between succeeding tones of a melody. Reference [21, p. 132] lists other descriptors: "[...] large intervals to represent joy, small intervals to represent sadness, ascending motion for pride but descending motion for humility, or disordered sequences of notes for despair."

The shape of the *melodic contour* can be extracted after pitch detection. For instance, [49] differs between initial (I), final (F), lowest (L), and highest (H) pitches in a melody. Then, contours can be grouped into several categories, e.g., I-L-H-F for four-stage contours with descending and then ascending movements, or (I,L)-H-F for three-stage contours whose initial pitch is also the lowest one. Further characteristics comprised statistics over pitch distribution in a melody (pitch deviation, highest pitch, etc.) and characteristics of vibrato. The integration of melodic audio features together with other groups of audio, MIDI, and lyric characteristics contributed to best achieved results for classification into emotions [45] (vibrato characteristics belonged to the most relevant melodic features selected by Relief) as well as for regression predicting arousal and valence [46].

A robust extraction of melody characteristics from audio is a challenging task and its success depends on the extraction of semitone spectrum. Further requirements are the robust onset detection and source separation for the identification of the main voice. Approaches to solve those tasks are presented in Sections 16.2 and 11.6.

21.5.3 Instrumentation and Timbre

Instrumentation can provide additional clues for emotion recognition as well. Organ can be associated with reverence and power, xylophone with dreaminess or joy, disturbed guitars with anger or sadness, and swarmatron[7] with mystery or anxiety. The same melody played by different real or synthesized instruments led to a measurable change in perceived emotion (happiness, sadness, anger, or fear) in [23].

Having the score at hand, instrumentation can be automatically analyzed. However, the scores are not always available and do not help to derive characteristics of a particular instrument or a playing style. Recognition of instruments from polyphonic audio recordings is a very complex task; see Chapter 18. From the signal perspective, an instrument timbre is characterized by a range of frequencies with varying amplitudes and their change over time. Timbral descriptors often applied in MER systems are Mel Frequency Cepstral Coefficients (MFCCs, cf. Section 5.2.3), spectral centroid (Definition 5.3), and spectral rolloff (Definition 5.8) [3]; others are discussed in Chapter 5. Spectral centroid had the best rank after the application of several feature selection methods in [6] for the classification into four regions of arousal/dimension space.

In particular, features which describe the *distribution of overtones* and balance between harmonic and non-harmonic partials (thus relating to perceived consonance) may be beneficial for emotion prediction. These features include tristimulus, inharmonicity, irregularity, and even/odd harmonics [71, p. 48]; see Section 5.3.4. The balance between even and odd harmonics was proved to be relevant for emotional perception of individual instrument tones with equal spectral centroid values and attack times in [67]. Here, the classes (categories) were happy, sad, heroic, scary, comic, shy, joyful, and depressed. The influence of timbre on listener perception was observed and visualized with dimensional models in [13]. The best seven features for the prediction of arousal and valence belonged to three groups: temporal (attack slope, envelope centroid), spectral (spectral skewness and regularity, ratio between energies of high and low frequencies), and spectro-temporal (spectral flux and flux of 6th sub-band). It is worth mentioning that partial descriptors depend on the previous successful estimation of fundamental frequency (cf. Section 4.8) and may be noisy for audio mixtures of many sources.

Example 21.6 (Relevant Timbre Features). *Similar to Example 21.5, the most individually relevant timbre features are estimated for the classes from Examples 21.1– 21.4. The 1st MFCC is the most relevant feature for the identification of anger and sadness, the 2nd MFCC for fear, and the 13th MFCC for joy. Again, anger and sadness are easier to identify than fear and joy.*

21.5.4 Dynamics

Indicators of loudness in the score may be helpful for the recognition of arousal or emotions related to arousal dimension, as in the Russell model (see Figure 21.2). We

---

[7]Analogue synthesizer based on multiple oscillators which was for instance used by Trent Reznor for the soundtrack of *The Social Network* movie.

may expect that pieces with annotations *pp* (pianissimo) and *p* (piano) may have a high positive correlation with emotions like calmness, sleepiness, and boredom. In contrast, *ff* (fortissimo) and *f* (forte) may indicate anger, fear, or excitement.

For audio recordings, an often applied estimator of a volume is the *root mean square* (RMS) of the time signal, Equation (2.48). Also the *zero-crossing rate* (Definition 5.1), which roughly measures the noisiness of the signal, can be helpful. The deviation of zero-crossing rate may correlate with valence [71, p. 42].

For the measurement of perceived loudness the spectrum can be transformed to scales adapted to subjective perception of volume like a sone scale, cf. Section 2.3.3. The temporal change of volume can be captured by statistics of energy over longer time windows. Examples of these features are lowenergy (Definition 5.4) and characteristics of RMS peaks such as overall number of peaks and number of peaks above half of the maximum peak [60]. Dynamics features contributed to the most efficient feature sets for the recognition of four emotions (anger, happiness, sadness, tenderness) in [48].

Example 21.7 (Relevant Dynamics Features). *Probability densities of the most individually relevant dynamics features for Examples 21.1–21.4 are provided in Figure 21.5. The number of energy peaks is higher for joy and lower for sadness as may be expected from music theory. The deviation of the 2nd sub-band energy ratio is higher for angry pieces. Fear seems to be particularly hard to predict using the most relevant dynamics feature, and distributions of the two classes are similar.*

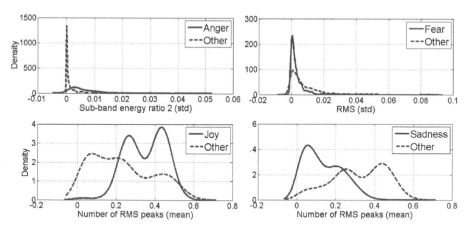

Figure 21.5: Probability densities for the most individually relevant dynamics features for the recognition of emotions from Examples 21.1–21.4.

21.5.5   Tempo and Rhythm

Fast *tempo* is often associated with high arousal, cf. Table 21.2. Manually annotated tempo (slow or fast) in [1] showed a significant correlation ($p < 0.05$) with 8 of 9 evoked emotions. Pieces which evoked joyful activation and amazement were characterized by fast tempo (Spearman's coefficient $r_{XY} = 0.76$ resp. 0.50); calmness, tenderness, and sadness by slow tempo ($r_{XY} = -0.64, -0.48, -0.45$). Algorithms for the extraction of tempo from audio are discussed in Chapter 20. A simple statistic of temporal musical progress is the *event density* (number of onsets per second), used for instance in [48]. It is also possible to save the density of beats and tatums.

Certain rhythmical patterns may be representative of a genre and emotions induced by corresponding music pieces; consider two syncopated rhythms typical for Latin American dances in Figure 3.43. The rhythmic periodicity of audio signal can be captured by the estimation of *fluctuation patterns* (Section 5.4.3) which had the largest regression β weights for the recognition of anger in [14]. Rhythmic characteristics can be calculated from onset curves (for onset detection see Section 16.2). Reference [39] defines the *rhythm strength* as the average onset strength, and *rhythm regularity* as the strength of peaks after autocorrelation of the onset curve. It is argued that the strength of a rhythm is higher for emotions with a high arousal and a low valence compared to emotions with a low arousal and a high valence. Rhythm regularity may have a high correlation with arousal [71, p. 41]. Both features along with tempo and event density improved the recognition of three out of four emotion regions with either low or high arousal and valence. The *pulse clarity* was the best individual predictor for arousal and the second best for valence in [11].

Other rhythm features showed, in [53], the largest individual correlation with arousal and valence, compared to descriptors of spectrum, chords, metadata, and lyrics. Here, signal energies of candidate tempi between 60 and 180 BPM were estimated after the application of comb filters. In the next step, the meter of music pieces was calculated. Rhythm features together with dynamic characteristics were suggested as the best descriptors to recognize 4 emotions in [48].

Example 21.8 (Relevant Tempo and Rhythm Features). *In the last example of feature comparison, best individually relevant tempo and rhythm features are estimated. The fluctuation patterns belong to the most relevant characteristics for three of four classes: the 1st dimension for anger, the 4th for fear, and the 5th for joy. Sad pieces tend to have a lower rhythmic clarity, which is the most relevant feature for this class.*

21.5.6 Lyrics, Genres, and Social Data

With the rapid growth of the Internet, new sources of information beyond score and audio became available for music classification, cf. Chapter 8. From textual sources used for MER systems, *lyrics* are probably investigated most frequently. A simple possibility is to count *occurrences of the most frequent words* [53]. In that study, words were reduced to their stems, e.g., "loved" and "loving" to the stem "love." The performance, however, was lower compared to audio features. Another commonly applied statistic is *TF-IDF*, Equations (8.1) and (8.2). Reference [9] distinguishes between three *categories of occurrences* of sentiment words which have a strong impact on their meaning: a word itself ("I love you"), a word with a negation ("I don't love you"), and with a modifier ("I love you very much").

Information about *genre or style* of a music piece, either provided by a music expert in a music web database or by listeners, may have a strong correlation with emotions. Asking listeners to rate emotions for classical, jazz, pop/rock, Latin American, and techno pieces revealed significant differences of emotion distributions over tested genres both for felt and perceived emotions [73]. In another study, the correlation between genres and emotions was measured as significant by χ^2-statistic for three large music data sets with 12 genres and 184 emotions [71, p. 199]. Reference [11] provides interesting insights on the impact of genre for the evaluation of MER systems. For example, the performance of regression model (R^2) for valence prediction dropped from 0.58 to 0.30 and further to 0.06 when the models were trained for film music and were validated on classical versus popular pieces. The same test scenario for arousal led to a decrease of R^2 from 0.59 to 0.54 and 0.37.

User-generated content in the social web can help to improve the performance of MER systems. Extension of audio features with last.FM *listener tags* sorted by their frequency improved the performance of mood clustering into four regions with high or low arousal/valence as well as into four MIREX[8] mood clusters [7]. To avoid an excessive number of tag dimensions, they can be mapped to sentiment words after the estimation of co-occurrences. Also, reduction techniques like clustering can be applied [37]; see Chapter 11. Another possibility to mine social data is to measure *co-occurrences of artists and songs* in the history of listening behaviour or user-crafted playlists from web radios [61].

21.5.7 Examples: Individual Comparison of Features

In the following we provide two examples comparing features for a categorical and a dimensional MER approach.

Example 21.9 (Feature Relevance for a Categorical Approach). *For the measurement of individual feature relevance, we may compare distributions of features from Examples 21.1–21.4 by means of the Wilcoxon rank-sum test. First, many audio descriptors were extracted with AMUSE [62]. Feature values from extraction frames between onset events were normalized and aggregated as mean and standard deviation for classification windows of 4 s with 2 s overlap. For each of four MER tasks, the Wilcoxon test was applied to compare features from windows either belonging or not belonging to a current class. The top three most relevant features were identified for each task; Table 21.4 lists corresponding p-values. The 1st–3rd places for tasks in column headers are given in brackets (bold font). For example, the 1st dimension of the fluctuation pattern characteristics was the best feature to distinguish between "angry" and "other" pieces.*

Most of the original feature implementations are from the MIR Toolbox [36], except for RMS peak number (implementation for [60] based on the MIR Toolbox peak function) and sub-band energy ratio (Yale implementation [42]). Characteristics of fluctuation patterns belong to the best descriptors. Energy features (RMS peak number and sub-band energy) are also relevant, followed by MFCCs and spectral

---

[8]http://www.music-ir.org/mirex/wiki/MIREX_HOME. Accessed 23 November 2015.

brightness. Note that the number of analyzed music pieces was rather low, so that outcomes of this example may be of less general significance.

Table 21.4: Comparison of Audio Features for the Recognition of Four Emotions. Number after Feature Name: Dimension; (m): Mean; (s): Standard Deviation. Table Entries Are p-Values from the Wilcoxon Test. Bold Values in Brackets Outline Top 1st, 2nd, and 3rd Features for the Corresponding Classification Task

Feature	Anger	Fear	Joy	Sadness
Fluct. pattern 1 (m)	8.09e-103 (1)	0.09	0.01	3.66e-54
Fluct. pattern 3 (m)	0.04	4.83e-35 (3)	3.14e-97 (3)	1.25e-30
Fluct. pattern 4 (m)	1.62e-06	4.02e-44 (1)	2.16e-111 (2)	4.63e-45
Fluct. pattern 5 (m)	1.08e-05	2.19e-25	4.73e-145 (1)	1.89e-88
MFCC 1 (m)	2.42e-85	3.65e-05	5.66e-25	3.00e-129 (1)
MFCC 2 (m)	2.26e-38	1.00e-35 (2)	1.04e-48	1.28e-16
No. of RMS peaks (m)	4.25e-40	5.41e-05	2.76e-53	8.12e-122 (3)
Spectral brightness (m)	3.16e-63	0.18	5.84e-25	2.49e-128 (2)
Sub-band energy 2 (m)	7.26e-98 (3)	0.82	0.07	7.46e-94
Sub-band energy 2 (s)	5.90e-100 (2)	0.30	0.14	8.28e-87

Example 21.10 (Feature Relevance for a Dimensional Approach). *For the evaluation of individual feature relevance in a dimensional approach, we randomly sampled 100 music pieces from the database 1000 Songs (further details of this and other databases are provided in Section 21.6 and Table 21.6). The same features were extracted as in Example 21.9 and aggregated for complete songs leading to 100 labeled data instances.*

Table 21.5 lists the five most relevant features for each task w.r.t. R^2 for linear regression. Both energy-related characteristics of RMS peaks can at best individually explain arousal, but the number of RMS peaks also has the highest individual contribution for valence prediction. Mean distance in the phase domain (Equation (5.38)) is the 3rd individually most relevant feature for arousal prediction. Phase domain characteristics were particularly successful for the classification of percussive versus "non-percussive" music in [42]. For valence prediction, three of the five most relevant features belong to rhythm descriptors: characteristics of fluctuation patterns and rhythmic clarity. As can be expected from music theory and previous studies (see, e.g., [11, p. 351]), arousal – which can also be described as a level of activation – is easier to predict than valence.

Original feature implementations are mostly from the MIR Toolbox, except for RMS peak characteristics (implementation from [60] based on the MIR Toolbox peak function) and distances in phase domain/spectral skewness (Yale implementation [42]).

In Figure 21.6, the values of the two most relevant features for the prediction of arousal and valence are plotted together with corresponding regression lines.

Table 21.5: Comparison of Audio Features for the Recognition of Arousal and Valence. Number after Feature Name: Dimension; (m): Mean. Table Entries Are R^2 after Linear Regression. Bold Values in Brackets Outline Top 1st–5th Features for the Corresponding Regression Task

Feature	Arousal	Valence
Distance in phase domain (m)	0.3637 **(3)**	0.0816
Fluct. pattern 5 (m)	0.0682	0.1907 **(3)**
Fluct. pattern 6 (m)	0.1829	0.2051 **(2)**
Number of RMS peaks (m)	0.3657 **(2)**	0.2105 **(1)**
Number of RMS peaks above mean amplitude (m)	0.4434 **(1)**	0.1826 **(4)**
Rhythmic clarity (m)	0.1496	0.1786 **(5)**
Spectral brightness (m)	0.3005 **(5)**	0.0993
Spectral skewness (m)	0.3331 **(4)**	0.1470

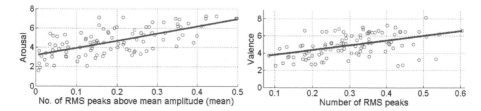

Figure 21.6: Best individually relevant tempo and rhythm features for the prediction of arousal and valence after linear regression model estimated for 100 randomly sampled music pieces from the 1000 Songs database.

21.6   Computationally Based Emotion Recognition

Probably the first MER system was introduced in [29]. Here, sentiment descriptors (melancholy, serious, pathetic, etc.) were predicted with heuristic rules based on music characteristics (chords, key, melody, and rhythm) estimated after audio transcription. Later, [19] described a method to extract four moods (anger, fear, happiness, sadness) from tempo and articulation features using a neural network. In the same year [38] presented results of a study on multi-label classification of 13 emotion groups using Support Vector Machines (SVMs, cf. Section 12.4.4) and low-level signal descriptors. The first dimensional MER system for the estimation of arousal and valence by means of fuzzy classification was probably proposed in [72].

The prediction goal of a MER system can be an expressed, perceived, or felt (evoked) emotion [71]. The recognition of an *expressed* emotion reveals the composer's intention of creating a music piece that transfers a certain emotion to a listener. Ideally, the expressed emotion is the same as the *perceived* emotion but this does not always hold. The prediction of a *felt* emotion may be hard or even impossible because personal experience also plays a role. For instance, 30% of all presented songs invoked autobiographical memories in [27]. Excellent examples illustrating the difference between induced and perceived emotions are provided in [21, p. 134].

A favorite tune "became associated with grief and tears," after being informed about the death of one's uncle during the playing. As another example, aggressive hard rock may induce calmness and relaxation for younger people who are fond of this genre. Most systems are designed to recognize perceived emotions.

For the evaluation and optimization of a MER system, annotated music data is necessary. Table 21.6 lists publicly available databases with annotated perceived emotions, sorted by publication year. The first half contains categorical annotations, the second half dimensional ones. Note that dimensional annotations can also be used for the classification into 4 regions with low or high arousal/valence.

Unfortunately, all data sets have individual drawbacks. One of the most accurate annotations, the Soundtrack database [15], was created with the help of music experts who carefully selected music according to "extremes of the three-dimensional model." However, this database only contains film music. Databases with annotations for many tracks often do not contain audio for copyright reasons. 1000 Songs [54] is a good alternative with a large freely distributed set of tracks and features, but contains probably less popular music from Free Music Archive.

After the merging of processed features with corresponding annotations, classification or regression models can be created and evaluated.

For *categorical MER systems*, supervised classification methods from Chapter 12 can be applied. Earlier studies often adopted a single classifier, like neural networks in [19] and SVMs in [38]. For better performance, it makes sense to compare several methods, as done in [45] for Naive Bayes (NB, introduced in Section 12.4.1), decision tree C4.5 (Section 12.4.3), k-Nearest Neighbors (k-NN, Section 12.4.2), and SVMs (Section 12.4.4). SVMs performed best achieving F-measure $m_F = 0.64$, cf. Equation (13.15), for the recognition of five emotional clusters. The best model in [6] across 3 classifiers (NB, k-NN, SVMs), 7 feature sets, and 3 feature selection methods was also built with SVMs (the accuracy $m_{ACC} = 0.65$, cf. Equation (13.11), for the recognition of four regions in arousal/valence space).

The choice of the "best" classifier strongly depends on data and a classification task. In [6], NB was the best method for three of seven feature sets, and in [45] the comparative performance of C4.5, k-NN, and NB varied for different feature sets. The choice of a classification method is usually harder when performance measures beyond classification performance are taken into account (see Section 13.3.6) and they are very seldom integrated in MER systems until now.

For *dimensional MER systems*, various regression methods can be considered. Linear regression (see Section 9.8.1) was inferior to k-Nearest Neighbors (k-NN) regression and Support Vector Regression (SVR) in [46] for the prediction of arousal and valence. The best performance w.r.t. R^2 was achieved using SVR. Similar results were reported in [26]: linear regression was the worst method and SVR the best one according to R^2 for the prediction of arousal and valence. In [14], best R^2 values for the prediction of valence, activity, and tension were achieved by means of Partial Least Squares Regression [66], compared to Multiple Linear Regression and Principal Component Regression (PCR).

Similar to the choice of a classifier for a categorical MER system, it is hard

Table 21.6: Databases with Annotated Emotions. No.: Number of Tracks; Goal: Prediction Goal; Genres: Genres of Database Tracks; Audio: Availability for Download; Features: Availability for Download; Ref.: Reference

No.	Goal	Genres	Audio	Features	Ref.
CATEGORICAL ANNOTATIONS					
CAL500, http://jimi.ithaca.edu/~dturnbull/data, annotations by at least three students for each track					
500	18 emotions	36 genres	low quality	MFCCs	[59]
Emotions, http://mlkd.csd.auth.gr/multilabel.htm, 3 annotations per track by experts					
593	6 emotions	7 genres	no	72 rhythmic and timbre features	[57]
MIREX-like Mood, http://mir.dei.uc.pt/resources/MIREX-like_mood.zip, annotations from AllMusicGuide					
193	5 clusters	mostly pop/rock	30s excerpts	no (lyrics and MIDIs included)	[45]
764				no (lyrics included)	
903				no	
Soundtracks, https://www.jyu.fi/hum/laitokset/musiikki/en/research/coe/materials/emotion/soundtracks, annotations by 6 experts					
110	5 emotions	soundtracks	10-30s excerpts	no	[15]
DIMENSIONAL ANNOTATIONS					
No.	Goal	Genres	Audio	Features	Ref.
Soundtracks, https://www.jyu.fi/hum/laitokset/musiikki/en/research/coe/materials/emotion/soundtracks, annotations by 6 experts					
110	valence, energy, tension	soundtracks	10-30s excerpts	no	[15]
NTUMIR-60, http://mac.citi.sinica.edu.tw/~yang/MER/NTUMIR-60, annotations by 40 students for each track					
60	arousal, valence	mostly pop/rock	no	252 features (timbre, pitch, rhythm, etc.)	[71, p. 92]
NTUMIR-1240, http://mac.citi.sinica.edu.tw/~yang/MER/NTUMIR-1240, annotations by 4.3 subjects for each track (online test)					
1240	arousal, valence	Chinese pop	no	213 features (timbre, pitch, rhythm, etc.)	[71, p. 92]
1000 Tracks, http://cvml.unige.ch/databases/emoMusic, annotations by 10 crowdworkers for each track					
744	arousal, valence	8 genres	45s excerpts	6669 features (spectrum, MFCCs, etc.)	[54]
AMG1608, http://mpac.ee.ntu.edu.tw/dataset/AMG1608, annotations by 665 subjects (students and crowdworkers); each track annotated by 15 crowdworkers					
1608	arousal, valence	mostly pop/rock	no	72 features (MFCCs, tonal, spectral, temporal)	[8]

to provide general recommendations. For example, *k*-NN regression outperformed SVR using a melodic feature set in [46].

21.6.1   *A Note on Feature Processing*

Carefully selected feature processing methods may help to avoid the failure of a MER system or even significantly improve the performance. In the following we list several techniques applied in studies on emotion recognition. For a general introduction to feature processing, see Chapter 14. According to Sections 14.3 and 14.4, one may distinguish between processing of feature and time dimensions.

Feature dimension processing in MER systems is almost always seen as a dimension reduction task. A prominent statistical approach to reduce the number of variables is Principal Component Analysis (PCA, Definition 9.48), used for instance in [11], where 9 components were created from 39 original features. For better interpretability, it may be meaningful to apply other methods that keep original feature dimensions and remove less relevant ones. This can be solved by means of feature selection; see Chapter 15. Sequential forward selection was applied in [5] (best found solution contained 32 features of 59), Relief in [45] (number of features reduced from 698 to 19).

To reduce the time dimension, simple statistics are often applied, like mean and standard deviation [11]. In [45], kurtosis and skewness were also estimated for short-frame features. These statistics contributed to the top-ranked features: the best melodic feature was the skewness of vibrato coverage, followed by its kurtosis. Autoregressive coefficients (cf. Definition 9.40) were used in [52] for the prediction of arousal and valence. For the same task in [40] more complex models were compared: Gaussian mixtures, autoregressive DAR and MAR (cf. Equations (14.5) and (14.6)), Markov, and vector quantization. However, only MFCCs were used as raw features. Thirteen energy, spectral, and vocal features were aggregated using more than 20 statistical functionals in [64]. These statistics included characteristics of percentiles, peaks, moments, modulation, temporal progress, and regression of feature time series.

The length of time windows for feature aggregation and classification plays an important role. Too short frames, typical for the extraction of signal low-level descriptors, may not be sufficient for the interpretation of emotional cues, and too long frames represent music with many musical events. On the other side, some features like descriptors of lyrics characterize complete songs. As for audio characteristics, often statistics over complete songs are calculated despite theoretical disadvantages: this is done in 12 of 22 studies listed in [3, Table 3]. 15-second frames were analyzed in [11]. The scope of [69] was to examine several window lengths, and the best results were achieved with 8- or 16-second frames.

The optimal length of classification windows may strongly depend on data. For example, in [69] only classical music was classified, and in [63] we observed that the optimal classification window length was around 24 s for a simpler music classification problem (close to "classic-against-pop" scenario) and 1.2–1.5 s for more complex classes. Recalling that genres and perceived emotions often have strong interrelationships (see discussion in Section 21.5.6), this strengthens the suggestion that the optimal length of a classification window depends on the concrete MER task.

21.6.2   Future Challenges

Even if a large number of approaches for computationally based emotion prediction were proposed in the last two decades, MER can be still described as being in its "infancy" [71, p. 6]. In the future many problems should be resolved.

In most studies a music piece is assigned as a whole to a single or more emotions, however this is not always the case. Only a few works address Music Emotion Variation Detection (MEVD) [4, p. 503], [71, p. 31], [30, p. 262].

Personal differences in emotion perception, unsharp boundaries between perceived and induced emotions, and the impossibility to learn induced emotions without intensive interaction with a user are particular problems for personalized MER systems. For instance, tracks tagged as "gruesome" may correspond to perceived, induced, or both emotions [21]. Further examples of differences between perceived and induced emotions are given in the beginning of Section 21.6.

Many approaches are limited to or are focussed on audio descriptors. Even if they provide advantages like extractability for any audio recording independent of its popularity and capture important characteristics not available in the score (timbre, style of the performer, etc.), the quality often suffers with an increasing number of playing sources. With better methods for audio transcription, the robustness of audio features may be significantly increased. In a recent study [1], a set of manually annotated music characteristics showed "the best performance as compared to all the features derived from signal-processing, demonstrating that our ability to model human perception is not yet perfect."

Multi-modal approaches bear great opportunities. Related studies report improvements of performance after the combination of features from different groups and sources, for example, audio and lyrics in [71, p. 179] or audio, melody, MIDI, and lyrics descriptors in [45]; see also sections on combinations of lyrics, tags, and images with audio characteristics in [30, p. 263].

For a better comparison of algorithms, publicly available databases with standardized annotations according to emotion models are necessary. At the moment it is not the case, as stated in [71, p. 23]: "there is still no consensus on which emotion model or how many emotion categories should be used;" see also the discussion of individual drawbacks of databases in Table 21.6. These databases should contain as different music as possible (classical, popular, non-Western, instrumental, etc.). As mentioned in Section 21.5.6, switching from one genre to another may significantly decrease the performance of a MER system, when it is trained on tracks with a poor variation of genres [11]. Finally, these databases should supply as many feature sources as possible (audio, score, lyrics, metadata, etc.), because multi-modal approaches were shown as superior to single feature groups (see references in the previous paragraph).

21.7   Concluding Remarks

Emotions and moods play an essential role in music composition and music perception. Almost all classical and popular pieces are created keeping a certain theme in mind which should invoke an associated emotional affect: calm relaxation in a lullaby, joy in a love song, fear in a thriller movie, etc. Automatic recognition of emotions in music may help to create more interpretable classification models and enable personal recommendations.

In this chapter, we introduced several emotional theories and models applicable to music, followed by lists with characteristics which can be used for automatic prediction of emotions and moods in music. We discussed further details of the implementation of music emotion recognition systems, such as the choice of classification and regression approaches, feature processing methods, and databases with categorical and dimensional annotations.

21.8 Further Reading

An extensive introduction into computationally based emotion recognition is provided in [71], followed by specific applications like fuzzy emotion recognition. Several references to enhanced topics like multi-modal approaches or music emotion variation detection are given in Section 21.6.2. Examples of MER tasks beyond categorical and dimensional approaches like prediction of personal preferences or emotional intensity are described as "miscellaneous emotion models" in [16].

Reference [3, Table 1] lists emotion models used in MER systems and [3, Table 3] lists statistics of various studies, including features, lengths of classification windows, classification and regression algorithms, etc. A brief but exhaustive overview of related studies is summarized in [30].

Studies which propose new feature sources for MER systems comprise vocal characteristics [71, p. 213] and listening context [12]. Preprocessing of lyric features for MER is addressed in [70], and the analysis of percussive and bass patterns in [58]. Reference [55] discusses the evaluation of MER systems, outlining weak points of present systems w.r.t. uncontrolled independent variables in data sets.

Bibliography

[1] A. Aljanaki, W. F., and R. C. Veltkamp. Computational modeling of induced emotion using GEMS. In *Proc. of the 15th International Society for Music Information Retrieval Conference (ISMIR)*, pp. 373–378. International Society for Music Information Retrieval, 2014.

[2] L. Balkwill and W. F. Thompson. A cross-cultural investigation of the perception in music: Psychophysical and cultural clues. *Music Perception*, 17(1):43–64, 1999.

[3] M. Barthet, G. Fazekas, and M. Sandler. Multidisciplinary perspectives on music emotion recognition: Implications for content and context-based models. In *Proc. of the 9th International Symposium on Computer Modelling and Retrieval (CMMR)*, pp. 492–507. Springer, 2012.

[4] M. Barthet, G. Fazekas, and M. Sandler. Music emotion recognition: From content- to context-based models. In *Proc. of the 9th International Symposium on Computer Modelling and Retrieval (CMMR)*, pp. 228–252. Springer, 2012.

[5] C. Baume, G. Fazekas, M. Barthet, D. Marston, and M. Sandler. Selection of audio features for music emotion recognition using production music. In *Proc. of the Audio Engineering Society 53rd International Conference (AES)*, pp. 54–62. Audio Engineering Society, 2014.

[6] S. Beveridge and D. Knox. A feature survey for emotion classification of Western popular music. In *Proc. of the 9th International Symposium on Computer Modelling and Retrieval (CMMR)*, pp. 508–517. Springer, 2012.

[7] K. Bischoff, C. S. Firan, R. Paiu, W. Nejdl, C. Laurier, and M. Sordo. Music mood and theme classification - a hybrid approach. In *Proc. of the 10th International Society for Music Information Retrieval Conference (ISMIR)*, pp. 657–662. International Society for Music Information Retrieval, 2009.

[8] Y. Chen, Y. Yang, J. Wang, and H. Chen. The AMG1608 dataset for music emotion recognition. In *Proc. of the 40th IEEE International Conference on Acoustics, Speech and Signal Processing (ICASSP)*, pp. 693–697. IEEE, 2015.

[9] T. Dang and K. Shirai. Machine learning approaches for mood classification of songs toward music search engine. In *Proc. of 2009 International Conference on Knowledge and Systems Engineering*, pp. 144–149. IEEE, 2009.

[10] H. de la Motte-Haber and G. Rötter, eds. *Musikpsychologie*. Laaber-Verlag, 2005.

[11] T. Eerola. Are the emotions expressed in music genre-specific? An audio-based evaluation of datasets spanning classical, film, pop and mixed genres. *Journal of New Music Research*, 40(4):349–366, 2011.

[12] T. Eerola. Modelling emotions in music: Advances in conceptual, contextual and validity issues. In *Proc. of the 53rd Audio Engineering Society Conference (AES)*, pp. 278–287. Audio Engineering Society, 2014.

[13] T. Eerola, R. Ferrer, and V. Alluri. Timbre and affect dimensions: Evidence from affect and similarity ratings and acoustic correlates of isolated instrument sounds. *Music Perception*, 30(1):49–70, 2012.

[14] T. Eerola, O. Lartillot, and P. Toiviainen. Prediction of multidimensional emotional ratings in music from audio using multivariate regression models. In *Proc. of the 10th International Society for Music Information Retrieval Conference (ISMIR)*, pp. 621–626. International Society for Music Information Retrieval, 2009.

[15] T. Eerola and J. K. Vuoskoski. A comparison of the discrete and dimensional models of emotion in music. *Psychology of Music*, 39(1):18–49, 2011.

[16] T. Eerola and J. K. Vuoskoski. A review of music and emotion studies: Approaches, emotion models, and stimuli. *Music Perception*, 30(3):307–340, 2013.

[17] P. Ekman. *Emotion in the Human Face*. Cambridge University Press, New York, 1982.

[18] P. Ekman. Moods, emotions and traits. In P. Ekman and R. Davidson, eds., *The Nature of Emotion: Fundamental Questions*, pp. 56–58. Oxford University

Press, New York, 1994.

[19] Y. Feng, Y. Zhuang, and Y. Pan. Music information retrieval by detecting mood via computational media aesthetics. In *Proc. of the IEEE/WIC International Conference on Web Intelligence (WI)*, pp. 235–241. IEEE, 2003.

[20] N. H. Frijda. *The Emotions*. Cambridge University Press, Cambridge, 1986.

[21] A. Gabrielsson. Emotion perceived and emotion felt: Same or different? *Musicae Scientiae*, 5(1):123–147, 2001.

[22] A. Gabrielsson and E. Lindström. The influence of musical structure on emotional expression. In P. N. Justin and J. A. Sloboda, eds., *Music and Emotion: Theory and Research*, pp. 223–248. Oxford University Press, London, 2001.

[23] J. C. Hailstone, R. Omar, S. M. D. Henley, C. Frost, M. G. Kenward, and J. D. Warren. It's not what you play, it's how you play it: Timbre affects perception of emotion in music. *The Quarterly Journal of Experimental Psychology*, 62(11):2141–2155, 2009.

[24] K. Hevner. Tests for aesthetic appreciation in the field of music. *Journal of Applied Psychology*, 14:470–477, 1930.

[25] K. Huber. *Der Ausdruck musikalischer Elementarmotive. Eine experimentalpsychologische Untersuchung*. Barth, Leipzig, 1923.

[26] A. Huq, J. Pablo Bello, and R. Rowe. Automated music emotion recognition: A systematic evaluation. *Journal of New Music Research*, 39(3):227–244, 2010.

[27] P. Janata, S. T. Tomic, and S. K. Rakowski. Characterisation of music-evoked autobiographical memories. *Memory*, 15(8):845–860, 2007.

[28] P. N. Juslin and P. Laukka. Communication of emotions in vocal expression and music performance: Different channels, same code? *Journal of Personality and Social Psychology*, 129(5):770–814, 2003.

[29] H. Katayose, M. Imai, and S. Inokuchi. Sentiment extraction in music. In *Proc. of the 9th International Conference on Pattern Recognition (ICPR)*, pp. 1083–1087. IEEE, 1988.

[30] Y. E. Kim, E. M. Schmidt, R. Migneco, B. G. Morton, P. Richardson, J. Scott, J. A. Speck, and D. Turnbull. State of the art report: Music emotion recognition: A state of the art review. In *Proc. of the 11th International Society for Music Information Retrieval Conference (ISMIR)*, pp. 255–266. International Society for Music Information Retrieval, 2010.

[31] S. Knobloch, K. Weisbach, and D. Zillmann. Love lamentation in pop songs: Music for unhappy lovers? *Zeitschrift für Medienpsychologie*, 16(3):116–124, 2004.

[32] S. Knobloch and D. Zillmann. Mood management via the digital jukebox. *Journal of Communication*, 52(2):351–366, 2002.

[33] V. Konecni. Review von: Patrick N. Juslin & John A. Sloboda (Eds.), Music and Emotion: Theory and Research, Oxford, U.K.: Oxford University Press, 2001. *Music Perception*, 1:332–340, 2002.

[34] V. J. Konecni. Social interaction and musical preference. In D. Deutsch, ed., *The Psychology of Music*, pp. 497–516. Academic Press, New York, 1982.

[35] G. Kreutz. Musik und emotion. In H. Bruhn, A. Lehmann, and R. Kopiez, eds., *Musikpsychologie*. Rowohlt, Reinbek, 2006.

[36] O. Lartillot and P. Toiviainen. MIR in MATLAB (II): A toolbox for musical feature extraction from audio. In *Proc. of the 8th International Conference on Music Information Retrieval (ISMIR)*, pp. 127–130. Austrian Computer Society, 2007.

[37] C. Laurier, M. Sordo, J. Serrà, and P. Herrera. Music mood representations from social tags. In *Proc. of the 10th International Society for Music Information Retrieval Conference (ISMIR)*, pp. 381–386. International Society for Music Information Retrieval, 2009.

[38] T. Li and M. Ogihara. Detecting emotion in music. In *Proc. of the 4th International Conference on Music Information Retrieval (ISMIR)*, pp. 239–240. The Johns Hopkins University, 2003.

[39] L. Lu, D. Liu, and H. Zhang. Automatic mood detection and tracking of music audio signals. *IEEE Transactions on Audio, Speech & Language Processing*, 14(1):5–18, 2006.

[40] J. Madsen, B. S. Jensen, and J. Larsen. Modeling temporal structure in music for emotion prediction using pairwise comparisons. In *Proc. of the 15th International Society of Music Information Retrieval Conference (ISMIR)*, pp. 319–324. International Society for Music Information Retrieval, 2014.

[41] G. Mandler. *Mind and Emotion*. John Wiley & Sons Inc, 1975.

[42] I. Mierswa and K. Morik. Automatic feature extraction for classifying audio data. *Machine Learning Journal*, 58(2-3):127–149, 2005.

[43] H. Motte-Haber. *Handbuch der Musikpsychologie*. Laaber-Verlag, 1985.

[44] M. B. Neiss, L. A. Leigland, N. E. Carlson, and J. S. Janowskya. Age differences in perception and awareness of emotion. *Neurobiology of Aging*, 30(8):1305–1313, 2009.

[45] R. Panda, R. Malheiro, B. Rocha, A. Oliveira, and R. P. Paiva. Multi-modal music emotion recognition: A new dataset, methodology and comparative analysis. In *Proc. of the 10th International Symposium on Computer Music Multidisciplinary Research (CMMR)*. Springer, 2013.

[46] R. Panda, B. Rocha, and R. P. Paiva. Dimensional music emotion recognition: Combining standard and melodic audio features. In *Proc. of the 10th International Symposium on Computer Music Multidisciplinary Research (CMMR)*. Springer, 2013.

[47] J. A. Russell. A circumplex model of affect. *Journal of Personality and Social Psychology*, 39(6):1161–1178, 1980.

[48] P. Saari, T. Eerola, and O. Lartillot. Generalizability and simplicity as criteria in feature selection: Application to mood classification in music. *IEEE Trans-*

actions on Audio, Speech, and Language Processing, 19(6):1802–1812, 2011.

[49] J. Salamon, B. M. M. Rocha, and E. Gómez. Musical genre classification using melody features extracted from polyphonic music signals. In *Proc. of 2012 IEEE International Conference on Acoustics, Speech and Signal Processing, (ICASSP)*, pp. 81–84. IEEE, 2012.

[50] K. R. Scherer. *Vokale Kommunikation: Nonverbale Aspekte des Sprachverhaltens*. Beltz, Weinheim/Basel, 1982.

[51] K. R. Scherer and M. R. Zentner. Emotional effects of music: Production rules. In P. N. Justin and J. A. Sloboda, eds., *Music and emotion: Theory and research*, pp. 361–392. Oxford University Press, London, 2001.

[52] E. Schubert. Modeling perceived emotion with continuous musical features. *Music Perception*, 21(4):561–585, 2004.

[53] B. Schuller, F. Weninger, and J. Dorfner. Multi-modal non-prototypical music mood analysis in continuous space: Reliability and performances. In *Proc. of the 12th International Society for Music Information Retrieval Conference (ISMIR)*, pp. 759–764. University of Miami, 2011.

[54] M. Soleymani, A. Aljanaki, Y. Yang, M. N. Caro, F. Eyben, K. Markov, B. W. Schuller, R. Veltkamp, F. Weninger, and F. Wiering. Emotional analysis of music: A comparison of methods. In *Proc. of the ACM International Conference on Multimedia*, pp. 1161–1164. ACM, 2014.

[55] B. Sturm. Evaluating music emotion recognition: Lessons from music genre recognition? In *Proc. of 2013 IEEE International Conference on Multimedia and Expo (ICME)*, pp. 1–6. IEEE, 2013.

[56] A. Tellegen, D. Watson, and L. A. Clark. On the dimensional and hierarchical structure of affect. *Psychological Science*, 10(4):297–303, 1999.

[57] K. Trohidis, G. Tsoumakas, G. Kalliris, and I. P. Vlahavas. Multi-label classification of music into emotions. In *Proc. of the 9th International Conference on Music Information Retrieval (ISMIR)*, pp. 325–330. Drexel University, 2008.

[58] E. Tsunoo, T. Akase, N. Ono, and S. Sagayama. Music mood classification by rhythm and bass-line unit pattern analysis. In *Proc. of 2010 IEEE International Conference on Acoustics Speech and Signal Processing (ICASSP)*, pp. 265–268. IEEE, 2010.

[59] D. Turnbull, L. Barrington, D. Torres, and G. Lanckriet. Semantic annotation and retrieval of music and sound effects. *IEEE Transactions on Audio, Speech, and Language Processing*, 16(2):467–476, 2008.

[60] I. Vatolkin. *Improving Supervised Music Classification by Means of Multi-Objective Evolutionary Feature Selection*. PhD thesis, Department of Computer Science, TU Dortmund, 2013.

[61] I. Vatolkin, G. Bonnin, and D. Jannach. Comparing audio features and playlist statistics for music classification. In A. F. X. Wilhelm and H. A. Kestler, eds., *Analysis of Large and Complex Data*, pp. 437–447. Springer, 2016.

[62] I. Vatolkin, W. Theimer, and M. Botteck. AMUSE (Advanced MUSic Explorer) - a multitool framework for music data analysis. In *Proc. of the 11th International Society on Music Information Retrieval Conference (ISMIR)*, pp. 33–38. International Society on Music Information Retrieval, 2010.

[63] I. Vatolkin, W. Theimer, and G. Rudolph. Design and comparison of different evolution strategies for feature selection and consolidation in music classification. In *Proc. of the IEEE Congress on Evolutionary Computation (CEC)*, pp. 174–181. IEEE, 2009.

[64] F. Weninger, E. F., B. W. Schuller, M. Mortillaro, and K. R. Scherer. On the acoustics of emotion in audio: What speech, music, and sound have in common. *Frontiers in Psychology*, 4(292), 2013.

[65] H. Werbik. *Informationsgehalt und emotionale Wirkung von Musik*. Schott, Mainz, 1971.

[66] S. Wold, M. Sjöström, and L. Eriksson. PLS-regression: A basic tool of chemometrics. *Chemometrics and Intelligent Laboratory Systems*, 58(2):109–130, 2001.

[67] B. Wu, A. Horner, and C. Lee. Musical timbre and emotion: The identification of salient timbral features in sustained musical instrument tones equalized in attack time and spectral centroid. In *Proc. of 40th International Computer Music Conference (ICMC) joint with the 11th Sound & Music Computing conference (SMC)*, pp. 928–934. Michigan Publishing, 2014.

[68] W. Wundt. *Grundriss der Psychologie*. Kröner, Leipzig, 1922. 15. Auflage.

[69] Z. Xiao, E. Dellandrea, W. Dou, and L. Chen. What is the best segment duration for music mood analysis? In *Proc. of 2008 International Workshop on Content-Based Multimedia Indexing (CBMI)*, pp. 17–24. IEEE, 2008.

[70] H. Xue, L. Xue, and F. Su. Multimodal music mood classification by fusion of audio and lyrics. In *Proc. of the 21st International Conference on MultiMedia Modeling (MMM)*, pp. 26–37. Springer, 2015.

[71] Y.-H. Yang and H. H. Chen. *Music Emotion Recognition*. CRC Press, 2011.

[72] Y.-H. Yang, C.-C. Liu, and H. H. Chen. Music emotion classification: A fuzzy approach. In *Proc. of the 14th Annual ACM International Conference on Multimedia*, pp. 81–84. ACM, 2006.

[73] M. Zentner, D. Grandjean, and K. R. Scherer. Emotions evoked by the sound of music: Characterization, classification, and measurement. *Emotion*, 8(4):494–521, 2008.

Chapter 22

Similarity-Based Organization of Music Collections

SEBASTIAN STOBER

Machine Learning in Cognitive Sciences, University of Potsdam, Germany

22.1 Introduction

Since the introduction of digital music formats such as MP3 (see Section 7.3.3) in the late 1990s, personal music collections have grown considerably. But not much has changed concerning the way we structure and organize them. Popular music players that also function as collection management tools like iTunes or AmaroK still organize music collections in tables and lists based on simple metadata, such as the artist and album tags. Even their integrated music recommendation functions, like iTunes Genius, often only rely on usage information, which is typically extracted from playlists and ratings – i.e., they largely ignore the actual music content in the audio signal.

Equipped with sophisticated content analysis techniques as described in earlier chapters, we are now ready to pursue new ways of organizing music collections based on music similarity. We can compare music tracks on a content-based level and group similar tracks together. Furthermore, we can generate maps that visualize the structure of a *similarity space*, i.e., the space defined by a set of objects and their pairwise similarities. Such maps easily allow to identify regions or neighborhoods of similar tracks. Moreover, they provide the foundation for new ways of interacting with music collections. For instance, users can find relevant music by starting from a general overview map, identify interesting regions and then explore these in more detail until they find what they are looking for. This way, they do not have to formulate a query explicitly. Instead, they can systematically and incrementally narrow down their region of interest and define their search goal implicitly during the exploration process. Note that this also works in scenarios where users just want to explore and find something new. Here, narrowing down the region of interest more and more, characterizes that "something."

However, such new approaches do not come without their own challenges. First, music similarity is not a simple concept to start with. In fact, various frameworks exist within the fields of musicology, psychology, and cognitive science and there are many ways of comparing music tracks. How we compare music may depend on our musical background, our personal preferences or our specific retrieval task. Hence, we have to accept the fact that there is no such thing as *the* music similarity. Instead, there are many ways of describing music similarity and we have to find out which one works best in each specific context – or better, we would like to have the computer figure this out for us automatically. Section 22.2 will present a way of achieving this.

Second, visualizing a similarity space as a two-dimensional map – often also called a *projection* – necessarily requires some sort of dimensionality reduction. Except for very trivial cases, it is generally not possible to perfectly capture the structure of the similarity space in two dimensions. Hence, there will be unavoidable projection errors that can have a negative impact on the users' experience. Specifically, some neighbors in the visualization may in fact not be similar whilst some similar music tracks may be positioned in very distant regions of the map. Section 22.3 will discuss ways to address this issue and utilize it for the task of exploring a collection.

As a third major challenge, music collections usually are not static. They change over time as we add new music or sometimes remove some tracks. With every change of the collection, we will also have to change the corresponding map visualization. But we want to change it as little as possible to maximize continuity in the visualization and not confuse the user. In Section 22.4, solutions for several projection algorithms are compared.

22.2 Learning a Music Similarity Measure

A proper definition of music similarity is crucial for many music information retrieval applications – not only for structuring music collections. In the context of this chapter, we are specifically interested in the similarity between music tracks whereas in different scenarios, the similarity of albums, artists or even genres could be of interest as well. Furthermore, we will focus on learning a *distance measure* which, from a mathematical perspective, can be considered as a dual concept to the less well defined concept of similarity.[1] Talking about distances also makes much more sense in the context of structuring a music collection on a map as we will see later.

There are many different ways of comparing two tracks based on their features when all of them represent equally valid views. However, depending on the circumstances, some of these views might be more appropriate than others. As an example, consider looking for a suitable background music track for a photo slide show. In this use case, we might want a structuring of the music collection that emphasizes similarity in tempo, rhythm and timbre whereas features related to tonality or harmonic progression might be much less helpful. When looking for cover versions of a song without metadata information such as the title, differences in timbre may be

---

[1] From the mathematical perspective, this relation of similarity and distance makes perfect sense. It may however be questioned from a psychological point of view. Such a discussion is beyond the scope of this chapter. An overview can, for instance, be found on *Scholarpedia* [1].

less interesting than the lyrics or the harmonic structure. Similarly, musicians might especially look after structures, tonality or instrumentation and possibly pay special attention, consciously or unconsciously, to their own instruments.

In order to build an application that can accommodate such diverse views on music similarity, we need three important ingredients. First, we need an adaptable *model* of music similarity. Each parameter setting of this model represents one possible view. We would like to have the computer find optimal parameter values for the desired outcome. Second, we require a way to express and capture *preferences* for choosing the right model parameters. Finally, we need an algorithm – the *adaptation logic* – that derives the model parameters from the preferences. We will address these points in the following subsections.

22.2.1 *Formalizing an Adaptable Model of Music Similarity*

As with all machine learning problems in general, we have to make a tradeoff between the model complexity and the ability to generalize (cp. Chapter 13). For a simple model with few parameters, only a small amount of training data (preferences) is usually needed to determine a good setting that generalizes well beyond the training data. Complex models generally require much more training data and are more prone to overfitting. They are also harder to comprehend and often impossible to tune manually if needed. But they can handle more complicated cases where simpler models might fail to accommodate all the preferences (underfitting).

For our model of music similarity, let us assume that we have a set of "atomic" distance measures – each computed on one or more features of the tracks. There could be one for timbre, one for tonality, and so on. We will call them *facet* distance measures and assume that they are purely *objective*, i.e., they are independent of the user and usage context. Formally, this can be defined as follows:

Definition 22.1 (Facet Distance Measure). Given a set of features F, let S be the space determined by the feature values for a set of music tracks T. A *facet f* is defined by a *facet distance measure* δ_f on a subspace $S_f \subseteq S$ of the feature space, where δ_f satisfies the following conditions for any $a, b \in T$:

- $\delta_f(a,b) \geq 0$ with $\delta_f(a,b) = 0$ iff a and b are identical w.r.t. facet f
- $\delta_f(a,b) = \delta_f(b,a)$ (symmetry)

Furthermore, δ_f is a *distance metric* if it additionally obeys the triangle inequality for any $a, b, c \in T$:

- $\delta_f(a,c) \leq \delta_f(a,b) + \delta_f(b,c)$ (triangle inequality)

In Section 11.2, various distance measures are discussed that can be used to compute facet distances. The choice depends on the features to be compared and the focus of the comparison. For instance, let us consider a feature that captures the frequency distribution of common major and minor chords within a track (cp. Chapter 19 on chord recognition) as a histogram vector. We could compare two of these vectors using the Euclidean distance. If we are not interested in the actual frequencies, but only which chords appear in the two tracks, we could derive the sets of chords with non-zero frequency and compare them using the Jaccard index (cp. Section 11.2). Finally, if we only want to compare the number of different chords used, we can compute the Manhattan distance (cp. Section 11.2) of the size of the chord sets.

In order to avoid a bias when aggregating several facet distance measures, the values should be normalized. The following normalization can be applied for all distance values $\delta_f(a,b)$ of a facet f:

$$\delta'_f(a,b) = \frac{\delta_f(a,b)}{\mu_f} \tag{22.1}$$

where μ_f is the mean facet distance w.r.t. f:

$$\mu_f = \frac{1}{|\{(a,b) \in T^2\}|} \sum_{(a,b) \in T^2} \delta_f(a,b). \tag{22.2}$$

As a result, all facet distances have a mean value of 1.0. Special care has to be taken, if extremely high facet distance values are present that express "infinite dissimilarity" or "no similarity at all." Such values introduce a strong bias for the mean of the facet distance and thus should be ignored during its computation. Further normalization methods for features that can also be applied to normalize distance values are introduced in Section 14.2.2.

The actual distance between objects $a,b \in T$ w.r.t. the facets $f_1, \dots, f_l$ is computed as the weighted sum of the individual facet distances $\delta_{f_1}(a,b), \dots, \delta_{f_l}(a,b)$:

$$d(a,b) = \sum_{i=1}^{l} w_i \delta_{f_i}(a,b). \tag{22.3}$$

This way, we introduce the facet weights $w_1, \dots, w_l \in \mathbb{R}$ which allow us to adapt the importance of each facet according to *subjective* user preferences or for a specific retrieval task. Note that the linear combination assumes the independence of the individual facets, which might be a limiting factor of this model in some settings. The weights obviously have to be non-negative and should correspond to proportions, i.e., add up to 1, thus:

$$w_i \geq 0 \qquad \forall 1 \leq i \leq l \tag{22.4}$$

$$\sum_{i=1}^{l} w_i = 1. \tag{22.5}$$

The resulting adaptable model of music similarity has a linear number of parameters – one weight per distance facet. The weights are intuitively comprehensible and can also be easily represented as sliders in a graphical user interface.

22.2.2 Modeling Preferences through Distance Constraints

So far, we have seen how music similarity can be modeled by introducing weight parameters that allow adaptations according to our preferences. Next, we need to describe these preferences so that they can be used to guide an optimization algorithm.

The algorithm will then eventually identify the values for the weight parameters that best reflect our preferences.

Distance or similarity preferences can be expressed in two ways, either through absolute (quantitative) statements or through relative (qualitative) statements. The former statements can be binary like *"x and y are / are not similar,"* which requires a hard decision criterion, or quantitative like *"the similarity / distance of x and y is 0.5,"* which requires a well-defined scale. Relative preference statements, on the contrary, do not compare objects directly but their pair-wise distances in the form *"x and y are more similar / less distant than u and v."* Usually, this is done relative to a seed object reducing the statements to the form *"x is more similar / less distant to y than z."* Such statements will be considered in the remainder of this chapter as they are much easier to express and thus more stable than absolute statements, i.e., when asked again, users are more likely to confirm the earlier expressed relative preference than stating the same absolute value for the distance / similarity again. We will concentrate on the simpler version that refers to a seed track and is easier to comprehend. However, all of the following can easily be modified to accommodate general relative distance constraints defined on two pairs of tracks without a seed.

Definition 22.2 (Relative Distance Constraint). A *relative distance constraint* (s, a, b) demands that object a is closer to the seed object s than object b, i.e.:

$$d(s,a) < d(s,b). \tag{22.6}$$

With Equation (22.3), this can be rewritten as:

$$\sum_{i=1}^{l} w_i(\delta_{f_i}(s,b) - \delta_{f_i}(s,a)) = \sum_{i=1}^{l} w_i x_i = \mathbf{w}^T \mathbf{x} > 0 \tag{22.7}$$

substituting $x_i := \delta_{f_i}(s,b) - \delta_{f_i}(s,a)$. As we will see later, such basic constraints can directly be used to guide an optimization algorithm that aims to identify weights that violate as few constraints as possible. But there is also an alternative perspective on the weight-learning problem as pointed out by Cheng and Hüllermeier [3] and illustrated in Figure 22.1. We can transform the optimization problem into a binary classification problem with positive training examples $(\mathbf{x}, +1)$ that correspond to satisfied constraints and negative examples $(-\mathbf{x}, -1)$ corresponding to constraint violations, respectively. In this case, the weights $(w_1 \dots w_l) = \mathbf{w}^T$ describe the model (separating hyperplane) to be learned by the classifier.[2]

The focus on relative distance constraints may seem like a strong limitation, but as we will see next, complex expressions of distance preferences can be broken down into "atomic" relative distance constraints. Let us, for instance, consider two very common user activities related to collection structuring: grouping and (re-)ranking. Grouping can be realized by assigning tags such that tracks in the same group share the same tag. It could also be done visually in a graphical user interface where tracks can be moved into folders or similar cluster containers by drag-and-drop operations.

---

[2]For an explanation of (binary) classification problems and the concept of the separating hyperplane, please refer to Chapter 12.

relative distance constraints ➡ linear classification problem

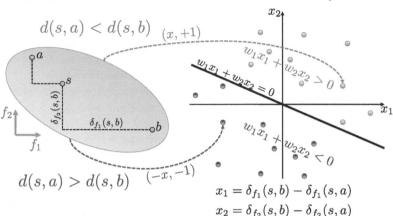

Figure 22.1: Transformation of a relative distance constraint (left) into two training instances of the corresponding binary classification problem (right). The negative example (bottom, dark gray) has the inverse relation sign of the positive example (top, light gray). For simplicity, this scenario only considers two facets. The axes in the diagram on the right refer to the differences for each facet distance as shown below the diagram. Every hyperplane through the origin that perfectly separates the (light gray) positive examples from the (dark gray) negative examples corresponds to a facet weight vector $\boldsymbol{w}$ that satisfies all distance constraints.

For ranking, a set of tracks is arranged as a list according to the similarity with a seed track. Groupings and ranking list, do not have to be created from scratch but can be pre-computed using default facet weights and then modified by the user. Once the user has modified the ranking or grouping, constraints can be derived. For a set G of tracks grouped together by similarity, every pair of tracks x, y within G should be more similar to each other than to any track o outside of G. This results in two relative distance constraints per triplet x, y, o:

$$\forall x, y \in G, o \notin G: \quad d(x,y) < d(x,o) \ \wedge \ d(y,x) < d(y,o). \tag{22.8}$$

For a ranked list of similar tracks $t_1, \ldots, t_n$ w.r.t. a seed track s, the track ranked first should be the one most similar to the seed, and in general, the track at rank i should be more similar than the tracks at ranks $j > i$. This results in the following set of relative distance constraints:

$$\forall i, j \in \{1, \ldots, n\} \ \text{with} \ i < j: \quad d(s, t_i) < d(s, t_j). \tag{22.9}$$

These two examples aim to demonstrate how atomic relative distance constraints can be inferred from a user's interaction with a system. Of course, more ways of interacting with a user interface are possible depending on the complexity of the application. Each will require a slightly different approach to model the expressed similarity preference. As long as an interaction relates to the (perceived) relative distance between at least three tracks, a similar representation by relative distance constraints can be found.

22.2.3 Dealing with Inconsistent Constraint Sets

Sometimes the set of constraints to be used for learning may be inconsistent because there are constraints that contradict each other. The reasons for this can be manifold – e.g., a user may have changed her or his mind or the constraints may be from different users or contexts in general. In such cases, it is impossible to learn a facet weighting that satisfies all constraints – regardless of the learning algorithm or the facets used. In order to obtain a consistent set of constraints, a constraint filtering approach described by McFee and Lanckriet [17] can be applied as follows:

1. A directed multigraph (i.e., a graph that may have multiple directed edges between two nodes) is constructed with pairs of objects as nodes and the distance constraints expressed by directed edges. For instance, for the distance constraint $d(b,c) < d(a,c)$, a directed edge from the node (b,c) to the node (a,c), would be inserted.

2. All cycles of length 2 are removed, i.e., all directly contradicting constraints. This can be done very efficiently by checking the graph's adjacency matrix.[3]

3. The resulting multigraph is further reduced to a directed acyclic graph (DAG) in a randomized fashion: Starting with an empty DAG, the edges of the multigraph are added in random order omitting those edges that would create cycles.

4. The corresponding distance constraints of the remaining edges in the DAG form a consistent set of constraints.

This can be repeated multiple times as the resulting consistent set of constraints may not be maximal because of the randomized greedy approach taken in step 3. An exhaustive search for a maximum acyclic subgraph would be NP-hard.

22.2.4 Learning Distance Facet Weights

With a consistent set of relative distance constraints, we are ready to take the next step and let a learning algorithm determine the optimal facet weights. Note that even though we removed all inconsistencies from the constraints set, it is still possible that there is no weighting that satisfies all constraints. This could happen, for instance, when the user's comparison of the tracks is based on features that are not or only partly covered by the distance facets. In order to deal with such situations, we need a learning algorithm that can tolerate constraint violations. As already pointed out in Section 22.2.1, it is possible to look at the problem of finding optimal facet weights from different perspectives – as an optimization or a classification problem. The following sections describe two optimization and one classification approach using different techniques.

---

[3]Let us assume there are m edges from node (a,c) to node (b,c) and n in the opposite directions. This can either be resolved by removing all $m+n$ edges or by just removing $min(m,n)$ edges in each direction and leaving $|m-n|$ edges in the direction of stronger preference.

22.2.4.1   Gradient Descent

One straightforward way of learning weights is to apply a gradient descent approach, which has been already introduced in Chapter 10. During learning, all constraint triples (s,a,b) are presented to the algorithm several times until convergence is reached. If a constraint is violated by the current distance measure, the weighting is updated by trying to maximize

$$obj(s,a,b) = \sum_{i=1}^{l} w_i(\delta_{f_i}(s,b) - \delta_{f_i}(s,a)), \tag{22.10}$$

which can be directly derived from Equation (22.7). This leads to the following update rule for the individual weights:

$$w_i = w_i + \eta \Delta w_i \quad \text{with} \tag{22.11}$$

$$\Delta w_i = \frac{\partial obj(s,a,b)}{\partial w_i} = \delta_{f_i}(s,b) - \delta_{f_i}(s,a), \tag{22.12}$$

where the learning rate η defines the step width of each iteration and can optionally be decreased progressively for better convergence. To enforce the bounds on w_i given by Equations (22.4) and (22.5), an additional step is necessary after the update, in which all negative weights are set to 0 and the weights are normalized such that we obtain a constant weight sum of 1. The algorithm can stop as soon as no constraints are violated anymore. Furthermore, a maximum number of iterations or a time limit can be specified as another stopping criterion.

This algorithm can compute a weighting, even if not all constraints can be satisfied. However, it is not guaranteed to find a globally optimal solution and no maximum margin is enforced for extra stability. Using the previous weight settings as initial values in combination with a small learning rate allows for gradual change in scenarios where constraints are added incrementally, but there may still be solutions with less change required.

22.2.4.2   Quadratic Programming with Soft Constraints

Another way to treat the weight learning as optimization is to model it as a quadratic programming problem for which very efficient solvers are readily available in various scientific computing packages. In general, a quadratic programming problem has an objective function of the form

$$\min_{x \in \mathbb{R}^n} \frac{1}{2} x^T G x + a^T x \tag{22.13}$$

subject to linear equality and inequality constraints

$$x^T C_e = b_e \tag{22.14}$$
$$x^T C_i \geq b_i. \tag{22.15}$$

Figure 22.2 illustrates how the equality and inequality constraints can be used to specify the weight-learning problem. The first l elements of the feature vector x are

$$
C_e = [\,\overbrace{1 \quad 1 \quad \cdots \quad 1}^{l} \quad \overbrace{0 \quad 0 \quad \cdots \quad 0}^{k}\,] \qquad\qquad b_e = [\,1\,]
$$

$$
C_i = \begin{bmatrix}
1 & 0 & \cdots & 0 & 0 & 0 & \cdots & 0 \\
0 & 1 & \cdots & 0 & 0 & 0 & \cdots & 0 \\
\vdots & \vdots & \ddots & \vdots & \vdots & \vdots & \ddots & \vdots \\
0 & 0 & \cdots & 1 & 0 & 0 & \cdots & 0 \\
c_{1,1} & c_{1,2} & \cdots & c_{1,l} & 1 & 0 & \cdots & 0 \\
c_{2,1} & c_{2,2} & \cdots & c_{2,l} & 0 & 1 & \cdots & 0 \\
\vdots & \vdots & \ddots & \vdots & \vdots & \vdots & \ddots & \vdots \\
c_{k,1} & c_{k,2} & \cdots & c_{k,l} & 0 & 0 & \cdots & 1
\end{bmatrix}
\qquad
b_i = \begin{bmatrix}
0 \\ 0 \\ \vdots \\ 0 \\ \varepsilon \\ \varepsilon \\ \vdots \\ \varepsilon
\end{bmatrix}
$$

Figure 22.2: Scheme for modeling a weight-learning problem with soft distance constraints through the equality and inequality constraints of a quadratic programming problem.

the facet weights we would like to optimize. We have to satisfy the weight constraints given in Equations (22.4) and (22.5). The single equality constraint corresponds to Equation (22.5) whereas Equation (22.4) is represented by the first l inequality constraints – one for each weight. Each of the remaining k inequality constraints models a relative distance constraint as formulated in Equation (22.7). For the i-th distance constraint (s,a,b), the value $c_{i,j}$ refers to the facet distance difference for the j-th facet, i.e., $\delta_{f_j}(s,b) - \delta_{f_j}(s,a)$. The constant ε in Figure 22.2 refers to a small value close to machine precision which is used to enforce inequality.

In order to allow each of the distance constraints to be violated, individual slack variables $\xi \geq 0$ are introduced such that:

$$
\sum_{i=1}^{l} w_i(\delta_{f_i}(s,b) - \delta_{f_i}(s,a)) + \xi > 0. \tag{22.16}
$$

A slack value greater than zero means that the respective constraint is violated. For the k constraints, we require k slack variables $\xi_1, \ldots, \xi_k$ that form the remaining k dimensions of the variable vector $x = (w_1, \ldots, w_l, \xi_1, \ldots, \xi_k)$. The objective is then to minimize the sum of the squared slack variables. This can be accomplished by setting the last k values on the diagonal of the matrix G to 1 and all other values of G and a to zero.

This approach will always find a globally optimal solution that minimizes the slack. However, it cannot be used directly in incremental scenarios. In order to support incremental learning, the objective function has to be modeled differently. In particular, we have to add a term for minimizing the weight change and balance it with the slack minimization objective. This, however, is beyond the scope of this chapter.

22.2.4.3 Maximum Margin Classifier

The third learning approach takes the classification perspective, i.e., all distance constraints are transformed into positive and negative training examples for a binary classifier as illustrated in Figure 22.1. With a maximum margin classifier, the "safety margin" (which does not contain any training examples) around the separating hyperplane is maximized. This leads to more stable solutions w.r.t. to noisy constraints. As a popular maximum margin classifier, a linear support vector machine (SVM) can be used. This is an SVM as introduced in Chapter 12 with a linear kernel, i.e., effectively without applying the kernel trick. The specific SVM implementation has to support hard constraints, i.e., constraints that must not be violated, to enforce non-negative weights, cp. Equation (22.4). Soft constraints could be violated by the optimization algorithms in favor of a larger margin or when not all distance constraints can be satisfied. Equation (22.5) can be accomplished through normalization.

If non-negative weights can be ensured, this approach finds a globally optimal solution. Moreover, there exist very efficient implementations like *LIBLINEAR* that can deal with a large number of training examples. They are especially suited for weight-learning problems with many constraints. Incremental learning can be accomplished by using incremental SVMs as, for instance, described by Cauwenberghs and Poggio [2].

22.3 Visualization: Dealing with Projection Errors

With a music similarity measure adapted to reflect the user's view using the techniques described in the preceding section, let us now focus on how to visualize the resulting similarity space.

22.3.1 Popular Projection Techniques

When it comes to visualizing a music collection by similarity, *neighborhood-preserving* projection techniques like self-organizing maps (SOMs) or multidimensional scaling (MDS) and the closely related principal component analysis (PCA) have become increasingly popular. The general objective of such techniques can be paraphrased as follows: Arrange the tracks in two or three dimensions (on the display) in such a way that neighboring tracks are very similar and the similarity decreases with increasing distance on the display.

Given a set of data points, classical MDS [11] finds an embedding in the target space (here $\mathbb{R}^2$) that maintains their distances (or dissimilarities) as far as possible – without having to know their actual values. Hence, it is also well suited to compute a layout for spring- or force-based visualization approaches. MDS is closely related to PCA (cp. Section 9.8.3), which projects data points simply onto the (two) axes of highest variance termed principal components. In contrast to SOMs, both are non-parametric approaches that compute a *globally* optimal solution, w.r.t. data variance maximization and distance preservation, respectively, in fixed polynomial time.

SOMs (cp. Section 11.4.2) are a special kind of artificial neural network commonly applied for structuring data collections by clustering similar objects into identical or neighboring cells of a two-dimensional grid. In the field of MIR, they have been used in a large number of applications for structuring music collections like the *Islands of Music* [23, 21], *MusicMiner* [19] (Figure 22.4), *nepTune* [9] (Figures 22.4 and 26.3), or the *Map of Mozart* (Figure 26.7). An overview on SOM-related publications in the field of MIR is given in [30]. As a major drawback, SOMs generally require that the objects they process are represented as vectors, i.e., elements of a vector space.[4] If the feature representation does not adhere to this condition, we need to vectorize it first. For instance, as proposed in [28], we can use MDS to compute an embedding of the tracks into a high-dimensional Euclidean space for vectorization. This vectorization step is exactly like using MDS directly for projection, but here the output space can have as many dimensions as needed to not lose any information about the distances between the tracks. Afterwards, we can use a regular SOM to project the vectorized data for visualization. Alternatively, there are also several special versions of SOMs that do not require vector input such as kernel SOMs and dissimilarity SOMs (also called median SOMs) [10].

22.3.2 Common and Unavoidable Projection Errors

The similarity space of the tracks to be projected usually has far more dimensions than the display space.[5] Therefore, the projection inevitably causes some loss of information, irrespective of which dimensionality reduction technique is applied. Consequently, this leads to a distorted display of the neighborhoods such that some tracks will appear closer than they actually are (type I projection errors). At the same time, some tracks that are distant in the projection may in fact be neighbors in the underlying similarity space (type II projection errors). Such neighborhood distortions are depicted in Figure 22.3. These projection errors cannot be fixed on a global scale without introducing new ones elsewhere as the projection is already optimal w.r.t. some criteria (depending on the technique used). Kaski et al. [8] define two measures, *trustworthiness* and *continuity*, that allow us to assess the *visualization quality* of a projection w.r.t. type I and II projection errors.

Definition 22.3 (Trustworthiness). A visualization can be considered *trustworthy* if the k nearest neighbors of a point on the display are also neighbors in the original space. The respective measure of trustworthiness can be computed by:

$$M_{trustworthiness} = 1 - C(k) \sum_{i=1}^{N} \sum_{j \in U_k(i)} (r_{ij} - k), \qquad (22.17)$$

where r_{ij} is the rank of j in the ordering of the distance from i in the original space,

---

[4]SOM cells are usually represented by prototype feature vectors and the SOM learning algorithm relies on vector operations to update these prototypes based on the assigned objects.

[5]In order to correctly display all pair-wise distances of N tracks, a space with at most N dimensions are needed. The *intrinsic dimensionality* of the collection refers to the smallest number of dimensions $1 \le m \le N$ where this is still possible. This value very much depends on the complexity of the similarity measure and the distribution of feature values in the collection. In all but trivial cases, it will be significantly higher than 2.

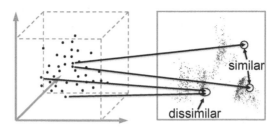

Figure 22.3: Possible problems caused by projecting music tracks represented in a high-dimensional similarity space (left) onto a low-dimensional space for display (right).

$U_k(i)$ is the set of i's false neighbors in the display, and $C(k) = 2/(Nk(2N - 3k - 1))$ is a constant for obtaining values in $[0, 1]$.

Definition 22.4 (Continuity). The measure of *continuity* considers the k nearest neighbors in the original space and captures how well they are preserved in the visualization:

$$M_{continuity} = 1 - C(k) \sum_{i=1}^{N} \sum_{j \in V_k(i)} (\hat{r}_{ij} - k), \tag{22.18}$$

where $\hat{r}_{ij}$ is the rank of j in the ordering of the distance from i in the visualization and $V_k(i)$ is the set of i's true neighbors missing in the visualized neighborhood.

Type I projection errors increase the number of dissimilar (i.e., irrelevant) tracks displayed in a local region of interest (low trustworthiness). While this might become annoying, it is much less problematic than type II projection errors. Type II projection errors are like "wormholes" connecting possibly distant regions in the two-dimensional display space through the underlying high-dimensional similarity space.[6] They result in similar (i.e., relevant) music tracks to be displayed far away from the region of interest – the neighborhood they actually belong to (low continuity). In the worst case, misplaced neighbors could even be off-screen if the display is limited to the currently explored region. This way, users could miss tracks they are actually looking for.

22.3.3 Static Visualization of Local Projection Properties

Several possibilities exist to statically visualize type I projection errors (trustworthiness) as shown in Figure 22.4. Here, mountain ranges are a popular metaphor. *MusicMiner* [19] draws mountain ranges between dissimilar music tracks that are

---

[6]As a metaphor for understanding projection wormholes, imagine crumpling a two-dimensional sheet of paper (the screen projection) to approximate the three-dimensional volume of a box (the actual similarity space). Each coordinate in the volume is mapped to the closest point on the crumpled paper. When the paper is flattened (visualization), not all mapped volume coordinates will be next to their actual neighbors. In reality, the similarity space usually has many more dimensions, which amplifies the problem.

Figure 22.4: Screenshots of approaches that use mountain ranges to separate dissimilar regions (left: *MusicMiner* [19], middle: *SoniXplorer* [15]) or to visualize regions with a high density of similar music tracks (right: *nepTune* [9], a variant of *Islands of Music* [23, 21]).

displayed close to each other. *SoniXplorer* [14, 15] uses the same geographical metaphor but in a 3D virtual environment that users can navigate with a game pad. The *Islands of Music* [23, 21] and related approaches [9, 20, 6] use the third dimension the other way around. Here, islands or mountains refer to regions of similar tracks (with high density) separated by water (with low density of similar tracks). All these approaches visualize local properties of the projection, i.e., neighborhoods of either dissimilar or similar music tracks.

22.3.4 Dynamic Visualization of "Wormholes"

Type II projection errors can be visualized as well. However, this is much harder as they are not confined to local regions. Figure 22.5 shows the *SoundBite* user interface [13] where lines are drawn to connect a selected seed track (highlighted by a circle) with its actual nearest neighbors. The rendering of neighborhood relationships is done dynamically for one seed track at a time. Visualizing all neighborhood connections statically would create too much visual clutter.

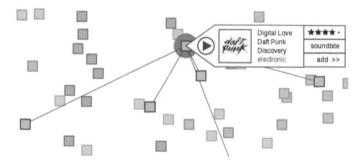

Figure 22.5: In the *SoundBite* user interface [13], a selected seed track and its actual nearest neighbors are connected by lines.

A different dynamic visualization technique is applied by the *MusicGalaxy* user

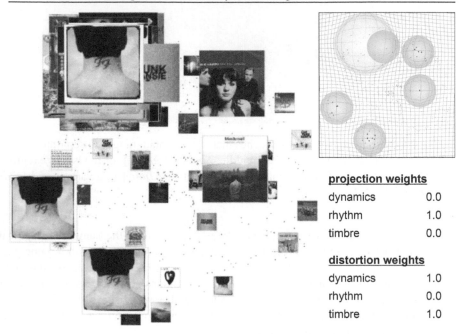

projection weights	
dynamics	0.0
rhythm	1.0
timbre	0.0

distortion weights	
dynamics	1.0
rhythm	0.0
timbre	1.0

Figure 22.6: Left: *MusicGalaxy* (inverted color scheme for print). Tracks are visualized as stars with brightness corresponding to listening frequency. For a well-distributed selection of popular tracks, the album cover is shown for better orientation. Same covers indicate different tracks from the same album. Hovering over a track displays the title. For tracks in focus, the album covers are shown with the size increased by the lens scale factor. Top right: corresponding SpringLens distortion resulting from (user-controlled) primary focus (large) and (adaptive) secondary lenses (small). Bottom right: facet weights for the projection and distortion distance measures (cp. Section 22.3.5).

interface [29] shown in Figure 22.6, which exploits the wormhole metaphor for navigation.[7] Instead of trying to globally repair errors in the projection (implemented through MDS), the general idea is to *temporarily* fix and highlight the neighborhood in focus through distortion. To this end, an adaptive mesh-based distortion technique called *SpringLens* is applied that is guided by the user's focus of interest. The SpringLens consists of a complex overlay of multiple fish-eye lenses divided into a primary and secondary focus (Figure 22.6, top right). The primary focus is a single large fish-eye lens used to zoom into regions of interest. At the same time, it compacts the surrounding space but does not hide it from the user to preserve overview. While the user can control the position and size of the primary focus, the secondary focus is automatically adapted. It consists of a varying number of smaller fish-eye lenses. When the primary focus is moved by the user, a neighbor index is queried

---

[7]Demo videos are available at http://www.dke-research.de/aucoma. Accessed 22 June 2016.

with the track closest to the new center of focus. If nearest neighbors are returned that are not in the primary focus region, secondary lenses are added at the respective positions. As a result, the overall distortion of the visualization temporarily brings the distant nearest neighbors back closer to the focused region of interest. This way, distorted distances introduced by the projection can, to some extent, be compensated whilst the distant nearest neighbors are highlighted. By clicking on a secondary focus region, users "travel through a wormhole" and the primary focus is changed respectively. This is like navigating an invisible neighborhood graph.

22.3.5 Combined Visualization of Different Structural Views

We can use the adaptive multi-focus distortion technique described above beyond its originally intended purpose of fixing projection errors. Specifically, we can combine two different views, a *primary view* and a *secondary view*, on the same music collection in a single visualization. These two views could, for instance, correspond to two distinct music similarity measures for comparing the tracks. In this scenario, we visualize the similarity space of the primary view directly by the map projection as described earlier. This time, however, the lens distortions do *not* visualize projection errors. Instead, we use the distortions to *indirectly* visualize the similarity space of the secondary view. For any given track in primary focus chosen by the user, we identify the nearest neighbors in the secondary view and highlight them with the lenses of the secondary focus. By moving the primary focus around, we can explore local neighborhood relations of the otherwise invisible secondary view. This becomes especially interesting, when orthogonal views defined by using non-overlapping facet sets are combined as shown in Figure 22.6. Here, the "rhythm" facet defines the similarity space for projection (primary view) and the other two facets, "dynamics" and "timbre," define the similarity space for the distortion of the secondary focus (secondary view). Consequently, in this example, the tracks in the secondary focus are very different in rhythm (large distance in the projection) but very similar in dynamics and timbre w.r.t. the track in the primary focus.

22.4 Dealing with Changes in the Collection

So far in this chapter, we have assumed a static music collection. However, this is not a practical assumption. Most music collections will change over time, e.g., as users add new music tracks or remove ones they do not like anymore. To what extent does a map (projection) change when tracks are added or removed? Could it even be necessary to re-compute the map from scratch? Such questions need to be answered as failing to support changes in the collection may significantly limit the usefulness of a MIR application in real-world scenarios. In this context, being able to add tracks to an existing map is more important than removal which is often trivial (at worst leading to blank spaces in the map) and also an uncommon use case. New tracks that are similar to existing ones should be embedded in the respective neighborhoods. At the same time, the map should also be able to deal with changes in music taste (adding tracks from new genres) and an increase of musical diversity.

Ideally, a map should be altered as little as possible and only as much as necessary to reflect the changes of the underlying collection. Too abrupt changes in the topology might confuse the user who over time will get used to the location of specific regions in the map.

22.4.1   Incremental Structuring Techniques

One strategy to deal with changing collections is to use an incremental technique that can start from an existing map and only changes it as much as needed to accommodate the new tracks. There are special variants for both MDS and SOMs that fall into this category.

Landmark multidimensional scaling as described in [4] is a computationally efficient approximation to classical MDS. The general idea of this approach is as follows. Given a sample set of *landmarks* or pivot objects, an embedding into a low-dimensional space is computed for these objects using classical MDS. Each remaining object can then be located within the output space according to its distances to the landmarks. Obviously, the quality of the projection depends on the choice of the landmarks – especially if the landmark sample set is small compared to the size of the whole collection. If the landmarks lie close to a low-dimensional subspace (e.g., a line), there is the chance of systematic errors in the projection. Landmark MDS can be applied to visualize growing music collections by using the initial tracks as landmarks. Consequently, the position of a track once added to the map never changes. However, the landmark set may become less and less representative with increasing collection size and possibly changing music taste. This may have a significant effect on the quality of the projection.

Growing self-organizing maps (GSOMs) are SOMs with a flexible cell grid for structuring. Their size does not have to be specified prior to training. Instead, they usually start with a small size and grow as needed and adapt incrementally to changes in the underlying collection whereas other approaches may always need to generate a new structuring when the grid becomes too small. There are flat and hierarchical variants of GSOMs that have both been applied for structuring music collections (e.g., [27] and [24, 22, 5] respectively). Flat GSOMs add cells at the outer boundary whereas hierarchical GSOMs grow in depth forming nested structures of SOMs within cells. As mentioned earlier in Section 22.3, SOMs usually require vectorization of the input data. For growing collections, we can vectorize the initial tracks using MDS and then use these as landmarks to embed new songs into the existing vector space. If many new tracks are added, the embedding quality might degrade and cause a significant drop in the nearest neighbor retrieval precision as observed in [28]. In that case, the vectorization and the SOM on top of it should be recomputed.

22.4.2   Aligned Projections

As an alternative to incremental structuring techniques, we can try to align a new map computed from scratch with its previous version – hoping that they share enough common structure for a meaningful outcome. This strategy worked surprisingly well

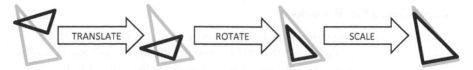

Figure 22.7: Procrustes superimposition of two triangles (sets of 3 points).

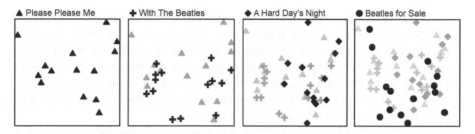

Figure 22.8: Aligned MDS projections computed after adding the first four Beatles albums to the collection.

in an experimental comparison with incremental structuring techniques. Here, a collection containing the 12 official albums of *The Beatles* was visualized adding one album at a time [26]. The approach liked best by the participants of this study was a combination of classic MDS for computing the projections and *Procrustes superimposition* for alignment.

MDS does not support incremental collection changes. Instead, a new map has to be computed every time the collection grows. Even with little change of the collection, the resulting map may look very different because it could be arbitrarily translated and rotated without affecting the pairwise distances. In order to remedy this issue, Procrustes superimposition [7] can be applied to align each newly generated map with the previous one. Procrustes superimposition involves a sequence of affine transformations – optimally translating, rotating, and uniformly scaling – with the goal to minimize the difference in placement and size between two shapes. This is illustrated in Figure 22.7 for two triangles (or sets of three points).

Figure 22.8 shows the MDS projections computed after adding the first four Beatles albums to the collection. Using transition animations to "cross-fade" between subsequent projections makes it easier for users to track individual position changes – as can be seen in the online demo.[8] Apart from the unavoidable scaling of the map to show new tracks outside of the old map's boundary, track positions are very stable. Stability further increases as the number of new tracks becomes relatively small compared to the size of the whole collection.

---

[8]An online demo of the *Beatles History Explorer* is available at http://demos.dke-research.de/beatles-history-explorer/. Accessed 22 June 2016.

22.5   Concluding Remarks

As we have seen, structuring a music collection by similarity poses several interesting challenges. Firstly, we learned that there is no such thing as "the" music similarity that is generally applicable. Instead, there are many individual views based on the user and the usage context that we need to adapt to, for instance, with the modeling approach and the corresponding learning algorithms for adaptation presented in this chapter.

Next, we turned towards a common problem of similarity-based structuring techniques that apply some form of dimensionality reduction to generate an overview map (projection) in which neighboring tracks should be similar. Such visualizations suffer from inevitable projection errors. Some tracks will appear closer than they actually are and some distant tracks may in fact be neighbors in the original similarity space. We discussed several static and dynamic visualization techniques that address these issues and can even be utilized for navigation.

Finally, we focused on the problem that music collections change over time and the challenge to update the corresponding visualizations accordingly. We compared two different strategies that both aim to minimize visual change for a maximum continuity between consecutive visualizations.

All the techniques covered in this chapter provide essential building blocks for applications that can structure a music collection by similarity. However, some issues and open questions still remain that have not been addressed. For instance, how can long-term adaptations of the music similarity be supported due to gradual change in preferences, how can music tracks be effectively visualized beyond simply showing their album covers, or how can we add semantics to axes of map projections to make them more meaningful?

22.6   Further Reading

An experimental comparison of the similarity adaptation approaches described in Section 22.2.4 is given in [31]. This paper also covers alternative formulations for the quadratic programming problem. There are also more complex ways to model and learn music similarity than covered here. A recent overview and comparison is given in [34].

Slaney et al. [25] apply several algorithms based on second-order statistics (such as whitening) and optimization techniques to learn Mahalanobis distance metrics (cp. Section 11.2) for clustering songs by artist, album or blog they appear on. For the optimization, an objective function that mimics the k-nearest neighbor leave-one-out classification error is chosen. Songs are represented as vectors containing various content-based acoustic features.

McFee et al. [17] apply a partial-order embedding technique with multiple kernels that maps artists into multiple non-linear spaces (using different kernel matrices), learns a separate transformation for each kernel, and concatenates the resulting vectors. The Euclidean distance in the resulting embedding space corresponds to the perceived similarity.

Further work by McFee et al. [16] focuses on adapting content-based song similarity by learning from a sample of collaborative filtering data. Here, they use the metric learning to rank (MLR) technique [18] – an extension of the Structural SVM approach – to adapt a Mahalanobis distance according to a ranking loss measure. This approach is also applied by Wolff et al. [33] whose similarity adaptation experiments are based on the *MagnaTagATune* dataset derived from the *TagATune* game [12]. Further experiments described in [32] compare the approaches that are using the more complex Mahalanobis distance to the weighted facet distance approach described in this chapter.

Bibliography

[1] F. G. Ashby and D. M. Ennis. Similarity measures. *Scholarpedia*, 2(12):4116, 2007. revision #142770.

[2] G. Cauwenberghs and T. Poggio. Incremental and decremental support vector machine learning. In *Advances in Neural Information Processing Systems (NIPS'00)*, pp. 409–415, Cambridge, MA, USA, 2000. MIT Press.

[3] W. Cheng and E. Hüllermeier. Learning similarity functions from qualitative feedback. In *Proceedings of the 9th European Conference on Advances in Case-Based Reasoning (ECCBR'08)*, pp. 120–134, Trier, Germany, 2008. Springer-Verlag.

[4] V. de Silva and J. B. Tenenbaum. Global versus local methods in nonlinear dimensionality reduction. In *Advances in Neural Information Processing Systems (NIPS'02)*, pp. 705–712, Cambridge, MA, USA, 2002. MIT Press.

[5] M. Dopler, M. Schedl, T. Pohle, and P. Knees. Accessing music collections via representative cluster prototypes in a hierarchical organization scheme. In *Proceedings of the 9th International Conference on Music Information Retrieval (ISMIR'08)*, pp. 179–184, 2008.

[6] M. Gasser and A. Flexer. FM4 Soundpark: Audio-based music recommendation in everyday use. In *Proceedings of the 6th Sound and Music Computing Conference (SMC'09)*, pp. 161–166, 2009.

[7] J. C. Gower and G. B. Dijksterhuis. *Procrustes Problems*. Oxford University Press, 2004.

[8] S. Kaski, J. Nikkilä, M. Oja, J. Venna, P. Törönen, and E. Castrén. Trustworthiness and metrics in visualizing similarity of gene expression. *BMC Bioinformatics*, 4(1):48, 2003.

[9] P. Knees, T. Pohle, M. Schedl, and G. Widmer. Exploring music collections in virtual landscapes. *IEEE MultiMedia*, 14(3):46–54, 2007.

[10] T. Kohonen and P. Somervuo. How to make large self-organizing maps for non-vectorial data. *Neural Networks*, 15(8–9):945–952, October–November 2002.

[11] J. B. Kruskal and M. Wish. *Multidimensional Scaling*. Sage, 1986.

[12] E. Law and L. von Ahn. Input-agreement: A new mechanism for collecting data

using human computation games. In *Proceedings of the 27th International Conference on Human Factors in Computing Systems (CHI'09)*, pp. 1197–1206, New York, NY, USA, 2009. ACM.

[13] S. Lloyd. *Automatic Playlist Generation and Music Library Visualisation with Timbral Similarity Measures*. Master's thesis, Queen Mary University of London, 2009.

[14] D. Lübbers. SoniXplorer: Combining visualization and auralization for content-based exploration of music collections. In *Proceedings of the 6th International Conference on Music Information Retrieval (ISMIR'05)*, pp. 590–593, 2005.

[15] D. Lübbers and M. Jarke. Adaptive multimodal exploration of music collections. In *Proceedings of the 10th International Society for Music Information Retrieval Conference (ISMIR'09)*, pp. 195–200, 2009.

[16] B. McFee, L. Barrington, and G. R. G. Lanckriet. Learning similarity from collaborative filters. In *Proceedings of the 11th International Society for Music Information Retrieval Conference (ISMIR'10)*, pp. 345–350, 2010.

[17] B. McFee and G. R. G. Lanckriet. Heterogeneous embedding for subjective artist similarity. In *Proceedings of the 10th International Society for Music Information Retrieval Conference (ISMIR'09)*, pp. 513–518, 2009.

[18] B. McFee and G. R. G. Lanckriet. Metric learning to rank. In *Proceedings of the 27th International Conference on Machine Learning (ICML'10)*, pp. 775–782, 2010.

[19] F. Mörchen, A. Ultsch, M. Nöcker, and C. Stamm. Databionic visualization of music collections according to perceptual distance. In *Proceedings of the 6th International Conference on Music Information Retrieval (ISMIR'05)*, pp. 396–403, 2005.

[20] R. Neumayer, M. Dittenbach, and A. Rauber. PlaySOM and PocketSOMPlayer: Alternative interfaces to large music collections. In *Proceedings of the 6th International Conference on Music Information Retrieval (ISMIR'05)*, pp. 618–623, 2005.

[21] E. Pampalk, S. Dixon, and G. Widmer. Exploring music collections by browsing different views. In *Proceedings of the 4th International Conference on Music Information Retrieval (ISMIR'03)*, pp. 201–208, 2003.

[22] E. Pampalk, A. Flexer, and G. Widmer. Hierarchical organization and description of music collections at the artist level. In *Proceedings of the 9th European Conference on Research and Advanced Technology for Digital Libraries (ECDL'05)*, pp. 37–49, Heidelberg / Berlin, 2005. Springer Verlag.

[23] E. Pampalk, A. Rauber, and D. Merkl. Content-based organization and visualization of music archives. In *Proceedings of the 10th ACM International Conference on Multimedia*, pp. 570–579, New York, NY, USA, 2002. ACM.

[24] A. Rauber, E. Pampalk, and D. Merkl. Using psycho-acoustic models and self-organizing maps to create a hierarchical structuring of music by musical styles.

In *Proceedings of the 3rd International Conference on Music Information Retrieval (ISMIR'02)*, 2002.

[25] M. Slaney, K. Q. Weinberger, and W. White. Learning a metric for music similarity. In *Proceedings of the 9th International Conference on Music Information Retrieval (ISMIR'08)*, pp. 313–318, 2008.

[26] S. Stober, T. Low, T. Gossen, and A. Nürnberger. Incremental visualization of growing music collections. In *Proceedings of the 14th International Society for Music Information Retrieval Conference (ISMIR'13)*, pp. 433–438, 2013.

[27] S. Stober and A. Nürnberger. Towards user-adaptive structuring and organization of music collections. In *Adaptive Multimedia Retrieval. Identifying, Summarizing, and Recommending Image and Music*, volume 5811 of *LNCS*, pp. 53–65. Springer Verlag, Heidelberg / Berlin, 2010.

[28] S. Stober and A. Nürnberger. Analyzing the impact of data vectorization on distance relations. In *Proceedings of 3rd International Workshop on Advances in Music Information Research (AdMIRe'11)*, pp. 1–6, Piscataway, NJ, USA, 2011. IEEE.

[29] S. Stober and A. Nürnberger. Musicgalaxy: A multi-focus zoomable interface for multi-facet exploration of music collections. In *Exploring Music Contents*, volume 6684 of *LNCS*, pp. 273–302. Springer Verlag, Heidelberg / Berlin, 2011.

[30] S. Stober and A. Nürnberger. Adaptive music retrieval: A state of the art. *Multimedia Tools and Applications*, 65(3):467–494, 2013.

[31] S. Stober and A. Nürnberger. An experimental comparison of similarity adaptation approaches. In *Adaptive Multimedia Retrieval. Large-Scale Multimedia Retrieval and Evaluation*, volume 7836 of *LNCS*, pp. 96–113. Springer Verlag, Heidelberg / Berlin, 2013.

[32] D. Wolff, S. Stober, A. Nürnberger, and T. Weyde. A systematic comparison of music similarity adaptation approaches. In *Proceedings of the 13th International Society for Music Information Retrieval Conference (ISMIR'12)*, pp. 103–108, 2012.

[33] D. Wolff and T. Weyde. Combining sources of description for approximating music similarity ratings. In *Adaptive Multimedia Retrieval. Large-Scale Multimedia Retrieval and Evaluation*, volume 7836 of *LNCS*, pp. 114–124. Springer Verlag, Heidelberg / Berlin, 2013.

[34] D. Wolff and T. Weyde. Learning music similarity from relative user ratings. *Information Retrieval*, 17(2):109–136, 2014.

Chapter 23

Music Recommendation

DIETMAR JANNACH
Department of Computer Science, TU Dortmund, Germany

GEOFFRAY BONNIN
LORIA, Université de Lorraine, Nancy, France

23.1 Introduction

Until recently, music discovery was a difficult task. We had to listen to the radio hoping one track will be interesting, actively browse the repertoire of a given artist, or randomly try some new artists from time to time. With the emergence of personalized recommendation systems, we can now discover music just by letting music platforms play tracks for us. In another scenario, when we wanted to prepare some music for a particular event, we had to carefully browse our music collection and spend significant amounts of time selecting the right tracks. Today, it has become possible to simply specify some desired criteria like the genre or mood and an automated system will propose a set of suitable tracks.

Music recommendation is however a very challenging task, and the quality of the current recommendations is still not always satisfying. First, the size of the pool of tracks from which to make the recommendations can be quite huge. For instance, Spotify,[1] Groove,[2] Tidal,[3] and Qobuz,[4] four of the currently most successful web music platforms, all contain more than 30 million tracks.[5] Moreover, most of the tracks on these platforms typically have a low popularity[6] and hence little information is available about them, which makes them even harder to process for the task of automated recommendation. Another difficulty is that the recommended tracks are

---

[1] http://www.spotify.com. Accessed 22 June 2016.
[2] http://music.microsoft.com. Accessed 22 June 2016.
[3] http://tidal.com. Accessed 22 June 2016.
[4] http://www.qobuz.com. Accessed 22 June 2016.
[5] This information can be obtained using the API's search services provided by these platforms.
[6] This information can be obtained using, for instance, the API of Last.fm or The Echo Nest, two of the currently richest sources of track information on the Web.

immediately consumed, which means the recommendations must be made very fast, and must, at the same time, fit the current context.

Music recommendation was one early application domain for recommendation techniques, starting with the *Ringo* system presented in 1995 [35]. Since then however, most of the research literature on recommender systems (RS) has dealt with the recommendation of movies and commercial products [17]. Although the corresponding core strategies can be applied to music, music has a set of specificities which can make these strategies insufficient.

In this chapter, we will discuss today's most common methods and techniques for item recommendation which were developed mostly for movies and in the e-commerce domain, and talk about particular aspects of the recommendation of music. We will then show how we can measure the quality of recommendations and finally give examples of real-world music recommender systems. Parts of our discussion will be based on [5], [8], and [22], which represent recent overviews on music recommendation and playlist generation.

23.2 Common Recommendation Techniques

Generally speaking, the task of a recommender system in most application scenarios is to generate a ranked list of items which are assumedly relevant or interesting for the user in the current context.[7] Recommendation algorithms are usually classified according to the types of data and knowledge they process to determine these ranked lists. In the following, we will introduce two common recommendation strategies found in the literature.

23.2.1 Collaborative Filtering

The most prominent class of recommendation algorithms in research and maybe also in industry is called *Collaborative Filtering* (CF). In such systems, the only type of data processed by the system to compute recommendation lists are *ratings* provided by a larger user community. Table 23.1 shows an example of such a rating database, where 5 users have rated 5 songs using a rating scale from 1 (lowest) to 5 (highest), e.g., on an online music platform or using their favorite music player.

In this simple example, the task of the recommender is to decide whether or not the *Song5* should be put in Alice's recommendation list and – if there are also other items – at which position it should appear in the list.

Many CF systems approach this problem by first *predicting* Alice's rating for all songs which she has not seen before. In the second step, the items are ranked according to the prediction value, where the songs with the highest predictions should obviously appear on top of the list.[8]

---

[7]In Section 23.4 we will discuss in more detail what relevance or interestingness could mean for the user.

[8]Taking the general popularity of items into account in the ranking process is, however, also common in practical settings because of the risk that only niche items are recommended.

Table 23.1: A Simple Rating Database, Adapted from [20]. When Recommendation Is Considered as a *Rating Prediction* Problem, the Goal is to Estimate the Missing Values in the Rating "Matrix" (Marked with '?')

	Song1	Song2	Song3	Song4	Song5
Alice	5	3	4	4	?
User1	3	1	2	3	3
User2	4	3	4	3	5
User3	3	3	1	5	4
User4	1	5	5	2	1

23.2.1.1 CF Algorithms

One of the earliest and still relatively accurate schemes to predict Alice's missing ratings is to base the prediction on the opinion of other users, who have liked similar items as Alice in the past, i.e., who have the same taste. The users of this group are usually called "neighbors" or "peers". When using such a scheme, the question is (a) how to measure the similarity between users and (b) how to aggregate the opinions of the neighbors. In one of the early papers on RS [32], the following approach was proposed, which is still used as a baseline for comparative evaluation today. To determine the similarity, the use of Pearson's correlation coefficient (Definition 9.20 in Chapter 9) was advocated. The similarity of users u_1 and u_2 can thus be calculated via

$$sim(u_1, u_2) = \frac{\sum_{i \in \hat{I}} (r_{u_1,i} - \overline{r_{u_1}})(r_{u_2,i} - \overline{r_{u_2}})}{\sqrt{\sum_{i \in \hat{I}} (r_{u_1,i} - \overline{r_{u_1}})^2} \sqrt{\sum_{i \in \hat{I}} (r_{u_2,i} - \overline{r_{u_2}})^2}}. \tag{23.1}$$

where $\hat{I}$ denotes the set of products that have been rated both by user u_1 and user u_2, $\overline{r_{u_1}}$ is u_1's average rating and $r_{u_1,i}$ denotes u_1's rating for item i.

Besides using Pearson's correlation, other metrics such as cosine similarity have been proposed.[9] One of the advantages of Pearson's correlation is that it takes into account the tendencies of individual users to give mostly low or high ratings.

Once the similarity of users is determined, the remaining problem is to predict Alice's missing ratings. Given a user u_1 and an unseen item i, we could for example compute the prediction based on u_1's average rating and the opinion of a set of N closest neighbors as follows:

$$\hat{r}(u_1, i) = \overline{r_{u_1}} + \frac{\sum_{u_2 \in N} (sim(u_1, u_2)(r_{u_2,i} - \overline{r_{u_2}}))}{\sum_{u_2 \in N} sim(u_1, u_2)}. \tag{23.2}$$

The prediction function in Equation (23.2) uses the user's average rating $\overline{r_{u_1}}$ as a baseline. For each neighbor u_2 we then determine the difference between u_2's average rating and his rating for the item in question, i.e., $(r_{u_2,i} - \overline{r_{u_2}})$, and weight the difference with the similarity factor $(sim(u_1, u_2))$ computed using Equation (23.1).

---

[9]For alternative distance measures, see Section 11.2.

When we apply these calculations to the example in Table 23.1, we can identify *User2* and *User3* as the closest neighbors to Alice ($sim(User2, User3)$ is 0.85 and 0.7). Both have rated *Song5* above their average and predict an above-average rating between 4 and 5 (exactly 4.87) for Alice, which means that we should include the song in a recommendation list.

While the presented scheme is quite accurate – we will see, later on, how to measure accuracy – and simple to implement, it has the disadvantage of being basically not applicable for real-world problems due to its limited scalability, since there are millions of songs and millions of users for which we would have to calculate the similarity values.

Therefore a large variety of alternative methods have been proposed over the last decades to predict the missing ratings. Nearly all of these more recent methods are based on offline data preprocessing and on what is called "model-building". In such approaches, the system learns a usually comparably compact model in an offline and sometimes computationally intensive training phase. At runtime, the individual predictions for a user can be calculated very quickly. Depending on the application domain and the frequency of newly arriving data, the model is then re-trained periodically. Among the applied methods we find various data mining techniques such as association rule mining or clustering (see Chapter 11), support vector machines, regression methods and a variety of probabilistic approaches (see Chapter 12). In recent years, several methods were designed which are based on *matrix factorization* (MF) as well as ensemble methods which combine the results of different learning methods [23].

In general, the ratings that the users assigned to items can be represented as a matrix, and this matrix can be factorized, i.e., it is possible to write this matrix R as the product of two other matrices Q and P:

$$R = Q^T \cdot P.$$

Matrix Factorization techniques determine approximations of Q and P using different optimization procedures (see Chapter 10). Implicitly, these methods thereby map users and items to a shared factor space of a given size (dimensionality) and use the inner product of the resulting matrices to estimate the relationship between users and items [23]. Using such factorizations makes the computation times much shorter and at the same time implicitly reveals some *latent* factors. A latent aspect of a song could be the artist or the musical genre the song belongs to; in general, however, the semantic meanings of the factors are unknown. After the factorization process with f latent factors (for example $f = 100$), we are given a vector $q_i \in \mathbb{R}^f$ for each item i and a vector $p_u \in \mathbb{R}^f$ for each user. For the user vectors, each value of the vector corresponds to the interest of a user in a certain factor; for item vectors, each element indicates the degree of "fit" of the item to the factor. Given a user u and an item i, we can finally estimate the "match" between the user and the item by using the dot product $q_i^T p_u$.

Different heuristic strategies exist for determining the values for the latent factor vectors p_u and q_i. The most common ones in RS, which also scale to larger-scale rating data bases, are stochastic gradient descent optimization and *Alternating Least Squares* [23]; see also Chapter 10.

In order to estimate a rating $\hat{r}_{u,i}$ for user u and item i, we can use the following general equation, where μ is the global rating average, b_i is the item bias, and b_u is the user bias.

$$\hat{r}_{u,i} = \mu + b_i + b_u + q_i^T p_u \tag{23.3}$$

The reason for modeling user and item biases is that there are items which are generally more liked or disliked than others, and there are, on the other hand, users who generally give higher or lower ratings than others. For instance, Equation (23.3) with $f = 2$ corresponds to the assumption that only two factors are sufficient to accurately estimate the ratings of users. These factors may be, for instance, the genre and the tempo of tracks, or any other factors, which are inferred during the factorization step.

The learning phase of such an algorithm consists of estimating the unknown parameters based on the data. This can be achieved by searching for parameters which minimize the squared prediction error (see Chapter 10), given the set of known ratings K:

$$\min_{q*,p*,b*} \sum_{(u,i)\in K} \left(r_{u,i} - (\mu + b_u + b_i + q_i^T p_u)\right)^2 + \lambda \left(\|q_i\|^2 + \|p_u\|^2 + b_u^2 + b_i^2\right). \tag{23.4}$$

The last term in the function is used for regularization and to "penalize" large parameter values.

Overall, in the past years much research in the field of recommender systems was devoted to such rating prediction algorithms. It however becomes more and more evident that rating prediction is very seldom the goal in practical applications. Finding a good ranking of the tracks based on observed user behavior is more relevant, which led to an increased application of "learning to rank" methods for this task or to the development of techniques that optimize the order of the recommendations according to music-related criteria such as track transitions or the coherence of the playlists [19].

23.2.1.2 *Collaborative Filtering for Music Recommendation*

As mentioned in the introduction, although music was one of the earliest application domains for recommender systems, music recommendation has until recently remained a niche topic. The first application, the *Ringo* system, actually used a CF technique [35], and modern online music services such as *Spotify*, *Last.fm*, or *iTunes Genius* use – among other techniques – collaborative filtering methods to generate playlists and recommend songs.

When compared to other approaches to (music) recommendation, collaborative filtering methods have some well-known advantages and limitations. From a system provider's perspective, one advantage of CF lies in the fact that besides the users' rating feedback, no additional information (about the musical genre, the authors or any sort of low-level data) has to be acquired and maintained. At the same time, CF methods are well understood and have been successfully applied in a variety of domains, including those where massive amounts of data have to be processed and a large number of parallel users have to be served. The inherent characteristic of CF-based algorithms can in addition lead to recommendations that are surprising and novel for the user, which can be a key feature for a music recommender in particular when the user is interested in discovering new artists or musical sub-genres.

On the down side, CF methods require the existence of a comparably large user community to provide useful recommendations. Related to that is the typical issue of data sparsity. In many domains, a large number of items in the catalog have very few (or even no) ratings, which can lead to the effect that they are never recommended to users. At the same time, some users only rate very few items, which makes it hard for CF methods to develop a precise enough user profile. Situations in which there are no or only a few ratings available for an item or a user are usually termed "cold start" situations. A number of algorithms have been proposed to deal with this problem in the literature. Many of them rely for example on hybridization strategies, where different algorithms or knowledge sources are used as long as the available ratings are not sufficient. Finally, as also discussed in [8], some CF algorithms have a tendency to boost the popularity of already popular items so that, based on the chosen algorithm, some niche items have a low chance of ever being recommended.

CF-based music recommendation has some aspects which are quite specific for the domain. Besides the fact that in the case of song recommendation it is plausible to recommend the same item multiple times to a user, it is often difficult to acquire good and discriminative rating information from the user. Analyses have shown that for example on *YouTube* users tend to give ratings only to items they like so that the number of "dislike" statements is very small. While this bias towards liked items can also be observed in other domains, it appears to be particularly strong for multimedia content as provided on *YouTube*, which "degrades" the user feedback basically to unary ratings ("like" statements). With respect to data sparsity as mentioned above, a common strategy in CF recommender systems is to rely on implicit item ratings, that is, one interprets actions performed by users on items as positive or negative feedback. In the music domain, such implicit feedback is often collected by monitoring the user's listening behavior, and in particular, listening times are used to estimate to which extent a user liked a song.

One of the most popular online music services that uses – among other techniques – collaborative filtering, is *Spotify*. Spotify provides several types of radios such as genre radios, artist radios, and playlist radios. Once the user has chosen a radio, the system automatically plays one recommended song after the other. The user can give some feedback (like, dislike, or skip) on the tracks and this feedback is taken into account and used to adapt the selection of the next recommendations. All these radios use collaborative filtering, and more precisely matrix factorization [4]. Another interesting feature of Spotify is the Discover weekly playlist, a playlist that is automatically generated each week and that the user can play to discover music he may like. This feature also uses collaborative filtering to select the tracks which are "around" the favorite tracks of the users in the similar users' listening logs [37].

An interesting aspect of Spotify is that it has a number of "social" or community features. Users can create a network of friends and follow the playlists they share.

23.2.2 Content-Based Recommendation

Content-based (CB) techniques are rooted in Information Retrieval (IR) and exploit additional information about the available catalog items to generate personalized recommendations. The "content" of an item (e.g., a song or album) can in principle be an arbitrary piece of information describing a certain aspect of the item. Historically, the term *content* was used to refer to the goal of many methods developed in the field of IR, which is the recommendation of text documents or web pages, whose content can be automatically extracted. In the context of music recommendation, however, we would also consider information about the artist, the musical genre or any other type of information that can be extracted with music analysis methods as content.

The rough idea of CB recommenders is to look at items which the current user has liked in the past and then scan the catalog for further items which are similar to these liked items. A CB recommender has to implement at least two functionalities: (A) The system first has to acquire and update a "user profile", which captures the user's interests and preferences.[10] (B) The system has to implement a retrieval function, which determines the estimated relevance of a given item for a certain user profile.

Regarding the maintenance of the user profile, one option for new users of an online music site would be to ask them to explicitly specify their favorite artists or genres or rate some songs. Alternatively, existing profile information taken, e.g., from social networks such as *Facebook* could be used as a starting point. After the initial ramp-up, the user profile should be continually updated based on implicit or explicit feedback.

How to represent and learn the user profile and how to retrieve suitable items depends on the available information. The most common approach is to represent the user profile and an item's content information in the same way and along the same dimensions.

Table 23.2: Content Information in a Song Database

Title	Artist	Genre	Feel	Liked?
Old man	Neil Young	Country	Melancholy	✓
Perhaps Love	John Denver	Country	Melancholy	✓
On the road again	Willie Nelson	Country	Driving Shuffle	
Harlem Shuffle	Rolling Stones	Rock	Use of Groove	✗
...	...	...	...	
Redemption Song	Bob Marley	Reggae	Reggae feel	✓

Table 23.2 shows an example for a content-enhanced music catalog, where the items marked with a tick (✓) correspond to those which the user has liked. A basic

---

[10]In contrast to CF methods, the user profile in CB approaches is not based on the behavior of the community but only on the actions of the individual user.

strategy to derive a user profile from the liked items would be to simply collect all the values in each dimension (artist, genre, etc.) of all liked items. The relevance of unseen items for the user can then be based, for example, on the overlap of keywords. In the example, recommending the Willie Nelson song appears to be a reasonable choice due to the user's preference for country music.

In the area of document retrieval, more elaborate methods are usually employed for determining the similarity between a user profile and an item which, for example, take into account how discriminative a certain keyword is for the whole item collection. Most commonly, the TF-IDF (term frequency - inverse document frequency) metric is used to measure the importance of a term in a document in IR scenarios. The main idea is to represent the recommendable item as a weight vector, where each vector element corresponds to a keyword appearing in the document. The TF-IDF metric then calculates a weight that measures the importance of the keyword or aspect, that is, how well it characterizes the document. The calculation of the weight value depends both on the number of occurrences of the word in the document (normalized by the document length) as well as how often the term appears in all documents, thus avoiding giving less weight to words that appear in most documents.

The user profile is represented in exactly the same way, that is, as a weight vector. The values of the vector, which represent the user's interest in a certain aspect can, for example, be calculated by taking the average vector of all songs that the user has liked.

In order to determine the degree of match between the user profile u and a not-yet-seen item i, we can calculate the cosine similarity as shown in Equation (23.5) and rank the items based on their similarity.

$$sim(u, i) = \frac{u \cdot i}{|u||i|}. \tag{23.5}$$

The cosine similarity between two vectors measures the distance (angle) between them and uses the dot product ($\cdot$) and the magnitudes ($|u|$ and $|i|$) of the rating vectors. The resulting values lie between 0 and 1.

Generally, the recommendation could be viewed as a standard IR ranked retrieval problem with the difference that we use the user profile as an input instead of a particular query. Thus, on principle, modern IR methods based, e.g., on Latent Semantic Analysis or classical ones based on Rocchio's relevance feedback can be employed; see [20]. Viewed from yet a different perspective, the recommendation problem can also be seen as a classification task, where the goal is to assess whether or not a user will like a certain item. For such classification tasks, a number of other approaches have been developed in the field of Information Retrieval, based, e.g., on probabilistic methods, Support Vector Machines, regression techniques, and so on; see Chapter 12.

23.2.2.1   *Content-Based Filtering for Music Recommendation*

Content-based techniques are particularly appropriate for textual document recommendation, as the content of the documents can be directly used to induce vectors of keywords. This is not possible for music (except maybe for recommending songs based on the lyrics), and other types of content features have to be acquired. On principle, all the various pieces of information that can be automatically extracted through automated music analysis, such as timbre, instruments, emotions, speed, or audio features, can be integrated into the recommendation procedure.

In general, the content features in CB systems have to be acquired and maintained either manually or automatically. In each case, however, the resulting annotations can be imprecise, inconsistent or wrong. When songs are, for example, labeled manually with a corresponding genre, the problem exists that there is not even a "gold standard" and that when using annotations from different sources the annotations may contradict.

In recent years, additional sources of information have become available with the emergence of Semantic Web technologies, see [10, 5], and in particular with the Social Web. In this frame, users can actively provide meta-information about items, for instance by attaching tags to items, thereby creating so-called folksonomies. This is referred to as Social Tagging, and it is becoming an increasingly valuable source of additional information. Since the manual annotation process of songs does not scale well, "crowdsourcing" the labeling and classification task is promising despite the problems of labeling inconsistencies and noisy tags. An important aspect here is that the tags applied to a resource not only tell us something about the resource, e.g., the song itself, but also about the interests and preferences of the person that tags the item.

Besides expert-based annotation and social tagging, further approaches to annotating music include Web Mining, e.g., from music blogs or by analyzing the lyrics of songs; automated genre classification; or similarity analysis [9].

Content-based recommendation methods have their pros and cons. In contrast to CF methods, for example, no large user community is required to generate recommendations. On principle, a content-based system can start making recommendations based on one single positive implicit or explicit user feedback action or based on a sample song or user query. More precise and more personalized recommendations can of course be made, if more information is available. The obvious disadvantage of content-based methods when compared with CF methods is that the content information has to be acquired and maintained. In that context, the additional problem arises that the available content information might not be sufficiently detailed or discriminative to make good recommendations.

From the perspective of the user-perceived quality of the recommendations, methods based on content features by design recommend items similar to those the user has liked in the past. Thus, recommendation lists can exhibit low diversity and may contain items that are too similar to each other. In addition, such lists might only in rare cases contain elements which are surprising for the user. Being able to make such "serendipitous" and surprising recommendations is however considered as an important quality factor of an RS. On the other hand, recommending at least a few familiar items – as content-based systems will do – can help the user to develop trust in the system's capability of truly understanding the user's preferences and tastes.

With respect to real-world systems, *Pandora Music* is most often cited as a content-based recommendation service and is based on the *Music Genome* project. The idea of the project was to codify every song as a vector of up to 400 to 500 different features (genes), relative to the genre. These features include both characteristics of the music itself (e.g., of structure, rhythm and meter, instrumentation, or tonality) as well as other information such as the roots of the style and other influences, the used recording techniques, characteristics of the lead vocals as well as information related to the lyrics.

The interesting aspect of *Pandora* is that these feature values are assigned manually by musical experts over years. Annotating a song can take an expert up to half an hour.[11] The similarity of tracks can then be easily computed based on a distance metric once a sufficient number of genes is available. A profile of a user is learned by collecting implicit and explicit feedback. The underlying assumption is that music can be classified in an objective way, and that the chosen set of genes is sufficient to capture all aspects that make musical tracks similar to each other.

Another limiting factor is that the manual annotation approach does not scale too well and new songs can only appear in the recommendations when the musical genes have been entered. As mentioned in [22], however, this particular aspect and the fact that the genre is not explicitly encoded in the genes, can also lead to surprising recommendations. As of 2014 [7], *Pandora Radio* has more than 250 million registered users and features more than 80,000 artists and 800,000 tracks in its library.

23.2.3   Further Knowledge Sources and Hybridization

Besides user-provided feedback and content data, different other types of knowledge sources can also be taken into account in the recommendation process. In so-called "demographic" approaches, information about the user's age, sex, education, or income group can be factored into the algorithms. Similarly, the user's personality, the hobbies or general interests and lifestyle might be relevant. Besides such demographic and "psychographic" systems, in e-commerce settings, knowledge-based approaches can be found. So-called critiquing-based systems are an example of such systems, where the user can interactively state and revise the requirements with respect to a certain set of item features. Other knowledge-based systems use explicit domain rules to match user requirements with product features. These types of systems however only play a minor role in music recommendation.

Besides the above-mentioned user-provided tags for resources, Social Web platforms can serve as a source for further knowledge to be exploited for music recommendation. One can, for instance, try to interpret direct friendship relations in a social network as an indicator of a user's trust in the recommendations of another person and amplify the neighborhood weight for such users in a collaborative approach. Explicit "Like" statements for certain artists or songs represent another natural source to learn about a user's musical preferences.

Finally, the incorporation of information about the user's current context appears to be a particularly important aspect for the music recommendation task. The term

---

[11]See http://en.wikipedia.org/wiki/Music_Genome_Project for details and examples of *genes*. Accessed 22 June 2016.

context may refer to user-independent aspects such as the time of the day or year but also to user-specific ones such as the current geographic location or activity. In particular, the second type of information, that is, including the information about whether the user is alone or part of a group, becomes more and more available thanks to GPS-enabled smartphones and corresponding Social Web applications.

Since the different basic recommendation techniques (e.g., collaborative filtering or content-based filtering) have their advantages and disadvantages, it is a common strategy to overcome limitations of the individual approaches by combining them in a hybrid approach. When, for example, a new user has only rated a small number of items so far, applying a neighborhood-based approach might not work well, because not enough neighbors can be identified who have rated the same items. In such a situation, one could therefore first adopt a content-based approach in which one single item rating is enough to start and switch to a CF method later on, when the user has rated a certain number of items. In [6], Burke identifies seven different ways that recommenders can be combined. Jannach et al. in [20] later on organize them in the following three more general categories:

- *Monolithic designs:* In such approaches, the hybrid system consists of one recommendation component which pre-processes and combines different knowledge sources; hybridization is achieved by internally combining different techniques that operate on the different sources (Figure 23.1).

- *Parallelized designs:* Here, the system consists of several components whose *output is aggregated* to produce the final recommendation lists. An example is a weighted design where the recommendation lists of two algorithms are combined based on some ranking or confidence score. The above-mentioned "switching" behavior can be seen as an extreme case of weighting (Figure 23.2).

- *Pipelined designs:* In such systems, the recommendation process consists of multiple stages. A possible configuration could be that a first algorithm pre-filters the available items which are then ranked by another technique in a subsequent step (Figure 23.3).

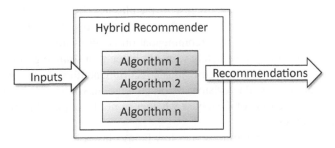

Figure 23.1: Monolithic hybridization design; adapted from [20].

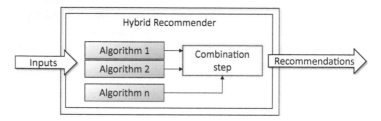

Figure 23.2: Parallelized hybridization design; adapted from [20].

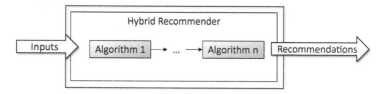

Figure 23.3: Pipelined hybridization design; adapted from [20].

23.3   Specific Aspects of Music Recommendation

The recommendation techniques discussed so far were typically not designed to work only for a certain class of products and can thus be applied to a variety of domains. There are, however, specific aspects in music recommendation which are slightly different from other domains and have to be considered in the specific system design. In particular, a number of works are devoted to the problem of "playlist generation", which can be considered a special form of music recommendation.[12]

Consider the following list of aspects, which is based on Paul Lamere's talk at the *ACM Recommender Systems 2012* conference "I've Got 10 Million Songs in My Pocket. Now What?" [25]; see also [5, 8].

- *Consumption-related aspects:* First, the recommended items can be either "consumed" immediately or not. In case of the classical book recommender of *Amazon.com*, the delivery of a book needs a couple of days. Thus, the current context of the user at the time of the recommendation is not as important as in situations where the customer immediately wants to listen to a song, e.g., in the case of an online radio station. In that sense, music recommendation shares similarities with video streaming (or IP television) recommendation, where considering the user's context (e.g., the time of the day or whether or not he enjoys the video alone or in a group) is crucial.

   Track consumption time is very low (songs last a few minutes). The systems thus have to generate a lot of recommendations. For instance, most users can listen to more than 20 songs a day, while they rarely read more than 20 books a year.

   A user can play the same song a hundred times, while, e.g., movies are more

---

[12]See [5] for an in-depth review of approaches for music playlist generation.

rarely watched again and again. Familiarity is a very specific and important feature of music. Users usually like to discover some new tracks, but at the same time like to listen to the tracks with which they are familiar. A very specific compromise thus exists between familiarity and discovery for user satisfaction.

Songs are often consumed in sequence. It is important that successive songs form a smooth transition regarding the mood, tempo or style. A good playlist thus not only balances possible quality features like coherence, familiarity, discovery, diversity and serendipity, but also has to provide smooth transitions.

Finally, in contrast to many other recommendation domains, tracks can be consumed when doing other things. One can listen to music while working, studying, dancing, etc., and each type of activity fits best with a different musical style.

- *Feedback mechanisms:* With respect to user feedback and user profiling, the *consumption times* (listening durations), track skipping actions, and volume adjustments can be used as implicit user feedback in the music recommendation domain. On some music websites, users can furthermore actively "ban" tracks in order to avoid listening to tracks which they actually do not like. At the same time, the *consumption frequency* can be used as another feedback signal. This repeated "consumption" of items seems to be particularly relevant for music, because it is intuitive to assume that the tracks that the users play most frequently are the tracks that the users like the most. This information about repeated consumption can also be used in other domains like web browsing recommendation, but it seems to be less relevant for these types of applications [13, 21].

    Another typical feature of many music websites is that their users can create and share playlists. Many users create such playlists,[13] and the tracks in these playlists usually correspond to the tracks the users like. Playlists can therefore represent another valuable source for an RS to improve the user profiles.

- *Data-related aspects:* Music recommendation deals with very large item spaces. Music websites usually contain tens of millions of tracks. Moreover, other kinds of musical resources can also be recommended, like for instance concerts. Some of these resources should, however, not appear in recommendation lists, for example karaoke versions, tribute bands, cover versions, etc.

    Furthermore, the available music metadata is in many cases noisy and hard to process. Users often misspell or type inappropriate metadata, as for instance "!!!" as an artist's name. At the same time, dozens of bands, artists, albums, and tracks can have identical names, which not only makes the interpretation of a user query challenging, but can also lead to problems when organizing and retrieving tracks based on the metadata.

- *Psychological questions and the cost of wrong recommendations:* From a psychological perspective, music represents a popular means of self-expression. With respect to today's Social Web sites, the question arises if one can trust that all the positive feedback statements on such platforms are true expressions of what users think and what they really like. Additionally, in the music domain, there exists a

---

[13] About 20% of *Last.fm* users have created at least one playlist.

certain number of "purist" enthusiasts. For such users, music recommendations have to be made very carefully and homogeneity with respect to the musical style and to the artists might be important.

23.4 Evaluating Recommender Systems

One challenging question after developing a novel recommendation algorithm or deploying a new music recommendation service is how to assess the quality or usefulness of the recommendations. In a system with real users, the way we measure the service's effectiveness depends on the underlying (business) goal and model. A typical evaluation setting would consist of conducting A/B tests. In such a test, the user community is split into two or more groups and each group receives recommendations using different algorithms. Based on a defined success metric, we can then compare, for example, how long users stay on the site, how many songs they skip, how many songs they download, how often they return, etc. An example of such an A/B test where different recommendation strategies were compared – although in a different domain – can be found in [18].

23.4.1 Laboratory Studies

In research and academic settings, unfortunately, such A/B tests can seldom be conducted as typically no real-world system is available with which such experiments could be made. Researchers therefore often rely on laboratory studies, which usually consist of a few dozen of participants and in which certain aspects of a recommendation system are analyzed.

Let us assume that the design of the recommender system's user interface has an effect on the perceived quality of the recommendations. To that purpose we can design a controlled experiment to test the hypothesis in which we let the subjects interact with a prototype system with two different interfaces. The possible experimental designs include *between-subject* and *within-subject*. In a between-subject design, each subject (participant) receives recommendations through only one of two implemented interfaces. In a within-subject design, each user will see both of them. When the subjects have ended their interactions with the system, they are asked, via a questionnaire, how they liked the recommendations (and other aspects) of the system.

Based on the answers of the participants, we can then check if any of the observed differences between the groups are statistically significant and support our initial hypothesis. Note that in some experiment designs the user behavior during the interaction, e.g., the listening times, can be automatically monitored or logged. In other designs, users are asked to think aloud when interacting with the system.

23.4.2 Offline Evaluation and Accuracy Metrics

Unfortunately, laboratory studies are costly, time-consuming and sometimes hard to reproduce. Since the participants of such studies are often students, they do not form a truly representative sample of a real population. Recommender systems research is therefore mostly based on offline experiments based on historical data sets. These datasets usually contain a set of user ratings for items (or purchase transactions or other forms of implicit feedback), which were collected on some real-world online platform.

The most common evaluation setting in recommender system research is based on the prediction of the relevance of a set of (hidden) items for a certain user, the application of different accuracy or ranking measures, and the repetition of the measurements using cross-validation, as described in Section 13.2.3 in Chapter 13. When the goal is to predict the value of the hidden ratings, the Mean Absolute Error (MAE) and the Root Mean Squared Error (RMSE) can be used; see Chapter 13. If the goal is to produce top-n recommendation lists, precision and recall are often applied.

While precision and recall represent the most popular evaluation metric in RS according to the study in [17], the absolute numbers reported in research papers should be considered with care as they might not reflect the "true" values very well. As mentioned in the survey paper on RS evaluation by Herlocker et al. [16], in RS evaluation scenarios the so-called ground truth for most user-item ratings is not known. Considering only the known ratings of the test set for the calculation of precision and recall results in unrealistically high values. In addition, it is not always clear how the set of relevant items is determined from the rating information.

Another shortcoming of classification accuracy metrics is that they only count the number of hits in a recommendation list but not at which position in the list the hits have been found. Intuitively, the relevant items should be placed on the top of the list as they have a higher chance to get the attention of the user. Therefore, ranking measures[14] are often applied in the IR field that take the position of an item into account. An example of such a measure is the "discounted cumulative gain" (DCG), which is applicable mostly for non-binary notions of relevance [26]. The cumulative gain (CG) corresponds to the sum of the relevance weights (ratings) of the items up to a certain list length. The idea of the DCG is to reduce the relevance value of items appearing later in the list, usually by a logarithmic factor. Let Rel_j be the relevance score for an item at position j (based on the known rating). The DCG of a ranked list of length k can be calculated as follows:

$$DCG_k = Rel_1 + \sum_{j=2}^{k} \frac{Rel_j}{log_2(j)}.$$ (23.6)

Variations of this scheme, e.g., concerning the logarithmic base, are also common in the literature. Usually, the DCG is also normalized and divided by the score of the "optimal" ranking so that finally the values of the normalized DCG lie between zero and one.

Note that other domain-specific or problem-specific schemes are possible. In the *2011 KDD Cup*,[15] the task was to separate highly rated music items from non-rated

---

[14] See also Chapter 9.

[15] http://www.kdnuggets.com/2011/02/kdd-cup-2011-recommending-music.html. Accessed 22 June 2016.

items given a test set consisting of six tracks, out of which 3 were highly rated and 3 were not rated by the user.

23.4.3 Beyond Accuracy: Additional Quality Factors

The above-mentioned accuracy metrics are relatively easy to measure given a set of historical rating data. The question, however, arises whether they really represent the best indicators for the quality of a recommender system. Consider, for example, a music recommender which has detected that all Rolling Stones songs have been rated very highly by a user. As the system is trained to minimize the prediction error, it would then probably recommend even more Rolling Stones tracks to him. However, presenting the user a list full of Rolling Stones songs might not represent a good recommendation even though the prediction was actually good, e.g., in terms of RMSE. Thus, such a list might be boring and not at all surprising for the user. In addition, such a recommender would probably also only focus on popular items (which are often rated highly), which leads to a possibly undesired effect that the major part of songs in the music collection will never be placed in a recommendation list. Therefore, in recent years, measures other than accuracy began to gain increasing attention in the research community.

23.4.3.1 Coverage, Cold Start, Popularity, and Sales Diversity

With the term coverage, either "user-space coverage" or "item-space coverage" can be meant in the literature of RS. User (-space) coverage is a measure that describes for how many of the known users a recommender system is capable of making a (useful) recommendation. When considering the basic neighborhood-based method described in Section 23.2.1, the system might be configured in a way that requires at least N neighbors whose similarity level exceeds a certain threshold. Thus, not for all users – in particular those who have only rated very few items or have a niche taste – recommendations will be calculated because the system's confidence in the recommendations might be too low. The cold start behavior of an RS is also related to coverage and can, for example, be measured by calculating user coverage and/or the predictive accuracy at different (artificially created) data set sparsity levels.

Item-space coverage (or catalog coverage), on the other hand, typically refers to the question of how many of the existing items can be, or more importantly are, actually ever recommended to users. Item coverage can be measured both in offline as well as in online experiments, for example, by analyzing how often each catalog item actually appeared in the first n elements (top-n) of the recommendation lists presented to the users.

Item coverage is also related to sales diversity and the popularity-bias of recommender systems. In [8], Celma discusses the skewed distribution of item popularity and the corresponding music "long tail" in detail. An example of such a long tail distribution is shown in Figure 23.4.

On the x-axis, the items (songs) are sorted according to their popularity, which is measured in terms of, e.g., the playcount on an online platform such as *Last.fm*,

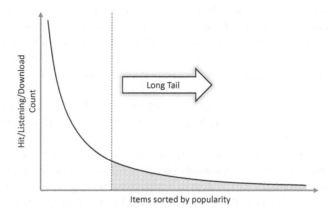

Figure 23.4: Long tail distribution.

sales or download numbers, or, when the goal is to measure the diversity of the recommendations, the number of appearances of a song in a recommendation list.

The term long tail refers to the fact that in many domains – and in particular the music domain – some very few popular items account for a large amount of the sales volume. In [8], Celma cites the numbers of a report from 2007 about the state-of-the-industry in music consumption, where 1% of all available tracks were reported to be responsible for about 80% of the sales or that nearly 80% of about 570,000 tracks were purchased fewer than 100 times.

Given this skewed distribution toward popular songs and mainstream artists, it could therefore be – according to marketing theory – a goal to increase sales of items in the long tail. Recommender systems are one possible method to achieve such a goal and studies such as [38] or [12] have analyzed how recommenders impact the buying behavior of customers and the overall sales diversity. On the one hand, one can observe that in some domains a recommender can help the user to better explore the item space and find new items he or she was not aware of. On the other hand, there is a danger that depending on the underlying strategy and algorithm, the usage of a recommender system can lead to the undesired effect of further boosting already very popular items as recommending blockbusters to everyone is a comparably safe strategy.

23.4.3.2 List Diversity, Novelty, Serendipity, and Familiarity

Besides the global sales diversity of a platform, the diversity of recommendations for an individual user can be an important quality factor for the customer. When given a Beatles seed song on an online music platform, recommending a playlist of 10 other Beatles songs or, even worse, 10 cover versions of the same song, might be technically plausible but perhaps not what the user would enjoy. Therefore, it is often advisable to make sure that the recommendation list (playlist) is not monotonous and that the items are not too similar to each other.

A possible strategy could therefore be to try to include items in the recommendation list that increase the diversity of the list, items that are supposedly novel for the user, as well as items which are to some extent surprising (but relevant) for the user. In order to control diversity, one needs a measure of item similarity. When following a content-based recommendation approach as described in Section 23.2.2, the underlying similarity metric can be used to compare two tracks. Alternatively, one could use metadata such as genre, artist, etc. Based on such a measure, an overall diversity metric for a list can be derived, e.g., by computing and aggregating the pairwise similarities.

Another related criterion is novelty. Novelty can be measured in user studies via a questionnaire. Some researchers also propose offline evaluation approaches which use the general popularity of an item to estimate the novelty of a recommendation list, assuming that highly popular items are not novel to the user. Alternatively, schemes that use the time stamps of the ratings to assess the novelty of an item are possible, see [33].

Serendipity is also often mentioned as a desirable playlist characteristic. This concept is often referred to as a measure of how surprising and unexpected, but accurate, the recommendations are. When a user is pointed through a recommendation list to a track of a genre, style, or artist she or he usually does not particularly like, and the user finds that he likes the track, then the recommendation can be considered as being serendipitous. This also corresponds to valuable recommendations given by friends or music enthusiasts. Serendipity can also be measured in a user study and approximated in offline experiments based on the similarity of items and the deviation of recommendations from obvious recommendations. Providing only serendipitous recommendations can, however, be dangerous, and it is also important that the recommendations include a set of items the user is familiar with as these items can help to increase the user's trust in the system.

An example of a user study on novelty and familiarity that includes 288 participants can be found in [8]. In that study the users were asked to rate recommended songs based on 30-second excerpts. The recommendations included tracks the users already knew and tracks which were novel to them. One of the observed results was that users rated the songs they knew much higher and the perceived quality of the system increased when familiar songs were recommended. One of the insights and conclusions of the study therefore was that a recommender should provide additional contextual information such as an explanation as to why a certain song had been recommended.

23.4.3.3 *Adaptivity, Scalability, and Robustness*

The quality measures discussed so far focus mainly on the utility of the recommendations for an individual user or a service provider. Also, other aspects can be relevant for the practical success of a recommender system.

Adaptivity refers to the capability of an RS to quickly adapt the recommendations based on very recent events. Such an event could, for instance, be that a track becomes popular overnight, e.g., due to its usage in a popular TV advertisement or its relation to some other event. In some domains such as news recommendation, items can become outdated very quickly as well. Another perspective on adaptivity is the system's rate of taking changes and additions in the user profile into account. When users rate tracks, they might expect that their preferences are immediately taken into account, which might not be the case if the underlying algorithm is based on a computationally expensive training phase and models are only updated, e.g., once a day.

Scalability is a characteristic of recommender systems which is particularly relevant for situations where we have to deal with millions of items and several million users as is the case in music recommendation. Techniques such as nearest-neighborhood algorithms do not scale even to problems of modest size. Therefore, only techniques which rely on offline pre-computation and model-building work in practice.

Robustness typically refers to the resistance of a system against attacks by malevolent users who want to push or "nuke" certain artists. Recent works such as [28] have shown that, for example, nearest-neighbor algorithms can be vulnerable to various types of attacks, whereas model-based approaches are often more robust in that respect.

23.5 Current Topics and Outlook

In this chapter we discussed the basic techniques for building and evaluating music recommender systems, which have already been successfully implemented in various domains. In this final section, we will briefly discuss three topics which have attracted increased interest in the recommender systems research community outside the music recommendation field: context-awareness, the incorporation of Social Web information in the recommendation process, and sequential recommendation.

23.5.1 Context-Aware Recommendation

When the recommended music is "consumed" immediately, the current situation or context of the listener can be of extreme importance. You might, for example, be interested in different types of music depending on the time of the day or depending on what you are currently doing. In the morning, on the way to work, you might enjoy a different type of music than when doing sports. But the term *context* in music recommendation can have even more facets. It can be the social environment (e.g., being part of a group or alone, one's geographical location, and even the current weather and your current emotional state). All these aspects may influence what type of music will be most appropriate as a recommendation. The context may finally refer to general and non-personal characteristics such as the time of year (think of recommending Christmas songs in May) or very specific ones like the set of tracks which you have been listening to during the current session.

When thinking about our main types of recommendation approaches (collaborative filtering and content-based filtering), the intuition is that CF methods are more suited to taking context into account. For instance, a CF-based recommender that takes into account what your friends have been recently listening to can help to alleviate at least some of the problems. On the contrary, traditional content-based methods are, for instance, not always capable of taking the cultural background of a track into account, except for cases in which cultural information can be extracted from tags or metadata annotations.

Kaminskas and Ricci classify the possible contextual factors in three major groups [22]:

- environment-related context, e.g., location, time or weather;
- user-related context, e.g., the current activity, demographical information (even though this can be considered part of the user profile, and provides information about the environment), and the emotional state; and
- multimedia context, which relates to the idea of combining music with other corresponding resources such as images or stories.

In the same work, the authors review a set of context-aware prototypical music recommenders and experimental studies in this area. They conclude that research so far is "data-driven" and that researchers often tend to fuse given contextual information into their machine learning techniques. Instead, the authors advocate a "knowledge-based" approach, where expertise, e.g., from the field of psychology, about the relationship between individual contextual factors and musical perception is integrated into the recommendation systems.

23.5.2 Incorporating Social Web Information

The emergence of what is called "Web 2.0" has dramatically changed how we behave in the online world. We are no longer pure consumers of edited content but we actively annotate, "like" and comment items on resource-sharing platforms, we voluntarily post information about our current situation, mood or interests on the Social Web, review items on e-commerce sites, or even write our own (micro-)blogs.

Given the discussions above, e.g., on contextual parameters that influence what music should be recommended, it is obvious that the Participatory Web opens new opportunities for music recommendation. On the one hand, more information about the users, in particular their preferences, tastes and current state, can be found online; on the other hand – through social annotations and tags – more information about the tracks in the catalog becomes available.

In recent years, the exploitation of social tagging data in the recommendation process was one of the key topics in RS research. In the context of music recommendation – but also in other domains – user-contributed tags can be simply seen as additional content information and, on principle, standard CB methods could be applied. However, in contrast to automatically extracted or manually annotated metadata, tagging data often contain lots of noise. While some tags for a certain track such as "classic jazz" or "dance music" carry potentially valuable information about the song, users may also tag an album in their personal and subjective view, e.g., with tags like "own it". Other problems with tags include malicious users who add various types of unusable or noisy tags to the data.

In [36], various methods for acquiring high-quality tags and addressing the problem of no common vocabulary is proposed; see also [22]. The methods include small-scale data collection within a defined user group and vocabulary, harvesting social tags from online music sites, and using tagging games or different strategies to automatically mine tags from other web sources. A complementary approach to harmonizing the vocabulary on tagging-enabled platforms is the use of "tag-recommenders". Such recommenders can already be found on today's resource sharing platforms such as *delicious.com* and make tagging suggestions to the users based on RS technology.

As discussed in [24], user-provided tags may carry different types of valuable information about tracks such as genre, mood or instrumentation that can help to address some challenging tasks in Music Information Retrieval such as similarity calculation, clustering, (faceted) search, music discovery and, of course, music recommendation. The major research challenges include however the detection and removal of noise and, as usual, cold start problems and the issue of lacking data for niche items.

23.5.3 Playlist Generation

As mentioned in Section 23.3, music is typically played in a sequential manner, according to playlists. This means that the relationships between the successive tracks are important. For instance, the transitions between the tracks may be important, as well as the overall diversity of the recommended tracks, the general topics or themes, the musical path from the first track to the last track, etc. For that reason, we can often consider the music recommendation problem as a playlist generation problem, i.e., the recommendation of an ordered collection of tracks.

Although playlist generation can be considered as a special case of track recommendation, it goes slightly beyond, as a playlist itself can be considered as an artistic resource. Moreover, as users like to listen to tracks they already know and rarely want to discover one single track at a time, providing static recommendation lists as done, for instance, for movies is not often relevant. For that reason, several track recommendation scenarios actually correspond to a form of playlist generation:

1. Simple playlist generation: a seed track is selected by the user, or some set of desired characteristics such as a minimum tempo, a set of genres, etc., and a whole playlist is generated. This scenario is used by iTunes Genius and the playlist generation service of The Echo Nest.

2. Repeated track recommendation for playlist construction: each time the user adds a track, a new list of recommended tracks is provided, and the user can use this list to add a new track to the playlist. This scenario was used in the Rush application [3].

3. Radios: recommended tracks are automatically played one after the other and the user can only skip them if he or she does not want to listen to them. This scenario requires little user effort and is the most frequent in commercial platforms (Spotify, Pandora, Last.fm, etc.).

Playlist generation has been part of the research literature since the early 2000s [29] but did not attract much attention until recently. Among the recent work, [15] proposed a frequent pattern mining approach where patterns of latent topics are extracted from user playlists and then used to compute recommendations. The authors of [14] exploited long-term preferences including artists liked on Facebook and usage data on Spotify to build playlists which consist of the most popular tracks of the artists who are the most similar to the artists the user likes. A statistical approach was presented in [27], where random walks on a hypergraph are used to iteratively select similar tracks. In the same spirit, [11] proposed a sophisticated Markov model in which tracks are represented as points in the Euclidean space and transition probabilities are derived from the corresponding Euclidean distances. The coordinates of the tracks are learned using a likelihood maximization heuristic. The authors of [19] went further by taking into account the characteristics of the tracks that are already in the playlist, and proposed a heuristic which tries to mimic these characteristics in the generation process.

23.6   Concluding Remarks

The way we consume music has dramatically changed during the last decade. Today, millions of tracks are instantly available through an ever-increasing number of online music services. Finding suitable music for a certain listening situation or discovering new music becomes more and more challenging given the millions of songs which are available for instant download. Music recommenders help users in different ways, e.g., discovering new songs or artists, exploring the catalog, creating personalized playlists, or recommending music that corresponds to the situation of the user.

23.7   Further Reading

In this chapter we introduced the basic techniques for building such systems, which have been successfully applied in industry in various domains over the last decade. We provided short explanations of several such techniques. For a comprehensive overview of other methods, see for instance [1, 23, 33, 20].

We have shown that music recommendation has some particularities which have to be taken into account. While some of them – like certain types of context – are already addressed in current research, we believe that in particular, the psychological aspects of music perception must be more carefully considered in future research on music recommendation. For an overview of context-aware music recommendation, see [22]. For an overview of context-aware recommendation in general, see also [2].

We also introduced some of the basic techniques for evaluating the recommendations and pointed out some limitations of the current evaluation strategies. More details about state-of-the-art user-centric evaluation procedures can be found in [31]. Further questions of experimental design, measurement and analysis, which are common in the social sciences, are covered in detail in [30], and comprehensive overviews of the common offline evaluation schemes for recommender systems can be found in [16, 34].

Bibliography

[1] G. Adomavicius and A. Tuzhilin. Toward the next generation of recommender systems: A survey of the state-of-the-art and possible extensions. *IEEE Transactions on Knowledge and Data Engineering*, 17(6):734–749, 2005.

[2] G. Adomavicius and A. Tuzhilin. Context-aware recommender systems. In F. Ricci, L. Rokach, B. Shapira, and P. B. Kantor, eds., *Recommender Systems Handbook*, pp. 217–253. Springer, 2011.

[3] D. Baur, S. Boring, and A. Butz. Rush: Repeated Recommendations on Mobile Devices. In *Proc. IUI*, pp. 91–100, New York, NY, USA, 2010. ACM.

[4] E. Bernhardsson. Systems and methods of selecting content items using latent vectors, August 18 2015. US Patent 9,110,955.

[5] G. Bonnin and D. Jannach. Automated generation of music playlists: Survey and experiments. *ACM Computing Surveys*, 47(2):1–35, 2014.

[6] R. Burke. Hybrid recommender systems: Survey and experiments. *User Modeling and User-Adapted Interaction*, 12(4):331–370, 2002.

[7] M. Burns. The Pandora One subscription service to cost $5 a month, `http://techcrunch.com/2014/03/18/the-pandora-one-subscription-service-to-cost-5-a-month/`, 2014, accessed February 2016.

[8] Ò. Celma. *Music Recommendation and Discovery - The Long Tail, Long Fail, and Long Play in the Digital Music Space*. Springer, 2010.

[9] Ò. Celma and P. B. Lamere. Music recommendation tutorial. ISMIR'07, `http://ocelma.net/MusicRecommendationTutorial-ISMIR2007/slides/music-rec-ismir2007-low.pdf`, September 2007, accessed February 2016.

[10] Ò. Celma and X. Serra. FOAFing the music: Bridging the semantic gap in music recommendation. *Journal of Web Semantics*, 6(4):250–256, 2008.

[11] S. Chen, J. Moore, D. Turnbull, and T. Joachims. Playlist prediction via metric embedding. In *Proc. KDD*, pp. 714–722, New York, NY, USA, 2012. ACM.

[12] D. M. Fleder and K. Hosanagar. Recommender systems and their impact on sales diversity. In *Proceedings of the 8th ACM Conference on Electronic Commerce (EC'07)*, pp. 192–199, New York, NY, USA, 2007. ACM.

[13] S. Fox, K. Karnawat, M. Mydland, S. Dumais, and T. White. Evaluating implicit measures to improve web search. *ACM Transactions On Information Systems (TOIS)*, 23(2):147–168, 2005.

[14] A. Germain and J. Chakareski. Spotify me: Facebook-assisted automatic playlist generation. In *Proc. MMSP*, pp. 25–28, Piscataway, NJ, USA, 2013. IEEE.

[15] N. Hariri, B. Mobasher, and R. Burke. Context-aware music recommendation based on latent topic sequential patterns. In *Proc. RecSys*, pp. 131–138, New York, NY, USA, 2012. ACM.

[16] J. L. Herlocker, J. A. Konstan, L. G. Terveen, and J. T. Riedl. Evaluating collaborative filtering recommender systems. *ACM Transactions on Information Systems (TOIS)*, 22(1):5–53, 2004.

[17] D. Jannach, M. Zanker, M. Ge, and M. Gröning. Recommender systems in computer science and information systems: A landscape of research. In *Proc. EC-WEB 2012*, Heidelberg / Berlin, 2012. Springer Verlag.

[18] D. Jannach and K. Hegelich. A case study on the effectiveness of recommendations in the mobile Internet. In *Proceedings of the 2009 ACM Conference on Recommender Systems (RecSys'09)*, pp. 41–50, New York, NY, USA, 2009. ACM.

[19] D. Jannach, L. Lerche, and I. Kamehkhosh. Beyond hitting the hits: Generating coherent music playlist continuations with the right tracks. In *Proceedings of the 9th ACM Conference on Recommender Systems*, pp. 187–194, New York, NY, USA, 2015. ACM.

[20] D. Jannach, M. Zanker, A. Felfernig, and G. Friedrich. *Recommender Systems: An Introduction*. Cambridge University Press, 2011.

[21] N. Jones, P. Pu, and L. Chen. How Users Perceive and Appraise Personalized Recommendations. *User Modeling, Adaptation, and Personalization (UMAP 2009)*, pp. 461–466, 2009.

[22] M. Kaminskas and F. Ricci. Contextual music information retrieval and recommendation: State of the art and challenges. *Computer Science Review*, Vol. 6(23):89–119, 2012.

[23] Y. Koren, R. Bell, and C. Volinsky. Matrix factorization techniques for recommender systems. *Computer*, 42(8):30–37, 2009.

[24] P. B. Lamere. Social tagging and music information retrieval. *Journal of New Music Research*, 37(2):101–114, 2008.

[25] P. B. Lamere. I've got 10 million songs in my pocket: Now what? In *Proceedings of the Sixth ACM Conference on Recommender Systems*, RecSys '12, pp. 207–208, New York, NY, USA, 2012. ACM.

[26] C. D. Manning, P. Raghavan, and H. Schütze. *Introduction to Information Retrieval*. Cambridge University Press, 2008.

[27] B. McFee and G. Lanckriet. Hypergraph models of playlist dialects. In *Proc. ISMIR*, pp. 343–348, 2012.

[28] B. Mobasher, R. Burke, R. Bhaumik, and C. Williams. Toward trustworthy recommender systems: An analysis of attack models and algorithm robustness. *ACM Transactions on Internet Technology*, 7(4):23, 2007.

[29] F. Pachet, P. Roy, and D. Cazaly. A combinatorial approach to content-based music selection. *Multimedia*, 7(1):44–51, 2000.

[30] E. Pedhazur and L. Schmelkin. *Measurement, Design, and Analysis: An Integrated Approach*. Lawrence Erlbaum Associates, 1991.

[31] P. Pu, L. Chen, and R. Hu. Evaluating recommender systems from the user's

perspective: Survey of the state of the art. *User Model. User-Adapt. Interact.*, 22(4-5):317–355, 2012.

[32] P. Resnick, N. Iacovou, M. Suchak, P. Bergstorm, and J. Riedl. Grouplens: An open architecture for collaborative filtering of netnews. In *Proceedings of the 1994 ACM Conference on Computer Supported Cooperative Work (CSCW'94)*, pp. 175–186, New York, NY, USA, 1994. ACM.

[33] F. Ricci, L. Rokach, B. Shapira, and P. B. Kantor, eds. *Recommender Systems Handbook*. Springer, 2011.

[34] G. Shani and A. Gunawardana. Evaluating recommendation systems. In *Recommender Systems Handbook*, pp. 257–297. Springer, 2011.

[35] U. Shardanand and P. Maes. Social information filtering: Algorithms for automating word of mouth. In *Proceedings of the SIGCHI Conference on Human Factors in Computing Systems*, CHI '95, pp. 210–217, New York, NY, USA, 1995. ACM.

[36] D. Turnbull, L. Barrington, and G. R. G. Lanckriet. Five approaches to collecting tags for music. In *Proc. ISMIR 2008*, pp. 225–230, Philadelphia, 2008.

[37] M. Vacher. Introducing Discover Weekly: Your ultimate personalised playlist, `https://press.spotify.com/it/2015/07/20/introducing-discover-weekly-your-ultimate-personalised-playlist`, accessed February 2016.

[38] M. Zanker, M. Bricman, S. Gordea, D. Jannach, and M. Jessenitschnig. Persuasive online-selling in quality & taste domains. In *Proceedings of the 7th International Conference on Electronic Commerce and Web Technologies (EC-Web'06)*, pp. 51–60, Heidelberg / Berlin, 2006. Springer Verlag.

Chapter 24

Automatic Composition

MAIK HESTER, BILEAM KÜMPER
Institute of Music and Musicology, TU Dortmund, Germany

24.1 Introduction

While most chapters of this book deal with analyzing given music, this one gives an introduction to synthesis tasks called automatic (or, in a broader sense, algorithmic) composition. We start with a discussion of what a composer does and what a composition actually is (or may be). There are some suggestions as to why the act of composing could (or should) be automatic. After a short outline of historical examples, we are going to show some of the basic principles composing computers work with, giving a broad rather than an in-depth overview. Particular software applications are numerous as well as subject to constant change and therefore not in the scope of this chapter.

24.2 Composition

24.2.1 What Composers Do

In order to understand how computers compose music, one should first try to find out how humans compose music. As early as 1959, in their book on composing with computers, Hiller and Isaacson state that "the act of composing can be thought of as the extraction of order out of a chaotic multitude of available possibilities" [8, p. 1]. These possibilities include all kinds of sounds, at least within the range of human audibility. The choice of sonic material and its required features is therefore one of the basic compositional activities.

If a composer decides to work mainly with pitched sounds, the next decision affects the number of pitches to be used. The continuous pitch range of the sound spectrum can be reduced to a finite set of pitches or pitch classes (cf. Section 3.5). These are commonly known as scales. Similarly, the time continuum may be subdivided into units to form a grid of sound durations, which can then be combined to rhythms or meters (cf. Section 3.6). These grids are necessary to write down music in notes. The musical notes commonly used today are a symbolic language. They do not store the music itself, "but rather instructions for musicians who learn which actions to perform from these symbols in order to play the music" [19, p. 4]. Notes can only describe a few features of the sound; others (e.g. the amplitudes of its overtones) cannot be represented, but they can be composed as additional declarations to the notes. The invention of MIDI (cf. Section 7.2.3) made it possible for computers to process music notation, and the notes could then be interpreted by musicians as well as by computers. As it is obviously easier to program a computer to compose with a finite set of, say, twelve pitches (e.g. the chromatic scale) than with an infinite number of possible pitches, most composing software works rather with notes than with sounds, although this is not essential (cf. [2]). But computers can do much more: they can control the production of any requested sound features (e.g. duration, pitch, overtone spectrum) on a microscopic level, which leads to possibilities that even the symphonic orchestra with its variety of instruments cannot provide. "But the larger the inventory the greater the chances of producing pieces beyond the human threshold of comprehension and enjoyment. For instance, it is perfectly within our hearing capacity to use scales of 24 notes within an octave but more than this would only create difficulties for both the composer and listener" [19, p. 13]. Moreover, "one of the most difficult issues in composition is to find the right balance between repetition and diversity" [19, p. 13].

On the whole, composing could be regarded as a chain of decisions and in this way be compared to the structure of computer programs, but this analogy raises some more questions. Certainly a human composer does not make every single decision fully consciously. Many of them may be made unconsciously from adopted traditions and paragons perceived in the past. Original ideas and new combinations, or clichés and stereotypes may arise – intended or not. Other decisions are made as a result of previously chosen rules and styles or with regard to the intended listeners or markets. If you want to compose a twelve-tone piece or decide to write a pop song, you do not have to care about quarter tones or white noise. After all, the order in which decisions are taken is obviously not always the same. The ways of composing may go bottom-up, from a detailed idea to a large-scale work, or top-down, from an overall plan to the single sound. In most cases, there will be a permanent intersection between these directions.

24.2.2 Why Automatic Composition?

If composing is that complex, why should one try to model this process in likely imperfect computer programs? Historically, this modeling is exactly the first reason for trying it, and an important one for sure. In programming, one could learn something about how humans compose music. From the conversion of century-old rules for human composers into machine-readable algorithms, one would get even more knowledge regarding music composed by humans.

There are many reasons for composers to use computer programs and thus have part of their work done automatically, or for computer scientists and musicologists to explore these programs. Conventionally, music is composed by a human being who applies traditional or self-made rules. As long as the music simply does what the composer wants it to do, there is little surprise in the resulting sound. But what if the composer tries to stand aside and "let the music do what *it* wants to do" or even "to let the music compose itself" [12, p. 2]? When Steve Reich lets microphones swing over loudspeakers, when Paul Panhuysen stretches long wires across a lake or Peter Ablinger plants a row of trees in the open landscape, the composer seems to vanish behind his work, which could somehow hardly be called a piece, and the music, as a result, is rather found than composed. "In all these cases the 'composers' are [...] simply letting music arise out of circumstances that they can not personally control" [12, p. 2]. This kind of music is not intended to be performed on traditional musical instruments and therefore there are no conventional notes. This kind of music has mainly to do with sound. Composers like Tom Johnson, on the other hand, try to find music in mathematical objects like Pascal's triangle. These objects deliver a set of numbers which the composer can transform into music, mapping them to concrete pitches and rests to make their inherent structure audible. This kind of music can be performed on traditional musical instruments, and it has mainly to do with notes (cf. [13]). Likewise, algorithmic composition is divided into two domains, generating scores or generating sounds. The generated scores may be written out in notes so that musicians can perform them on classical instruments while the generated sounds can only be heard through loudspeakers (cf. [24]).

Computers can be used at any step during the process of composing. They can perform precise calculations just to spare the composer time and effort. They can help to learn rules typical of a certain time period, musical style or composer's handwriting. Or they can assist in creating an automatic accompaniment or even completely new music based on a mixture of rules from different genres or periods. Moreover, computer programs could be set up to compose automatically or even autonomously. In doing so, computers can serve as supporting devices, they can produce compositions or meta compositions (cf. [1]). "In its purest form, (computer-based) algorithmic music is the output of a stand-alone program, without user controls, with musical content determined by the seeding of the random number generator [...]. Most systems allow for some sort of control, however, through inputs to the algorithm, or live controls to a running process. Interactive music systems take advantage of algorithmic routines to produce output influenced by their environments, while live coders burrow around inside running algorithms, modifying them from within" [1, p. 300]. One application could be live composition in real time, interactive or on demand, for computer games or web sites, to influence the user's current mood or to create a new experience (cf. [26]).

Algorithmic composition makes use of mathematical statements and production rules, which are used within (computer) programs (cf. [28]). Algorithms may be influential from the micro to the macro structures. They can cover the whole range from the generation of sounds or notes over aspects like timbre or expression to the musical form (cf. [28]). If an algorithm models traditional composition rules the result will also sound traditional. Inventing new algorithmic rules or taking over those from outside the realm of music, on the other hand, can lead to completely new musical forms and structures (cf. [24]). Adapting algorithms from the natural sciences may simply arise from the curiosity to find new means of musical expression. But in a Platonic sense, it is also about investigating and explaining the world around us in musical terms, and thus is linked with historical concepts like the music of the spheres. "In any case, the modeling of natural processes requires the use of computers owing to the large number of mathematical calculations needed. The computer is not only used as a compositional facilitator – it becomes a necessity" [24, p. 53].

24.2.3 A Short History of Automatic Composition

The use of algorithms in composition is neither new in the history of music nor is it restricted to computers. A very early example often cited is Guido d'Arezzo's method (1026) to find melodies corresponding to vowels in a given text (see, e.g. [18]). Later, particularly polyphonic music such as isorhythmic motets or complex canons tend to show some kind of "computational thinking in music" [6]. Even some pieces employing random processes exist in earlier history; Mozart's dice game (1793) became one of the most famous (see [21]). The second Viennese school around Arnold Schoenberg in the 1920s and the serialism which was derived from Schoenberg's approach in the 1950s as well as John Cage's principle of indeterminacy as a counter-movement to serialism were further important milestones on the way to automatic composition (cf. [19]).

While some ingenious people have been experimenting with (mechanical) automation in composition since the 17th century (cf. [21]), electronic computers made things easier from the 1950s on (cf. [21]). Harry F. Olson and Herbert F. Belar, the inventors of the *Olson-Belar Composing Machine* (1950), tried to find new melodies in the style of Stephen Foster. After observing relative frequencies of consecutive notes and groups of notes in several Foster songs (such as "Oh Susannah"), they programmed the machine to find new ones with the same distribution (cf. [10]). In 1957 Lejaren Hiller and Leonard Issacson wrote the *ILLIAC Suite*, named after the University of Illinois' *ILLIAC* computer and commonly called the very first "real piece" composed by a computer (cf. [8]).

Around the same time, the Greek composer and architect Iannis Xenakis began to work with stochastic distributions of sound in his music, controlling not the single sound, but the density of sound masses. While starting to calculate manually, he soon developed computer programs to help himself (cf. [27]). For instance, he made use of the Maxwell–Boltzman distribution in his piece *Pithoprakta*. In mapping the movements of individual gas molecules to the glissando movements of string instruments he was able to create a music "in which separate voices cannot be discerned, but the shape of the sound mass which they generate is clear" [28].

Some effort has been made to generate music in historic styles, modeling rules or patterns of the paragons. David Cope's work *Experiments in Musical Intelligence* has attracted attention since the 1990s and still lead to CD productions in different styles, while the Turkish composer Kemal Ebcioğlu built a system to compose chorales sounding like those of J.S. Bach, using a large number of rules (cf. [5]).

A quick search of the Internet today would show you dozens of automatic composition programs, often written using quite special techniques, such as fractals or evolutionary algorithms. Commercial software is available, e.g. for composing band accompaniments to a given melody or jazz solos over a chord scheme. After all, computer music languages that were written for studio or academic use decades ago have improved and are available for everybody nowadays.

24.3 Principles of Automatic Composition

24.3.1 Basic Methods

In computer-based composition, humans define a set of rules to achieve a certain aim. The computer follows these rules generating data which can be transformed into classical notes or directly into sounds (cf. [28]). According to the underlying processes, computer composition can be divided into five domains: deterministic, stochastic, rule-based or grammar-based processes, as well as methods of artificial intelligence.

In any case, there are some basic methods, which can be used generally: you could transform a structure by reversing or mirroring, transposing pitches or augmenting durations. These are methods of classical counterpoint as well as of serial music (cf. Section 3.2.1). Transposing and augmentation could be seen as adding a certain constant value to those values of the structure. The structure could be cut into several pieces and be recombined or permuted in different orders.

Another method is the mapping of external data to musical parameters. If you find an appropriate mapping, you could use whatever data you want, be it color data from JPG pictures, text data from Twitter messages, or topographic height structures of the earth, measured by satellites (cf. [9]).

24.3.1.1 Deterministic Processes

Composing with deterministic processes means that you have total control over the outcome. If you have the same input, your algorithm will always calculate the same result. Obviously this is not a proper model for the human composing process as a whole. On the other hand, the algorithm could compute much more data in the same time and it does so without errors. If you like "errors" or surprises in music, this approach is probably not to your taste.

If you compose with a finite set of values for a parameter (e.g. with a bunch of pitches), you could use *finite automata*. These iterative rules describe how to get more values out of a few starting values. The composer Tom Johnson, for instance, makes use of so-called L-systems (named after the biologist Aristid Lindemeyer). In biology, L-systems are used to describe the growth of plants by a restricted set of simple transformation rules like

$a \rightarrow b$
$b \rightarrow ab$

which result in rapidly growing self-similar structures:

a
b

ab
bab
abbab
bababbab and so forth.

These structures cannot be derived from traditional compositional rules (cf. [24] and [14]).

Bear in mind that one may map these variables not only to notes, but to any musical structures. Think for instance of the first two bars of a famous German children's song, as shown in Figure 24.1. This is just to keep things simple and continues a tradition of using folk songs as starting material for manageable experiments.[1]

Figure 24.1: The first two bars of "Hänschen klein".

Applying the above L-system to the motifs in Figure 24.1, the rather dull, however quite long melody[2] in Figure 24.2 results.

Figure 24.2: A simple L-system applied to the two motifs.

Of course there are *infinite automata* too, which go beyond those finite sets to infinity. If you apply them, for instance, to pitches, the music will at some point not be playable or even audible anymore, so you may want to stop the algorithm somewhere.

---

[1]Music pedagogue Fritz Jöde remarked in the 1920s: "Proceeding beyond this kind of folklore should only happen when the child has developed so far that its perceiving organs have grown far enough for further designs which exceed the simple folk-like structure that it could really perceive the substance" [11, translation by the authors, p. 108].

[2]If you want to work in this direction, compare Helmut Lachenmann's "Ein Kinderspiel" [17], which employs the same song.

These automata could, e.g., be fractal processes which have been adapted from chaos theory (cf. [20]). The term fractal was coined by Benoit Mandelbrot to describe self-similar patterns. Common examples are the growth of trees or snowflakes.

Cellular automata are another example of algorithms that have been adapted by composers from the natural sciences. Cellular automata have been designed to describe the behavior of particles in liquids. "Individual particles influence each other by knocking against one another or changing places, etc., thereby defining the movement of the fluid as a whole. The easily comprehensible, elementary effects within a cell's immediate area can have an unexpected effect on the system as a whole, because each cell can simultaneously influence and be influenced by several of its neighbors" [24, p. 52].

24.3.1.2 Stochastic Processes

In stochastic music, the events are determined by chance operations, i.e. we can use random numbers and map them to any musical parameters like pitch, duration, order, rhythm, dynamics, overtone spectrum, vibrato, etc. Random variables can be *discrete*, taking any of a list of possible values (like flipping a coin or rolling a perfect dice), or *continuous*, taking any numerical value in an interval (like measuring the duration of sound events via MIDI). Random numbers can easily be generated by computers. If we map them to musical parameters, the result is random music within the boundaries of the parameters (cf. [28]).

It takes three basic steps to generate a computer-based composition: the generation of raw material (e.g. random numbers), the modification of this raw material (e.g. permutation), and finally the selection from the modified material. Hiller and Isaacson called this process the *random sieve method* (cf. [28]). If the selection is done according to test criteria, this process is called *generate and test* (cf. [28]). In any case, a certain amount of *indeterminacy* remains in stochastic music. It is quite certain to reach different results if you run an algorithm twice.

There can be different types of probability distributions as well as different density functions (e.g. linear, exponential, Gaussian). Certain notes (e.g. those which do not belong to a major scale) can be omitted from the output by assigning them the probability value 0. To find suitable values, e.g. for the probabilities of the notes of a melody, we could make use of the numerous collections of digitalized music and analyze a number of melodies in the desired style.

Looking back to our children's song example, we try different distributions. The whole original melody is shown in Figure 24.3.

To keep things simple, we only change pitches and leave the rhythm untouched. A random melody containing the five notes of the song (C, D, E, F, G) with equal probabilities may render Figure 24.4, while in Figure 24.5 we have exponentially rising probabilities, which means that G is much more probable than C.

Obviously, the note G is the most frequent one in this melody. If you are not really satisfied, you could try again and again, and pick the best results. In fact, there are 5^{49} possible five-note melodies in the rhythm of the original song. Another idea is to identify frequencies (in the statistical sense) of the notes in given songs. If we count the absolute frequencies of the five notes in "Hänschen klein", we would get

Figure 24.3: "Hänschen klein"

Figure 24.4: Five notes with equal probability in the rhythm of the original song.

the values in Table 24.1. Letting this feature of the old song be the distribution of notes in a new song, we could get a melody like the one in Figure 24.6.

Table 24.1: Frequencies of Notes in "Hänschen klein."

note	absolute frequency	relative frequency
C	5	0.102
D	12	0.245
E	15	0.306
F	6	0.122
G	11	0.224

   The next step in building a model for melody generation is to observe not the frequencies of single notes, but those of two-note combinations instead. We count

Figure 24.5: Five notes with exponentially rising probabilities.

Figure 24.6: Melody with a distribution of notes as observed in the original melody.

the absolute frequencies of pairs of notes and get the second-level feature shown in Table 24.2. The columns represent the first note, and the rows the second note of a two-note pair.

Table 24.2: Relative Frequencies of Note Pairs Observed in the Original Song

	C	D	E	F	G
C	0	0.042	0.042	0	0
D	0.063	0.146	0.042	0	0
E	0	0	0.146	0.125	0.042
F	0	0.063	0.020	0	0.042
G	0.042	0	0.063	0	0.125

In this way, we could model a series of events that are linked with probability

Figure 24.7: Melody with a distribution of note pairs as observed in the original song.

values using Markov chains (cf. Example 9.24). To emulate a human composer with the computer, we have to take the balance of probabilities into consideration. Since the pitches within a diatonic scale are not uniformly distributed, we could not use standard uniform random numbers to pick them. A perfect dice gives equal chances for every number, and no event has any influence on any of the following events. But within a diatonic scale, the probabilities differ for each note, and the selection of a note has an influence on the following ones. The tonic, for instance, will be found more often than the dominant. In a C major scale, we may find the note G quite often since it belongs to both the tonic and the dominant chord [1]. In order to go beyond independent draws we could create a discrete state space where each state has an associated probability. Since each outcome must represent one state, the sum of all probabilities is always 1. If the current state is dependent on 0 prior states, we have a zero-order Markov chain, the result being pure chance. In Markov chains of higher order, the calculations may require very large tables.

If we want to generate a melody to be performed on a certain instrument, we have to take care that the pitches remain within the pitch range of this instrument. From the lowest pitch, we can only move upwards, and from the highest pitch we can only move downwards. We could define that most of the notes should be in the middle of the pitch range by assigning them higher probability values than the notes close to the boundaries, or we could advise the program to stop once a boundary is met (cf. [28]).

Stochastic processes can even be used for sound generation in *Granular Synthesis*. Here, sounds are made from *grains* – sound particles too short to be perceived by ear separately. Probability functions control the density of grains in clouds, and hereby the spectra of resulting sounds (cf. [22]). Xenakis's *GENDY3* forms another approach to stochastic synthesis. Here he directly calculates segments of waveforms in order to get new, unexpected sounds (cf. [23]).

24.3.2 Advanced Methods

In the following section we achieve a more advanced approach by using and combining the basic methods.

24.3.2.1 Rule-Based Processes

In rule-based processes, the composer defines a set of rules which the computer has to follow. Rules can be applied to randomly generated sets of musical data, as described above for Hiller and Isaacson's approach in their *ILLIAC Suite*. These may be the rules of classical counterpoint or any other rules that map features of certain musical styles (e.g. rhythms in a reggae, chords in a blues, imitation in a canon). Doing so requires knowledge about the different styles, therefore we call this approach *knowledge based*.

As an example, we create an algorithm that checks the melody for intervals that are difficult to sing (cf. Figure 24.8). In this case it compares a newly composed note N_n with its predecessor $N\_(n-1)$ and drops all intervals larger than a fourth.

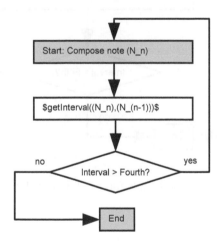

Figure 24.8: Rule avoiding larger intervals.

In a more complex example, a second voice is introduced avoiding parallel fifths. Let one voice already be composed and a second one to be automatically composed by an algorithm as seen above. For more simplicity, both voices shall have the same rhythm. You can model the rule "no parallel fifths" (cf. Section 3.5.2) as shown in Figure 24.9. For any note (V_2,N_n) in the second voice, the rule checks for a fifth compared to the first voice (V_1,N_n). If so, it also checks the note before the new one. If two fifths follow each other, the rule refuses the new note and starts to compose another one.

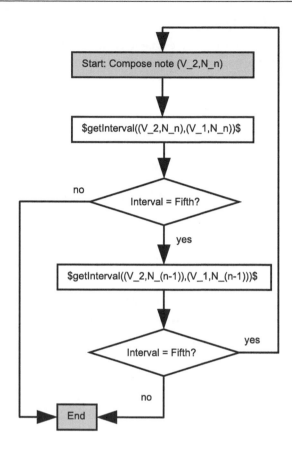

Figure 24.9: Rule for avoiding parallel fifths.

24.3.2.2   *Grammar-Based Processes*

Think of a musical work as a hierarchy of structural parts – a sonata movement consists of exposition, development, recapitulation; a melody consists of phrases, phrases of motifs, motifs of notes, and so on. If one describes this hierarchy in a more abstract way, one could refer to the methods of formal grammar as established by Noam Chomsky in the 1950s. In a common language, you could fill the formal structure with units from a "lexicon". All possible words lead to correct sentences – although not all of them are really meaningful. In music we do not have to care about meaning, so we could decide to calculate a bunch of "correct" structures by filling in our lexicon units.

In our example of "Hänschen klein" (Figure 24.10) we have four phrases forming a bar form with recapitulation (A-A'-B-A') (cf. Figure 3.46). Parts A and B end with the dominant on G, while A' ends with the tonic C. Basically, there are four motifs. Two of them (a and d) are transformed by transposition up or downwards.

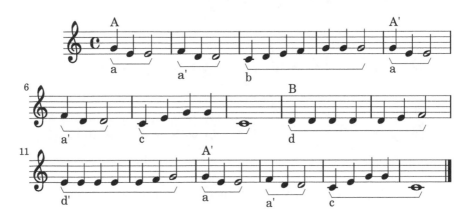

Figure 24.10: Analysis of phrases and motifs in "Hänschen klein".

Let's pick some more motifs from the children's songs "Frère Jacques" and "Twinkle, Twinkle, Little Star"[3] (cf. Figure 24.11).

Figure 24.11: Some more motifs from other children's songs.

Now we could recombine all these following the structure of "Hänschen klein", choosing motifs randomly out of the following lists:

$a \rightarrow \{ a, e, f \}$
$b \rightarrow \{ b, g \}$
$c \rightarrow \{ c, h \}$
$d \rightarrow \{ b, c, d, g, h \}$

One example of how these rules can be applied to musical notes is shown in Figure 24.12.

Of course, not only transposition but all basic transformation methods can be applied. As you can see from our example, formal grammars could be described in terms of finite automata, which help getting them into computer code (cf. [19]).

---

[3]If you want to use different melodies as a database, you may have to transpose them into the same key.

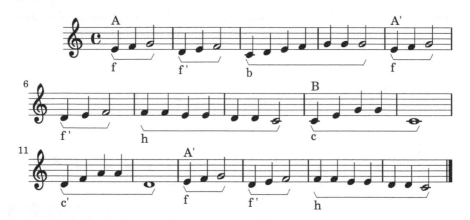

Figure 24.12: New melody from old motifs using a simple formal grammar.

24.3.2.3   Evolutionary Algorithms

If you set up an algorithm to compose a melody, the output may be excellent, quite interesting or just boring. To get a melody that is somewhat familiar but not too similar to one of your favorite melodies, it is necessary to optimize the output of your algorithm. Instead of evaluating every output individually and starting all over again to hopefully get a better result, you can make use of evolutionary algorithms.

The first task is to define mutation operators to get mutations – new versions of the original musical material. In our simple melody-building example, mutations could be any translation in time or pitch (cf. Section 3.3). Some possible offspring are shown in Figure 24.13. In Example 1 the intervals are inverted (diatonic) while in Example 2 one note in each bar is replaced randomly by another. Example 3 finally replaces each half note with a quarter rest and a quarter note of the same pitch. Of course several different parents could be used for greater diversity (cf. [15]).

Figure 24.13: Some mutations of a melody.

Evolutionary algorithms make use of artificial intelligence (AI) to simulate the evolution of a population on the computer. AI means that algorithms can generate an output but also evaluate this output according to underlying rules, and they can learn how to optimize their output. To reach this aim, they can compare their output with data from databases (sound or MIDI libraries) via rule-based or machine listening processes. David Cope, for instance, uses AI to extract databases of a musical style, on which processes of formal grammar work to (re)compose a new work (cf. [4]).

24.3.3 Evaluation of Automatically Composed Music

Evaluating the system's output is a difficult but important task, since it has direct influence on the operation's creation, mutation and recombination. Since music is a quite complex matter consisting of sounds and rests, of pitches, rhythms, and chords, these features have to be extracted from the musical data and evaluated independently, e.g. by use of machine learning, neural networks, or decision trees (cf. [15]). To save the composer from evaluating the quality of the automatically composed music himself, algorithms could be set up that evaluate the fitness of the composing algorithms' output. This could be achieved by means of feature selection (cf. Chapter 15), classification (cf. Chapter 12), and optimization (cf. Chapter 10).

24.4 Concluding Remarks

As we have shown, computers can compose music in many different ways. They can help human composers save time and effort, but a computer would not begin to compose without being told. A computer does not need music, has no interest in music and has no emotional reference to music. At least the first decision, whether there should be music at all, has to be taken by a human being.

24.5 Further Reading

For a thorough and in-depth survey of automatic composition, please refer to Eduardo Miranda's book *Composing Music with Computers* [19]. Nick Collins's *Introduction to Computer Music* [1] provides many inspiring examples and exercises, even if there are only a few really automatic ones.

Various examples of self-similar structures like automata, foldings, or weaving patterns can be found in Tom Johnson's book *Self-similar Melodies* [14]. If you want to know more about granular synthesis, read the basic book by Curtis Roads [22].

David Cope gives more insight into his concepts and methods for experiments in artificial creativity in various books, e.g. [4], [3]. In his thesis [7], Jonathan Gillick uses AI to emulate jazz solos of famous jazz artists.

Several programming languages are especially designed for musical applications. Johannes Kreidler [16] explains basic programming and acoustic principles using *Pure Data*, an open-source graphical environment, while Heinrich Taube [25] gives an introduction to composing music in the also open-source functional programming language *Lisp*.

Bibliography

[1] N. Collins. *Introduction to Computer Music*. Wiley, Chichester, 2010.

[2] N. Collins. Automatic composition of electroacoustic art music utilizing machine listening. *Computer Music Journal*, 3(36):8–23, 2012.

[3] D. Cope. *Virtual Music*. MIT Press, Cambridge, MA, 2001.

[4] D. Cope. *Computer Models of Musical Creativity*. MIT Press, Cambridge, MA, 2005.

[5] K. Ebcioğlu. An expert system for chorale harmonization. In *Proceedings of AAAI-1986*, volume 2, pp. 784–788. AAAI Press, 1986.

[6] M. Edwards. Algorithmic composition: Computational thinking in music. *Communications of the ACM*, 54(7):58–67, 2011.

[7] J. Gillick. *A Clustering Algorithm for Recombinant Jazz Improvisations*. Wesleyan University, Middletown, CT, 2009. (Doctoral dissertation).

[8] L. A. Hiller and L. M. Isaacson. *Experimental Music*. McGraw-Hill, New York, 1959.

[9] S. Himmelsbach, ed. *Jens Brand. Book*. Edith-Ruß-Haus für Medienkunst, Oldenburg, 2009.

[10] T. Holmes. *Electronic and Experimental Music: Technology, Music, and Culture*. Routledge, New York, NY, 2012.

[11] F. Jöde. *Das schaffende Kind in der Musik*. Möseler, Wolfenbüttel, 1962.

[12] T. Johnson. Automatic music. http://editions75.com/Articles/Automatic%20music.pdf. Accessed: 2014-07-29.

[13] T. Johnson. Found Mathematical Objects. http://editions75.com/Articles/Found%20Mathematical%20Objects.pdf. Accessed: 2014-07-29.

[14] T. Johnson. *Self-similar Melodies*. Editions 75, Paris, 1996.

[15] R. Klinger and G. Rudolph. Automatic composition of music with methods of computational intelligence. *WSEAS Transactions on Information Science and Applications*, 4(3):508–515, 2007.

[16] J. Kreidler. *Loadbang*. Wolke, Hofheim am Taunus, 2009.

[17] H. Lachenmann. *Ein Kinderspiel*. Breitkopf & Härtel, Wiesbaden, 1982.

[18] G. Loy. *Musimathics: The Mathematical Foundations of Music*. MIT Press, Cambridge, MA, 2006.

[19] E. R. Miranda. *Composing Music with Computers*. Focal Press, Oxford, 2001.

[20] G. Nierhaus. *Algorithmic Composition*. Springer, Wien, 2009.

[21] C. Roads. Algorithmic Composition Systems. In *The Computer Music Tutorial*, pp. 819–852. MIT Press, Cambridge, MA, 1996.

[22] C. Roads. *Microsound*. MIT Press, Cambridge, MA, 2001.

[23] M.-H. Serra. Stochastic composition and stochastic timbre: GENDY3 by Iannis Xenakis. *Perspectives of New Music*, 31(1):236–257, 1993.

[24] M. Supper. A few remarks on algorithmic composition. *Computer Music Jour-*

nal, 1(25):48–53, 2001.

[25] H. K. Taube. *Notes from the Metalevel*. Taylor & Francis, New York, NY, 2004.

[26] J. Togelius et al. Procedural content generation: Goals, challenges and action-
able steps. `http://drops.dagstuhl.de/opus/volltexte/2013/4336/`
`pdf/7.pdf`. Accessed: 2015-07-08.

[27] I. Xenakis. *Formalized Music: Thought and Mathematics in Composition*. Pendragon Press, Stuyvesant, NY, 1992.

[28] H. J. Yoon. *Stochastische und fraktale Modelle in der Algorithmischen Kom-
position*. Electronic Publishing Osnabrück, Osnabrück, 2002.

Part IV

Implementation

Part IV

Implementation

Chapter 25

Implementation Architectures

MARTIN BOTTECK

Fachhochschule Südwestfalen, Meschede, Germany

25.1 Introduction

This chapter's author started to investigate music signal processing whilst being a member of a mobile phone manufacturer's corporate research unit. During the early 2000s, mobile phone devices were beginning to be equipped with music players and large enough flash storage in order to accommodate music collections that were hard to sort and navigate. Our idea was to utilize music signal processing techniques to help present the contents of a music collection on a mobile device somehow taking account of the listening experience connected to each track, i.e. provide the user with a personalized set of track lists that are generated automatically. This will be the topic of this chapter.

Since the computational effort required to execute the algorithms exceeded a mobile device's capabilities during the early 2000s by far, Linux computing grids were used instead.[1] Despite much effort in research on processing concepts demanding less computation power, this concept still has not found its way into today's products. A concise description of the considerable achievements can be found in Chapter 27. The research work was complemented by activities designed to separate a major part of the calculations and perform them on dedicated servers outside the mobile device. These attempts up to now seem to have faltered. Sorting your private music collection by a dedicated service on the network still is not a very popular application despite the fact that several attempts for commercialization have indeed been made (cf. references given in Section 25.3). There exist, however, a few services on the Internet that offer small applications based on music signal processing techniques. Some of them may be used in combination with each other, thereby offering more complex functionality. This concept deserves a closer look.

---

[1] The experiments were carried out as part of the Music Information Retrieval Evaluation eXchange (MIREX) coordinated by the Graduate School of Library Information Science, University of Illinois at Urbana-Champaign, www.music-ir.org. Accessed 22 June 2016.

This chapter intends to demonstrate the challenges when determining a suitable implementation architecture by assessing a few examples for applications that utilize music signal processing techniques. It will, however, not deal with the challenges connected with creating and transporting massive stream network traffic and also does not look into details of integrated circuit design. Instead, we will consider implementation architectures in the following way.

Definition 25.1 (Implementation Architecture). An *Implementation Architecture* in connection with software-intensive systems is "the fundamental organization of a system embodied in its components, their relationships to each other, and to the environment, and the principles guiding its design and evolution" [1].

This definition is intended to help control system design and understand the underlying requirements including functionality, cost, and risk. Notably, the definition does not include criteria to decide between "good" and "bad" partitioning of the system into components. It has nonetheless been adopted as an ANSI standard defining the state of the art in engineering practice. Such partitioning, however, constitutes one of the most crucial tasks in the system design process.

On first sight, implementation architectures for music signal processing are strongly determined by the algorithms to be executed: the winning concept would be the one most "efficiently"[2] executing the tasks. On the other hand, we seem to like computation architectures that allow straightforward implementation and installation of applications despite not being specifically tailored to the tasks at hand: desktop PCs for a long time have been the most popular computing environment, only recently outrun by mobile computing platforms (smartphones), both offering, though, rather "inefficient" performance for specific computational tasks.

Hence, the dependency between applications and implementation architectures is bi-directional: algorithms determine the requirements on architectures, but in turn available platforms (with their specific properties) determine the limits for algorithmic complexity and functional extent. Chapter 27 elaborates on opportunities for reflecting algorithmic needs in computer hardware. This chapter, however, will give a little more insight into the bi-directional nature of dependency between algorithms and architectures.

25.2 Architecture Variants and Their Evaluation

Definition 25.2 (Music Classification). *Music classification* constitutes one kind of computer music processing that attempts to associate music tracks with pre-defined contexts. Its processing chain is depicted in Figure 25.1: Perceptional features are computed for music tracks from a music data base. A selected and processed subset of these features are used to classify the tracks into classes (i.e. categories). Classification is then performed according to parameters suitably defined in order to achieve

---

[2]"Efficiency" seems to not be uniquely defined though; in some cases it seems reasonable to relate to specific "effort"-determining measures which may include chip size, number of electronic components, amount of data exchange, power consumption, computation time, or any combination of these.

the desired association of tracks. The classes themselves may be defined in several ways in order to create track lists according to the desired listening experience.

Please refer to Section 12.3 for further explanation of this process and the meaning of key terms.

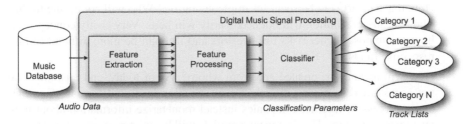

Figure 25.1: Processing chain for music classification [4].

With respect to music classification, the requirements that algorithms impose on implementation concepts will be separated into

- processing complexity,
- data complexity / data volume, and
- network connectivity.

Specifically, computing and processing of features constitute far more demanding requirements than subsequent classification. An example of a detailed study is given in [4].[3]

Reassigning the feature processing tasks to remote elements of the implementation architecture, however, might result in massive data transfer including the music tracks themselves as well as pre-processed features.

At this point it should be noted that a suitable architecture is determined by technical considerations as outlined so far, but maybe even more so by economic aspects. Any solution will require substantial amounts of effort for its implementation, and these efforts need to be compensated by market revenues at some point in time. Consequently, an implementation will not only be judged by the effort needed for its implementation but more so by the market revenues it promises to achieve. During past years the markets for recorded music have undergone tremendous change. Due to substantially improved technology for digitalization and data communication it has become increasingly difficult to obtain revenues from music listeners; drastically reduced cost and effort for music delivery and exchange undermine existing copyright protection agreements. Future developments in these markets are very hard to predict, specifically since all successful solutions will require a mass market uptake in order to create positive returns on their related investment. Investigating the mass market aspects themselves will go far beyond this book's focus, but it should be noted that a basic requirement for suitable architectures will include aspects like

---

[3]Many music features rely on the spectrum of the musical signal. Then, simply the number of calculations required for Fourier transform (DFT/FFT) already constitutes a substantial computational challenge.

ease of use, scalability, decreasing marginal cost, plus a concept to establish a critical minimum market presence to begin with ([8], [6]).

For the time being, the influence of economic considerations on implementation architectures can be summarized by a list of questions to be addressed in evaluating implementation proposals:

- How much effort is to be spent for implementation? Who will contribute to this? The "cheapest" solution will not necessarily win here. Very costly concepts will however impose higher risks for investors and it might be more difficult to raise respective funds.

- How will revenue be created and from whom? As of today, end customers are not accustomed to paying directly for information in digital format. Commercially successful Internet companies instead monetarize information about these customer's behavior. Hence, revenue models will be quite more complicated than they used to be in the past.

- Which legal or commercial regulations are required? Copyright laws were established in order to provide revenue opportunities for creative artists. They are not a law of nature but rather a mere agreement between people in our society. In view of technological advances, these copy protection rules might seem not to be technically enforceable any longer. Consequently, the law might change and different regulations might be established in order to retain some commercial incentive for creating art.

- What happens in case one of these crucial regulations ceases to exist?

Specifically with respect to the latter aspect, architectures with ample potential for ad hoc change, rapid update, and further development provide clear benefits.

Three different approaches for implementing a music processing solution as proposed in previous chapters can be identified and will be discussed.

25.2.1 Personal Player Device Processing

In a device-centric approach, processing of music data and presentation of results resides on the same computation device that also stores the music data to be classified. Playback of this music will also happen on this very same device based on a track list provided by the classifier (cf. Figure 25.2). The definition of classes will be based on direct user input. Hence, these classes will be very personal to start with.

Music feature processing and recommendation imposes several demanding requirements on the implementation hardware:

- large enough storage space for music, features and (less demanding) classification results,

- sufficient computational power to perform complex feature processing, feature selection and classifier training, and

- versatile graphics and sound capabilities in order to present recommendations in a meaningful way.

Mobile devices have limits with respect to several of these requirements. Their com-

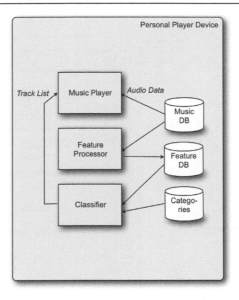

Figure 25.2: Personal player device-based processing.

putational capabilities will remain limited due to limits in power consumption and even more when compared to dedicated computing servers that may be located somewhere in the IP network.

Considerations on suitable device architectures for this approach will be discussed in more detail in a subsequent chapter (Chapter 27), thereby addressing a wide range of alternative processor types and integrated circuit solutions.

25.2.2 Network Server-Based Processing

In network server-based processing, all processing happens on a remote server and the music itself will be streamed to the personal player device in the end customer's hands (cf. Figure 25.3). In this case, the task of the playback device is reduced to mere audio rendering (i.e. converting compressed digital audio data into an analogue signal). In this scenario, network connectivity is of crucial importance: although the data volume to be communicated to the mobile device may be considered moderate (approx. 200 kbit/s for top-quality compressed audio) this stream needs to be available constantly and without delay variations.[4] Specifically, the latter constitutes a requirement difficult to support in packet switching (IP) networks due to its conflicts with other traffic of largely varying statistical properties [10].

As of today, the most popular music recommendation services are built on this architecture. More details about such services are presented in a previous chapter

---

[4]Although it is indeed possible to limit the requirements on maximum delay and delay variations by implementing playback buffers or progressive download techniques, it should be noted here that these compensation techniques will degrade user experience to some extent: at least when changing tracks the user cannot avoid rather long reaction times of the user interface.

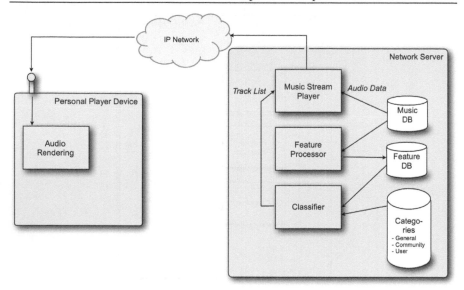

Figure 25.3: Networked server-based processing.

on music recommendation (Chapter 23). In contrast to straightforward applications purely based on personal player devices, these services offer classes that are based on recommendations from other users as well as being purely personal. Advantages of such multi-modal class definition are discussed in Chapter 26.

25.2.3   Distributed Architectures

Implementations on mobile devices may benefit from distributing selected tasks to other nodes on the network, specifically if these are computationally intensive (cf. Figure 25.4). Essentially, this concept allows more or less all processing components required for music classification to be available on the arbitrary nodes involved. Depending on the application to be realized, these components will be used or not, taking account of specific limits (bandwidth, processing power, data capacity). Hence, this "hybrid" type of architecture allows us to find a good compromise between technical requirements and capabilities provided by the actual end customer hardware. This will not only help overcome shortcomings in the hardware platform, but in turn offer a high potential to develop use cases and applications that were not already taken into account in the beginning. Furthermore, distributed concepts promise to provide substantial advantages concerning acceptance of rapid updates and ad hoc changes. The latter will be helpful for easier load balancing, allow more robust behavior in case of failures, and provides room for scalable improvements by adding more computation nodes in the network.

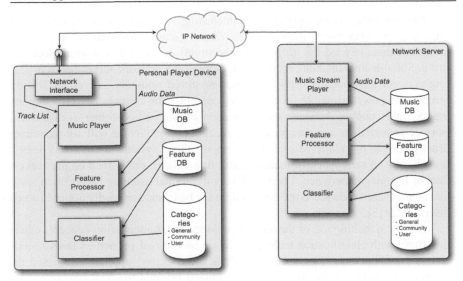

Figure 25.4: Distributed architecture for music classification.

25.3 Applications

The following sections present architectural concepts for a selected set of applications.

25.3.1 Music Recommendation

As outlined in Chapter 23, the task of an application for music recommendation is to provide the user with a ranked list of music tracks which are interesting with respect to the current context. A range of algorithms has been developed for this task. These not only provide an equally large range of difference in complexity, but also vary in their understanding of "interesting" and "context".

A straightforward solution will provide the track most similar to the one currently playing on the end customer device. "Interesting" in this case will imply similarity, "context" simply denotes the track currently selected. This scenario directly maps to a personal player device processing architecture. Respective applications will help customers navigate through their personal music data base. Network connectivity for data exchange is not required. However, all the processing power needs to be provided by the playback device. For mobile devices, this constitutes a challenging requirement despite possibilities to conduct a large part of the processing with background priority. Listening experience will be limited to the tracks already stored. Although commercial products for this scenario were released (mostly for personal computers), these seemingly have not been successful on the market.[5]

---

[5]A music player for PCs released by *mufin* in 2011 was discontinued. The company since then provides products for music recognition instead: www.mufin.com. Accessed 22 June 2016.

More popular systems understand "interesting" in a much wider way: they take into account other users' music choice. *iTunes Genius*[6] and *last.fm*[7] provide recommendations beyond the scope of music stored in the end customer's local data base. Suitable implementation architectures rely on a network server-based processing concept thus making use of information stored remotely as well as abundant processing power available on centralized servers. The concept however implies a rather tight coupling between the application on the end customer device and the service provider's music data base. Specifically, *iTunes Genius* will not recommend tracks outside the *iTunes Store*. Furthermore, these systems as of today seem to mostly rely on metadata information (artist, title, album) rather than on musical content or listening experience. Users with very individual listening habits often remain unsatisfied since their preferred music might not be labeled or hardly noticed by other users so far [5].

A system making use of the already provided recommendation concepts extending these with classification techniques based on musical properties leads to a distributed architecture. One such example is the *Shazam*[8] application: based on a snippet of audio, it presents a list of recognized original material (the most likely metadata) including hyperlinks to the *iTunes Store*. Coupling of this information is possible across an interface agreed on between the application providers, in this case a solution proprietary to *Shazam* and *Apple*.

25.3.2   Music Recognition

During recent years several attempts were made to automatically recognize music. The development to a large degree followed paths similar to speech and voice recognition systems. The general idea is to compare audible input to elements of a data base (containing music tracks instead of text fragments) and provide a list of best matches. End customers should thereby be able to hum sections of a song and – alas – the music player starts to play back that very song performed by the original artist. This scenario (often referred to as "Query by Humming", Chapter 26) is more challenging than it might appear on first sight since people hum tunes typically in incorrect pitch, they frequently miss a few notes completely and the input signal is often heavily polluted with noise and – maybe even more often – other music is playing in the background. Consequently, early implementations of such applications (provided by, e.g., *Vodafone* in Germany based upon the *Teleca* technology) in the early 2000s did not rely on the processing power available in the end customer device only but performed all calculations on *Vodafone's* server in the mobile network.

This network server-based processing architecture comes with another advantage: end customers do not necessarily expect the desired song to reside in their local music data base already. Instead, they welcome the search to be extended across a

---

[6]http://www.apple.com/legal/internet-services/itunes/de/genius.html. Accessed 22 June 2016.

[7]www.last.fm. Accessed 22 June 2016.

[8]www.shazam.com. Accessed 22 June 2016.

wider range of music tracks. Further developments of applications for music recognition consequently focused on mobile devices. Music recognition is seen as a welcome extension of the limited user interface of such devices. Specifically, the *Shazam* app has gained wide acceptance on the market. A popular use case is to let the app recognize music played in commercials in order to subsequently purchase the track from the *iTunes Store* by just one further click.[9] In this case, the system uses a distributed architecture thus developing a new use case and revenue opportunity. It shall be noted here that *Shazam* and *Apple* do not share their revenues; each application provider collects fees for its service separately.

Only the distributed architecture approach has survived on the market in this case. The service originally presented by Vodafone was discontinued only months after its introduction due to lack of customer interest.[10] Seemingly, the music data base provided by these services so far did not provide an offer tempting enough for people to start browsing through it. Connecting the well-established *iTunes Store* to another application (music recognition by *Shazam*) instead raised enough interest to exceed a critical minimum market acceptance.

25.4 Novel Applications and Future Development

The distributed architecture approach has mainly survived since it supported applications that were not specifically intended in the beginning. In the future, therefore, we will see more and more connections of distributed modules through service mashups. Hence, the definition of interfaces related to music recognition constitutes a crucial task: data formats for various types of information need to be agreed upon. This not only concerns musical data (cf. "Digital Representation of Music", Chapter 7) but also specific information about

- musical features (cf. Chapter 5),
- classification parameters (cf. Chapter 12), and
- listening context (cf. Chapter 26).

Figure 25.5 shows the various types of information that need to be made available between the devices participating in a distributed architecture for music classification. The most obvious interface for data exchange is between the music data interface and the music database on the device or server. Several common standards for this exist and are described in Chapter 7. Public exchange of such data is however legally restricted. Consequently, devices might rely on the exchange of metadata (artist, title, etc.) information as included in, e.g., MP3 playlists or with reference to tag values of *.mp3* files. This reference may however easily be broken when music tracks are copied or might be completely unavailable for a large range of tracks. Future applications therefore will need to rely on more general references

---

[9]There are examples of several tracks or artists that became hits or stars through just this scenario: e.g. "1,2,3,4" by Feist was boosted by its use in an *Apple* commercial.

[10]*Vodafone* has made several attempts since to relaunch music services. A collaboration with *Ampya* found only 200,000 German customers in 2013 [9]. The service was relaunched in 2014 after *Ampya's* acquisition by *Deezer*.

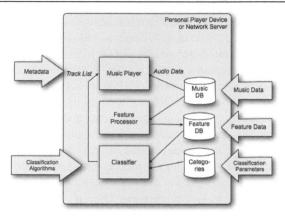

Figure 25.5: Data exchange and interfaces for distributed architectures.

denoted by class information, preferably generated through music classification. In order to reproduce this classification on remote devices, the extracted features need to be transported. Exchange formats for these might be defined in alternative ways:

- A straightforward method lists the features computed for a music track sorted somehow with reference to the track's timeline in a set of files associated with the music track. AMUSE [11] describes the .arff file type and has demonstrated the suitability of such approach. It has become apparent, though, that the amount of feature data involved in successful classification is quite large, often even larger than the music data itself.

Therefore, a more compact data format will be desired, preferably specifying the features to be computed in a *formal* way. Again, alternative approaches might be taken:

- Relying on the definitions given in Chapters 5 and 8, an XML tagged list could be provided listing a selection of features needed for the classification task at hand as well as some configuration parameters like, e.g., frame lengths or windowing functions. Such representations would be rather compact. Obviously, the algorithms associated with this reference list need to be available at the point of processing. In networked devices, this does not necessarily imply the device itself; feature processing might be executed at yet another remote location invoked through common web-based methods (e.g. Remote Function Calls (RFC) embedded in SOAP messages).[11]

- The latter approach necessitates maintaining reference links between XML tags and algorithms, a similar reference we have experienced to be hardly maintainable between *.mp3* tag values and music tracks already. So, why not transport the feature extraction algorithm itself? Such exchange would merely require the

---

[11]This "yet another remote location" might be present in a completely different network segment. Imagine a moderately powerful server in your home network rather than something out on the public Internet.

definition of a suitable computation platform. "Suitable" in this case will go little beyond what has led to existing and well-known engines like Java or other pre-packaged execution environments: a platform for music classification will need to possess respective libraries and has to come with implementations of basic functions in a very efficient way. Otherwise, as outlined in Chapter 27, the amount of processing might become excessive for some of the algorithms. Besides, Chapter 27 identifies those algorithms that are most demanding in terms of processing power.

Given the availability of a generic computation platform, we then could not only exchange algorithms for feature extraction but also for the rest of the classification task: parameters for classification and entire classification algorithms. Any user having classified a set of tracks in the personal collection to belong to a favorite personal class could communicate the algorithms and their parameters that have led to this classification. We may call this set a "Musical Taste" since it will enable other users to apply a similar classification to any music track accessible for processing.

```
<feature_space>
  <General>
     <class_ref>Summer Music</class_ref>
        <!-- class of music these features are relevant for -->
     <created_by>AutoTrackLister</created_by>
     <date>ddmmyyy</date>
     <feature_reference>www.featureparadise.edu/def_tables</
        feature_reference>
        <!-- location of acronyms and algorithmic definitions -->
  </General>
  <feature_proc>
     <win_len>512</win_len>
        <!-- Window length in samples -->
     <win_func>RECT</win_func>
        <!-- choose from RECT, HAM, HANN, or UDEF -->
  </feature_proc>
  <feature_list>
        <!-- feature acronym; temporal weighting; reference
           period -->
     Z_Cross; GMM1; Window
        <!-- Zero-Crossing Rate; 1st order Gaussian Mixture
           Model; one value per Window -->
     ACF; GMM3; Window
        <!-- Auto Correlation Function; 3rd order GMM; one
           value per Window -->
     LDN; GMM1; Track
        <!-- Loudness; 1st order GMM; one value per track -->
  </feature_list>
</feature_space>
```

Listing 25.1: Example of feature reference listing in exchangeable XML format.

An example of an XML-style notation of a reference to features that shall be used for a specific classification is shown in the grey box Listing 25.1. The comments give some explanation of the parameters set therein. The above example constitutes a rather straightforward approach; only references to features computed elsewhere are given. A much more elaborate concept was presented by Mackay in [7]. Its complexity, however, tends to produce rather large data sets. Mackay intends to directly specify algorithmic behavior or list the computation results (feature values). Behavioral description in languages intended for execution (like e.g. Java, Python, etc.) promises to be much more compact than this approach. Anyhow, with the availability of universally agreed feature references, an abundant range of novel applications and scenarios would become possible:

- Music recommendation services already existing could eventually be extended to provide track lists based on perceptional features, not only on metadata (similar to the *Shazam-iTunes* case).

- People could trade "Musical Taste". For example, why shouldn't a number of customers be interested in celebrities' musical preferences?

- Individual users' "Musical Taste" could provide them with recommendations from virtually any sort of music collection. The amount of music somehow published per month continuously grows, so there is a lot of good music out there for everyone to match. Discovering it will be the problem of the future.

The next chapter on user interaction (Chapter 26), amongst other things, covers opportunities to utilize context information. Depending on the end user's listening situation, recommendations for music or maybe the behavior of the user interface as such might change: why not propose a different type of music when commuting back from work than during physical exercise at the gym? Maybe even choose tracks with a beat matching to your current heartbeat or workout rhythm?

Mobile devices typically provide several types of sensors (GPS position, accelerometer, compass, ...) which deliver basic signals. To determine the "listening context" from these signals constitutes a classification task in itself, which shall not be discussed here. However, "listening context" needs to be available in a machine-readable form in order to adapt classification parameters or recommendation settings. In its most straightforward realization, a specific listening context could be described by a reference to a specific "Musical Taste". Again, we face the challenge of maintaining link references here (as we already have seen in the context of MP3 files (see Section 7.3.3) and metadata information). However, if a formal description, eventually as part of the generic computing platform or script engine as described above would become available, a further set of use cases might be developed, providing an even more intriguing listening experience.

25.5 Concluding Remarks

Many of the music analysis processes presented throughout this book are computationally demanding. They easily exceed the processing resources of typical office computers or mobile devices. Hence, implementation architectures have been developed to distribute selected tasks to other more powerful nodes in computer networks. In several cases these partial processing tasks have developed into self-contained network applications addressing a specific user community. Some of them were later combined, resulting in even another specific application which was not foreseen in the beginning. With the advent of formal notations to exchange algorithms and/or processing results between computing nodes, a rapid development of an abundant range of novel applications for music analysis can be foreseen. Its pace will by far outrun the development of concepts that require all processing to reside on a single office computer or mobile device.

25.6 Further Reading

As mentioned already, the definition of a "good" architecture not only depends on technical criteria, but equally on market dependencies. Readers interested in relations of these will find further relevant insight through a large range of literature; hence, this section intends to suggest publications as an entry point for further reading. With respect to novel applications in the world of consumer electronics and information technology, G. A. Moore [8] provides an almost "classic" foundation on *market uptake mechanisms*. R. G. Cooper [6] has compiled a meaningful set of considerations for most commercial aspects to be regarded for the *innovation of technology products*. For those interested in actually *programming mobile applications*, an introduction to basic principles and procedures is provided by R. B'Far in [3]. An in-depth investigation of technical *principles of software architecture* including distributed architectures is provided by L. Bass and R. Kazman in [2]. This book also makes a quite elaborate attempt at understanding criteria for deciding between "good" and "bad" system partitions.

Bibliography

[1] I. Architecture Working Group. *IEEE Recommended Practice for Architectural Description of Software-Intensive Systems*. IEEE, New York, 2000.

[2] L. Bass and R. Kazman. *Software Architecture in Practice*. Addison Wesley Pearson, New York, 2012.

[3] R. B'Far. *Mobile Computing Principles*. Cambridge University Press, 2004.

[4] H. Blume, M. Botteck, M. Haller, and W. Theimer. Perceptual Feature based Music Classification: A DSP Perspective for a New Type of Application. In *Proceedings of the 8th International Workshop SAMOS VIII; Embedded Computer Systems: Architectures, Modeling, and Simulation*. IEEE, 2008.

[5] O. Celma. *Music Recommendation and Discovery*, pp. 5–6. Springer Science and Business Media, 2010.

[6] R. G. Cooper. *Winning at New Products*. Perseus HarperCollins, New York, 2001.

[7] C. McKay and I. Fujinaga. Expressing musical features, class labels, ontologies, and metadata using ace xml 2.0. In J. Stein, ed., *Structuring Music through Markup Language: Designs and Architectures*, pp. 48–79. IGI Global, Hershey, 2013.

[8] G. A. Moore. *Crossing the Chasm*. Perennial HarperCollins, New York, 1999.

[9] J. Stüber. Was Berliner Musikstreamingdienste besser als Spotify können (in German). *Berliner Morgenpost*, 12(2), 2013.

[10] A. S. Tanenbaum and D. J. Wetherall. *Computer Networks*, chapter The Medium Access Control Sublayer, pp. 257–354. Prentice Hall Pearson, Boston, 2011.

[11] I. Vatolkin, W. M. Theimer, and M. Botteck. AMUSE (Advanced MUSic Explorer): A multitool framework for music data analysis. In *Proc. Int. Conf. Music Information Retrieval (ISMIR)*, pp. 33–38. International Society for Music Information Retrieval, 2010.

Chapter 26

User Interaction

WOLFGANG THEIMER
Volkswagen Infotainment GmbH, Bochum, Germany

26.1 Introduction

Humans use their senses such as sight, hearing, touch, taste, and smell to perceive
the environment. These modalities also serve to respond to input signals. A technical
system can interact with its environment with the same modalities, but is not limited
to them. Think for example about other (electronic) data channels or new sensors and
actuators which extend the classical "senses". Therefore, user interfaces are often
categorized according to which input and output modalities are used in the system.
In the context of performing music, the tactile and the audio channels dominate the
artist's interaction with the instruments. Technical music processing systems extend
the interaction to all input and output modalities and typically rely on visual user
feedback.

Music is to a large extent a social activity. Musicians are, for example, performing music together in an orchestra and music listeners are gathering in concert
halls. Thus, it is important to take into account the interaction and communication
among musicians, listeners and novel technical systems for music processing. In order to generalize the following argument, all objects with which the user interacts,
be it a physical instrument, electronics, or a software implementation, are defined as
music processing systems. Figure 26.1 gives a top-level overview of how the user
interfaces of technical music processing systems can be characterized from an architectural point of view. The figure uses a system engineering approach to describe an
entity, which in our case is a system to process musical signals: A complete system is
decomposed into self-contained subsystems which are characterized as blocks with a
set of input signals, processing of the input inside the subsystem, and a set of output
signals. The functional block *Music processing system - User X* can represent a musical instrument or a technical system which processes an input signal in the context
of music under the control of a user. The input can be a direct user input, for example
to play a musical instrument, or it could be a musical signal which is manipulated by
the user for music editing.

The user input to a music processing system in essence has four purposes:

1. Generation of music (via musical instruments or vocals)
2. Modification of existing music (music editing)
3. Definition of a music query (i.e. specifying the search parameters)
4. Navigation through a music collection

The functions of a music processing system are mainly generating, modifying, finding and exploring music. Thus the output of a musical system covers four different areas as well:

- Music performance (playback of music and related multimedia)
- Presentation of music query results
- Representation of a music collection
- Feedback for the user to confirm the input and state of the system

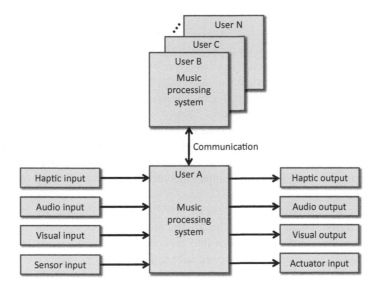

Figure 26.1: Top-level view of user interaction.

This chapter is an overview of interaction modalities. It cannot and does not intend to give an exhaustive analysis of interaction methods since electronic devices, communication, and processing quickly evolve. These developments provide new opportunities and extend the interaction space. Nevertheless, examples are given to illustrate the user interaction concepts.

26.2 User Input for Music Applications

26.2.1 Haptic Input

Haptic Input for Music Generation

Traditionally, haptic user input has been the most important input modality in music. All non-electronic musical instruments rely on sound wave generation within mechanical structures (see Chapter 2). The user holds and plays an instrument by adapting the haptic input to the musical score and expression. In the case of wind instruments, the user also provides a modulated air flow to achieve the desired sound output. Even in the case of electronic instruments, the main interaction is based on tactile input. Additional sensors can, however, also pick up more indirect input and the option of an electronic remote control exists (see also Section 26.2.3).

While users provide their input, additional processing and information exchange can be performed in electronic systems at the same time. The relationship between haptic input and system functions does not have to be direct any longer. An abstraction and mapping of the input can be made based on the context and system state. An example is an automatic error correction during a music performance, based on the knowledge of the musical score and the current position in the song. The same tactile input, entered via strings, buttons, knobs, sliders, touch, or other input elements, can be used in different ways, depending on the system logic. While the first electronic music systems had individual buttons and knobs for each and every parameter leading to a complex interaction surface, newer systems purposely abstract the functionality and offer programmable input elements for a cleaner user interface (see Figure 26.2).

Figure 26.2: Early synthesizer (Studio 66 system) on the left compared to a modern version (Korg Kronos X88) on the right [pictures under CC license].

Haptic Input for Querying Music

In former times, music could only be experienced directly by playing instruments and/or listening to the performance. With the advent of mass storage and the availability of large music databases locally or in the Internet, music has become immediately accessible and reproducible. The user faces the challenge of how to find the appropriate music for a specific mood and context. Haptic input can also help in this respect by defining music search parameters or by navigating through a larger set of music.

The simplest approach in electronic music players is a textual search for music titles, albums, artists, genres, and the option to travel through lists of matching music. This is a suitable approach if the metadata of a piece of music is at least partially known so that the user can formulate a search query. But in many cases the user does not recall song- or artist-specific textual information. It is often easier to remember the melody line of a song. In this situation, the user can play the melody (or any other characteristic notes) on an electronic instrument. The melody is recorded and sent to a music database to perform an error-tolerant search for the melody string, for example, by comparing it with a database of music in MIDI format. As a result, a ranked list of possible song candidates is returned and the user can select the suitable song from a short list of alternatives.[1] This approach can be generalized to a music query based on audio input, as will be seen in Section 26.2.3.

Haptic Input for Navigation in Music Collections

Often music listeners are not specifically interested in a certain piece of music, but rather would like to explore a collection of music. Of course this can be done simply by listening to music channels or skipping forward in a (random) playlist. The problem with this approach is the sequential nature of the audio signal: While listening to a piece of music, it is difficult to keep the overview of the complete collection since the concentration is focused on the current song. In a more sophisticated approach, the music navigation borrows concepts from computer games and follows the gamification trend by transferring gaming elements to other application domains [4]. Envision that all songs are placed in a virtual music library building (House of Music concept; not published). Each room in this library contains a certain music genre. Artists, albums, and titles are arranged on shelves in the respective rooms. The user navigates through this House of Music with game-like haptic controls, can listen to ambient music representing the different music styles while walking, and is able to select music, for example by climbing up a shelf and grasping an album. For this concept it is very important to obtain timely haptic, audio, and visual feedback.

A second more abstract concept along those lines is *nepTune* [8]: The user is represented as a pilot, flying his aeroplane over a musical landscape. Gaining altitude gives a better overview of all the music and descending towards the earth shows more details (like, for example, music metadata) and allows pre-listening to the music. It should be mentioned that the visualization of the musical landscape plays an important role in *nepTune*; the concept is mentioned in this section due to the pilot-specific haptic 3D navigation.

Haptic Input for Music Editing

Editing music is mainly done by haptic input. Changes can be performed by manipulating the recorded music, such as changing the sound characteristics or changing the notes and chords of electronic instruments. Alternatively, the musicians can replay

---

[1]http://www.musipedia.org. Accessed 22 June 2016.
[2]http://www.cp.jku.at/projects/nepTune. Accessed 22 June 2016.

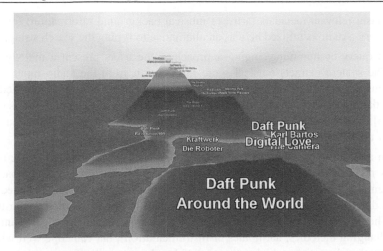

Figure 26.3: User interface screenshot of nepTune for an exemplary music collection (Department of Computational Perception, Johannes Kepler University, Linz).[2]

a passage with their instruments and replace an existing time interval in a recording. Modern sound recording solutions allow multi-channel recordings (one or several separate voices for each instrument) and enable channel-specific editing and synchronization operations before the complete opus is assembled again from the individual channels. Music editing can be supported by intelligent systems which are able to align different recordings in terms of timing and absolute pitch.

26.2.2 Audio Input

Besides the haptic input (for playing instruments) the direct audio input of music is another relevant input modality. Vocals play an important role in music, from classical operas to modern pop or rock. From a system perspective, a musician might play an instrument, but the output of the instrument is an audio signal which mixes with direct audio input such as vocals.

Audio Input for Music Generation
Audio input is available when recording a musical performance such as a concert. It can be used more selectively to correct and replace passages with wrong notes. Different channels can be recorded separately and can be integrated either directly or one after the other by a sequencer program. This topic will not be discussed in more detail since it is only loosely related to the topics of this book, but in general, this is a very active and diverse field of research.

Audio Input for Music Query
Similar to the haptic domain, audio has been used in the past mainly for music creation (vocal music). In an electronic music analysis system, the audio signal can fulfill additional new roles:

• A speech recognition algorithm can interpret the voice input, which provides music-relevant metadata (artist or musical background information) for a search query. Audio is utilized here as an alternative to typing the search string.

- In query-by-humming/singing approaches, the users present a melody vocally. The system extracts musical score information from the audio input to start a music query. In most approaches the recognized string of notes is matched against a database of music, allowing an error-tolerant search for similar pieces of music [6]. Due to the symbolic nature of notes, different performances of the same piece of music can be found.

- In audio fingerprinting approaches, the playback of an unknown piece of music is recorded to identify its origin (metadata like title, artist). A compact representation of the music, the so-called fingerprint, is extracted and matched against a database of music. The most similar candidates are returned to the user [3]. The fingerprint only identifies a piece of music without any musical changes (like changing melodies, harmonies or instrumentation), but can cope well with noise or other distortions which are not related to the music itself.

Audio Input for Music Navigation

Most users use some kind of haptic input (for example a scroll wheel) to browse through a music collection. This approach is not suitable for disabled people who lack coordination in their hands. If these people are not deaf and can provide audio input, they can use their voice to explore a music collection. Consider for example a music collection where each piece of music is mapped as a point in a three-dimensional coordinate system consisting of the axes of time, genre, and title. In order to address an individual song (point in 3D) the following three signals could be extracted from the user's audio input: audio volume, pitch and similarity of the audio input to a certain vowel. Varying those parameters independently or in combination allows one to navigate in this three-dimensional music space. The basic concept is outlined in [15].

Simpler solutions can also be conceived. Think, for example, of a music playlist which can be scrolled via the user's voice by mapping the pitch frequency to the position in this list. Alternatively, the initial pitch frequency could serve as reference and increasing the pitch would initiate a scrolling to the top of the list; decreasing the pitch relative to the starting pitch frequency would initiate a scroll down in the list.

Audio Input for Music Editing

Music editing based on audio only is beyond reach at the moment since it is difficult to specify the editing parameters only by audio. Rather a combination of audio and haptic input is applied. It is, for example, perfectly possible to provide audio input to replace music passages which are specified beforehand. This can, for example, be done by marking the interval which should be changed and playing the original music signal in the background so that the user can synchronize with the music. Alternatively, the user could mark the channels in a multi-channel recording that should be modified and whenever the user provides an audio input on those channels, the previous signal is overwritten.

26.2.3   Visual and Other Sensor Input

Visual and other sensory input for music-specific tasks is novel and made possible by real-time image and signal processing. It can be considered as indirect music input compared to playing a physical instrument, singing, or manipulating a piece of music via haptic or audio input.

Visual and Sensor Input for Music Generation

One of the first electronic musical instruments which has been played touchlessly with hands in the air is the Theremin, developed in 1919 by Lew Termen [1]. The instrument contains a high-frequency LC oscillator (several hundred kHz up to several MHz resonance frequency) which is mixed down via a fixed second oscillator into the audible frequency range. If the user's hand approaches the Theremin, the hand's capacity influences the resonance frequency of the first oscillator and thus also modulates the mixed-down frequency. The position of the hand relative to the Theremin is translated into tones with different pitch and volume. Many alternative technologies emerged which sense the musician's hand in front of an instrument based on other measurement principles, e.g. from the optical or acoustical domain.

Another method to generate or better modulate music is the concept of the electronic conductor: When the user moves the electronic conductor's baton, he/she can initiate a playback of a pre-defined piece of music and synchronize the music with the tempo and rhythm of the baton. Technologically, this is enabled by sensors such as accelerometers, magnetic sensors, and gyroscopes. User gestures can also be detected by dedicated sensors in the surroundings or by body sensors which are worn by the user. One of the most elaborate examples in this category is a *conductor's jacket* equipped with a plurality of sensors to estimate the motion and engagement of a musician who conducts a virtual orchestra [9]. This solution is used for training to conduct orchestras. The conductor is synthesizing orchestra music by his activity and can also generate very false music if he is doing it incorrectly. The user can listen to music which serves as direct feedback.

The majority of classical instruments have been portable, except for heavier and larger instruments (organ, piano, ...). In the past, most electronic instruments have been stationary due to the bulkier electronics and the constant need for electrical power. In recent years mobile devices have taken over the innovation lead from stationary computers and are also making inroads into the area of electronic music instruments. Mobile devices contain a plurality of sensors which enable them to become musical instruments in combination with haptic interaction. Due to their communication capabilities they can also form electronic orchestras. An exploration of the mobile music device design space is given in [5].

Visual and Sensor Input for Music Query

Sensors are mostly used to determine the context of the user, i.e. the physical activities and the user reaction in a specific situation. They need to be interpreted for meaningful input. It is not very likely that the sensory information alone determines a music query, but it can be used efficiently in conjunction with direct (mainly haptic and audio) input. It helps to disambiguate the query, taking into account user habits for a certain spatial-temporal context. An example to map visual information from an image into musical properties for music generation purposes is given in [17]. Image properties such as contours, colors, and textures are mapped to musical features like pitch, duration, and key. Instead of using those musical properties for music generation, they could specify music query parameters as well and create a query for music which is compatible in its mood with the imagery.

Visual and Sensor Input for Music Navigation
Visual and other sensor signals can be used directly for music navigation. This is done by sensing the body, arm, and hand postures and translating them into navigation input similar to a haptic input. An example of this approach, using a Wii Remote, is outlined in [13].

But sensors can also be used as indirect input for music navigation: A video analysis can be used to synchronize the music to the user or adapt the tempo of the music (for example in music games). A typical simplification of the analysis can be done when the user carries markers or sends out signals such as the positions of infrared light sources as is, for example, used in game consoles. Image processing can concentrate on the user, for example, by identifying the user mood from the facial expression. This information is used for a suitable selection of music or as general user feedback. Alternatively, a complete visual scene analysis can be made. This leads to future applications which perform a music audience analysis and use the results for adapting the music playback.

Visual and Sensor Input for Music Editing
The difference between body sensors and other sensor devices is their "wearable" nature, i.e. they are integrated into the clothing, are worn (for example like a watch) or are implanted due to a medical indication. Interesting concepts emerge which make the body signals audible and give real-time feedback to the user. The "motion sonification" project [2] is an approach to optimize motion coordination in rehabilitation or sports training by picking up body limb orientation and acceleration to create an audio feedback in relation to how far the motion pattern deviates from the optimal coordination. In the future, more music-like (pleasant) feedback signals might help to achieve a long-term user acceptance of this effective method.

26.2.4   Multi-Modal Input

For complex interactions between man and machine it is often advantageous to use multiple input modalities, either at the same time or sequentially. An audio input can be a reliable input in silent environments, but might fail if the background noise level is too high. In these situations a tactile input (for example entering the information via a keyboard) or the detection of facial expressions for supporting speech recognition with visual cues can resolve ambiguities in the interpretation. The technical system can be designed to expect parallel alternative inputs all the time or switch between the input modalities based on user choices.

26.2.5 Coordination of Inputs from Multiple Users

Telecommunication systems make it possible to collaborate with remote communities in real time and share experiences via worldwide networks. Especially in the music domain and for music creation, novel services emerge when different participants at distant places all contribute to a piece of music. A typical use case is a distributed orchestra, where the members all play their own instruments and want to listen to the joint orchestra sound as feedback (see also Figure 26.1). Different challenges exist in coordinating the multiple users: The transmission from the originator to all others should be done with low latency (on the order of 10ths of milliseconds) so that the different musicians can also listen while the others are playing. The different inputs have to be synchronized and aligned. This can be assisted by technology, but is ultimately only possible if the other musicians hear their own input relative to the other music components.

26.3 User Interface Output for Music Applications

26.3.1 Audio Presentation

In music it is natural that the most significant output modality is audio. As a simple application of audio it is used as feedback for user input. When the user presses a button or performs a touch action, often an acoustic feedback is given to the user. In a musical application, the audio channel is used for music playback and can represent a mix of audio signals (for example in music editing). But audio can also convey more complex musical parameters such as timbre, tempo, rhythm, harmony, and melody (refer to Chapters 2 and 3 for the physical and psychological fundamentals and Chapter 5 for the signal processing perspective). While a single user can only provide monophonic audio input, the resulting audio output can carry multiple signals and voices at the same time.

Audio can also give subtle feedback. One example is an Internet radio application which allows the user to tune in to a dedicated station by turning a frequency knob like in a classic AM/FM radio. In one concept the audio feedback during the tuning interval is typical radio noise to indicate the ongoing tuning activity.

26.3.2 Visual Presentation

Today's user interfaces are dominated by visual elements since screens with higher graphical resolution have become standard in most technical systems. The visual modality can provide high-dimensional representations of data. The information can be elements of a typical music player such as playback controls, music cover and visuals, playlists, and the like. But, in addition, it is possible to represent the style of music via animations, providing the score (for musicians) or representing the complete music collection.

An in-depth analysis how a music collection can be structured and presented visually can be found in Chapter 22. Similarity metrics for songs based on musical features are introduced in Section 22.2. Based on these metrics, complete music collections can be visualized as two-dimensional maps (see Section 22.3). In addition, concepts are presented for dealing with projection errors when the high-dimensional space of features is reduced to a 2D representation.

Many other novel concepts for music analysis and navigation systems exist which demonstrate the expressive power of visual information in parallel to the music:

- In *MusicRainbow* [10], music artist names are projected onto a circle. Artists whose music is similar based on the audio signal are placed close to each other. A traveling salesman algorithm is used to optimize the placement of artists relative to each other. The user rotates a knob to select an artist and pushes it to listen to the artist's music. The user interface is shown in Figure 26.4.

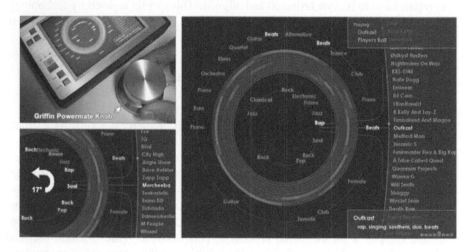

Figure 26.4: User interface of MusicRainbow music navigation.

- *Sourcetone*[3] [7] provides an automatic music classification system which ranks music according to attributes such as mood, activity, and health. Users can select a desired set of attribute properties in the user interface and listen to the matching songs. In Figure 26.5, a circle with four quadrants represents the different categories of moods, where neighboring segments show similar moods and opposite segments represent inverse moods.

- The *GlassEngine*[4] was developed to provide graphical access to the work of the composer Philip Glass. Each vertical stripe represents a piece of music. Any piece of music can be found on different sliding bars which are used to sort the complete music according to different criteria such as title and time period. The user clicks and slides a bar for navigation to select a piece of music for playback.

- *Musicroamer*[5] is a visualization of the musical relationships among artists. A

---

[3]http://www.sourcetone.com. Accessed 22 June 2016.
[4]http://www.philipglass.com. Accessed 22 June 2016.
[5]http://musicroamer.com. Accessed 11 July 2016.

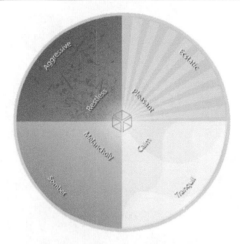

Figure 26.5: Sourcetone emotion wheel for music selection.

graph structure shows the artists most similar to an initially provided artist based on user tag data retrieved from LastFM.[6] These artists can be used as starting points for a new similarity search. A similar service is offered by *LivePlasma*[7] which provides audio streaming for playlists of the selected artists. An exemplary visualization of LivePlasma is shown in Figure 26.6.

- *Map of Mozart*[8] is another type of visualization of a musical landscape (see Figure 26.7). Rhythm patterns are extracted from each piece of music. A self-organizing map [12] groups acoustically similar pieces of music close to each other (cp. Section 11.4.2). A user can listen to a certain type of music by selecting a region on the map.

26.3.3 Haptic Presentation

A haptic response can give feedback to a user action. This is a classical property of user interfaces whenever a user touches a surface. A simple example is the mechanical click of a button as confirmation that it is switched on. A more sophisticated case is a programmable action such as a vibra feedback on a mobile device. Haptic feedback can also reveal music content-related properties in an eyes- and ears-free fashion. Think, for example, of a playlist of pieces of music or a two-dimensional music map on a touchscreen (like those described in the previous section). Whenever the finger passes a relevant piece of music fitting to a music query on the screen, the device can respond with a vibration signal, indicating a match with the search criteria. Haptic feedback in general simplifies blind, i.e. eyes-free, operation of devices.

---

[6]http://www.last.fm/api. Accessed 22 June 2016.
[7]http://www.liveplasma.com. Accessed 22 June 2016.
[8]http://www.ifs.tuwien.ac.at/mir/mozart. Accessed 22 June 2016.

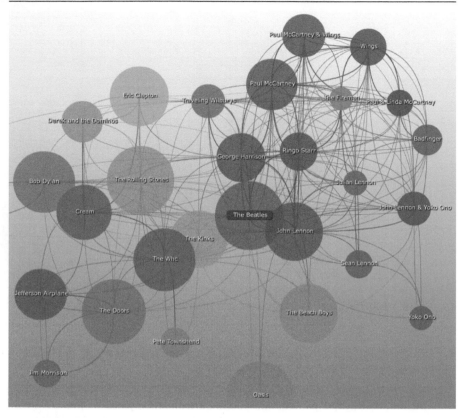

Figure 26.6: Example of LivePlasma output for *The Beatles* (screenshot from Live-Plasma web site, October 3rd, 2014).

Another similar interaction concept is implemented in Shoogle [16]. It reveals content in a mobile device through shaking. The device is used as a virtual container for pieces of content which are represented as (virtual) balls. If the user shakes the device, the balls are accelerated and collide with the virtual walls of the container, causing them to bounce back. Each collision with the walls can be made audible and can be felt by the user as a short vibra pulse. The number of collisions is proportional to the number of balls which represent the information objects. If, for example, the user queries the system for a certain type of music, the number of matching songs could be mapped to the number of balls rattling in the box. Shorter songs could be represented as lighter balls with weaker vibra and audio collision signals.

26.3.4 Multi-Modal Presentation

A multi-modal representation of music utilizes different modalities of information to balance the shortcomings of the various individual modalities or to enrich the

Figure 26.7: Map of Mozart: Self-organizing map for Mozart's music.

presentation. A typical multi-modal music presentation combines the different output channels described in Sections 26.3.1–26.3.3.

A web-based example for the trend towards multi-modal input and output is Musipedia.[9] It allows music queries by recording voice input, but also allows haptic input by clicking notes via a computer mouse on a virtual keyboard or by recording input from a connected MIDI keyboard. The output of a music search is presented visually as melody lines, metadata of the song, and as musical output.

26.4 Factors Supporting the Interpretation of User Input

26.4.1 *Role of Context in Music Interaction*

Music preferences are certainly person-specific. But even for the same person the music perception and the favored music can depend on the context. The context of a person is defined by a variety of properties:

- *Location*: A home environment can lead to different music preferences than a more public environment, such as being abroad.
- *Time*: Often the music selection strongly depends on the time of the day and the

---

[9]http://www.musipedia.org. Accessed 22 June 2016.

type of day (for example work day vs. weekend), reflecting different attention levels of a music listener.

- *Mood*: The user's mood is a more subtle parameter influencing situation-specific music preferences, but it certainly impacts the more emotional music genres.

- *Activity pattern*: Physical activity influences music listening habits, but conversely, music can also support an activity. Motivational music in sports is a typical example of the latter.

- *Social interaction*: The social interaction with friends and other persons in a certain situation also shapes the behavior of a user. For example, it can directly influence the music preferences during a party or other events.

- *Environment*: The atmosphere created by the surroundings, such as climate, audiovisual and haptic stimuli, in short everything that a user can perceive via his or her senses, is an external input for the user's mood. Indirectly the environment has an impact on the music preferences in a certain situation.

These context parameters are additional indirect input signals. They can be estimated by a technical system by evaluating information from a positioning system, a clock, user input patterns, and sensors. These measurements help to characterize the physical environment as well as how the user reacts to it. Thus, it is possible to implement a context engine which tries to infer the user's music preferences. The system can suggest suitable music for certain contexts or confine itself to an existing selection, see also Section 23.2.3.

26.4.2   Impact of Implementation Architectures

User interfaces have progressed significantly during the last decades and users have become more demanding as well. Two important acceptance factors are the responsiveness of the user interface, i.e. a low-latency system response, and clear user feedback indicating the system behavior. In some cases the system feedback should be given within milliseconds, for example, when musicians are playing together (locally or via the Internet). The generated music must be kept in synchronization with the other artists. Latency in the range of seconds, while tolerable in a pure playback scenario, is seldom tolerated in most other use cases.

In Chapter 25 the various solutions to partition a music processing system were presented (see Section 25.2). Especially the distributed architecture solutions pose a challenge in this respect due to the time required for communication and processing with the remote parts of a networked system. Therefore, the user interaction activities are separated into foreground and background activities which are executed concurrently, but on different time scales.

The foreground activities typically react to user input within milliseconds by providing acoustic, visual, and haptic feedback. Playing music on an electronic instrument is also such a foreground activity, which should provide musical sound with low latency. In Chapter 27 various hardware approaches for efficient and high-performance device architectures are described. In the software domain real-time operating systems, such as RT Linux or QNX for example, provide low-latency multi-tasking capabilities by distributing tasks to multiple cores and offer elaborate scheduling mechanisms for managing time slice parallelism in software threads and processes [14][11]. The combination of hardware and software concepts enables a fast feedback loop for user input. The background activities are related to the communication with other parts of a distributed system and are not as predictable as the local threads and processes. Response times can vary significantly from milliseconds to seconds or even minutes due to two factors: Data transfer times depend on the available channel bandwidth and are typically time variant. Secondly, the processing time at the receiving end might also vary, for example, due to different server loads.

Therefore, user interfaces in networked devices often cache relevant data (for example user interface representation information and music content) to provide a fluent and uninterrupted user experience. Especially for music, this caching of data is appropriate since the music consumption is mainly sequential and continuous by listening to tracks from a playlist. See also Section 23.3 for more details on music recommender systems. The positive consequence of data caching is a responsive user interface since the fast local interaction between user and device (latencies on a millisecond time scale) and the slow (distributed) processing of music data are decoupled. In general, a balance between device-centric computations and remote processing has to be found.

Another aspect related to the distributed and networked architectures available today can be derived from the fact that a user is able to access the entire music available via the Internet. The scalability of a user interface is very important given the fact that people are no longer dealing with small local music collections only, but have access to millions of songs. Making this huge quantity accessible means being able to partition the collection on-the-fly and to show a subset in the user interface based on user preferences. This poses a heavy load on the device hardware (memory, processing, graphics subsystem). Scalability also has another meaning since multiple interaction devices can be used for musical purposes. It is quite typical to interchangeably use classic home stereo systems (with network interface), PCs/laptops, mobile devices such as tablets, smartphones, wearables, car headunits, and possibly further devices in the future. Nowadays, it is expected that the content follows the user into different device contexts. As a consequence, the user interfaces have to be adapted to the different interaction devices. Each of these devices poses specific challenges in terms of user interface elements and user attention.

26.4.3 *Influence of Social Interaction and Machine Learning*

In the past, user interaction with technical devices has been a one-to-one relationship. The user provided explicit and unambiguous input to a device, for example, by pressing a button whose purpose was clear. Today, social interaction with other users provides an additional input source so that multiple entities influence the user interface content. In a recommender system with collaborative filtering (see Section 23.2.1) it becomes relevant to also show the social interaction in the user interface presentation.

Another emerging trend of today's user interface is the use of implicit information. In contrast to explicit user input (for example pressing a button with a defined meaning), implicit inputs have to be interpreted by a technical system since they are extracted from a complex user behavior or the user's context.

- The user context can be estimated through a variety of sensors and influences the user habits; therefore, the user interface output is often adapted to those changing needs. Examples are a recommender system which proposes different music based on the user's mood or adapted representations of the user interface for mobile and stationary devices.

- Implicit information can also be generated directly by the user during the interaction with the technical system in two ways: One option is that the user is certain about his input (for example when issuing a speech command or performing a gesture). However, for the machine it is implicit input which has to be interpreted via feature extraction, processing, and classification. Alternatively, the user input is explicit, but only the sequence of explicit inputs provides new information and has to be interpreted. An example is a music recommendation system where users are often reluctant to provide explicit ratings. However, their listening behavior, i.e. how long they listen to a song, skip forward or revisit a piece of music, gives a good indication of their music preferences; see Section 23.3.

In both cases machine learning algorithms are typically used to create a relationship between implicit user interface input and user interface output (for example recommended music).

26.5   Concluding Remarks

In this chapter the description of user interaction has been structured according to the modalities of the human senses. While in the past, single input and output modalities have been used in the interaction between a music device and its user, it is increasingly common to find multi-modal systems, both for the input and output side. A clear message from user studies is that the latency between (music-related) user input and device output should be as low as possible in order to make the music device controllable and maintain synchronization with other musicians. The use of sensors as additional interaction channels increases the potential of electronic music systems and complements classical input modalities. The user interface should be responsive and provide clear feedback to the user with minimum latency.

In the future, we will see multi-modal user interfaces for distributed systems, which will increase the demand for high-speed communication and processing. Distributed systems for music could emerge from a network of musicians who want to make music via their network of electronic instruments. In distributed architectures, processing can be offloaded from the local device to other cloud- or server-based machines, as shown in Chapter 25. Another trend, not only for music user interfaces, is the application of machine learning algorithms to interpret the user context and implicit inputs. This allows us to build electronic music systems which are intelligent enough to understand the user and possibly also reduce the number of music-related input errors.

Bibliography

[1] I. Aldshina and E. Davidenkova. The history of electro-musical instruments in Russia in the first half of the twentieth century. In *Proceedings of the Second Vienna Talk, University of Music and Performing Arts Vienna, Austria*, pp. 51–54, 2010.

[2] H. Brückner, W. Theimer, and H. Blume. Real-time low latency movement sonification in stroke rehabilitation based on a mobile platform. In *International Conference on Consumer Electronics (ICCE)*, pp. 262–263, Las Vegas, January 2014. IEEE.

[3] P. Cano, E. Batlle, T. Kalker, and J. Haitsma. A review of audio fingerprinting. *J. VLSI Signal Processing*, 41:271–284, 2005.

[4] S. Deterding, R. Khaled, L. Nacke, and D. Dixon. Gamification: Toward a definition. In *CHI 2011 Workshop Gamification: Using Game Design Elements in Non-Game Contexts*, Vancouver, May 2011. ACM.

[5] G. Essl and M. Rohs. Interactivity for mobile music-making. *Organised Sound*, 14(2):197–207, 2009.

[6] A. Ghias, J. Logan, V. Chamberlain, and B. Smith. Query by humming: Musical information retrieval in an audio database. *Proc. ACM Multimedia*, pp. 231–236, 1995.

[7] A. Huq, J. Bello, A. Sarroff, J. Berger, and R. Rowe. Sourcetone: An automated music emotion recognition system. In *Proceedings; International Society for Music Information Retrieval (ISMIR 2009); Late breaking papers / demo session*, Kobe, Japan, 2009. International Society for Music Information Retrieval.

[8] P. Knees, M. Schedl, T. Pohle, and G. Widmer. Exploring music collections in virtual landscapes. *IEEE Multimedia*, 14(3):46–54, 2007.

[9] T. Nakra. Synthesizing expressive music through the language of conducting. *Journal of New Music Research*, 31(1):11–26, 2001.

[10] E. Pampalk and M. Goto. MusicRainbow: A new user interface to discover artists using audio-based similarity and web-based labeling. In *Proceedings International Society for Music Information Retrieval*, pp. 367–370, 2006.

[11] QNX Software Systems Limited. QNX Neutrino RTOS System Architecture, 2014. http://www.qnx.com.

[12] H. Ritter, T. Martinetz, and K. Schulten. *Neural Computation and Self-Organizing Maps: An Introduction*. Addison-Wesley, Boston, MA, USA, 1992.

[13] R. Stewart, M. Levy, and M. Sandler. 3D Interactive environment for music collection navigation. In *Proc. of the 11th Int. Conference on Digital Audio Effects (DAFx-08)*, pp. 1–5, Espoo, Finland, September 2008.

[14] A. Tanenbaum and H. Bos. *Modern Operating Systems*. Prentice Hall Press, Upper Saddle River, NJ, USA, 4th edition, 2014.

[15] W. Theimer, U. Görtz, and A. Salomäki. Method for acoustically controlling an electronic device, in particular a mobile station in a mobile radio network,

US patent 6321199, 2001. filed 13-Apr-1999, published 20-Nov-2001, `http://www.lens.org/lens/patent/US_6321199_B1`.

[16] J. Williamson, R. Murray-Smith, and S. Hughes. Shoogle: Excitatory multi-modal interaction on mobile devices. In *Proceedings of CHI 2007*, pp. 121–124. ACM, 2007.

[17] X. Wu and Z. Li. A study of image-based music composition. In *IEEE International Conference on Multimedia and Expo*, pp. 1345–1348, Piscataway, NJ, June 2008. IEEE.

Chapter 27

Hardware Architectures for Music Classification

Ingo Schmädecke, Holger Blume

Institute of Microelectronic Systems, Leibniz Universität Hannover, Germany

27.1 Introduction

Music classification is a very data intensive application evoking high computation effort. Dependent on the hardware architecture utilized for signal processing, this may result in long computation times, high energy consumption, and even in high production cost. Thus, the utilized hardware architecture determines the attractiveness of a music classification-enabled device for the user. This multitude of requirements and restrictions a hardware architecture should meet cannot be covered simultaneously. This is why a hardware designer must know the quantitative properties of all suitable architectures regarding the application for music classification in order to develop a successful media device.

In this chapter, we discuss challenges a system designer is confronted with when creating a hardware system for music classification. Several requirements like preferred short computation times, low production costs, low power consumption, and programmability cannot be covered by one hardware architecture. Instead, each hardware architecture has its advantages and disadvantages.

We will present several hardware architectures and their corresponding performance regarding computation times and efficiency when utilized for music classification. In detail, General Purpose Processors (GPP), Digital Signal Processors (DSP), an Application Specific Instruction Processor (ASIP), a Graphics Processing Unit (GPU), a Field Programmable Gate Array (FPGA), and an Application Specific Integrated Circuit (ASIC) are examined. Special care is given to the extraction of short-term features which can be accelerated in different ways and represents the most time-consuming step in music classification besides decoding. Finally, a practical example of energy cost-limited end-consumer devices demonstrates that this design space exploration of hardware architectures for music classification can support the design phase of stationary and mobile end-consumer devices.

Media devices like smartphones and stationary home entertainment systems with built-in music classification techniques provide several benefits compared to Internet-based solutions. First, no Internet connection is required to transfer data between a media device and a server offering music classification services. This also reduces energy costs since power-consuming wireless connections can remain switched off. This is a very important aspect considering the operating time of mobile devices. Finally, new music files, even unpublished music, can be processed offline. Hence, the approach of media devices with built-in music classification support promises high flexibility.

This chapter is structured as follows: First, different metrics are presented that are suitable to evaluate hardware architectures. Then, feature extraction-specific approaches to utilize available hardware resources are explained. Hardware architecture basics are provided afterwards. Finally, a comparison of architectures regarding their efficiency in terms of performing music classification as well as expected costs is presented.

27.2 Evaluation Metrics for Hardware Architectures

Hardware architectures can be compared using various metrics that are related to the cost aspects of a device. They can be differentiated into measurable (e.g. silicon area) and non-measurable (e.g. flexibility) metrics. In general, a hardware designer is interested in a subset of such metrics to choose a suitable architecture approach for a target product. In the following, fundamental cost factors relevant for music classification systems are presented. In addition, combined cost metrics are introduced that are utilized to effectively evaluate an architecture.

27.2.1 Cost Factors

Silicon area (A): The silicon area of an architecture primarily impacts the overall production costs for a high number of units. For an objective comparison of different architecture approaches, the impact of the feature size of the applied technology, which is the transistor size used for production, must be eliminated. Therefore, silicon areas have to be scaled to the same transistor size to ensure that only the architectures are compared regardless of the technology used.

Power consumption (P): The power consumption significantly influences operation costs as well as battery lifetimes. Hence, it affects the attractiveness of a product. Furthermore, the resulting heat determines the required cooling method. This is why power consumption is one of the most important aspects of music classification systems.

Computation time per result (T) (resp. throughput rate (η/T)): This application-specific cost factor requires the definition of a task to be performed or a result to be computed. In terms of music classification, a task can be the classification of one music file including computation steps like feature extraction, feature processing, and classification. Then, T corresponds to the time needed to identify a matching music label for the examined music file. A related metric is the throughput rate, which specifies the number of music files (η) that can be classified within a defined time (T).

Energy consumption per result (E_{result}): The energy consumption per result is related to the power consumption and the computation time per result for a given architecture. The advantage of this cost metric is that information about the expected battery lifetime can be directly estimated if the energy capacity is known.

System quality: This cost factor has a limited dependency on the utilized hardware architecture. For music classification systems, the system quality results from the interaction between hardware and software. A suitable system quality metric is the classification rate.

Flexibility: An architecture's flexibility is hard to measure since it comprises many facets like development costs, reusability, adaptability to further tasks, and signal processing methods. Thus, this factor is frequently evaluated qualitatively regarding the implementation effort of a signal processing chain.

27.2.2 Combined Cost Metrics

An effective consideration of several cost factors can be achieved by combined cost metrics that merge multiple cost factors with scalar metrics. The area efficiency

$$E_{Area} = \frac{\eta/T}{A}$$

and energy efficiency

$$E_{Energy} = \frac{\eta/T}{P}$$

represent two typical combined cost metrics that are useful for music classification systems. These cost metrics relate the application-specific throughput rate of a hardware architecture to its silicon area or its power consumption, respectively.

The throughput rate in the context of music classification depends on the amount of music content per file. An evaluation metric, which does not depend on content per file, is required. This metric can incorporate an alternative definition of the throughput rate, which is based purely on the time taken to extract a series of feature sets within a prescribed time window. This approach can be used since the feature extraction step is by far the most time-consuming step of the signal processing chain. Hence, the efficiency metrics are related to the feature extraction and are used to estimate the efficiency of performing music classification. In this case, the computation time per result is related to the time for extracting a feature set from a frame.

Finally, the ATE product

$$C_{ATE} = A \cdot T \cdot E_{result}$$

considers the silicon area, the computation time, and the energy consumption of a hardware architecture at the same time. This combined cost metric can also be used to estimate a hardware architecture's efficiency. In addition, it can be used together with the achievable classification rate in order to explore most of the discussed cost factors in a two-dimensional design space.

27.3 Specific Methods for Feature Extraction for Hardware Utilization

The efficient usage of hardware architectures requires that the available computation resources are utilized as much as possible. In this context, one possible optimization criterion is the reduction of the computation time that can be achieved by parallel computation for example. The extraction of short-term features allows an accelerated execution by parallel processing in several ways. Considering the feature extraction procedure itself, a single feature can be extracted from multiple frames concurrently and several independent features can be extracted from the same frame at once. Beside this, the actual extraction of a feature can benefit from parallel processing capabilities, too. A typical approach is to process several data with the same instruction, which is called Single Instruction Multiple Data (SIMD). Moreover, a concurrent execution of different instructions can be applied at once to extract a single feature, which is denoted as Multiple Instruction Multiple Data (MIMD). However, the quality of these approaches strictly depends on the particular feature. Another approach to increase the efficiency is to reduce hardware resources while keeping the computational performance at the same level. Of course, this approach is only possible if the hardware architecture design itself can be modified. All these approaches are covered by the architectures that are investigated in the next section.

27.4 Architectures for Digital Signal Processing

As a consequence of various demands on hardware systems, several hardware architectures exist today. They cover a broad field of approaches starting from low-cost standard designs for general purposes, up to highly specialized solutions. This section introduces basic hardware architecture designs. Afterwards, design concepts are presented that are more application specific but may still adopt elements of basic designs. The challenges of designing combined hardware architectures, also called heterogeneous architectures, will not be discussed in this chapter. The benefits of combining two architectures, one for extracting features and the other one for performing remaining signal processing steps, are discussed in Section 27.5. The following brief introduction to architecture fundamentals is based on [4] and [10], which provide a more detailed in-depth discussion of processor designs.

27.4.1 General Purpose Processor

General Purpose Processors (GPPs) are not designed for specific tasks but to successfully perform any arbitrary application as the term "General Purpose" indicates. GPPs can be found in server, desktop, and embedded systems and therefore, dif-

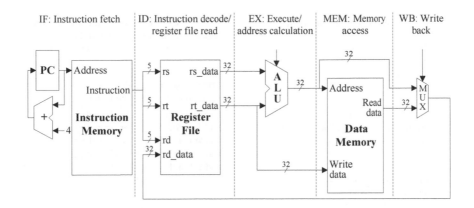

Figure 27.1: Simplified 32-bit RISC processor core architecture.

ferent design philosophies are followed to achieve high performance and low power demands. However, the baseline of the architecture design of GPPs remains the same and is based on two main components: datapath and control unit. The datapath performs arithmetic operations. The control unit tells the datapath and other elements of a processor and system what to do, according to the instructions to be executed. GPPs normally include internal memory like caches that can reduce data access times resulting from bigger but also slower external memories. A recent example of these general purpose processors is the Intel Core i7-2640M.

For data-intensive processing steps like the extraction of short-term features, the datapath limits the runtime performance more than the control path and is therefore a subject for further consideration. The fundamental structure of a typical GPP datapath is illustrated in Figure 27.1. For simplification, an architecture of a reduced instruction set computer (RISC) is shown without the control path and without additional datapath elements required for program branches.

Typically, datapaths are subdivided into five pipeline stages. Within the instruction fetch (IF) stage, instructions are read from an appropriate memory. Instructions are coded by 32 bit and therefore, a program counter (PC) pointing to the current memory address is incremented by four bytes. In more detailed datapath illustrations, instructions are also able to modify the PC value. During the instruction decode (ID) stage, the two independent source register addresses (rs, rt) and the destination register address (rd) of an instruction are identified. These addresses are related to an array of registers, which is called a register file. This is a set of registers, to which the datapath has direct access. The actual operation of an instruction is performed within the execute (EX) stage. The operation is executed by an arithmetic logic unit (ALU) which can process up to two 32-bit operands to compute one result. Moreover, the ALU is used to calculate a memory address from which data is read or written to memory. This is done within the latter memory access (MEM) stage. Finally, data read from memory is executed by the ALU and results are written into the datapath's register within the write-back (WB) stage.

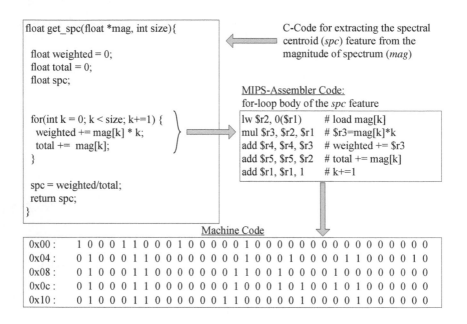

Figure 27.2: Translation flow from high-level C-language to machine code; example program code is based on the spectral centroid feature.

In the following, the program code for the extraction of the spectral centroid (*spc*) feature is explained (cp. Definition 5.3). A C-language-based function for its extraction is depicted in Figure 27.2. It is assumed that this feature is extracted from the magnitude of the spectrum (declared as *mag*) which has already been previously computed from an audio frame. The number of spectral components to be considered is specified by the function parameter *size*.

An essential element of such a feature extraction function is the for-loop in which spectral components, respectively audio samples, are sequentially processed. This procedure is typical for most of the short-term features and generally corresponds to the most time-consuming part of a feature extraction function. For an execution of this function, the C-code must be translated into machine code. Therefore, an intermediate step is performed that translates the architecture-independent high-level program code into the so-called assembler code. On the right-hand side of Figure 27.2, the assembler code of the for-loop body is shown. These are all instructions executed during one for-loop turn. This assembler syntax is applicable for "Microprocessor without Interlocked Pipeline Stages" (MIPS) [10]. At this level, source and destination registers within the register file are directly addressed (denoted by $r). Furthermore, the C-code is split into instructions that can be executed by the ALU. At the end, the assembler code can be translated into machine code that is readable by the processor. An example of machine code is also presented at the bottom of Figure 27.2.

Program execution order (in instructions)	Cycle	1	2	3	4	5	6	7	8	9	10
lw $r2, 0($r1)		IF	ID	EX	MEM	WB					
mul $r3, $r2, $r1			IF	ID	EX	MEM	WB				
add $r4, $r4, $r3				IF	ID	EX	MEM	WB			
add $r5, $r5, $r2					IF	ID	EX	MEM	WB		
add $r1, $r1, 1						IF	ID	EX	MEM	WB	

Figure 27.3: Execution flow of pipelined GPPs.

It takes a specific amount of time to process an instruction since each datapath stage implies a delay until an input signal of a stage affects its output. This means that the sequence of instructions must be triggered in time, which is done by a clock. Usually, a clock is a periodic signal and therefore can be specified by the length of one cycle in time units and by the number of cycles per second. Therefore, the cycle time must not be lower than the maximum time of the datapath to perform an instruction. In any case, the subdivision of the datapath into stages allows an architecture implementation technique called pipelining in which multiple instructions are overlapped in execution. This is a key approach to make processors fast. On the one hand, the clock cycle time can be increased because the overall delay of a datapath is split by the number of stages. On the other hand, the time to execute the for-loop is decreased which is illustrated in Figure 27.3.

The examined pipelined processor requires five clock cycles per instruction while the same processor with no pipelining performs one instruction within one cycle. It can be assumed that both processor variants process one instruction within the same time since the processor with pipelining has an approximately five times higher clock rate than the one without pipelining (assuming that any data and control dependencies are resolved). However, the pipelining approach allows the overlap of instruction executions. This means that the for-loop body of the *spc* function is executed after nine clock cycles while a processor with no pipelines takes 2.8 times longer to execute the same instructions although it requires only five cycles. This demonstrates that pipelining is a suitable concept for processors and for other hardware architectures. In practice, only additional registers between the stages must be integrated into the datapath design in order to extend an architecture for pipelining.

27.4.1.1 SIMD Instruction Set Extensions

Recent GPPs provide Single Instructions Multiple Data (SIMD) extensions that are able to process arrays of operands with the same type of instruction concurrently. Therefore, an application with SIMD instructions requires that multiple operands to be processed are arranged in one register. Besides an enhanced ALU, an extension of GPPs for SIMD support can be realized either by reusing the available register file or by adding an SIMD exclusive register file.

One suitable approach to use SIMD instructions for the extraction of features is to reduce the number of for-loop turns by parallel processing. This is demonstrated

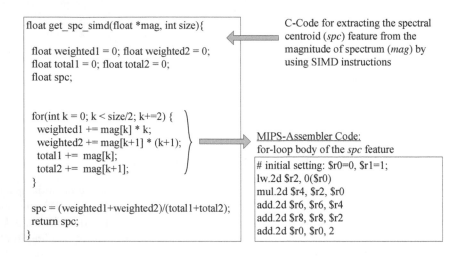

Figure 27.4: Modified program code of the spectral centroid feature (*spc*) for SIMD execution.

by the *spc* feature code in Figure 27.4. In this example, a GPP is supposed to execute two single instructions concurrently using SIMD instructions, so that the for-loop counts can be reduced by a factor of two. The for-loop body is expanded in order to execute twice the number of instructions per turn. This partition requires a certain data independency. Moreover, a final step must be performed after the for-loop structure to merge the partitioned results and to extract the *spc* feature. As a result, the instruction count to extract the *spc* feature is reduced overall.

The corresponding assembler code is similar to the single instruction-based program code. One obvious change is the identifier appended to the instruction names that indicates the respective array size of each operand. This is also reflected in the selection of register addresses. In detail, only the first register address of an array is specified while the next upper address is indirectly used. That is why only even addresses are used within the assembler code.

27.4.2 Graphics Processing Unit

Graphics Processing Units (GPUs) are designed to meet the high computation effort of modern graphic applications that can significantly take advantage of parallel computations. Current GPUs offer high flexibility in terms of programmability and are thereby also suitable for general purpose computations. Recently this has led to an increased use of GPUs in several research areas. The general GPU hardware architecture is illustrated in Figure 27.5 based on an NVIDIA GPU. The available hardware resources are hierarchically organized and allow concurrent processing at different levels. In detail, an NVIDIA GPU consists of a set of so-called Streaming Multiprocessors (SMs). An SM includes several CUDA [8] (Compute Unified De-

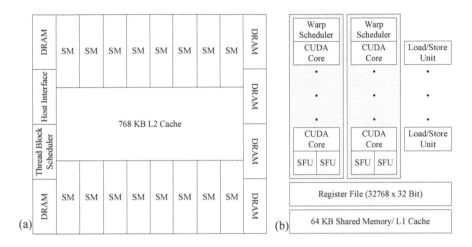

Figure 27.5: Regular structure of a graphics processing unit (GPU): (a) Top-level GPU structure, (b) Streaming Multiprocessor (SM) structure.

vice Architecture) cores each containing an ALU and register file. The computation flow of a CUDA core is similar to the flow of a GPP. However, all CUDA cores of an SM execute the same instructions concurrently on different data, which constitutes SIMD-like processing. In contrast to real SIMD computation, CUDA cores are also able to execute unconditional and conditional branches. Since all cores must execute the same instructions, the output of each core can be masked. For example, if an *if*-condition has to be executed, all cores, where the condition is not true, are disabled by masking. Afterwards, an optional *else* branch is executed by only letting the remaining cores of an SM being enabled. This procedure demonstrates that such programming structures must be avoided to increase the hardware utilization and the efficiency of a GPU. Besides, each SM provides its own shared memory to which all CUDA cores of the same SM have access, allowing a certain degree of data dependency. However, the same is not valid for data dependencies between other SMs since they cannot be synchronized. The usage of shared memory is recommended because it provides higher data rates than the external memory of a GPU card that must contain all data to be processed and that offers higher storage capacities than the shared memory available.

The GPU concept can be applied to feature extraction in order to extract one feature from several frames simultaneously. In detail, the extraction of a feature from one frame is performed by exactly one SM and the extraction itself is additionally accelerated by the available CUDA cores. NVIDIA offers a special programming model for their GPUs that is based on the C/C++ language. Since a GPU cannot be used alone, it must be integrated within a system, also called the host, that includes a CPU. The CPU must execute additional program code in order to call GPU specific functions to be executed. The general programming concept and workflow is explained in the example of the *spc* feature in Figure 27.6.

```
int main() {                                                    Host program code
  int size=512, frames = music_length/(2*size);
  int numBlocks = 4, numThreads = size/numBlocks;
  float mag[frames*size], spc[frames];
  ...
  get_spc<<<numBlocks, numThreads>>>(mag, spc); // gpu function call
  get_mfcc<<<numBlocks, numThreads>>>(mag, mfcc); // gpu function call
}
```

```
__global__ void get_spc(float *A, float* result) {              GPU function code
  int tid  = threadIdx.x;
  int bid = blockIdx.x;
  int K   = blockDim.x

  extern __shared__ realv varArray[];
  float *weighted = &varArray[0];
  float *total = &varArray[K];
```

```
  total[tid] = A[K*bid+tid];
  weighted[tid]=tid*total[tid];          ( data transfer to shared memory )
  __syncthreads();
```

```
  sum_reduction<256>(weighted, tid);         ( parallel reduction )
  sum_reduction<256>(total, tid);
```

```
  if(tid == 0) result[bid] = weighted[0]/total[0];  ( single thread execution )

}
```

Figure 27.6: GPU specific program code to extract the *spc* feature.

The NVIDIA programming model is based on a hierarchical thread organization. A thread corresponds to the instructions executed by one CUDA core. Furthermore, threads are clustered to thread blocks. A thread block is handled by one SM and cannot be shared by different SMs. In this way, a data dependency between SMs' respective thread blocks is avoided. Within the host-specific program code, the number of threads per block and number of blocks to be utilized is specified for each GPU function within angle brackets. The GPU function call is asynchronous. Hence, the host system is able to continue program execution parallel to the GPU data processing.

A GPU function containing the program code of a single thread is illustrated in Figure 27.6 and can be structured into three basic steps as shown by the *spc* feature. Data access to global memory and shared memory is managed by identification numbers that are unique for each thread. In this way, data indexing and parallel processing as well as a distinction in program execution is realized. The extraction of the *spc* feature begins by loading required frame data into shared memory, which provides fast data access for further processing. During the extraction of a low-level feature, the frame data is reduced to only a few values. This implies that the utilization of available hardware resources of an SM gradually decreases, which impacts the GPU resource utilization.

Accumulation is a typical computation step for many low-level features and offers only limited concurrent execution. A common and efficient approach for such processing tasks is the "parallel reduction" method, which reduces a data set to a

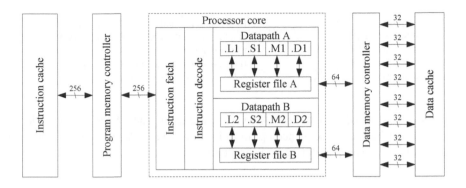

Figure 27.7: Digital signal processor with Very Long Instruction Word architecture for up to eight instructions per instruction word.

single value by utilizing as many parallel computation units as possible. Therefore, NVIDIA provides highly detailed programming examples that also describe optimization techniques to maximize the GPU's efficiency [3]. Within the presented program code, the parallel reduction is called *sum_reduction* and is programmed as a GPU function that can only be called by other GPU functions but not by the host. This function utilizes $x/2$ threads for the accumulation of x data and requires $log_2(x)$ steps instead of $x - 1$ steps in the case of a sequential accumulation. The amount of data to be accumulated is reduced by half within each step because of pairwise additions. After this reduction step, the remaining computation steps cannot be accelerated by concurrent processing. In case of the *spc* feature, only one CUDA thread can be utilized to execute the last instructions and to finally extract the feature.

27.4.3 Digital Signal Processor

Digital Signal Processors (DSPs) are designed to provide sufficient computation power while keeping the power consumption low. Therefore, they normally provide special instructions that are frequently required by most signal processing tasks. For high-performance applications, they are additionally based on a very long instruction word (VLIW) architecture. Such designs offer heterogeneous function units which support different instructions and can be used simultaneously. This means that instructions can be processed in parallel even if they are not of the same type. Such an instruction parallelism requires an extended instruction decoding because several instructions can be included within a single instruction word. An example VLIW DSP design is presented in Figure 27.7.

The VLIW design can execute up to eight instructions concurrently by eight independent heterogeneous function units (.L1/2, .S1/2, .M1/2, .D1/2). Thus, an instruction word has a size of 256 bits with up to eight 32-bit instructions. The function units are organized into two separate datapaths, each with its own register file. Because of this separation, the included data memory controller can read and write, respectively, up to two 64-bit words from data memory. Although the register files are separated

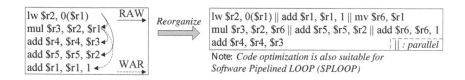

Figure 27.8: Read after write (RAW) and write after read (WAR) conflict identification and instruction sequence optimization for reduced VLIW instruction count; symbol || separates parallel 32-bit instructions.

from each other, data access between the datapaths is possible by special cross-paths that impact computation performance. Finally, the DSP utilizes pipelining as introduced in the GPP-related Section 27.4.1.

The degree of instruction-level parallelism is not only dependent on the hardware architecture but also on data dependencies between instructions. Considering the sequence of MIPS instructions of the *spc*'s for-loop body as shown in Figure 27.8, two different data dependencies exist that limit a concurrent instruction execution. The first one is a Read after Write (RAW) conflict that occurs if a register value is read after it has been updated one or more cycles before. In this case, the respective instruction must not be executed before this subsequent instruction or at the same time. The second dependency is a Write after Read (WAR) conflict. In detail, an instruction updates a register content that has to be read before. Both conflict types must be respected if the program code should be optimized for an instruction parallel execution. A reasonable rescheduling of the instruction sequence can enable a reduction of the VLIW instruction count as depicted in Figure 27.8.

The first and the last instruction of the sequential program code (lw and add) are merged together in order to be executed at once. Although a WAR conflict between these two instructions exists, a parallel execution is possible because of the register file implementation. Thereby, read and write access is performed in two phases beginning with a read access. In this way, a register content is read before a register content can be updated which avoids WAR conflicts that could occur between instructions of the same cycle.

It has to be noted that an additional instruction is inserted and can be found within the second line of the optimized DSP code. This instruction is duplicated from the last line of the original program code in order to further optimize DSP-specific program execution. The reason for this optimization is explained by examining the sequence of parallel instructions of successive for-loop runs, which is illustrated in Figure 27.9.

The instructions of the for-loop body are executed concurrently as far as possible while the instructions of different for-loop runs are executed sequentially. However, a further increase in computation performance can be achieved if the instructions of different for-loops could be overlapped, too. This requires that RAW and WAR conflicts between for-loop runs are considered. From a software point of view, com-

| | | | loop | | |
cycle	1	2	3	4	5
1	lw, add				
2	mul, add, add				
3	add				
4		lw, add			
5		mul, add, add			
6		add			
.				.	.
.				.	.
.				.	.
13					lw, add
14					mul, add, add
15					add

Figure 27.9: VLIW instruction sequence of for-loop turns without SPLOOP optimization.

piler methods that implement overlapped for-loop runs for VLIW architectures are called Software Pipelined LOOPs (SLOOPs). Such methods may utilize several code optimization approaches like instruction reordering, the insertion of No OPeration (NOP) instructions, or additional instructions as shown in Figure 27.8 in order to reduce the overall runtime. The instruction sequence resulting from SPLOOP utilization is shown in Figure 27.10. The optimized program code can be subdivided into three parts: prolog, kernel, and epilog. The prolog code corresponds to the initial phase of the for-loop execution during which the maximum number of overlapping for-loop turns is not reached. The kernel part of the program code possesses a repeating sequence of parallel instructions and the maximum number of for-loop turns are overlapped. In case of the *spc* feature, this means that all instructions are executed concurrently in each cycle. At the end of the for-loop execution, the number of overlapped turns decreases and requires another sequence of parallel instructions named epilog. The requirements of these three parts illustrates that the SPLOOP concept

| | | | loop | | | |
cycle	1	2	3	4	5	
1	lw, add					Prolog
2	mul, add, add	lw, add				
3	add	mul, add, add	lw, add			Kernel
4		add	mul, add, add	lw, add		
5			add	mul, add, add	lw, add	
6				add	mul, add, add	Epilog
7					add	

Figure 27.10: VLIW instruction sequence of for-loop turns with SPLOOP optimization.

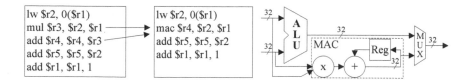

Figure 27.11: Integrated multiply-accumulate instruction set extension for an application specific instruction set processor.

increases the program size. Anyhow, DSPs may provide special hardware support for SPLOOPs that automatically generates prolog, kernel, and epilog program code from SPLOOP prepared program code. Thus, the additional DSP hardware allows the program size increase to be limited.

27.4.4 Application-Specific Instruction Set Processor

Application-Specific Instruction set Processors (ASIPs) are adaptable hardware architectures that are optimized to meet application requirements. They are normally based on a basic processor design like a GPP or a VLIW architecture and at least allow the instruction set and the configuration of the on-chip memory to be customized. In order to give an example of an ASIP adoption, an instruction set extension is presented that accelerates the *spc* feature specific for-loop execution as shown in Figure 27.11. A Multiply-ACcumulate (MAC) software instruction is introduced that allows the multiplication of two operands and adds the product to an accumulated result within one step. Therefore, the new instruction can merge the instructions from the second and third lines of the original for-loop code. In addition, this instruction is suitable to accelerate further feature extraction algorithms as well as Fourier transformations. From a hardware point of view, additional hardware elements like a multiplier, an adder, and a register are required in order to support the new software instructions that are summarized as hardware instructions. Thereby, a cost-efficient hardware instruction design supports various software instructions for higher flexibility and efficiency. For example, the hardware multiplier of the hardware instruction design could additionally be utilized to support software multiplication instructions which only requires simple hardware modifications.

27.4.5 Dedicated Hardware

A dedicated hardware design is the hardware implementation of a digital signal processing algorithm or function. By respecting application constraints like high throughput rates at low hardware resources, the dedicated hardware approach promises to achieve the maximum hardware efficiency of all design concepts. Therefore, this approach is additionally used to extract multiple features from a frame at once by implementing a dedicated hardware for every intended feature. The disadvantage of a dedicated hardware is that it limits flexibility since it can only perform the particular signal processing task for which it is designed. Figure 27.12 presents a simplified

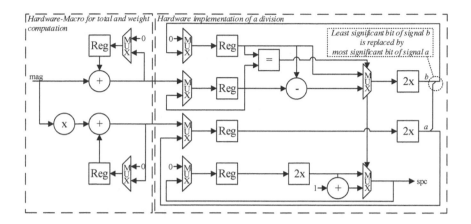

Figure 27.12: Simplified dedicated hardware implementation for extracting the *spc* feature from a continuous data stream.

example of a dedicated hardware design that is able to extract the *spc* feature from a continuous data stream.

In order to provide a better overview, only the datapath is presented while most of the control elements and signals are omitted. Only relevant REGisters (REG), MUltipleXers (MUX) for signal selection, and a signal comparator are illustrated as they are needed to describe the algorithm. Moreover, arithmetic elements (depicted as circles) are assumed to require one clock cycle for signal processing. The input signal (*mag*) is shown on the left side of the figure and provides a continuous data stream. Thus, it must be permanently processed and the *total* and *weight* values are concurrently processed by separate hardware elements. The division operator is performed by a long-division-like algorithm that computes the quotient (*spc*) with several iterations. It utilizes two times the elements that correspond to binary shift operations. In practice, such elements do not require any hardware resources because they can be realized through modified wiring. Although the complete division hardware takes several cycles to compute the result, it can operate concurrently with the computation of the total and weight values. Thus, all hardware elements can be utilized at the same time which further increases hardware utilization and efficiency. The hardware design presented can be physically realized by various technologies. Two popular approaches are presented in the following section that additionally affect computation performance, efficiency, and flexibility.

27.4.5.1 FPGA

Field Programmable Gate Arrays (FPGAs) are complex logic devices that can include millions of programmable logic elements. These logic elements can implement simple logic functions. By connecting several of these logic elements together, even complex hardware designs can be mapped onto the regular structure of FPGAs. An example "island style" FPGA structure is depicted in Figure 27.13.

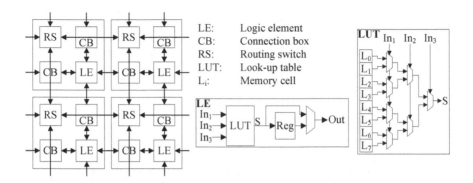

Figure 27.13: Fundamental "island style"–based FPGA structure.

The FPGA consists of various "islands" that include a Logic Element (LE), Connection Boxes (CB), and a Routing Switch (RS). The connection boxes are used to link adjacent logic elements to a global connection network. Thereby, horizontally and vertically routed connections of the global network are managed by the available routing switches, each containing several connection points to flexibly establish signal connections between signal lines. More advanced FPGA structures may provide further logic elements per "island" and additional elements like dedicated hardware multipliers, for example, that can be used to accelerate signal processing or to utilize available hardware resources more effectively.

Dedicated hardware elements can be described by Boolean functions, which means they can be specified by truth value expressions. Thus, logic elements normally include a Look-Up Table (LUT) with up to six input ports in order to implement such Boolean functions. Moreover, a one-bit storage element called Flip Flop is included that may be utilized to implement clock synchronous signals. In the following, the implementation of a hardware adder on an FPGA is demonstrated as it is also required to extract the *spc* feature by its dedicated hardware design. Because of the complexity of floating point hardware, a 4-bit adder design for signed integer operands is examined instead to present the essential implementation steps. These steps are shown in Figure 27.14.

The binary addition of two signed operands can be segmented into computation elements called Full Adders (FA). A full adder is designed to compute the sum of three input bits: one bit of the input operands with the same significance (a_i and b_i) and an additional bit (c_{i-1}) from a prior full adder. The decimal result of three input bit ranges between zero and three. Hence, two bits are required to represent the result. The lower Significant result bit (s_i) is used directly as an output signal while the upper significant result bit, also called the Carry bit (c_i), is routed to the successive full adder. Because each full adder is dependent on the prior full adder elements, this adder design is called a ripple carry adder.

Considering the presented FPGA design, two logic elements are required to implement a full adder circuit. One logic element is used to compute s_i and the other one to generate c_i. Therefore, truth tables of the output signals are created that in-

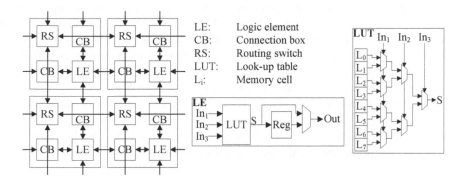

Figure 27.14: FPGA-based implementation flow of a 4-bit ripple carry adder.

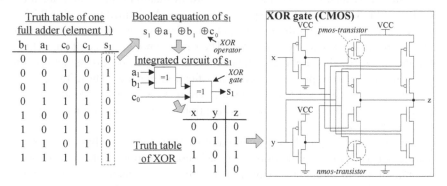

Figure 27.15: ASIC-based implementation flow of a 4-bit ripple carry adder.

clude the resulting output signals for each possible input signal combination. These tables can be easily stored within the look-up tables of the corresponding logic elements. Through this work flow, the complete ripple adder and even more complex hardware designs can also be mapped on FPGAs.

27.4.5.2 ASIC

With Application Specific Integrated Circuits (ASICs), dedicated hardware is implemented on transistor level. This approach is even more complex than FPGA-based designs because additional implementation steps have to be performed, like physical restrictions and effects, in order to get the final hardware design into production. Therefore, a convenient approach to reduce development effort is to use standard cells provided by chip manufacturers like TSMC. Standard cells define transistor placement and dimension of logic gates, which corresponds to hardware implementations of Boolean operators (e.g. AND, OR, NOT, etc.) or basic arithmetic functions (ADD, SUB, etc.). Thus, dedicated hardware must be described by Boolean, respectively, logic functions before it can be implemented on the basis of such logic gates. For example, the result signal of a full adder (s_i) is investigated again in Figure 27.15.

The output signal s_i can be described by two XOR operators (exclusive or) that combine a_i, b_i, and c_{i-1}. The truth table of an XOR shows that the combination of two variables is one when exactly one of the two variables is one. Otherwise the result is zero. An XOR gate corresponds to the hardware implementation of such an XOR operator that is typically included in standard cell libraries. The presented XOR gate implementation requires twelve transistors. In total, 24 transistors are required to implement the sum computation of a full adder element which is less than the number of transistors required to implement a logic element of an FPGA. Thus, an ASIC-based hardware design is typically more efficient compared to an FPGA but less flexible as well because of the fixed hardware design after production.

27.5   Design Space Exploration

Combined cost metrics are suitable hardware evaluation criteria and are the main subject of the subsequent design space exploration. As defined in Section 27.2, these metrics are dependent on the throughput rate respectively on the runtime required for the analysis of a selected music file. The runtime performance of an architecture and the system quality are mainly related to the extracted feature set and secondary to the utilized feature processing and classification method. Thus, Table 27.1 lists four different signal-level feature sets that are used to explore the design space of hardware architectures for music classification. For the definition of signal-level features, cp. Chapter 5.

Feature sets 1, 2, and 4 are composed of frequency domain–based features while feature set 3 requires a time domain–based representation of the audio signal. In combination with a running mean- and deviation-based feature processing method and an SVM classifier (cp. Section 12.4.4), the extraction time per window as well as the overall runtime are proportional to the achievable classification rates that result from cross-validating the GTZAN database [7]. That is, the system quality can be increased by increasing the computational effort. For these music classification methods, the percentage computation effort for extracting features from 30 seconds of music content in relation to the complete processing ranges from 98% to 99% on a desktop PC GPP. This confirms that the feature extraction is the step consuming most of the processing time. Next, seven different hardware architectures are introduced that are representative of the discussed hardware architecture approaches. These are applied for the investigated music classification methods. The results are shown in Table 27.2 including essential architecture properties.

In order to perform a fair comparison, architecture parameters which are technology independent, like power consumption and silicon area, are technology scaled [13]. Furthermore, two GPPs are considered that offer high performance (x86) and low power for mobile systems (RISC). Based on these architecture specifications, the design space regarding the energy and area efficiency of the investigated architectures is presented in Figure 27.16.

Both energy and area efficiency cover five orders of magnitude. Moreover, each architecture comprises a particular field of efficiency which results from the difference in computation effort related to the examined feature sets. This is why the area

Table 27.1: Investigated Feature Sets and Classification Rates Achieved by Music Classification Experiments Including the GTZAN Music Database that Includes 10 Different Genres and 100 Music Files per Genre. Extraction Times per Frame are Measured on an ARM Cortex A8 RISC Processor Featuring a Clock Frequency of 1 GHz

Set	Feature	Classification rate	Extraction time
1	Mel Frequency Cepstral Coefficients Spectral Centroid Spectral Flux Spectral Rolloff	73.4 %	86.9 μs
2	Mel Frequency Cepstral Coefficients Octave Spectral Contrast Normalized Audio Spectrum Envelope	80.8 %	196.2 μs
3	Zero-Crossing Rate Root Mean Square Low-Energy Window	50.6 %	45.3 μs
4	Sub Band Energy Ratio Spectral Crest Factor Maximum Amplitude in Chromagram	63.7 %	54.7 μs

Table 27.2: Investigated Hardware Architectures with an Applied Technology Scaling of Respected Architecture Properties to 40 nm

Name	Type	Frequency [MHz]	Est. Power Cons. [W]	Est. Area [mm^2]
Intel Core i7-2640M	x86	3500	43.00	233.00
ARM Cortex-A8	RISC	1000	0.23	3.50
NVIDIA GF100	GPU	1150	120	529.00
TI C674x	DSP	800	0.62	4.00
Synopsys ARC 600	ASIP	550	0.024	0.16
Virtex 5 LX220T	FPGA	126-205	0.54	200.30
TSMC (Low Power)	ASIC	151-602	0.002-0.024	0.007-0.179

and energy efficiency of an architecture is so high when extracting feature set 3. The lowest efficiency values of all architectures are offered by the x86-based GPP which in contrast provides high flexibility at moderate performance. Although the area and power consumption is 2.3 respectively 2.8 times higher compared to the x86 processor, the GPU provides an efficiency increase of about one order of magnitude because of the significantly higher computational performance. The RISC processor possesses a similar area efficiency as the GPU. However, the energy efficiency of this low-power GPP is almost two orders of magnitude higher compared to the x86 GPP. The examined DSP provides an even higher area and energy efficiency than the RISC

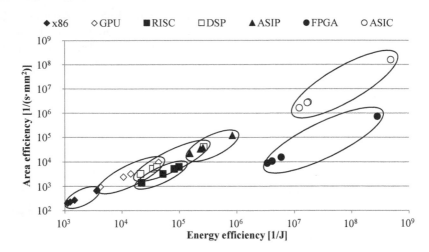

Figure 27.16: Energy and area efficiency of investigated hardware architectures determined by extracting one of the four respected feature sets.

processor. Thereby, the power consumption is 2.7 times higher compared to the RIS, which limits the increase in energy efficiency. Because of its very low power consumption and silicon area, the ASIP offers a twice higher area and energy efficiency than the Intel GPP. Thereby, the achievable extraction rate is of the same magnitude as the ARM GPP. The ASIP therefore is a very attractive architecture approach for low-power devices. It has to be mentioned that the high energy efficiency results are the result of intensive architecture optimization that requires long development times.

The highest efficiency results are achieved by the dedicated hardware architecture approach that extracts different features from a frame at the same time. The FPGA-based solutions offer lower area efficiency than the ASIC implementation because the available FPGA resources are not completely utilized for all implemented feature sets. Finally, the ASIC offers the highest efficiency values and concurrently the lowest silicon areas and power consumptions as expected. However, these results come along with the very low flexibility of the ASIC concept. The ASIC is therefore a suitable approach for low-power and high-performance devices that require a fixed set of acoustic features.

A practical example that helps to interpret these results is given by a Samsung Galaxy S2 smartphone. This mobile device utilizes a GPP that is comparable to the ARM Cortex-A8. Based on its battery, which is implemented by default, 1.5% of the overall battery capacity is consumed if a database of 1000 music files, each with 3 minutes of music content, is classified. This corresponds to a reduction of 22 minutes in operating time if the overall operating time is assumed to be 1 day. The ATE costs of flexibly programmable processors as well as suitable combinations of hardware architectures are presented in Table 27.3.

Table 27.3: ATE Costs of Programmable Processors and Heterogeneous Architecture Approaches (Physical Unit: $[mm^2 sJ]$)

	Set	No Coprocessor	NVIDIA GF100 (GPU)	Xilinx V5LX220T (FPGA)	TSMC LP40nm (ASIC)	TI C674x (DSP)
Intel	1	6.0×10^0	1.3×10^{-1}	3.0×10^{-1}	2.3×10^{-1}	—
Core i7	2	6.9×10^0	7.9×10^{-1}	3.6×10^{-1}	2.8×10^{-1}	—
2640M	3	7.0×10^{-1}	7.8×10^{-3}	1.4×10^{-1}	1.0×10^{-1}	—
(x86)	4	4.2×10^0	7.1×10^{-2}	3.2×10^{-1}	2.5×10^{-1}	—
ARM	1	1.1×10^{-2}	3.2×10^0	1.3×10^{-3}	2.0×10^{-5}	2.3×10^{-3}
Cortex	2	5.4×10^{-2}	1.6×10^1	4.0×10^{-3}	6.5×10^{-5}	6.0×10^{-2}
A8	3	2.8×10^{-3}	2.1×10^{-1}	5.9×10^{-4}	8.3×10^{-6}	3.9×10^{-4}
(RISC)	4	4.2×10^{-3}	1.4×10^0	1.4×10^{-3}	2.2×10^{-5}	1.7×10^{-2}
Synopsys	1	2.3×10^{-4}	7.5×10^0	4.7×10^{-3}	1.6×10^{-6}	9.8×10^{-3}
ARC600	2	5.0×10^{-4}	4.3×10^1	2.7×10^{-2}	1.1×10^{-5}	2.5×10^{-2}
(ASIP)	3	1.7×10^{-5}	5.2×10^{-1}	4.2×10^{-4}	4.8×10^{-8}	1.6×10^{-4}
	4	1.8×10^{-4}	2.7×10^0	1.7×10^{-3}	6.0×10^{-7}	7.2×10^{-3}

The ATE costs are related to the classification of a music file with a length of 30 seconds as reference. Thereby, the Intel processor is reasonably combined with a GPU as the degree of ATE cost decrease compared to the single processor solution. In contrast, a GPU is not a suitable extension for the examined ARM Cortex A8 GPP and the Synopsys ARC600 ASIP because of the resulting increase in ATE costs. However, an ASIC-based coprocessor applied for the feature extraction step can significantly reduce the ATE costs for both processors. This demonstrates which combination of architecture approaches are suitable in order to efficiently classify music. Finally, the computation time of heterogeneous architectures to analyze a database of 1000 music files, each with a length of 3 minutes, is shown in Table 27.4. By utilizing the measured ATE costs, the related computation time results during the early design phase of hardware systems, cost-efficient hardware systems can be designed that are very suitable for realizing the applications as they are presented in this book.

27.6 Concluding Remarks

In this chapter, we have evaluated a variety of different hardware architectures for music classification systems. A comparison of these architectures regarding their efficiency in terms of performing music classification as well as expected hardware-related costs such as silicon area and power consumption were presented. The results of this comparison show that dedicated architectures and FPGAs offer the highest area and energy efficiency for music classification tasks whereas more general purpose solutions like GPUs and CPUs offer more flexibility. With the presented cost

Table 27.4: Computation Time in Seconds to Classify a Complete Music Database with 1000 Music Files Each with Three Minutes of Music Content

	Set	No Copro- cessor	NVIDIA GF100 (GPU)	Xilinx V5LX220T (FPGA)	TSMC LP40nm (ASIC)	TI C674x (DSP)
Intel	1	147.06	6.09	28.65	20.18	—
Core i7	2	157.54	15.07	31.53	26.33	—
2640M	3	50.35	1.51	19.40	6.59	—
(x86)	4	123.01	4.54	29.68	20.93	—
ARM	1	696.72	22.68	28.65	22.68	362.96
Cortex	2	1572.41	50.52	50.52	50.52	583.27
A8	3	356.82	5.76	19.40	6.59	46.75
(RISC)	4	439.02	14.82	29.68	20.93	310.81
Synopsys	1	1444.49	71.01	71.01	71.01	362.96
ARC600	2	2119.03	170.55	170.55	170.55	583.27
(ASIP)	3	399.72	18.76	19.40	18.76	46.75
	4	1276.32	42.75	42.75	42.75	310.81

metrics, suitable combinations of these so-called heterogeneous architectures can be selected for future music classification systems.

27.7 Further Reading

A variety of literature is available which deals with the problem of identifying the most suitable architecture for a given signal processing task. This problem is referred to as Design Space Exploration (DSE). The following list of references is recommended for further reading dealing with the DSE problem as well as describing features and properties of single available architectures [1, 2, 4, 5, 6, 9, 11, 12].

Bibliography

[1] H. Blume. *Modellbasierte Exploration des Entwurfsraumes für heterogene Architekturen zur digitalen Videosignalverarbeitung.* Habilitation thesis, RWTH Aachen University, 2008.

[2] M. Gries and K. Keutzer. *Building ASIPs: The Mescal Methodology.* Springer US, 2006.

[3] M. Harris and et al. Optimizing parallel reduction in CUDA. *NVIDIA Developer Technology,* 2(4):1–39, 2007.

[4] J. L. Hennessy and D. A. Patterson. *Computer Architecture: A Quantitative Approach.* Elsevier, 2012.

[5] H. Kou, W. Shang, I. Lane, and J. Chong. Efficient MFCC feature extraction on graphics processing units. *IET Conference Proceedings,* 2013.

[6] C.-H. Lee, J.-L. Shih, K.-M. Yu, and H.-S. Lin. Automatic music genre classification based on modulation spectral analysis of spectral and cepstral features. *IEEE Transactions on Multimedia*, 11(4):670–682, 2009.

[7] MARSYAS. Music analysis, retrieval and synthesis for audio signals: Data sets. http://marsyas.info/downloads/datasets.html. [accessed 09-Jan-2016].

[8] J. Nickolls and W. J. Dally. The gpu computing era. *IEEE Micro*, 30(2):56–69, 2010.

[9] Y. Patt and S. Patel. *Introduction to Computing Systems: From Bits & Gates to C & Beyond*. Computer Engineering Series. McGraw-Hill Education, 2003.

[10] D. A. Patterson and J. L. Hennessy. *Computer Organization and Design: The Hardware/Software Interface*. Newnes, 2013.

[11] E. M. Schmidt, K. West, and Y. E. Kim. Efficient acoustic feature extraction for music information retrieval using programmable gate arrays. In *Proceedings of the 10th International Society for Music Information Retrieval Conference*, pp. 273–278, Kobe, Japan, October 26-30 2009.

[12] G. Schuller, M. Gruhne, and T. Friedrich. Fast audio feature extraction from compressed audio data. *IEEE Journal of Selected Topics in Signal Processing*, 5(6):1262–1271, 2011.

[13] H. J. Veendrick. *Nanometer CMOS ICs: From Basics to ASICs*. Springer, 2010.

Notation

Unless otherwise noted, we use the following symbols and notations throughout the book.

Abbreviations

Abbreviation	Meaning
iff	if and only if

Basic symbols

Symbol	Meaning
$:=$	equal by definition, defined by
$\mathbb{R}$	set of real numbers
$\mathbb{C}$	set of complex numbers
$\mathbb{Z}$	set of integers
$\mathbb{N}$	set of natural numbers
$[x_i]$	vector with elements x_i
$[x_{ij}]$	matrix with elements x_{ij}
$\boldsymbol{x}$	vectors are represented using bold lower case letters
$\boldsymbol{X}$	matrices are represented by upper case bold letters

Mathematical functions

Symbol	Meaning
z^*	complex conjugate of a complex number $z = x + iy$
cov	covariance
f_0	fundamental frequency
f_s	sampling frequency
f_μ	Fourier frequency, center frequency of the DFT bins
log	natural logarithm (base e)
$\log_{10}$	base-10-logarithm
$\log_2$	base-2-logarithm
$\boldsymbol{X}^T, \boldsymbol{x}^T$	transpose of $\boldsymbol{X}$, $\boldsymbol{x}$. Transposing a vector results in a row vector.
$\bar{x}$	(arithmetical) mean of observations x
med	median
mod	mode
$P(\cdot)$	probability (function)

Index